The 2nd Workshop on Natural Language Processing Meets Climate Change (ClimateNLP 2025)

Vienna, Austria
31 July 2025

ISBN: 979-8-3313-2405-6

Printed from e-media with permission by:

Curran Associates, Inc.
57 Morehouse Lane
Red Hook, NY 12571

Some format issues inherent in the e-media version may also appear in this print version.

Association for Computational Linguistics
209 N. Eighth Street
Stroudsburg, Pennsylvania 18360

Phone: 1-570-476-8006
Fax: 1-570-476-0860

acl@aclweb.org

Additional copies of this publication are available from:

Curran Associates, Inc.
57 Morehouse Lane
Red Hook, NY 12571 USA
Phone: 845-758-0400
Fax: 845-758-2633
Email: curran@proceedings.com
Web: www.proceedings.com

ClimateNLP 2025

The 2nd Workshop on Natural Language Processing Meets Climate Change

Proceedings of the Workshop

July 31, 2025

Introduction

We are happy to welcome you to ClimateNLP 2025, the second ACL workshop on Natural Language Processing Meets Climate Change. The workshop takes place on the 31st of July 2025 in the wonderful city of Vienna, Austria.

ClimateNLP aims to be the premier publication venue for research in the intersection of Natural Language Processing (NLP) and climate change. The workshop aims to discuss how NLP methods can be incorporated into climate change science and climate change action. This year, the program includes four keynote talks by Frida Berry Eklund (Klimatkollen), Emily Kormanyos (Bundesbank), Ruth Schmidt (German Corporation for International Cooperation), and Naomi Oreskes (Harvard University). Furthermore, we hold two panel discussions on the role of ClimateNLP in the industry and future research directions of ClimateNLP. A group discussion, four paper presentations, and two poster sessions round up the day.

We received 35 submissions this year, and recruited 45 active Program Committee (PC) who are distinguished experts in the field of NLP, climate change, or both. Every submission received at least two reviews. When making our selections for the program, we carefully considered the reviews, and conducted extensive debate and discussion among the organizing committee members. The members of the Program Committee did an excellent job in reviewing the submitted papers, and we thank them for their essential role in selecting the accepted papers and helping produce a high-quality program for the conference. In line with our purpose of discussing and learning about the intersection of NLP and Climate Change, our aim has been to create an inclusive program that accommodates as many favourably rated papers as possible. We accepted 22 papers (acceptance rate 62.8

We thank our program committee members for committing their time to help us select an excellent technical program.

We thank all the authors who submitted to the workshop and all workshop participants for making ClimateNLP 2025 a success and for growing the research areas of NLP for climate change with their fine work.

Gaku Morio, Tobias Schimanski, Jingwei Ni, and Organizing Committees

Organizing Committee

Program Chairs (by Last Name Alphabetical Order)

Kalyan Dutia, Climate Policy Radar, the UK
Peter Henderson, Princeton University, the US
Markus Leippold, University of Zurich, Switzerland
Christoper Manning, Stanford University, the US
Gaku Morio, Stanford University, the US
Veruska Muccione, University of Zurich, Switzerland
Jingwei Ni, ETH Zurich, Switzerland
Tobias Schimanski, University of Zurich, Switzerland
Dominik Stammbach, Princeton University, the US
Alok Singh, University of Oxford, the UK
Alba (Ruiran) Su, University of Oxford, the UK
Saeid A. Vaghefi, University of Zurich, Switzerland

Program Committee

Reviewers

Marco Bronzini, University of Trento
Janelle Cai, Massachusetts Institute of Technology
Thomas Corringham, University of California, San Diego
Hari Prasanna Das, University of California, Berkeley
Lukas Ebeling, ETH Zurich
Yu Fan, ETH Zurich
Henry Franks, Climate Policy Radar
Elizabeth Gallagher, Nesta
Nupoor Gandhi, Carnegie Mellon University
Dario Garigliotti, University of Bergen
Sanjay Girija, Google
Andre Graubner, Tsinghua University
Lavanya Gupta, J.P. Morgan Chase
Yifan Hou, Department of Computer Science, Swiss Federal Institute of Technology
Aditya Jain, Applied Research Scientist
Charlott Jakob, Technische Universität Berlin
Elphin Joe, Pennsylvania State University
Matyas Juhasz, Climate Policy Radar
Lynn Kaack, Hertie School of Governance
Shashank Kapoor, Google
Ken Kawamura, Independent
Shima Khanehzar, Cisco
Imene Kolli, University of Zurich
Sai Koneru, Pennsylvania State University
Ambar Nag, Scetti
Wilhelmina Nekoto, Masakhane
Poli Nemkova, University of North Texas
Vincent Nguyen, Cisco
Harrison Pim, Climate Policy Radar
Jakob Prange, Universität Augsburg
Harri Rowlands, InfluenceMap
Diya Saha, Tata Consultancy Services Limited, India
Thoudam Doren Singh, National Institute of Technology Meghalaya
Nick Sorros, MantisNLP
Anna Steinberg, Ludwig-Maximilians-Universität München
David Thulke, RWTH Aachen University and AppTek
Tommy Tran,
Mark Tyrrell, GFA Consulting Group GmbH
Adrian Ulges, RheinMain University of Applied Sciences
Aida Usmanova, Leuphana Universität Lüneburg
Junling Wang, ETHZ - ETH Zurich
Jakob Wedemeyer, Potsdam Institute for Climate Impact Research
Zukang Yang, Eonum Inc
Yongan Yu, University of McGill
Zhengyuan Zhu, University of Texas at Arlington

Table of Contents

Program

Thursday, July 31, 2025

09:00 - 09:05	*Opening Remarks*
09:05 - 09:30	*Keynote Speech 1: Frida Berry Eklund*
09:30 - 10:00	*Oral Presentation Session 1*
10:00 - 10:45	*Poster Session 1*
10:45 - 11:00	*Coffee Break*
11:00 - 11:05	*Session 2 Introduction*
11:05 - 11:30	*Keynote Speech 2: Emily Kormanyos*
11:30 - 12:00	*Panel Discussion 1 - ClimateNLP in practice with Frida Berry Eklund, Emily Kormanyos, Ruth Schmidt*
12:00 - 13:30	*Lunch Break*
13:30 - 13:35	*Session 3 Introduction*
13:35 - 14:00	*Keynote Speech 3: Ruth Schmidt*
14:00 - 14:30	*Oral Presentation Session 2*
14:30 - 15:00	*Panel Discussion 2 - Future of Research in ClimateNLP with Markus Leippold, Christopher D. Manning, Peter Henderson*
15:00 - 15:45	*Poster Session 2*
15:45 - 16:00	*Coffee Break*
16:00 - 16:05	*Session 4 Introduction*

Thursday, July 31, 2025 (continued)

16:05 - 16:30 *Keynote Speech 4: Naomi Oreskes*

16:30 - 17:15 *Question-Guided Open Discussion: Needs of ClimateNLP*

17:15 - 17:30 *Closing Remarks by Markus Leippold*

17:30 - 17:30 *Closing Remarks*

Enhancing Retrieval for ESGLLM via ESG-CID: A Disclosure Content Index Finetuning Dataset for Mapping GRI and ESRS

Shafiuddin Rehan Ahmed **Ankit Parag Shah** **Quan Hung Tran** **Vivek Khetan**[†]
Sukryool Kang **Ankit Mehta** **Yujia Bao** **Wei Wei**
Center for Advanced AI, Accenture, Mountain View, CA, USA
{shafiuddin.r.ahmed,ankit.parag.shah,yujia.bao,wei.wei}@accenture.com
[†] Accenture Labs, San Fransisco, CA, USA
vivek.a.khetan@accenture.com

Abstract

Environment, Social and Governance (ESG) reporting provides a diagnostic lens for evaluating a company's alignment with sustainability goals and stakeholder expectations, while also serving as an expression of its corporate identity and values. Frameworks like the Global Reporting Initiative (GRI) and the new European Sustainability Reporting Standards (ESRS) aim to standardize ESG reporting, yet generating comprehensive reports remains challenging due to the considerable length of ESG documents and variability in company reporting styles. To facilitate ESG report automation, Retrieval-Augmented Generation (RAG) systems can be employed, but their development is hindered by a lack of labeled data suitable for training retrieval models. In this paper, we leverage an underutilized source of weak supervision—the disclosure content index found in past ESG reports—to create a comprehensive dataset, ESG-CID, for both GRI and ESRS standards. By extracting mappings between specific disclosure requirements and corresponding report sections, and refining them using a Large Language Model as a judge, we generate a robust training and evaluation set. We benchmark popular embedding models on this dataset and show that fine-tuning BERT-based models can outperform commercial embeddings and leading public models, even under temporal data splits for cross-report style transfer from GRI to ESRS[1].

1 Introduction

ESG reporting serves as a diagnostic tool that enables structured self-assessment of a company's alignment with long-term sustainability goals and stakeholder expectations. It also is a comprehensive narrative that articulates the company's corporate identity, values, and, its impact to the world. The accelerating global climate crisis and increasing societal demands for corporate accountability have made ESG reporting a critical aspect of modern business. Natural Language Processing plays a pivotal role in understanding and drafting these long documents. Recent advancements in Large Language Models (LLMs) enable the analysis of vast amounts of textual data related to climate policies, sustainability reports, and environmental impact assessments (Vaghefi et al., 2023; Schimanski et al., 2024). By extracting insights from ESG reports, LLMs enhance transparency and inform stakeholders, driving data-driven decision-making in sustainability practices.

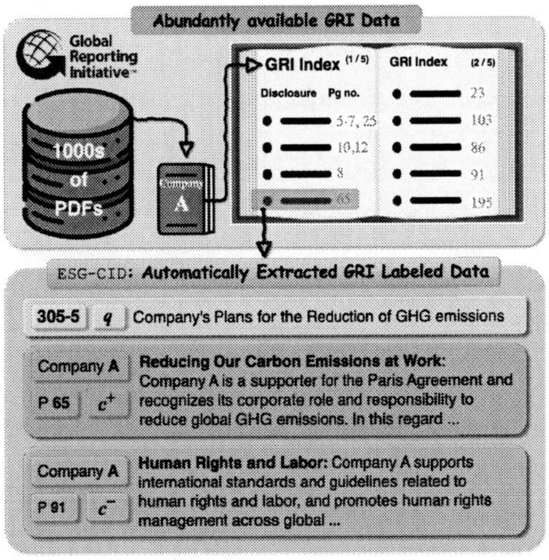

Figure 1: We extract content indices from GRI-compliant sustainability PDFs to create an ESG relevance dataset: ESG-CID. Each entry consists of a disclosure query (q), a relevant chunk (c^+) from the indexed page, and a randomly selected irrelevant chunk (c^-) from the rest of the document

Despite these advancements, generating comprehensive and standardized ESG reports remains a significant challenge. ESG documents are extensive—averaging 120 pages—and exhibit variability in reporting styles and structures among organizations. The lack of standardized and accessible

[1] huggingface.co/datasets/airefinery/esg_cid_retrieval

Proceedings of the 2nd Workshop on Natural Language Processing Meets Climate Change (ClimateNLP 2025), pages 1–16
July 31, 2025

ESG data can lead to greenwashing, obscures true risks, and impedes the effective allocation of resources toward sustainable investments and practices. Frameworks like the Global Reporting Initiative (GRI) and the new European Sustainability Reporting Standards (ESRS) aim to standardize ESG reporting, but automating this process requires effective Retrieval-Augmented Generation (RAG) systems. The development of such systems is hindered by a lack of labeled data suitable for training and evaluating retrieval models in the ESG domain.

The scarcity of labeled data arises mainly due to two factors: First, the considerable length of ESG reports makes manual annotation labor-intensive and time-consuming. Second, the lack of uniformity in reporting styles across different companies presents a challenge in creating datasets that generalize well. The combination of these factors makes it difficult to develop robust retrieval models needed for automating ESG reporting tasks.

In this paper, we leverage an underutilized yet readily available source of weak supervision: the **disclosure content index** found in past reports. We observed that GRI-compliant reports often include a content index linking specific disclosure requirements to corresponding sections or page numbers within the report. By extracting these mappings, we can generate large amounts of weakly supervised data that associates ESG disclosure queries with relevant text passages. To enhance the quality of this data, we use an LLM-as-a-judge to refine and validate the mappings. Additionally, it allows for an in-depth analysis of the standards' inter-relations providing insights on effectively using abundantly available past ESG data.

Using this dataset, we benchmark popular embedding models on the ESG retrieval task and explore the impact of fine-tuning. Our findings reveal that finetuning smaller BERT-based embedding models (`gte-large-en-v1.5`, `bge-large-en-v1.5`, `roberta-large`) can outperform commercial embedding models (`text-embedding-3-small`, `text-embedding-3-large`) and top-performing public models (`gte-Qwen2-1.5B-instruct`, `gte-Qwen2-7B-instruct`). Notably, our benchmark evaluates model performance under temporal data splits and cross-report style transfer from GRI to ESRS, demonstrating the generalizability of the fine-tuned models.

In summary, our contributions are as follows:

- We create the ESG-Content Index Dataset (`ESG-CID`), a dataset leveraging disclosure content indices from ESG reports to facilitate research in the ESG domain and support the development of retrieval models for standardized ESG reporting.
- We benchmark state-of-the-art embedding models on `ESG-CID`, highlighting their limitations in the ESG retrieval task out of the box and demonstrating the benefits of domain-specific fine-tuning.
- We conduct detailed analyses of model performance under temporal splits and cross-report style transfer, offering insights into the challenges and solutions for automating ESG report generation, particularly in the context of the new ESRS standards.

Metric	Value
Unique Topics	11
Unique Sections	112
Total Datapoints	1230
Avg. Sections/Topic	10
Avg. Dataponts/Section	11
Sections with GRI Overlap	99
Sections without GRI Overlap	13
Sections GRI Overlap ratio	0.88
Datapoints with GRI Overlap	648
Datapoints without GRI Overlap	582
Datapoints GRI Overlap ratio	0.53

Table 1: ESRS Statistics and Overlap with GRI. The table presents counts for unique topics, sections, and datapoints, along with their averages in the ESRS guidelines from the official GRI-ESRS interoperability data[2]. Section overlap is counted if at least one datapoint in the section overlaps with a GRI datapoint

2 Related Work

In our research, we build on the foundational work of GRI and the European Financial Reporting Advisory Group (EFRAG), which demonstrates the interconnection between the two standards—GRI and ESRS. Using their preliminary mapping, we illustrate the overlap between ESRS and GRI in Table 1. The table also presents statistics on unique topics, sections, and data points within ESRS, with significant overlaps highlighted in green. This overlap forms the basis of our approach, which is to leverage GRI data to meet ESRS standards.

The ESG domain has abundant public sustainability reports but lacks labeled data. Recent ad-

[2]GRI-ESRS-Mapping.xlsx

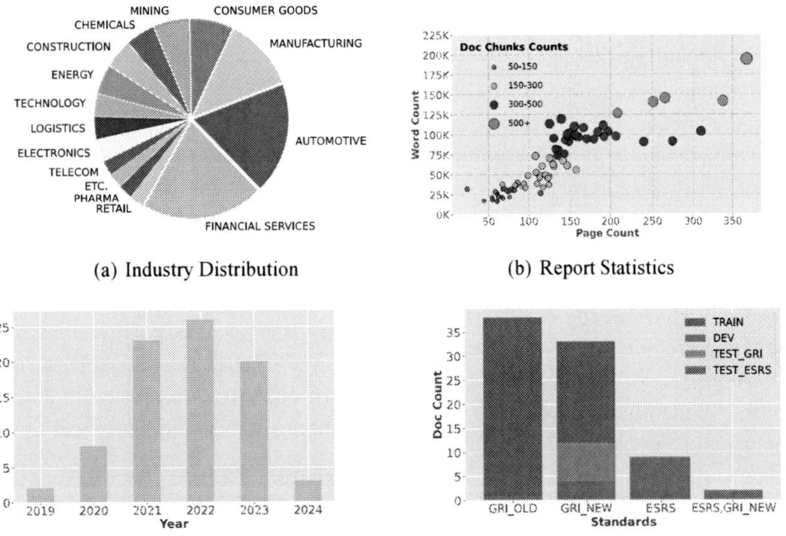

(a) Industry Distribution

(b) Report Statistics

(c) Documents by Year

(d) Dataset Split

Figure 2: Dataset characteristics and challenges: (a) Industry distribution, showcasing the diversity of reporting sectors. (b) Report statistics (page count vs. average word count per chunk, sized by chunk count), highlighting the variability in report length and chunk size, which pose challenges for retrieval models. (c) and (d): Dataset splits (Train, Dev, Test GRI, Test ESRS), illustrating the chronological approach and the out-of-domain ESRS test set.

vancements in LLMs and PDF ingestion are bridging this gap. Vaghefi et al. (2023) demonstrates the potential of LLMs to transform the ESG domain with a Climate-change query specific chat interface called *ChatClimate* powered by LLMs. More recent studies, such as *ChatReport* (Ni et al., 2023) and *ClimRetrieve* (Schimanski et al., 2024), focus on Question Answering within this domain through RAG. These studies, however, are limited by their focus on a narrow set of queries and evaluations based on only 10-20 documents. In contrast, our approach covers a broad spectrum of ESG framework requirements and queries, supported by extensive training and evaluation data.

Distant supervision is a key concept in low-resource model training (Quirk and Poon, 2017; Qin et al., 2018). Polignano et al. (2022) first proposed using the GRI content index as distant supervision for ESG annotations, focusing on table identification via Optical Character Recognition and its role in sentiment analysis. Our work extends this by linking ESRS and GRI frameworks and advancing representation learning through RAG-based automated content index creation.

RAG is a framework that enhances text generation by retrieving relevant external information, improving accuracy and contextual relevance in NLP tasks (Lewis et al., 2020; Jiang et al., 2023). However, most works on ESG domain rely on proprietary embeddings such as OpenAI, which are difficult to adapt to specific needs and pose privacy risks for company data. We enhance retrieval by fine-tuning on ESG-specific content indexes, exploring whether cost-efficient fine-tuning with high-quality data and smaller models can match more resource-intensive methods. We fine-tune various BERT-based models (both base and large) (Devlin et al., 2019; Liu et al., 2019; Li et al., 2023; Zhang et al., 2024; Xiao et al., 2023), leveraging the Model Test Evaluation Benchmark (MTEB; Muennighoff et al. (2022)) to identify the best-performing ones. Additionally, our study also evaluates ModernBERT (Warner et al., 2024) to further understand the impact of domain-specific fine-tuning on retrieval.

3 `ESG-CID`: Dataset Construction

In line with our goal to enhance ESG-specific retrieval systems, we first collected a comprehensive set of sustainability and annual reports from companies across various industries and regions. Utilizing a combination of automated web crawling and manual collection techniques, we gathered over 10,000 reports from 2018 to 2023. The automated collection leveraged databases such as the now-decommissioned GRI database and the SRN database (Donau et al., 2023). After filtering out duplicates and non-English reports, we retained approximately 2,500 unique reports.

Out of these, around half adhered to the GRI standards, with a subset including the disclosure content index in a machine-readable format. We manually curated 73 GRI reports containing detailed content indices to form the primary dataset for our study. Additionally, we identified 11 reports from early adopters of the ESRS standards, which included ESRS content indices, enriching our dataset with cross-standard representations. The collected reports cover a diverse array of industries[3], predominantly from the financial, automotive, and manufacturing sectors (see Figure 2(a)).

3.1 Leveraging Content Indices for Weak Labeling

The disclosure content index serves as a structured bridge between the ESG standard requirements and the report content, providing an opportunity to create weakly labeled data without extensive manual annotation. Each content index lists the standard disclosure requirements (e.g., GRI or ESRS IDs and descriptions), along with references to the pages in the report where these disclosures are addressed.

As illustrated in Figure 2(b), the sustainability reports are significantly lengthy, averaging around 120 pages each, with the longest document exceeding 350 pages. Annotating such extensive documents is labor-intensive and impractical, especially when fine-grained annotations at the chunk or sentence level are considered. To address this challenge, we manually extracted only the content indices from the reports focusing only on these specific but crucial sections. Two experienced annotators, well-versed in ESG reporting and familiar with both GRI and ESRS standards, undertook this task. Their expertise ensured the accuracy and consistency of the extracted content indices.

Using the extracted content indices, we align the disclosure requirements with their corresponding page numbers in the reports. By automatically associating each standard query q (i.e., the disclosure requirement) with the relevant sections of the report indicated by the page numbers, we generate a set of query-document pairs. The query is a standard disclosure requirement, and the document is the corresponding page content addressing that requirement. Leveraging this inherent structure allows us to create a weakly labeled dataset suitable for training and evaluating retrieval models.

3.2 Creating Triplets for Embedding Models

To train and evaluate retrieval models in a contrastive learning framework, we construct triplets consisting of a query q, a positive (matched) chunk c^+, and a negative (unmatched) chunk c^-.

Positive Chunks We preprocess the PDF documents to segment them into manageable chunks (details in §D). The positive chunks c^+ are extracted from the pages referenced in the content index for each disclosure requirement. This ensures that c^+ contains information pertinent to the query q.

Negative Chunks For the negative samples c^-, we randomly sample chunks from the same report that are not associated with the given disclosure requirement. This assumes that these chunks are less relevant or irrelevant to the query, providing a contrastive signal for training.

3.3 Refining Labels with LLM Judgments

While the content indices provide page-level references, not all text within the referenced pages may directly address the disclosure requirement. To enhance the quality of our dataset, we employ Large Language Models (LLMs) as automated judges to assess the relevance of each chunk to the corresponding query.

We define a scoring function $s = \texttt{LLMScore}(q, c)$ that assigns a relevance score between 0 and 5 to each query-chunk pair. The LLM evaluates whether the chunk c sufficiently addresses the disclosure requirement q. By applying a relevance threshold (e.g., $s \geq 3$), we filter out positive chunks that are not sufficiently relevant, thus improving the quality of the triplets.

This refinement step ensures that our dataset contains high-quality, relevant query-document pairs, enhancing the effectiveness of retrieval models trained or evaluated on this data[4].

3.4 Dataset Splitting for Real-World Evaluation

To simulate real-world scenarios, particularly the temporal evolution of ESG standards and the adoption of new reporting requirements, we strategically split our dataset based on report release years and reporting standards.

[3] We provide the company name and year information of the reports of the dataset in §B

[4] Details on the LLM prompts and scoring criteria are provided in the §C

Temporal Splitting The 73 GRI reports are ordered chronologically. We allocate the 10 most recent reports released after 2020, which adhere to the updated `GRI-NEW` standards, to form the test set (`TEST − GRI`). The next 5 most recent reports are designated as the development set for hyperparameter tuning. The remaining 58 reports, primarily following the older `GRI-OLD` standards, constitute the training set as shown in Fig 2(d). This split emulates a scenario where models trained on earlier data are evaluated on newer standards, testing their ability to generalize over time.

Cross-Standard Transfer The 11 ESRS reports form a separate test set (`TEST − ESRS`), allowing us to assess the models' performance on a different but related standard. This setup facilitates the evaluation of cross-standard transferability and the models' adaptability to new reporting frameworks.

Organizing the dataset this way ensures our evaluations reflect the challenges faced in real-world applications, such as adapting to evolving standards and handling reports from different time periods.

4 Experimental Setup

4.1 Embedding Models

We benchmark the retrieval performance of several state-of-the-art embedding models, including both LLMs and lightweight BERT-based models (< 1B Params). The LLM-based embeddings comprise open-source models such as `gte-Qwen2-1.5B-instruct` and (Li et al., 2023), `gte-Qwen2-7B-instruct` (Li et al., 2023), which are known for their strong capabilities in capturing complex language representations. We also include commercial models from OpenAI, namely `text-embedding-3-small` and `text-embedding-3-large`.

In addition to the LLMs, we evaluate lightweight BERT-based models suitable for deployment in resource-constrained environments. These include `roberta-large` (Liu et al., 2019), `bge-large-en-v1.5` (Xiao et al., 2023), `ModernBERT-Large` (Warner et al., 2024) and `gte-large-en-v1.5` (Li et al., 2023; Zhang et al., 2024). We also compare their smaller base models thus offering balance between performance and computational efficiency. By comparing these models, we aim to understand the trade-offs between large-scale embeddings and more efficient alternatives in the ESG retrieval context.

4.2 Fine-tuning on `ESG-CID`

To enhance the domain-specific performance of the lightweight BERT-based models, we fine-tune them on the training split of our constructed dataset (`ESG-CID`). We utilize the standard Multiple Negatives Ranking Loss (Reimers and Gurevych, 2019) for contrastive learning using triplets consisting of a query, a positive chunk, and a negative chunk $((q, c^+, c^-))$. Each query is associated with one relevant positive chunk and one irrelevant negative chunk, as detailed in Section 3.

The fine-tuning process spans five epochs and we pick the best checkpoint that achieves the lowest evaluation loss. Further training details are provided in the Appendix. The fine-tuned models using the entire training set are referred to by adding the suffix–FT to the model card (e.g., `roberta-large`–FT, `gte-large-en-v1.5`–FT, etc). Fine-tuned models trained by only using the `LLMScore`-curated training data have the suffix–FT_{LLM}. We hypothesize that fine-tuning will imbue these models with ESG-specific knowledge, improving their retrieval capabilities on domain-specific queries.

4.3 Evaluation Metrics

We evaluate the models using standard retrieval ranking metrics to assess their ability to retrieve relevant document chunks given a query. Since we do not directly label the relevant chunks for the disclosure and some chunks within the indexed page can be irrelevant, we slightly modify the evaluation. Given that the ground-truth is provided in the form of page numbers[5], we conduct the final ranking assessment based on relevant pages instead of chunks. This involves creating the assessment in a way that ranks page numbers using the metadata of the retrieved chunks.

The metrics calculated using the `ranx` library (Bassani, 2022) include:

Recall@10: Measures the proportion of relevant document pages retrieved in the top 10 chunks. We use '@10' to reflect the typical RAG use case that retrieves 10 documents.

Mean Reciprocal Rank at 50 (MRR@50): Indicates how early the first relevant document page appears.

Mean Average Precision at 50 (MAP@50): Averages precision scores at ranks where relevant document pages are found.

[5]assuming companies report their content index accurately and comprehensively

Model	Size	TEST – GRI REC @10	MRR @50	MAP @50	NDCG @50	TEST – ESRS REC @10	MRR @50	MAP @50	NDCG @50
gte-Qwen2-1.5B-instruct	1.5B	0.667	0.437	0.385	0.528	0.566	0.355	0.307	0.459
gte-Qwen2-7B-instruct	7B	0.713	0.469	0.412	0.551	0.597	0.403	0.347	0.495
text-embedding-3-small		0.684	0.459	0.405	0.545	0.546	0.336	0.284	0.439
text-embedding-3-large		0.730	0.540	0.471	0.602	0.617	0.439	0.379	0.524
Frozen BERT-based Models									
roberta-base	125M	0.045	0.054	0.032	0.109	0.055	0.048	0.029	0.106
BAAI/bge-base-en-v1.5	109M	0.542	0.278	0.242	0.404	0.351	0.213	0.174	0.336
Alibaba-NLP/gte-base-en-v1.5	137M	0.603	0.366	0.313	0.465	0.461	0.277	0.225	0.390
answerdotai/ModernBERT-Base	150M	0.112	0.078	0.056	0.165	0.157	0.103	0.072	0.194
roberta-large	355M	0.146	0.107	0.08	0.203	0.161	0.110	0.077	0.189
BAAI/bge-large-en-v1.5	335M	0.608	0.373	0.325	0.475	0.435	0.257	0.212	0.374
Alibaba-NLP/gte-large-en-v1.5	434M	0.635	0.382	0.333	0.485	0.492	0.291	0.247	0.408
answerdotai/ModernBERT-Large	396M	0.101	0.075	0.053	0.160	0.108	0.105	0.064	0.177
Fine-tuned BERT-based Models on entire data (FT)									
roberta-base		$0.77_{\pm.03}$	$0.57_{\pm.02}$	$0.51_{\pm.02}$	$0.64_{\pm.02}$	$0.59_{\pm.02}$	$0.42_{\pm.02}$	$0.35_{\pm.02}$	$0.50_{\pm.02}$
BAAI/bge-base-en-v1.5		$0.79_{\pm.01}$	$0.61_{\pm.01}$	$0.54_{\pm.01}$	$0.66_{\pm.01}$	$0.63_{\pm.01}$	$0.45_{\pm.01}$	$0.38_{\pm.00}$	$0.53_{\pm.00}$
Alibaba-NLP/gte-base-en-v1.5		$0.78_{\pm.01}$	$0.60_{\pm.02}$	$0.53_{\pm.02}$	$0.65_{\pm.02}$	$0.64_{\pm.03}$	$0.45_{\pm.03}$	$0.39_{\pm.02}$	$0.53_{\pm.02}$
answerdotai/ModernBERT-Base	–"–	$0.75_{\pm.01}$	$0.54_{\pm.03}$	$0.47_{\pm.02}$	$0.61_{\pm.02}$	$0.54_{\pm.02}$	$0.37_{\pm.02}$	$0.31_{\pm.02}$	$0.46_{\pm.02}$
roberta-large		$0.78_{\pm.02}$	$0.59_{\pm.03}$	$0.52_{\pm.02}$	$0.65_{\pm.02}$	$0.60_{\pm.02}$	$0.43_{\pm.02}$	$0.36_{\pm.02}$	$0.51_{\pm.02}$
BAAI/bge-large-en-v1.5		$0.79_{\pm.02}$	$0.59_{\pm.03}$	$0.53_{\pm.03}$	$0.65_{\pm.03}$	$0.63_{\pm.03}$	$0.46_{\pm.04}$	$0.39_{\pm.04}$	$0.54_{\pm.03}$
Alibaba-NLP/gte-large-en-v1.5		$0.79_{\pm.01}$	$0.59_{\pm.02}$	$0.52_{\pm.02}$	$0.65_{\pm.02}$	$0.64_{\pm.02}$	$0.45_{\pm.03}$	$0.38_{\pm.03}$	$0.53_{\pm.02}$
answerdotai/ModernBERT-Large		$0.78_{\pm.02}$	$0.57_{\pm.02}$	$0.50_{\pm.02}$	$0.63_{\pm.02}$	$0.57_{\pm.03}$	$0.41_{\pm.03}$	$0.34_{\pm.02}$	$0.48_{\pm.02}$
Fine-tuned BERT-based Models on LLMScore *filtered data* (FT_{LLM})									
roberta-base		$0.79_{\pm.01}$	$0.59_{\pm.03}$	$0.53_{\pm.03}$	$0.65_{\pm.02}$	$0.61_{\pm.03}$	$0.43_{\pm.03}$	$0.36_{\pm.03}$	$0.51_{\pm.03}$
BAAI/bge-base-en-v1.5		$0.79_{\pm.01}$	$0.59_{\pm.02}$	$0.53_{\pm.02}$	$0.65_{\pm.02}$	$0.63_{\pm.01}$	$0.45_{\pm.02}$	$0.39_{\pm.02}$	$0.53_{\pm.01}$
Alibaba-NLP/gte-base-en-v1.5		$0.79_{\pm.01}$	$\mathbf{0.62}_{\pm.02}$	$0.54_{\pm.02}$	$0.66_{\pm.01}$	$0.65_{\pm.02}$	$0.46_{\pm.02}$	$0.40_{\pm.02}$	$0.54_{\pm.02}$
answerdotai/ModernBERT-Base	–"–	$0.76_{\pm.04}$	$0.56_{\pm.05}$	$0.49_{\pm.05}$	$0.62_{\pm.04}$	$0.57_{\pm.06}$	$0.39_{\pm.06}$	$0.33_{\pm.05}$	$0.48_{\pm.05}$
roberta-large		$\mathbf{0.80}_{\pm.01}$	$0.61_{\pm.02}$	$0.54_{\pm.03}$	$0.66_{\pm.02}$	$0.62_{\pm.03}$	$0.45_{\pm.03}$	$0.38_{\pm.03}$	$0.53_{\pm.02}$
BAAI/bge-large-en-v1.5		$\mathbf{0.80}_{\pm.01}$	$\mathbf{0.62}_{\pm.02}$	$\mathbf{0.55}_{\pm.01}$	$\mathbf{0.67}_{\pm.01}$	$0.65_{\pm.02}$	$0.47_{\pm.03}$	$0.40_{\pm.03}$	$0.55_{\pm.02}$
Alibaba-NLP/gte-large-en-v1.5		$\mathbf{0.80}_{\pm.01}$	$\mathbf{0.62}_{\pm.02}$	$0.55_{\pm.02}$	$\mathbf{0.67}_{\pm.01}$	$\mathbf{0.66}_{\pm.02}$	$\mathbf{0.48}_{\pm.02}$	$\mathbf{0.41}_{\pm.02}$	$\mathbf{0.55}_{\pm.01}$
answerdotai/ModernBERT-Large		$0.79_{\pm.02}$	$0.58_{\pm.04}$	$0.52_{\pm.04}$	$0.64_{\pm.03}$	$0.59_{\pm.05}$	$0.42_{\pm.05}$	$0.35_{\pm.04}$	$0.50_{\pm.04}$

Table 2: Overall effectiveness of the models on ESG-CID comparing the mean and std of the ranking metrics for the finetuned models on 5 different runs. The row corresponding to Alibaba-NLP/gte-large-en-v1.5 is highlighted as our best performing finetuned model, while OpenAI 's text-embedding-3-large serves as the best available baseline. Our best model outperforms the baseline by 7-8% on TEST – GRI and 3-4% on TEST – ESRS.

Normalized Discounted Cumulative Gain at 50 (NDCG@50): Emphasizes the ranking positions of relevant document pages.

Performance is reported on both the GRI test split (TEST – GRI) and the ESRS test split (TEST – ESRS). It is noteworthy that the fine-tuned models were trained exclusively on the GRI training data and have not been exposed to any ESRS data, allowing us to evaluate their generalization capabilities across different ESG reporting standards.

4.4 Real-world Applicability: ESRS Content Indexing

Beyond standard retrieval metrics, we assess the practical utility of the models in constructing the ESRS content index within a company's report. According to ESRS, companies are required to provide structured disclosures in a tabular format. Our objective is to automate the extraction and indexing of relevant information from PDF reports according to each disclosure requirement.

In this task, given a document D and a set of ESRS disclosure queries $Q = \{q_1, q_2, \ldots, q_n\}$, we aim to map each query q_i to its corresponding page numbers in D. We experiment with reports from two companies—one in the automotive industry and one in agriculture—to capture diversity in reporting styles. We report the precision, recall and F1 of these mappings.

Each report D is segmented into chunks, and for each disclosure query q_i, the model retrieves the *top-10* most relevant chunks from D. The retrieved chunks are then mapped back to their page numbers, using the LLMScore effectively constructing the content index. Evaluation is based on the accuracy of these mappings, reflecting the models' effectiveness in automating the ESRS content indexing process.

5 Results and Analysis

5.1 Benchmarking Pre-trained Embedding Models

Table 2 presents the retrieval performance of various state-of-the-art embedding models on the GRI

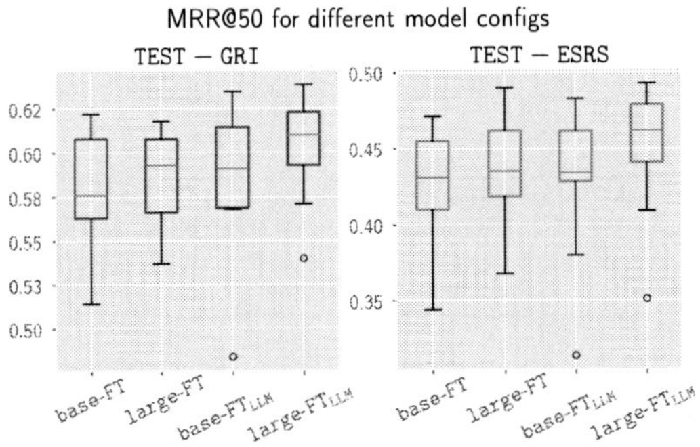

Figure 3: Box plot of the MRR@50 results from various fine-tuning runs (FT, FT_{LLM}) using `base` and `large` models. Each box represents the results from 20 different runs, comparing small and large BERT-based models in our experiments, with and without the use of `LLMScore` for filtering the training data.

and ESRS test sets. We show each finetuned model's aggregate performance on 5 different runs.

Firstly, we observe that most of the LLM-based embedding models demonstrate strong performance out of the box. For instance, the 1.5B parameter `gte-Qwen2-1.5B-instruct` embedding model achieves a Recall@10 of 0.667 without any domain-specific fine-tuning. Additionally, the open-source model `gte-Qwen2-7B-instruct` performs comparably to the commercial model `text-embedding-3-large`, highlighting the competitiveness of open-source solutions.

Secondly, LLM-based embedding models (listed in the first section of the table) significantly outperform the BERT-based embedding models (listed in the second section). This difference is attributed to the higher representational power and larger pretraining datasets of the LLM-based models, which enable better capture of semantic relationships in the ESG domain.

Thirdly, we note that the ESRS dataset presents a much greater challenge compared to GRI. There is a substantial performance degradation across models when evaluated on ESRS, indicating that ESRS retrieval tasks are more difficult.

5.2 Benchmarking Fine-tuned Embedding Models

We present the performance of our fine-tuned models in the last two sections of Table 2. While the original BERT-based models perform significantly worse than the LLM-based embeddings in their pre-trained state, fine-tuning on our dataset results in substantial performance improvements. After fine-tuning, the BERT-based models not only close the gap but, in most cases, outperform the larger LLM-based embeddings.

Specifically, for the GRI test set, `gte-large-en-v1.5`–FT achieves improvements of over 5-6 percentage points across all ranking metrics. The other BERT-based models, both small and large, demonstrate consistent gains, outperforming the LLM-based models despite having fewer parameters. This showcases the effectiveness of fine-tuning on `ESG-CID` for enhancing model performance.

When evaluating the transfer performance to the ESRS test set, the fine-tuned models continue to perform significantly better than their pre-trained counterparts. Notably, the fine-tuned `gte-large-en-v1.5`–FT model outperforms the commercial baselines across all ranking metrics, despite not having been trained on any ESRS data. This suggests that fine-tuning on GRI data imparts transferable knowledge that generalizes to ESRS retrieval tasks to a great extent.

5.3 Impact of `LLMScore` Filtering

To understand the contribution of the `LLMScore` filtering step and see the difference in performance between the `base` and the `large` models, we plot the `MRR@50` grouping the common runs. As shown in Figure 3, there is a consistent overall improvement when using the filtered data when compared to finetuning with entire data. This confirms that

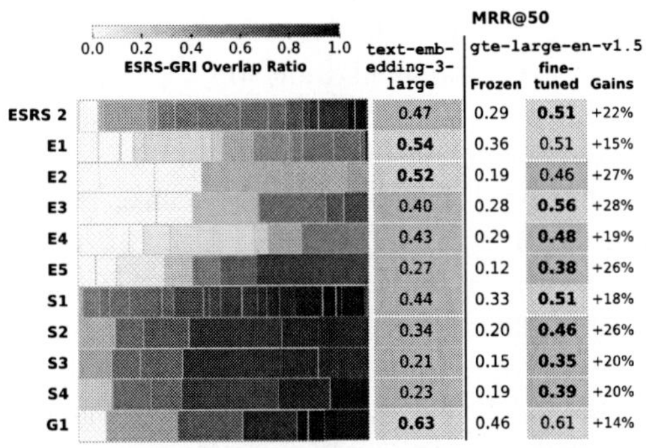

Figure 4: ESRS-GRI overlapping datapoints grouped by topics (top to bottom). Sections within each topic are ordered by their overlapping ratio (left to right). The table on the right displays ranking scores, using the MRR@50 metric, comparing OpenAI embeddings, the frozen and the fine-tuned `gte-large-en-v1.5` model. Scores from the better-performing model are boldened. Positive results (with MRR > 0.5) are highlighted in green, while negative results are highlighted in red.

the LLM filtering helps to remove noise and improve the quality of the training data, leading to a more effective retrieval model. We also observe consistent (albeit small) improvements when using larger counterparts justifying their higher capacity for this GRI/ESRS retrieval task.

5.4 Interplay between ESRS and GRI

To investigate the lower baseline scores observed in the ESRS test set, we conducted a detailed analysis of the overlap between ESRS topics and GRI standards. The heatmap in Figure 4 illustrates the overlapping sections, paired with the MRR@50 scores achieved by our best-performing model, `gte-large-en-v1.5–FT`$_{\text{LLM}}$, compared to the OpenAI baseline for each ESRS topic. We also include scores from the frozen counterpart to evaluate the performance gains from fine-tuning.

Our analysis reveals that the fine-tuned model consistently outperforms its frozen counterpart, with the most significant improvements observed in the `E2`, `E3`, `E5`, and `S2` topics, achieving gains of 26-27%. When compared to OpenAI's `text-embedding-3-large`, the fine-tuned model performs better in all but the `E1`, `E2`, and `G1` topics, with the maximum improvement of 16% observed in the `E3` topic, pushing the performance over the 50% MRR threshold.

However, certain topics, such as `E4` and `E5` (focusing on Biodiversity and Resource Use) remain challenging, as neither the large general-purpose model nor the fine-tuned model surpasses the 50% performance threshold. Similarly, topics from the Social category (`S2`, `S3`, and `S4`) show significant improvements from fine-tuning but still do not cross the threshold. In contrast, topics such as `ESRS 2` (General Disclosures), `E1`, `E3`, `S1`, and `G1` (Governance) demonstrate strong performance, indicating their suitability for automation. These topics exhibit high overlap with GRI, highlighting the potential to leverage existing GRI data to fine-tune retrieval systems for ESRS/CSRD-compliant reporting.

The problematic topics, highlighted in red, underscore areas where additional data collection and methodological refinement are necessary to improve mapping accuracy. Future work should focus on enhancing the GRI-ESRS correspondence or incorporating additional standards into the training set to further boost ESRS performance.

Company	Model	Prec	Rec	F1
Auto	`text-embedding-3-large`	0.36	0.34	0.35
	`gte-large-en-v1.5` ❄	0.36	0.27	0.31
	`gte-large-en-v1.5–FT`	0.39	0.36	0.38
	`gte-large-en-v1.5–FT`$_{\text{LLM}}$	0.39	**0.40**	**0.40**
Agri	`text-embedding-3-large`	0.62	0.42	0.50
	`gte-large-en-v1.5` ❄	0.67	0.40	0.50
	`gte-large-en-v1.5–FT`	**0.69**	0.43	0.53
	`gte-large-en-v1.5–FT`$_{\text{LLM}}$	0.63	**0.51**	**0.56**

Table 3: Comparison of GTE and OpenAI models for content index generation on an Automotive (Auto) and an Agricultural (Agri) companies.

5.5 ESRS Content Indexing

Table 3 presents the results of ESRS content indexing, comparing the performance of our fine-tuned `gte-large-en-v1.5–FT` model with OpenAI embeddings. Our analysis reveals that `gte-large-en-v1.5–FT`$_{\text{LLM}}$ outperforms OpenAI embeddings in both the automotive and agricultural domains. Notably, our training set contains a substantial amount of automotive data but very few agricultural company reports, as illustrated in Figure 2(a). Despite this imbalance, `gte-large-en-v1.5–FT`$_{\text{LLM}}$ demonstrates emergent properties, generalizing well to the agricultural domain despite limited training data.

Interestingly, the inclusion of `LLMScore` reduces the precision of the RAG system. This suggests that models trained with LLM filtering may introduce hard relevant-looking false positives, thereby confusing the RAG system. Future work could address this issue through finer prompt tuning.

6 Conclusion

This paper addresses the critical need for scalable ESG information retrieval by leveraging disclosure content indices to align GRI and ESRS frameworks. By using content indices as a source of weak supervision, we developed a novel benchmark for ESG retrieval finetuning and showed our ESG models outperform strong baselines, such as OpenAI. Our results demonstrate GRI indices can effectively bootstrap models for ESRS compliance, achieving moderate transferability despite limited ESRS-specific data. The `LLMScore` filtering process further enhanced training data quality, enabling our models to generalize across evolving ESG standards. These findings highlight the practical benefits of structured indices in automating ESG reporting and compliance tasks. By harmonizing the GRI and ESRS frameworks, this research establishes a robust foundation for future inquiries into standard-agnostic capabilities, adaptability across regulatory frameworks, and holistic ESG reporting solutions.

Limitations & Future Work

While our work lays a strong foundation for automated inter-framework ESG reporting and auditing, there are several limitations and areas for future research that we aim to address.

Firstly, the modest improvements between larger and smaller models suggest that our dataset may lack the size and diversity to fully exploit the capabilities of more complex models or the chosen samples for finetuning could be refined further being too noisy. Future research should focus on expanding and diversifying the dataset. This could include the incorporation of advanced techniques in automatic content index extraction from documents, leveraging recent advancements in PDF parsing and layout analysis on long documents (Saad-Falcon et al., 2023; Morio et al., 2024; Xie et al., 2025). Also, table reasoning through multi-agent refinement (Wang et al., 2024; Yu et al., 2025) could be explored to handle the diverse ESG reporting standards across different companies and frameworks more effectively. To address learning with noise, future work could investigate iterative training methodologies, such as multi-step training with hard negatives (Zhang et al., 2024) or using a cross-encoder as a re-ranker (Han et al., 2020) to filter out noise and harness a larger model's full potential.

Secondly, while retrieval is a crucial component of our RAG approach, it is not an endpoint. Future work should explore the automated generation of comprehensive sustainability reports from a wide array of a company's source documents. Current research (Ni et al., 2023; Wu et al., 2024), including ours, limits ESG analysis to a single document. Expanding this to include multiple documents such as financial reports, proxy statements, and annual reports would provide a more holistic and realistic approach to ESG reporting, reflecting the multifaceted nature of real-world data.

Lastly, our current work is restricted to the English language, which limits its applicability, especially given the diverse linguistic landscape of ESG reporting, particularly in Europe (Gutierrez-Bustamante and Espinosa-Leal, 2022). Future efforts should aim to extend this work to other languages, leveraging the availability of parallel corpora where companies report in multiple languages. This would not only enhance the accessibility and applicability of our models but also open up exciting avenues for analyzing the multilingual dependencies and nuances in ESG reporting.

Ethics Statement

We highlight the ethical aspects related to the participation of annotators in research activities. We are committed to ensuring that our approach to data annotation is humane, respectful, and inclusive, as this not only enhances the quality of the datasets but also respects and preserves the dignity and rights of all participants.

Acknowledgments

We thank the anonymous reviewers for their valuable comments which greatly helped improving our work. We would also like to extend our gratitude to Rozga Rhett, Dhruv Malik A, Kapil Mahajan and Jens Laue for their assistance throughout this project.

Disclaimer

This content is provided for general information purposes and is not intended to be used in place of consultation with our professional advisors. This document may refer to marks owned by third parties. All such third-party marks are the property of their respective owners. No sponsorship, endorsement or approval of this content by the owners of such marks is intended, expressed or implied.

References

Elias Bassani. 2022. ranx: A blazing-fast python library for ranking evaluation and comparison. In *ECIR (2)*, volume 13186 of *Lecture Notes in Computer Science*, pages 259–264. Springer.

Jacob Devlin, Ming-Wei Chang, Kenton Lee, and Kristina Toutanova. 2019. BERT: Pre-training of deep bidirectional transformers for language understanding. In *Proceedings of the 2019 Conference of the North American Chapter of the Association for Computational Linguistics: Human Language Technologies, Volume 1 (Long and Short Papers)*, pages 4171–4186, Minneapolis, Minnesota. Association for Computational Linguistics.

Charlotte-Louise Donau, Fikir Worku Edossa, Joachim Gassen, Gaia Melloni, Inga Meringdal, Bianca Minuth, Arianna Piscella, Paul Pronobis, and Victor Wagner. 2023. SRN Document Database. Accessed: 2023.

Marcelo Gutierrez-Bustamante and Leonardo Espinosa-Leal. 2022. Natural language processing methods for scoring sustainability reports—a study of nordic listed companies. *Sustainability*, 14(15):9165.

Shuguang Han, Xuanhui Wang, Mike Bendersky, and Marc Najork. 2020. Learning-to-rank with bert in tf-ranking. *Preprint*, arXiv:2004.08476.

Zhengbao Jiang, Frank Xu, Luyu Gao, Zhiqing Sun, Qian Liu, Jane Dwivedi-Yu, Yiming Yang, Jamie Callan, and Graham Neubig. 2023. Active retrieval augmented generation. In *Proceedings of the 2023 Conference on Empirical Methods in Natural Language Processing*, pages 7969–7992, Singapore. Association for Computational Linguistics.

Patrick Lewis, Ethan Perez, Aleksandra Piktus, Fabio Petroni, Vladimir Karpukhin, Naman Goyal, Heinrich Küttler, Mike Lewis, Wen-tau Yih, Tim Rocktäschel, et al. 2020. Retrieval-augmented generation for knowledge-intensive nlp tasks. *Advances in Neural Information Processing Systems*, 33:9459–9474.

Zehan Li, Xin Zhang, Yanzhao Zhang, Dingkun Long, Pengjun Xie, and Meishan Zhang. 2023. Towards general text embeddings with multi-stage contrastive learning. *arXiv preprint arXiv:2308.03281*.

Yinhan Liu, Myle Ott, Naman Goyal, Jingfei Du, Mandar Joshi, Danqi Chen, Omer Levy, Mike Lewis, Luke Zettlemoyer, and Veselin Stoyanov. 2019. Roberta: A robustly optimized bert pretraining approach. *Preprint*, arXiv:1907.11692.

Gaku Morio, Soh Young In, Jungah Yoon, Harri Rowlands, and Christopher Manning. 2024. Reportparse: A unified nlp tool for extracting document structure and semantics of corporate sustainability reporting. In *Proceedings of the Thirty-Third International Joint Conference on Artificial Intelligence, IJCAI-24*, pages 8749–8753. International Joint Conferences on Artificial Intelligence Organization. Demo Track.

Niklas Muennighoff, Nouamane Tazi, Loïc Magne, and Nils Reimers. 2022. Mteb: Massive text embedding benchmark. *arXiv preprint arXiv:2210.07316*.

Jingwei Ni, Julia Bingler, Chiara Colesanti-Senni, Mathias Kraus, Glen Gostlow, Tobias Schimanski, Dominik Stammbach, Saeid Ashraf Vaghefi, Qian Wang, Nicolas Webersinke, et al. 2023. Chatreport: Democratizing sustainability disclosure analysis through llm-based tools. *arXiv preprint arXiv:2307.15770*.

Marco Polignano, Nicola Bellantuono, Francesco Paolo Lagrasta, Sergio Caputo, Pierpaolo Pontrandolfo, and Giovanni Semeraro. 2022. An NLP approach for the analysis of global reporting initiative indexes from corporate sustainability reports. In *Proceedings of the First Computing Social Responsibility Workshop within the 13th Language Resources and Evaluation Conference*, pages 1–8, Marseille, France. European Language Resources Association.

Pengda Qin, Weiran Xu, and William Yang Wang. 2018. Robust distant supervision relation extraction via deep reinforcement learning. In *Proceedings of the 56th Annual Meeting of the Association for Computational Linguistics (Volume 1: Long Papers)*, pages 2137–2147, Melbourne, Australia. Association for Computational Linguistics.

Chris Quirk and Hoifung Poon. 2017. Distant supervision for relation extraction beyond the sentence boundary. In *Proceedings of the 15th Conference of the European Chapter of the Association for Computational Linguistics: Volume 1, Long Papers*, pages 1171–1182, Valencia, Spain. Association for Computational Linguistics.

Nils Reimers and Iryna Gurevych. 2019. Sentence-BERT: Sentence embeddings using Siamese BERT-networks. In *Proceedings of the 2019 Conference on Empirical Methods in Natural Language Processing and the 9th International Joint Conference on Natural Language Processing (EMNLP-IJCNLP)*, pages 3982–3992, Hong Kong, China. Association for Computational Linguistics.

Jon Saad-Falcon, Joe Barrow, Alexa Siu, Ani Nenkova, David Seunghyun Yoon, Ryan A. Rossi, and Franck Dernoncourt. 2023. Pdftriage: Question answering over long, structured documents. *Preprint*, arXiv:2309.08872.

Tobias Schimanski, Jingwei Ni, Roberto Spacey, Nicola Ranger, and Markus Leippold. 2024. Climretrieve: A benchmarking dataset for information retrieval from corporate climate disclosures. *Preprint*, arXiv:2406.09818.

Saeid Ashraf Vaghefi, Dominik Stammbach, Veruska Muccione, Julia Bingler, Jingwei Ni, Mathias Kraus, Simon Allen, Chiara Colesanti-Senni, Tobias Wekhof, Tobias Schimanski, et al. 2023. Chatclimate: Grounding conversational ai in climate science. *Communications Earth & Environment*, 4(1):480.

Zilong Wang, Hao Zhang, Chun-Liang Li, Julian Martin Eisenschlos, Vincent Perot, Zifeng Wang, Lesly Miculicich, Yasuhisa Fujii, Jingbo Shang, Chen-Yu Lee, and Tomas Pfister. 2024. Chain-of-table: Evolving tables in the reasoning chain for table understanding. *Preprint*, arXiv:2401.04398.

Benjamin Warner, Antoine Chaffin, Benjamin Clavié, Orion Weller, Oskar Hallström, Said Taghadouini, Alexis Gallagher, Raja Biswas, Faisal Ladhak, Tom Aarsen, Nathan Cooper, Griffin Adams, Jeremy Howard, and Iacopo Poli. 2024. Smarter, better, faster, longer: A modern bidirectional encoder for fast, memory efficient, and long context finetuning and inference. *Preprint*, arXiv:2412.13663.

Qilong Wu, Xiaoneng Xiang, Hejia Huang, Xuan Wang, Yeo Wei Jie, Ranjan Satapathy, Bharadwaj Veeravalli, et al. 2024. Susgen-gpt: A data-centric llm for financial nlp and sustainability report generation. *arXiv preprint arXiv:2412.10906*.

Shitao Xiao, Zheng Liu, Peitian Zhang, and Niklas Muennighoff. 2023. C-pack: Packaged resources to advance general chinese embedding. *Preprint*, arXiv:2309.07597.

Xudong Xie, Hao Yan, Liang Yin, Yang Liu, Jing Ding, Minghui Liao, Yuliang Liu, Wei Chen, and Xiang Bai. 2025. Pdf-wukong: A large multimodal model for efficient long pdf reading with end-to-end sparse sampling. *Preprint*, arXiv:2410.05970.

Peiying Yu, Guoxin Chen, and Jingjing Wang. 2025. Table-critic: A multi-agent framework for collaborative criticism and refinement in table reasoning. *Preprint*, arXiv:2502.11799.

Xin Zhang, Yanzhao Zhang, Dingkun Long, Wen Xie, Ziqi Dai, Jialong Tang, Huan Lin, Baosong Yang, Pengjun Xie, Fei Huang, et al. 2024. mgte: Generalized long-context text representation and reranking models for multilingual text retrieval. *arXiv preprint arXiv:2407.19669*.

A Hyperparameter settings

This section provides detailed information on the hyperparameter settings and training procedures used for fine-tuning the retrieval models (RoBERTa-large and GTE-large).

A.1 Hyperparameter Optimization

We used a combination of prior work, best practices for transformer fine-tuning, and empirical evaluation on a small validation set (carved out from the training set) to select the hyperparameters. Specifically, we held out five documents from the training set to form a validation set. This validation set was used solely for checkpoint selection and is distinct from the development set used for model evaluation. The primary metric for checkpoint selection was ‘dev_cosine_accuracy‘, defined below.

A.2 Training Arguments

Table 4 summarizes the key hyperparameters used for training. These settings were largely consistent across both RoBERTa-large and GTE-large, with the primary difference being the batch size due to GPU memory constraints.

We use saving and evaluation strategy based on the number of steps we take.

We used the ‘SentenceTransformerTrainingArguments‘ class from the ‘sentence-transformers‘ library to manage the training process. The key parameters are as follows:

Hyperparameter	RoBERTa-large	GTE-large
Training Epochs	5	5
Train Batch Size	32	8
Eval Batch Size	32	8
Warmup Ratio	0.05	0.05
FP16	False	False
BF16	False	False
Batch Sampler	No Duplicates	No Duplicates
Eval Steps	50	50
Save Steps	50	50
Save Total Limit	5	5
Logging Steps	20	20
Learning Rate	5e-5	5e-5
Load Best Model	True	True
Weight Decay	0.01	0.01
Metric for Best Model	‘cosine accuracy‘	‘cosine accuracy‘
DDP Find Unused Params	False	False

Table 4: Hyperparameter settings for fine-tuning RoBERTa-large and GTE-large.

- ‘output_dir‘: The directory where the trained models and checkpoints are saved. - ‘overwrite_output_dir‘: If ‘True‘, overwrites the contents of the output directory. - ‘num_train_epochs‘: The number of training epochs. We chose 5 epochs based on preliminary experiments, observing that performance plateaued after this point. - ‘per_device_train_batch_size‘: The batch size per GPU during training. We used a batch size of 32 for RoBERTa-large and 8 for GTE-large due to GPU memory limitations. - ‘per_device_eval_batch_size‘: The batch size per GPU during evaluation. - ‘warmup_ratio‘: The proportion of training steps used for a linear warmup of the learning rate. - ‘fp16‘ and ‘bf16‘: These were set to false due to hardware constraints. - ‘batch_sampler‘: We used the ‘NO_DUPLICATES‘ batch sampler, which ensures no duplicate examples within a batch. - ‘eval_strategy‘ and ‘eval_steps‘: Evaluation was performed every 50 training steps. - ‘save_strategy‘ and ‘save_steps‘: Model checkpoints were saved every 50 training steps. - ‘save_total_limit‘: Limited to 5 checkpoints to conserve disk space. - ‘logging_steps‘: Training statistics were logged every 20 steps. - ‘learning_rate‘: The initial learning rate for the AdamW optimizer was set to 5e-5. - ‘load_best_model_at_end‘: If ‘True‘, loads the model checkpoint with the best performance on the validation set at the end of training. - ‘weight_decay‘: The weight decay parameter for the AdamW optimizer. - ‘metric_for_best_model‘: The metric used for best model checkpoint selection was ‘eval_gri-chunk-dev_cosine_accuracy‘. - ‘ddp_find_unused_parameters‘: Set to ‘False‘ since distributed data parallel (DDP) training was not used.

A.3 Loss Function and Evaluation

The loss function used was 'MultipleNegatives-RankingLoss' from the 'sentence-transformers' library. This loss function is designed for contrastive learning, ensuring that similar pairs (query and positive chunk) have higher similarity scores than dissimilar pairs (query and negative chunk). Each batch considered all other examples as negatives.

For development set evaluation, we used the 'TripletEvaluator' from 'sentence-transformers'. The 'TripletEvaluator' takes three lists as input:

- 'anchors': A list of query examples. - 'positives': A list of relevant chunks. - 'negatives': A list of irrelevant chunks.

The evaluator computes the cosine similarity between anchor-positive and anchor-negative embeddings and calculates the 'cosine_accuracy' metric.

A.4 Cosine Accuracy Metric

The 'eval_gri-chunk-dev_cosine_accuracy' metric is calculated as follows:

1. Compute the cosine similarity between the query embedding and the positive chunk embedding: 'sim_pos = cosine_similarity(M(q), M(c+))'. 2. Compute the cosine similarity between the query embedding and the negative chunk embedding: 'sim_neg = cosine_similarity(M(q), M(c-))'. 3. Count the number of triplets where 'sim_pos > sim_neg'. 4. Compute 'cosine_accuracy' as the percentage of triplets where the positive chunk has a higher cosine similarity to the query than the negative chunk.

This metric reflects the model's ability to rank relevant chunks higher than irrelevant chunks.

A.5 Training Procedure

The models were trained using 'MultipleNegatives-RankingLoss', which is well-suited for contrastive training. Triplets of (query, positive chunk, negative chunk) were constructed, ensuring each query had one associated positive and one negative chunk. No significant overfitting was observed during the five training epochs.

B Company Information

See Table 5 for the company names and publication years of the ESG reports used in `ESG-CID`.

C `LLMScorePrompt` Details

Below is the prompt used for 'LLMScore', which leverages a Large Language Model (LLM) to assess the relevance of a text chunk to a given query, both extracted from an ESG report. The LLM is instructed to provide a numerical score on a scale of 0 to 5, reflecting the degree of relevance. See Figure 5 for further details.

LLMScore Prompt

Given the following [query], and a [text chunk] from an ESG report, please rate the relevancy of the chunk to the disclosure on a scale of 0-5, in terms of being able to provide evidence for the disclosure. Provide higher rating if the chunk has enough evidence to answer the query.

- The output should be a single number between 0 and 5. 0 means not relevant at all, 5 means highly relevant.
- The output should be an integer

[query]
{disclosure}
[text chunk]
{chunk}
Relevancy Score (1-5): <YOUR ANSWER HERE>

Figure 5: Prompt for `LLMScore`

D PDF Preprocessing

For the ingestion of long sustainability PDF documents, we adopt the popular `PyMUPdfLoader` library with scalability in mind. After extracting the text from each page of the report we perform the following steps:

1. **Newline Removal:** Remove newline characters to produce continuous text.
2. **Chunking:** Partition the text on a pagewise basis into segments of 2048 characters.
3. **Overlap:** Apply an overlap of 512 characters between contiguous chunks to preserve context.

Formally, for a given PDF document $d \in \mathcal{D}$, the loader produces a set of text chunks:

$$\mathcal{C}(d) = \{c_1, c_2, \ldots, c_n\},$$

where each chunk c_i is a sequence of 2048 characters (with a 512-character overlap with c_i and c_{i+1}). These chunks serve as the basic units for further processing in our pipeline.

E Dataset Example

In this section, we provide examples of the GRI index and the ESRS index from the HYUNDAI 2024 sustainability report. This communicates the complexity of the existing pdf data and why generating an ESRS report from the the GRI format report is challenging. Additionally, once relevent ESRS index and GRI index are identified; collating related content is non-trivial. See Figures 6, 7, and 8 for example content indices both in ESRS and GRI standards.

ESRS (European Sustainability Reporting Standards)

ESRS 2. General Disclosures

Indicator No.	Title	Page
ESRS 2 BP-1	General basis for preparation of the sustainability statements	124
ESRS 2 BP-2	Disclosures in relation to specific circumstances	28, 36, 42, 43, 97, 98, 100, 117-122
ESRS 2 GOV-1	The role of the administrative, management and supervisory bodies	9, 21, 81-85
ESRS 2 GOV-2	Information provided to and sustainability matters addressed by the undertaking's administrative, management and supervisory bodies	82, 85
ESRS 2 GOV-3	Integration of sustainability-related performance in incentive schemes	9, 17, 20, 37, 59
ESRS 2 GOV-4	Statement on sustainability due diligence	50-53, 67-69
ESRS 2 GOV-5	Risk management and internal controls over sustainability reporting[1)]	-
ESRS 2 SBM-1	Market position, strategy, business model(s) and value chain	6-7, 25-26
ESRS 2 SBM-2	Interests and views of stakeholders	11-13
ESRS 2 SBM-3	Material impacts, risks and opportunities and their interaction with strategy and business model(s)	15-17
ESRS 2 IRO-1	Description of the processes to identify and assess material impacts, risks and opportunities	14
ESRS 2 IRO-2	Disclosure Requirements in ESRS covered by the undertaking's sustainability statements	110-112

Figure 6: ESRS 2. General Disclosures Content Index of Hyundai found on page 110 of their 2024 sustainability report. The Indicator No. represents the standard's identifier, Title is used as the query text for our RAG system, and Page gives us the gold standard location of the relevant pages for the query within the report.

ESRS E1. Climate Change

Indicator No.	Title	Page
ESRS E1-1	Transition plan for climate change mitigation	32
ESRS E1-2	Policies related to climate change mitigation and adaptation	23-32
ESRS E1-3	Actions and resources in relation to climate change policies	32, 37
ESRS E1-4	Targets related to climate change mitigation and adaptation	24-26, 30-32, 38
ESRS E1-5	Energy consumption and mix	98
ESRS E1-6	Gross Scopes 1, 2, 3 and Total GHG emissions	36, 98
ESRS E1-7	GHG removals and GHG mitigation projects financed through carbon credits	16, 31
	Avoided emissions of products and services	15, 27
ESRS E1-8	Internal carbon pricing[2)]	-
ESRS E1-9	Potential financial effects from material physical and transition risks and potential climate-related opportunities	22, 33-35

Figure 7: ESRS E1. Climate Change: Content index of the climate change related topics found on page 110 of the Hyundai 2024 sustainability report. The Indicator No. represents the standard's identifier, Title is used as the query text for our RAG system, and Page gives us the gold standard location of the relevant pages for the query within the report.

GRI Index

Topic Specific Standards - Environmentall

GRI Standards		Page
No.	Title	
301-1	Materials used by weight or volume	42, 98
301-2	Recycled input materials used	42, 98
301-3	Reclaimed products and their packaging materials	42
302-1	Energy consumption within the organization	98
302-2	Energy consumption outside of the organization	36
302-3	Energy Intensity	98
302-4	Reduction of energy consumption	23-24
303-1	Interactions with water as a shared resource	42-43, 99
303-2	Management of impacts related to wastewater	43, 100
303-3	Water withdrawal	99
303-4	Water discharge	99
303-5	Water consumption	20, 42, 99
304-1	Operational sites owned, leased, managed in, or adjacent to, protected areas and areas of high biodiversity value outside protected areas	46-48
304-2	Significant impacts of activities, products and services on biodiversity	46-48
304-3	Habitats protected or restored	46-48
304-4	IUCN Red List species and national conservation list species with habitats in areas affected by operations	48

GRI Standards		Page
No.	Title	
305-1	Direct (Scope 1) GHG emissions	36, 98
305-2	Energy indirect (Scope 2) GHG emissions	36, 98
305-3	Other indirect (Scope 3) GHG emissions	36, 98
305-4	GHG emissions intensity	36, 98
305-5	Reduction of GHG emissions	23-32
305-7	Nitrogen oxides (NOx), sulfur oxides (SOx), and other significant air emissions	100
306-1	Waste generation and significant waste-related impacts	40-43
306-2	Management of significant waste-related impacts	40-43
306-3	Waste generated	100
306-4	Waste diverted from disposal	43, 100
306-5	Waste directed to disposal	100
308-1	New suppliers that were screened using environmental criteria	67-68
308-2	Negative environmental impacts in the supply chain and actions taken	69

Figure 8: GRI Content Index for GRI 300: Topic Specific Standards - Environmental.

DOCUMENT NAME	COMPANY	YEAR	INDUSTRY_CLUSTER	STANDARDS	SPLIT
FORD_2024	FORD	2024	AUTOMOTIVE	ESRS	TEST_ESRS
HYUNDAI_2019	HYUNDAI	2019	AUTOMOTIVE	GRI_OLD	TRAIN
HYUNDAI_2020	HYUNDAI	2020	AUTOMOTIVE	GRI_OLD	TRAIN
HYUNDAI_2021	HYUNDAI	2021	AUTOMOTIVE	GRI_OLD	TRAIN
HYUNDAI_2022	HYUNDAI	2022	AUTOMOTIVE	GRI_OLD	DEV
HYUNDAI_2022_A	HYUNDAI	2022	AUTOMOTIVE	GRI_OLD	TRAIN
HYUNDAI_2023	HYUNDAI	2023	AUTOMOTIVE	ESRS, GRI_NEW	TEST_ESRS
HYUNDAI_2024	HYUNDAI	2024	AUTOMOTIVE	ESRS, GRI_NEW	TEST_ESRS
KIA_2024	KIA	2024	AUTOMOTIVE	GRI_NEW	TEST_GRI
SKODA_2023	SKODA AUTO	2023	AUTOMOTIVE	ESRS	TEST_ESRS
TOYOTA_2023	TOYOTA	2023	AUTOMOTIVE	GRI_NEW	TEST_GRI
TRAIN_18	Nissan Motor Corporation	2022	AUTOMOTIVE	GRI_OLD	TRAIN
TRAIN_186	Nissan Motor Corporation	2021	AUTOMOTIVE	GRI_OLD	TRAIN
TRAIN_25	Geely Automobile Holdings	2022	AUTOMOTIVE	GRI_NEW	TRAIN
TRAIN_22	Benteler Group	2022	AUTOMOTIVE	GRI_NEW	TRAIN
TRAIN_123	SKC	2023	CHEMICALS	GRI_NEW	TRAIN
TRAIN_294	NOVA Chemicals	2021	CHEMICALS	GRI_OLD	TRAIN
TRAIN_306	NOVA Chemicals	2022	CHEMICALS	GRI_NEW	TRAIN
CTP_2023	CTP	2023	CONSTRUCTION	ESRS	TEST_ESRS
HELVAR_2023	HELVAR OY AB	2023	CONSTRUCTION	ESRS	TEST_ESRS
HH_2023	H+H	2023	CONSTRUCTION	ESRS	TEST_ESRS
TRAIN_242	Heidelberg Materials	2022	CONSTRUCTION	GRI_NEW	TRAIN
TRAIN_119	NESTE	2021	ENERGY	GRI_OLD	TRAIN
TRAIN_218	Fortis Inc.	2022	ENERGY	GRI_NEW	TRAIN
TRAIN_228	FortisBC	2022	ENERGY	GRI_NEW	DEV
SANTADER_2023	SANTADER BANK POLSKA GROUP	2023	FINANCIAL SERVICES	ESRS	TEST_ESRS
TRAIN_191	YUANTA FINANCIAL HOLDINGS	2021	FINANCIAL SERVICES	GRI_OLD	TRAIN
TRAIN_194	Banca Transilvania	2020	FINANCIAL SERVICES	GRI_OLD	TRAIN
TRAIN_239	Gulf International Bank	2022	FINANCIAL SERVICES	GRI_NEW	DEV
TRAIN_307	Taishin Financial Holding	2021	FINANCIAL SERVICES	GRI_OLD	TRAIN
TRAIN_71	Capital One	2021	FINANCIAL SERVICES	GRI_NEW	TRAIN
TRAIN_127	LOOMIS	2022	FINANCIAL SERVICES	GRI_NEW	TRAIN
TRAIN_155	Loomis	2021	FINANCIAL SERVICES	GRI_OLD	TRAIN
TRAIN_0	ALLY FINANCIAL	2021	FINANCIAL SERVICES	GRI_OLD	TRAIN
TRAIN_2	Energy Recovery	2021	TECHNOLOGY	GRI_OLD	TRAIN
TRAIN_77	Motorola Solutions	2021	TECHNOLOGY	GRI_NEW	TRAIN
TRAIN_3	Meta	2021	TECHNOLOGY	GRI_OLD	TRAIN
KPN_2023	KPN	2023	TELECOMMUNICATIONS	GRI_NEW	TEST_GRI
TRAIN_153	NTT DOCOMO	2020	TELECOMMUNICATIONS	GRI_OLD	TRAIN
ARLA_2023	ARLA	2023	CONSUMER PACKAGED GOODS	ESRS	TEST_ESRS
TRAIN_81	Ryanair	2022	AVIATION	GRI_NEW	TRAIN
TRAIN_124	HITEJINRO	2023	CONSUMER PACKAGED GOODS	GRI_NEW	DEV
TRAIN_212	Molson Coors Beverage Company	2022	CONSUMER PACKAGED GOODS	GRI_OLD	TRAIN
TRAIN_197	Illumina	2021	BIOTECH	GRI_OLD	TRAIN
TRAIN_181	CWT	2022	LOGISTICS	GRI_OLD	TRAIN
KERRY GROUP_2023	KERRY GROUP	2023	CONSUMER PACKAGED GOODS	GRI_NEW	TEST_GRI
LACTALIS_2023	LACTALIS	2023	CONSUMER PACKAGED GOODS	GRI_NEW	TEST_GRI
TRAIN_138	LS ELECTRIC	2023	ELECTRONICS	GRI_NEW	TRAIN
TRAIN_245	TAIFLEX	2023	ELECTRONICS	GRI_NEW	TRAIN
TRAIN_185	KONE	2022	MANUFACTURING	GRI_NEW	TRAIN
TRELLEBORG_2019	Trelleborg AB	2019	MANUFACTURING	GRI_OLD	TRAIN
TRELLEBORG_2020	Trelleborg AB	2020	MANUFACTURING	GRI_OLD	TRAIN
TRELLEBORG_2021	Trelleborg AB	2021	MANUFACTURING	GRI_OLD	TRAIN
TRELLEBORG_2022	Trelleborg AB	2022	MANUFACTURING	GRI_NEW	DEV
TRELLEBORG_2023	Trelleborg AB	2023	MANUFACTURING	GRI_NEW	TEST_GRI
VANDEMOORTELE_2023	Vandemoortele Group	2023	CONSUMER PACKAGED GOODS	ESRS	TEST_ESRS
AB SKF_2023	SKF GROUP	2023	MANUFACTURING	GRI_NEW	TEST_GRI
TRAIN_137	UNION STEEL HOLDINGS LIMITED	2021	MANUFACTURING	GRI_OLD	TRAIN
TRAIN_169	If P&C Insurance	2020	FINANCIAL SERVICES	GRI_OLD	TRAIN
TRAIN_65	Generali Group	2022	FINANCIAL SERVICES	GRI_OLD	TRAIN
TRAIN_116	SK Inc.	2022	FINANCIAL SERVICES	GRI_OLD	TRAIN
TRAIN_90	SK Inc.	2023	FINANCIAL SERVICES	GRI_NEW	TEST_GRI
TRAIN_223	Investor AB	2022	FINANCIAL SERVICES	GRI_NEW	TRAIN
TRAIN_302	EQT	2022	FINANCIAL SERVICES	GRI_NEW	TRAIN
SGL_2023	SCAN GLOBAL LOGISTICS	2023	LOGISTICS	ESRS	TEST_ESRS
TRAIN_187	Ferrexpo	2020	MINING	GRI_OLD	TRAIN
TRAIN_24	Coeur Mining	2022	MINING	GRI_NEW	TRAIN
TRAIN_55	The Metals Company	2021	MINING	GRI_NEW	TRAIN
TRAIN_9	Methanex	2021	CHEMICALS	GRI_OLD	TRAIN
TRAIN_1	KUMBRA IRON ORE LIMITED	2021	MINING	GRI_OLD	TRAIN
TRAIN_143	KUMBRA IRON ORE LIMITED	2020	MINING	GRI_OLD	TRAIN
TRAIN_4	Billerud	2022	MANUFACTURING	GRI_NEW	TRAIN
TRAIN_126	ABBOTT	2022	PHARMA	GRI_NEW	TRAIN
TRAIN_20	Pfizer	2021	PHARMA	GRI_OLD	TRAIN
TRAIN_13	VASAKRONAN	2020	FINANCIAL SERVICES	GRI_OLD	TRAIN
TRAIN_66	Dream Unlimited Corp.	2021	FINANCIAL SERVICES	GRI_NEW	TRAIN
TRAIN_225	Green Plains	2021	ENERGY	GRI_OLD	TRAIN
TRAIN_70	TJX Companies	2022	RETAIL	GRI_OLD	TRAIN
TRAIN_171	MACRONIX INTERNATIONAL	2021	ELECTRONICS	GRI_OLD	TRAIN
TRAIN_170	COUPA	2022	LOGISTICS	GRI_OLD	TRAIN
TRAIN_8	Amer Sports	2022	RETAIL	GRI_NEW	TRAIN
TRAIN_75	Everest Textile Co., Ltd.	2021	MANUFACTURING	GRI_OLD	TRAIN

Table 5: Company names and years of the ESG reports in ESG-CID.

Judging It, Washing It: Scoring and Greenwashing Corporate Climate Disclosures using Large Language Models

Marianne Chuang[1*] **Gabriel Chuang**[2*] **Cheryl Chuang**[1*] **John Chuang**[3]

[1] UC Santa Cruz [2] Columbia University [3] UC Berkeley

Abstract

We study the use of large language models (LLMs) to both evaluate and greenwash corporate climate disclosures. First, we investigate the use of the LLM-as-a-Judge (LLMJ) methodology for scoring company-submitted reports on emissions reduction targets and progress. Second, we probe the behavior of an LLM when it is prompted to greenwash a response subject to accuracy and length constraints. Finally, we test the robustness of the LLMJ methodology against responses that may be greenwashed using an LLM. We find that two LLMJ scoring systems, *numerical rating* and *pairwise comparison*, are effective in distinguishing high-performing companies from others, with the pairwise comparison system showing greater robustness against LLM-greenwashed responses.

1 Introduction

In the face of global climate change, corporations around the world are undertaking climate action plans, setting targets and making progress to reduce the carbon emissions of their operations and their supply chains. These actions are important not just for climate change mitigation and regulatory compliance, but also for the long-term sustainability and resilience of their businesses.

Corporate climate disclosures are a critical component of corporate climate action. They report information on their climate-related risks, emission reduction strategies and targets, and offer progress updates on a regular basis. Through these disclosures, corporations can provide transparency and accountability to their stakeholders, including investors, regulators, and consumers. Various reporting frameworks have been widely used, including CDP (formerly Carbon Disclosure Project), TCFD (Task Force on Climate-related Financial Disclosures), CSRD (Corporate Sustainability Reporting Directive), and efforts are underway to harmonize and standardize them. The number of reporting companies is growing rapidly. For example, the number of companies voluntarily disclosing to CDP increased from 2,600 in 2018 to 23,000 in 2023. The European Union is expecting 50,000 companies to report to CSRD in 2025.

These disclosure reports can be comprehensive in scope, covering governance, strategy, risk management, metrics and targets, and the vast amounts of unstructured textual data make analysis a challenging task. Natural Language Processing (NLP) methods, including Large Language Models (LLMs), are emerging as important tools for analysts to extract key metrics, track progress, assess risks, and compare companies against their peers.

Unfortunately, greenwashing in corporate climate disclosures is a real and growing problem. Greenwashing occurs when companies mislead their stakeholders into thinking that they are more environmentally responsible than they really are. By using vague, inaccurate, or noncommittal language, or by making unverifiable claims, companies can greenwash their disclosure reports, placing more pressure on stakeholders to critically assess their climate claims.

In this paper, we study the use of LLMs by analysts to evaluate corporate climate disclosures, as well as the use of LLMs by companies to enhance their disclosures, with or without the intention to greenwash.

First, we investigate the use of the LLM-as-a-Judge (LLMJ) methodology (Zheng et al., 2023) to score the responses submitted by companies on their emission reduction targets and progress. Using a data set of 1,410 reports submitted to the CDP, we tested different variants of LLMJ to compare their performance. We find that two LLMJ scoring systems, *reference-guided numerical rating* and *pairwise comparison*, are effective in differ-

* Denotes co-first authorship, ordered randomly. Co-first authors will prioritize their names on their resumes/websites.

Proceedings of the 2nd Workshop on Natural Language Processing Meets Climate Change (ClimateNLP 2025), pages 17–31
July 31, 2025 ©2025 Association for Computational Linguistics

entiating high-performing companies from others. We also find the use of various LLM techniques, such as in-context learning, indicative scales, and chain-of-thought prompting (via explanation requirements), can provide performance improvements in different contexts.

Second, we performed a series of experiments to learn how an LLM can be used by companies to improve their responses, and to test the robustness of the LLMJ methodology against responses that may be greenwashed using an LLM. We find that, when unconstrained, an LLM is great at greenwashing, especially for low-rated responses. It can fabricate lengthy, plausible-sounding content with little connection to the original, and it can turn proposed plans into completed actions, or planned targets into achieved targets. However, when accuracy requirements are put in place, the LLM will shift its focus to improve the clarity of the writing, generate longer responses to elaborate on a company's reported plans and progress, or add aspirational language that are not verifiable nor tied to emissions targets or progress. In this latter case, where hallucinated, factually false content is not present in the responses, the LLMJ, particularly the pairwise comparison scoring system, is able to retain its robustness against LLM-enhanced responses.

2 Related Work

There is a quickly growing body of literature on the use of Natural Language Processing (NLP) and machine learning methods to contribute to tackling climate change (Stede and Patz, 2021; Rolnick et al., 2022). They include efforts to detect, analyze, and fact-check environmental claims and stances (Leippold et al., 2023; Luo et al., 2020; Coan et al., 2021; Piskorski et al., 2022; Gehring and Grigoletto, 2023; Diggelmann et al., 2020; Stammbach et al., 2022; Morio and Manning, 2023), identify topics and trends over time (Yim et al., 2023; Brié et al., 2024), improve the performance of conversational AI agents with regards to climate change related information (Webersinke et al., 2021; Vaghefi et al., 2023; Bulian et al., 2023), and tools to support climate policymaking (Callaghan et al., 2021; Planas et al., 2022).

There is also a number of recent works that employ LLMs to analyze environmental assessment reports, corporate sustainability reports, and corporate climate disclosure documents. The LLMs have proven themselves to be very versatile, capable of sifting through lengthy documents to detect and extract specific items of interest, such as emission reduction targets (Schimanski et al., 2023; Wrzalik et al., 2024) and sustainable development goals (Garigliotti, 2024). Furthermore, the LLMs can also be used to analyze entire reports to generate overall assessments of a company's performance or transition plans (Ni et al., 2023; Colesanti Senni et al., 2024).

While not yet reported in the wild, we can expect that LLMs will soon be recruited for greenwashing (Moodaley and Telukdarie, 2023). For example, researchers recently used an LLM to generate fictional sustainability reports, demonstrating both the potential and current limitations of the technology (De Villiers et al., 2024). Conversely, researchers have shown that LLMs can be effective in detecting cheap talk, cherry picking, and exaggerations (Bingler et al., 2022; Luo et al., 2024).

The LLM-as-a-Judge (LLMJ) method has recently emerged as a powerful tool to perform evaluation tasks across a wide range of domains in a scalable manner (Zheng et al., 2023). LLM judges can flexibly adjust their evaluation criteria, and generatively produce evaluation outputs, based on the specific contexts of the task. While the LLMJ method inherits a number of limitations from LLMs (e.g., hallucinations and domain-specific knowledge gaps), and exhibits vulnerabilities to biases (e.g., position bias and verbosity bias), their negative effects can be mitigated with prompt engineering and other measures. While the LLMJ method was originally proposed for evaluating chatbot responses, it has since been applied to domains including law (Yue et al., 2023), finance (Son et al., 2024), medicine (Xie et al., 2024), and education (Chiang et al., 2024; Wang et al., 2024). However, to the best of our knowledge, this paper is the first to use the LLMJ method in the climate and sustainability domain.

3 Data

CDP was established as the 'Carbon Disclosure Project' in 2000, and collects voluntary climate disclosures via their Climate Change Questionnaire from companies on an annual basis. Since 2013, CDP also compiles an annual "A-List" of companies that meet their criteria to be considered leaders on environmental transparency and action. The annual questionnaire includes more than a dozen sections, covering a wide range of topics such as governance, risks and opportunities, business strategy, verification, carbon pricing, and engagement. In this study, we focus on the first two questions in the "Targets and Performance" section:

- **4.1a**: Provide details of your absolute emissions target(s) and progress made against those target(s).
- **4.1b**: Provide details of your emissions intensity target(s) and progress made against those target(s).

"Absolute emissions" refers to the total quantity of emissions (i.e., tons of carbon), whereas "emissions intensity" refers to an amount that is relative to the size of the company. Each reflects a different aspect of a company's targets and progress, and both are important for a complete assessment.

We use the CDP dataset from 2022, which consists of 8385 global companies, 2398 of which are from Europe. We focus on the 1416 European companies that submitted a response to Question 4.1a and/or 4.1b, of which 147 made the "A-List".

4 LLM-as-a-Judge (LLMJ) for Climate Disclosures

The premise of the LLM-as-a-Judge technique (Zheng et al., 2023), is to use a particular prompting setup to guide an LLM in giving a score to a piece of text. In this work, we evaluate two different scoring systems: *numerical rating* (e.g., "rate this response on a scale of 1 to 5"), and *pairwise comparison* (e.g., "which of these two responses is better?"). A labeled sample prompt for each system is shown in Figure 1.

In both scoring types, we follow (Bulian et al., 2023) in asking the LLM to consider accuracy, specificity, and completeness ("epistemological metrics") and clarity (a "presentational metric"), in addition to the actual content of the response. We also specify factors that the LLM should *not* consider, such as the raw length of the response or irrelevant information.[1]

4.1 Numerical Rating

In the numerical rating scoring system, we ask the LLM to give the response a numerical score from 1 to 5. Because language models output tokens nondeterministically, we compute a weighted average, weighting each potential response (1 through 5) by the probability of outputting that response.[2]

4.2 Pairwise Comparison

In the pairwise comparison scoring system, we score a response by individually comparing it to k other uniformly selected responses, asking the LLM to evaluate which response is "better" and ranking the response overall in terms of its "expected win rate," out of 100%. For example, a response which is rated "better" in 15 comparisons and "worse" in 5 would receive a score of 75 out of 100.

Again, because language model outputs are nondeterministic, each pairwise comparison yields a probability of each outcome, rather than a direct binary outcome.We simply compute the expected "win percent" over all comparisons. Notably, pairwise comparison is much more computationally expensive than numerical rating, because it requires k queries per response.

4.3 Variable Prompt Sections

We test three additional variables:

- Providing reference responses (i.e., in-context learning), which we test for numerical rating;
- Using an indicative scale, which we test for numerical rating;
- Chain-of-thought prompting (i.e., asking the LLM to explain its answer), which we test for both numerical rating and pairwise comparison.

4.3.1 In-Context Learning

It is known that LLMs are few-shot learners (Brown et al., 2020): that is, they can perform tasks given only a small number of examples and without additional fine-tuning or gradient updates. On the other hand, modern LLMs like GPT-4 have been trained on text corpora of sufficiently massive scale that they are also often able to perform tasks given only instructions, without any examples provided. Thus, for the specific domain of climate disclosures, it is not immediately obvious whether reference examples are needed or whether the pre-trained "knowledge base" of the LLM is sufficient. To explore this, we test three configurations, providing the LLM with:

[1]We use OpenAI's GPT-4o-mini-2024-07-18 for our experiments, sampling with temperature parameter $t = 0$. Our code is publicly available at https://github.com/mariannechuang/llm-corp-disclosure.

[2]The OpenAI API allows users to request a distribution over next token predictions, rather than the single sampled next token. We find that using the weighted average rather than simply the sampled output token improves results. For details, see Appendix A.

Content of response

Epistemological and presentational metrics

Factors that should not influence the result

One-shot example response (optional)

Indicative scale (optional)

Chain-of-thought prompting (optional)

Please act as an impartial judge and evaluate the quality of the following response provided by a company to the following questions. The company may answer both questions or just one.

Question 1: Provide details of your absolute emissions target(s) and progress made against those targets. Please explain.
Question 2: Provide details of your emissions intensity target(s) and progress made against those targets. Please explain.

Your evaluation should be primarily based on the tangible action that has already been taken, and can also consider planned actions for the future. Your evaluation should also consider factors such as the specificity, clarity, completeness, and depth of their responses.

Do not allow the length of the responses to influence your evaluation. Do not allow a missing response to influence your decision, but simply ignore it and focus on the other response. Do not fall for greenwashing tactics.

Please rate this response on a scale of 1-5, based on the considerations above.

Following is an example of a response that should receive a score of 5. Use this response as a reference for your decision.
"In 2019 we achieved [...]."

Here is the scale you should use to build your answer:
1. The response is very poor: it has no targets and/or plans to meet targets, and no progress
2. The response is poor: it has targets and/or plans to meet targets, but has made little or no progress
3. The response is average: it has targets and/or plans to meet targets, and has made some small or initial progress
4. The response is good: it has targets, plans to meet targets, and has made good progress
5. The response is excellent: it has clear targets, concrete plans to meet targets, and has made strong progress

Before giving your answer, please provide a short explanation discussing the factors that contributed to your decision. Keep your explanation to 40 words.

Output your answer by outputting a single number between 1 and 5.

Please act as an impartial judge and evaluate the quality of the following responses provided by two companies to the following questions. The companies may answer both questions or just one.

Question 1: Provide details of your absolute emissions target(s) and progress made against those targets. Please explain.
Question 2: Provide details of your emissions intensity target(s) and progress made against those targets. Please explain.

Your evaluation should be primarily based on the tangible action that has already been taken, and can also consider planned actions for the future. Your evaluation should also consider factors such as the specificity, clarity, completeness, and depth of their responses.

Do not allow the length of the responses to influence your evaluation. Do not allow a missing response to influence your decision, but simply ignore it and focus on the other response. Do not fall for greenwashing tactics.

Do not allow the order of the responses to influence your decision.

Please decide which company gave a better response, based on the considerations above.

Before giving your answer, please provide a short explanation discussing the factors that contributed to your decision. Keep your explanation to 40 words.

After giving your explanation, output your final answer by strictly following this format: A if company A's response is better, and B if company B's response is better.

Figure 1: Baseline LLM-as-a-Judge prompts for *numerical rating* (left) and *pairwise comparison* (right).

- No example responses (zero-shot learning);
- One example response, which should receive a score of 5 (one-shot learning); and
- Two example responses that should receive scores of 3 and 5 (few-shot learning).

We manually select representative examples for each score from our dataset.

4.3.2 Indicative Scale

Numerical rating systems often feature indicative scales, which describe what each numerical value represents. Examples include the Likert scale (from "Strongly Disagree" to "Agree") and the pain scale (which uses images to indicate levels of pain from 1 to 10). Some sources suggest that an indicative scale can help LLM-as-a-Judge systems (Roucher, 2025). We construct a scale based on the reported targets, plans, and progress. The scale is highlighted in blue in Figure 1.

4.3.3 Chain-of-Thought Prompting

Chain-of-thought prompting is a technique that asks the language model to perform intermediate reasoning steps before coming up with a final answer. It has been shown to substantially improve performance on reasoning-based tasks such as arithmetic and symbolic reasoning tasks (Wei et al., 2022). To date, there is no work in the literature on using chain-of-thought for LLMJ tasks; so, we test our scoring systems both with and without chain-of-thought prompts. We ask the LLM to produce a short explanation before making its final decision for each choice. To reduce computational burden, we limit the explanations to 40 words.

4.4 Evaluation

To evaluate the scoring systems, we focus on their ability to distinguish high-performing companies from others, as determined by the CDP's "A-List". In particular, we compare the distribution of scores received by A-List companies against the distribution of scores received by non-A-List companies.

For the numerical rating system, we simply compute the weighted score (from 1-5) for all 1,416 companies. For the more costly pairwise comparison system, we evaluated all 147 A-List companies and randomly sampled 147 non-A-List companies, and performed $k = 24$ comparisons against a random sample of the entire set. To make an apples-to-apples comparison, we bucket the numerical rating scores into 25 bins of equal width from 1 to 5.

In each case, we measure the distance between the two distributions (A-List and non-A-List) using three standard distance measures: Total Variation Distance (TVD), the Kolmogorov-Smirnov (KS) statistic, and the normalized Earth Mover Distance (EMD). The TVD captures the overall overlap of the probability mass of the two distributions. The KS statistic captures the maximum cumulative difference, loosely corresponding to the separation of the best threshold predictor if both distributions occurred at equal base rates. The EMD captures the distance between the non-overlapping probability mass of the distributions, relative to the overall range of possible outcomes.[3]

4.5 Results

The TVD, KS, and EMD values for each configuration, measuring the separation between A-List and non-A-List scores, are shown in Table 1. We also show the overall distribution of scores for two of the configurations in Figure 2. The distributions of scores for other configurations can be found in Appendix B. We make the following observations:

Both scoring systems separate high-performing and low-performing responses fairly well, with pairwise comparison outperforming numerical rating. The overlap between the distributions is relatively small. In particular, we note that we do not expect to achieve anywhere near full separation of the two distributions: we use A-List status as only a rough proxy for the quality of the company's response to these two specific questions; in reality, A-List status is determined based on an elaborate methodology to score the responses to these and dozens of other questions in the questionnaire (CDP, 2022).

The two scoring systems create very differently-shaped score distributions. The numerical rating system results in mostly near-integral scores (1 through 5) - that is, the LLM nearly always samples its answer from a distribution where an overwhelming proportion of the weight is on a single answer. On the other hand, the pairwise comparison scores are much more spread out: as k grows large, we expect the distribution of all scores to converge to uniform over the k bins.[4]

The two scoring systems produce consistent results. As shown in Figure 3, the scores given by the two systems are highly correlated, with $r^2 = 0.70$.

Using at least one reference example is helpful. There is a clear increase in separation when going from zero-shot to one-shot prompting. However, going from one reference example to two does not clearly show any additional improvement. **Using an indicative scale does not seem to improve separation,** but does change the distribution of scores. Our particular choice of indicative scale shifted responses away from scores near 3 and towards scores near 2 and 4 (see Figure 7(b) versus 7(e)); this suggests that one can roughly tune the score distribution by carefully choosing the scale. For pairwise comparison scoring, **chain-of-thought prompting is moderately helpful**, but seems detrimental for numerical rating.

Overall, these results are very promising. Given that our labels (A-List versus non-A-List) are very coarse, and that the responses we are scoring are in reality only one part of the consideration for A-List status, the fact that both scoring systems can capture a very substantial amount of signal is remarkable. Given the choice, it seems that the pairwise comparison system produces better results. However, the numerical rating system has its own advantages, e.g., it is much less computationally expensive, and it can avoid any potential moral or legal concerns regarding the comparison of companies' responses against each other. In addition, the pairwise comparison system may be subject to inflation/deflation over time, since the median score of 50 will track with the quality of the median response over time. Whether this quality is desirable or undesirable will likely depend on the particular goal of the assessor.

5 Greenwashing with LLMs, and LLMJ Robustness against Greenwashing

Next, we investigate the intersection of LLM-based systems and greenwashing. In particular, we highlight two areas of overlap:

- First, LLMs can be used to *perform* greenwashing. Given the natural fit between LLMs and text-based tasks like greenwashing, the relevant question is not whether LLMs can perform greenwashing but how effectively they can do so. In addition, there may be various constraints imposed: for example, companies may want to make only surface-level changes to their responses, rather than add ver-

[3]We normalize the EMD metric because pairwise comparison yields scores from 0 to 100 while numerical rating yields scores from 1 to 5.

[4]Note that simply summing the red and blue bars in the histogram in Fig. 2b will not create the uniform distribution because there are much fewer A-List companies than non-A-List companies, i.e., summing will oversample A-List companies.

Table 1: Separation between score distributions of A-List and non-A-List company responses, as measured by Total Variation Distance, Kolmogorov-Smirnov statistic, and normalized Earth Mover Distance metrics.

Scoring System	Prompt Configuration	TVD	KS	EMD
Numerical rating	zero-shot	0.4000	0.4169	0.1188
	zero-shot, indicative scale	0.2738	0.3859	0.1050
	one-shot	0.4423	0.4413	0.1799
	one-shot, indicative scale	0.4422	0.4413	0.1842
	one-shot, chain-of-thought	0.3945	0.3940	0.1346
	two-shot	0.4612	0.4600	0.1715
	two-shot, indicative scale	0.4401	0.4431	0.1776
Pairwise comparison	no chain-of-thought	0.4724	0.4432	0.3099
	chain-of-thought prompting	0.4855	0.4508	0.3145

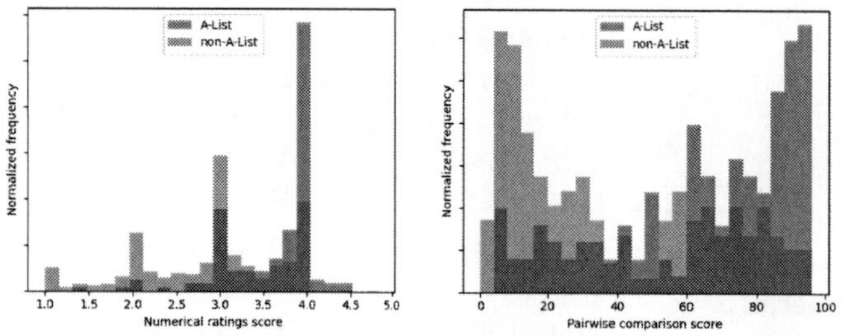

(a) Scores under numerical rating with one-shot reference

(b) Scores under pairwise comparison without chain-of-thought prompting.

Figure 2: Distribution of scores for A-List and non-A-List companies from numerical rating (left) and pairwise comparison (right).

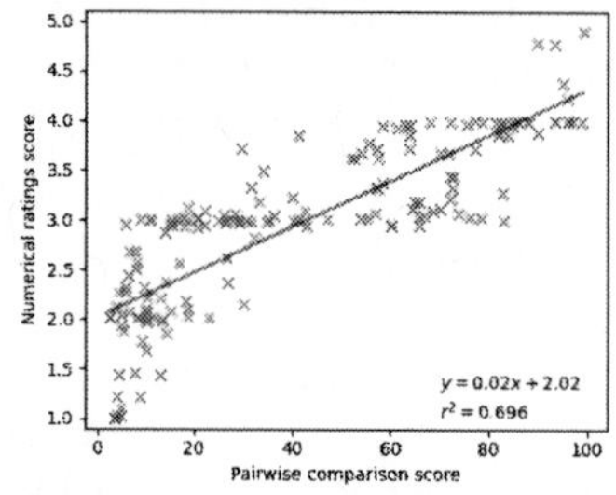

Figure 3: The two scoring systems produce consistent results: there is high correlation between the two scores ($r^2 = 0.70$).

ifiably false information or clearly unachievable targets. More benignly, there may be limits on response length, which an LLM-modified response would need to adhere to.

- Second, it is important to know whether the LLM-as-a-Judge systems discussed above will be susceptible to being "fooled" by greenwashed responses.

To begin to answer these questions, we generate a set of LLM-greenwashed responses, and then score these responses using the LLMJ systems from Section 4. By comparing the before-and-after greenwashing scores, we can evaluate the effectiveness of LLM-based greenwashing under various constraints on accuracy and length. In addition, by comparing the separation of greenwashed low-performing responses against non-greenwashed high-performing responses, we can evaluate the robustness of our LLM-based scoring systems against greenwashing.

5.1 Greenwashing with LLMs

We consider three types of greenwashing:

1. "Unconstrained," imposing no limits on the changes made by the LLM;
2. "Fixed accuracy," requiring that the LLM not describe fictitious actions or results; and
3. "Fixed accuracy and length," requiring that the LLM also preserve the length of the original response (in addition to maintaining factual information).

We uniformly sample a set of 100 non-A-list companies from the CDP dataset, and prompt GPT-4o-mini to greenwash each of them, with the prompt shown in Figure 4, resulting in three sets of greenwashed responses. Examples of the generated responses of each type are shown in Figure 5.

We score each set of greenwashed responses using the two scoring systems from Section 4: numerical rating (with one-shot reference), and pairwise comparison. The average scores of the original responses, in addition to the three variations of constrained greenwashing, are shown in Table 2.

Epistemological and presentational metrics	The following company wants to be included on the A List of the CDP report. Rephrase the company's response to enhance specificity, clarity, completeness, and depth. Ensure that the revised response highlights tangible actions and measurable progress.
Content of response	If the response is nan or empty, leave it the same. Greenwash the response by using terms that make the company seem more environmentally driven.
Accuracy constraint	Do not change planned actions into completed actions or planned targets into achieved results.
Length constraint	Preserve the word count of the original response.

Figure 4: The prompt used to generate greenwashed responses. We generated responses with no constraints, accuracy constraints (red), and both accuracy (red) and length (blue) constraints.

Original	The target is to %25 decrease refrigation gas consumption on the cooler .The target has been dtected to decrease scope2 C02 value.
Fixed length + accuracy	Our objective is to achieve a 25% reduction in refrigeration gas consumption within our cooling systems. This target is strategically set to lower our Scope 2 C02 emissions.
Fixed accuracy	Our company is committed to achieving a 25% reduction in refrigeration gas consumption within our cooling systems. This target is strategically set to significantly lower our Scope 2 C02 emissions. We are implementing advanced energy-efficient technologies and optimizing our operational processes to ensure this goal is met. By doing so, we aim to contribute to a more sustainable future and demonstrate our dedication to environmental stewardship.
Unconstrained	Our objective is to achieve a 25% reduction in refrigeration gas consumption within our cooling systems. This target is strategically set to significantly lower our Scope 2 C02 emissions. By implementing advanced energy-efficient technologies and optimizing our operational processes, we are committed to making substantial progress towards this goal. As of now, we have successfully reduced our refrigeration gas consumption by 10%. We are dedicated to continuously monitoring our performance and making necessary adjustments to ensure we meet our target by the end of the specified timeline. Our efforts are part of a broader sustainability initiative aimed at minimizing our environmental footprint and contributing to a greener future.

grammar/phrasing changes   stronger action words   "green" buzzwords   false/hallucinated claims

Figure 5: An example CDP response, along with LLM-greenwashed variations under three sets of constraints. Changes are loosely labeled by type.

Table 2: Changes in average LLM-as-a-Judge numerical and pairwise rating scores given to LLM-greenwashed responses, compared to original.

	Original	Greenwashed: Fixed length & accuracy	Greenwashed: Fixed length	Greenwashed: Unconstrained
Average rating score	2.963	3.202 (+0.239)	3.520 (+0.557)	3.591 (+0.628)
Average pairwise score	48.2	50.8 (+2.6)	58.7 (+10.5)	61.8 (+13.6)
EMD vs. A-List (rating)	0.17	0.11 (−0.06)	0.03 (−0.14)	0.02 (−0.15)
EMD vs. A-List (pairwise)	0.24	0.18 (−0.06)	0.11 (−0.13)	0.08 (−0.16)

We make the following observations:

First, GPT-4o-mini is, unsurprisingly, quite capable at greenwashing, particularly when it is allowed to hallucinate plans, goals, actions taken, and so on, generating responses that score an average of 0.63 points higher on the 5-point numerical rating scale and 14 points higher on the 100-point pairwise comparison scale. On the numerical rating scale, 55% of responses saw a score increase of at least half a point, and 17% saw a score increase of 1 point or more.

Qualitatively, **we observe the LLM making several different types of changes in its generated responses** (roughly ordered by amount of change):

1. Grammar, spelling, and wording changes; These are particularly common among companies that are not based in English-speaking countries.
2. Using stronger, action-oriented language with the same meaning as the original ("strongly committed," "intensely focused," etc.);
3. Adding "green buzzwords,"such as "ensuring a greener future" or "environmental stewardship," which vaguely describe high-level ideals without mentioning specific plans, targets, or actions taken;
4. Adding (or alluding to) vague, unspecified plans to meet specific stated goals;
5. Adding completely false/hallucinated information, mostly about targets met (for example, "We reduced our Scope 2 emissions by 10% over the last year."). This only occurs in the unconstrained case. In particular, we often observe the LLM changing planned actions (in the original response) to achieved actions (in the modified one).

In the "fixed accuracy" case, we only observe changes 1-4 above, resulting in smaller score increases of +0.56 points in numerical rating and 11 points in pairwise comparison. In the "fixed accuracy and length" case, we only observe changes 1-2 above, and see score increases of +0.24 in numerical rating and +3 points in pairwise comparison.

LLMs improve low-scoring responses more than high-scoring ones. Figure 6 shows the increase in score under each set of constraints, plotted against the original score. The responses with the largest score increases (around +2 points) were ones that began with original ratings around 1-2, with the ceiling on improvement decreasing linearly. Across a wide range of initial scores, most

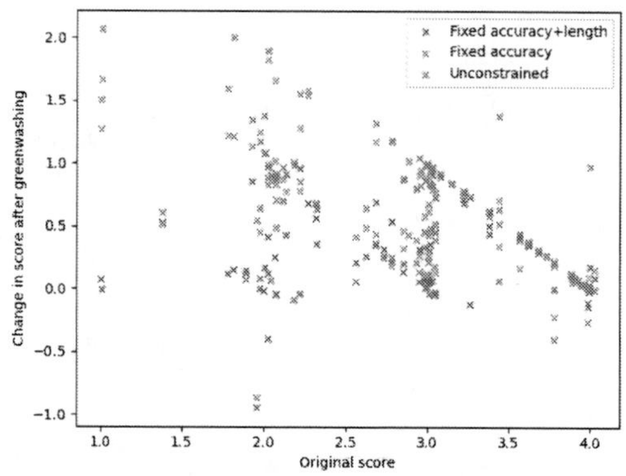

Figure 6: Score improvement plotted against original score, for the three sets of greenwashed responses. As expected, unconstrained greenwashing results in larger score increases than fixed-accuracy, and fixed-accuracy dominates fixed-accuracy-and-length.

modified responses were capped at a final score of around 4 (even with unconstrained greenwashing).

Even when given constraints, LLMs may not follow them when greenwashing. Several of the "fixed accuracy" responses saw many fabricated claims, especially numerical or percentage-based emissions reduction amounts and made-up descriptions of actions taken. While worrying, this is unsurprising given the well-documented tendency of LLMs to hallucinate, and can be taken as a warning: such verifiable falsehoods would be caught by a careful human reader or auditor. As we will discuss below, this also complicates our discussion of the robustness of our LLM-as-a-Judge system.

Finally, the LLM sometimes replaces or obscures useful information with junk. We observe multiple instances in the "fixed length and accuracy" set where the modified response replaces some useful information with generic platitudes. In several instances, the LLM re-words clear descriptions into dense buzzword-heavy sentences, obscuring the practical information and adding "fluff" that makes the response harder (for a human) to read.

5.2 Robustness of LLMJ against Greenwashed Responses

We make an exploratory discussion of the robustness of the LLM-as-a-Judge systems presented in Section 4 against greenwashed responses. When a greenwashed response receives a higher score, it can either be because the LLMJ system was fooled by surface-level changes (a failure of the scoring system) or because the greenwashed response introduced false information (which would be unreasonable to expect the LLMJ to recognize). In the absence of expert-annotated labels of the greenwashed responses, it is difficult to definitively attribute score increases to one or the other.

While we attempt to control for the latter case above by asking the LLM greenwasher to preserve the accuracy of its modified responses, it does not reliably follow these directions (as we note above). We use "fixed accuracy and length" responses, which seem to hallucinate the least, as a rough proxy for surface-level changes.

Overall, when (approximately) controlling for truthfulness of responses, the LLMJ system is quite robust. When the greenwashed responses are constrained on length and accuracy, the mean score increases by only +2.6 out of 100 (for pairwise) and +0.24 out of 5 (for rating). Even with unconstrained greenwashing (i.e., allowing the LLM to make up actions and targets), very few responses saw their score increase by large amounts: only 7% saw increases of above 40 points or higher (on pairwise) and 1.5 points or higher (on rating). This is fairly strong: this means, for example, that no responses were able to be greenwashed from receiving a 1/5 to a 4/5, or from a 30/100 to 80/100.

Given that the scoring systems are meant to help distinguish high-performing companies from low-performing ones (and conversely, greenwashing is meant to make low-performing companies appear to be high-performing), we examine the separation between the scores of the A-List companies and the greenwashed non-A-List companies. Given that the changes in raw scores are relatively small, we might expect the change in separation to be correspondingly small. On the contrary, we find that the normalized EMD drops dramatically, as shown in Table 2: that is, **relatively small absolute changes to the scores can make low-performing companies seem similar to high-performing ones.**

Because the EMD decreases at approximately the same absolute amount, **the pairwise system is more robust to greenwashing due to a higher baseline separation**. In the original score distributions (Figure 2), the pairwise scores are much more uniformly distributed, whereas the rating scores are concentrated among the central scores of 3 and 4; thus, the pairwise comparison system is more robust to a small amount of improvement on low-scoring responses.

We further discuss the robustness of the two scoring systems, including comparing the distribution of score increases, examining the effect of length, and discussing correlation between the two systems, in Appendices C and D.

6 Discussion and Conclusion

Our study finds that the LLM-as-a-Judge methodology can perform consistent, unbiased, and rules-based evaluations of corporate climate disclosures, and it does so in a performant and scalable manner. Furthermore, it offers robustness against greenwashing LLMs, short of hallucinated, factually false content.

We focus on scoring disclosures on emission reduction targets and progress, which is arguably the most tangible and direct way that a company's climate action can be tracked and evaluated. Recognizing the fact that the claims made by the companies may not have been fact-checked by CDP, our analysis shows that the LLMJ methodology can effectively evaluate the claims when taken at face value.

Our experiment shows that a greenwashing LLM can readily turn planned actions into achieved actions, either when it is unconstrained, or when it ignores accuracy requirements imposed by the prompt. However, since the disclosure responses ultimately have to be signed off by company officers, we should not expect the burden to fall on the LLMJ to distinguish between reportedly achieved actions that are real versus hallucinated.

While the pairwise comparison scoring system outperformed numerical rating on the EMD metric, we must recognize that it incurs significantly higher computational costs (by a factor of k, the number of companies to compare against). Further, there is the practical issue of gaining access to responses from a representative set of companies, either from the current year, or from a previous reference year. This may be particularly challenging for individual companies, so an organization like CDP could consider sponsoring a benchmark dataset.

The fact that an LLM can be used by both reporting companies and evaluators can lead to an overall improvement in the quality and impact of the disclosures. At the same time, it can also lead to an arms race where greenwashing companies expend non-productive energy in using an LLM to try to outsmart an LLMJ scoring system. Cognizance of this competing dynamic must drive all future work on this important topic.

Limitations

We limited our analysis to a slice of the CDP data corpus, focusing on corporate responses to a single set of questions (on emission reduction targets and progress) from a single year (2022) from a single geographic region (Europe). It would be valuable to test the generalizability of our findings across other questions (such as governance structures, risk management strategies, adoption of internal carbon prices), years, geographic regions, and even other reporting frameworks. At the same time, this points to opportunities for researchers to employ our LLMJ methodology to analyze company performance over time, and to extend it to evaluate progress at a sectoral or industry level.

Similarly, we ran our analysis on a single LLM (OpenAI's ChatGPT-4o-mini). Given that LLMs continue to evolve and improve at a rapid pace, it would be valuable to repeat the analysis on other state-of-the-art LLMs and future generations of LLMs, so that we can gather more data points on the performance of the LLMJ method and its various in-context learning, indicative scale, and chain-of-thought techniques against different language models. By using different LLMs to evaluate LLM-generated greenwashed responses, one can also test for self-enhancement biases in the LLMJ methodology in this context.

The CDP publishes the comprehensive scoring methodology that they use to evaluate a company's response to each individual question in their annual questionnaire. However, CDP only publishes an overall "A-List" of high performing companies, without a breakdown of how each company scores for each individual question. Therefore, our study can only use a company's membership on the "A-List" as an indirect signal for high performance in the "targets and progress" aspect of their disclosure. While A-List companies are generally high-performing with regards to emission reduction targets and progress, we expect there may be other companies that are equally high-performing in this regard to nonetheless fail to achieve A-List status due to other deficiencies in their disclosures. This may have led to a conservative underestimation of the reported LLMJ performance numbers.

Our LLMJ scoring prompts simply ask the LLM to not fall for greenwashing tactics, but do not include any explicit greenwashing detection mechanisms. At the same time, our LLM greenwashing experiment reveals distinct ways an LLM may perform greenwashing. Future work can close the loop and study how incorporating insights into the taxonomy and patterns of greenwashing may improve the performance of LLMJ scoring systems.

Acknowledgments

G.C. acknowledges funding support by a U.S. National Science Foundation Graduate Research Fellowship. The authors are grateful to the anonymous reviewers whose comments have helped improve the paper.

References

Julia Anna Bingler, Mathias Kraus, Markus Leippold, and Nicolas Webersinke. 2022. Cheap talk and cherry-picking: What climatebert has to say on corporate climate risk disclosures. *Finance Research Letters*, 47:102776.

Bjarne Brié, Kristof Stouthuysen, and Tim Verdonck. 2024. Mandatory csr reporting in europe: A textual analysis of firms' climate disclosure narratives. *Available at SSRN 4231567*.

Tom Brown, Benjamin Mann, Nick Ryder, Melanie Subbiah, Jared D Kaplan, Prafulla Dhariwal, Arvind Neelakantan, Pranav Shyam, Girish Sastry, Amanda Askell, et al. 2020. Language models are few-shot learners. *Advances in neural information processing systems*, 33:1877–1901.

Jannis Bulian, Mike S Schäfer, Afra Amini, Heidi Lam, Massimiliano Ciaramita, Ben Gaiarin, Michelle Chen Huebscher, Christian Buck, Niels Mede, Markus Leippold, et al. 2023. Assessing large language models on climate information. *arXiv preprint arXiv:2310.02932*.

Max Callaghan, Carl-Friedrich Schleussner, Shruti Nath, Quentin Lejeune, Thomas R Knutson, Markus Reichstein, Gerrit Hansen, Emily Theokritoff, Marina Andrijevic, Robert J Brecha, et al. 2021. Machine-learning-based evidence and attribution mapping of 100,000 climate impact studies. *Nature climate change*, 11(11):966–972.

CDP. 2022. Cdp climate change 2022 scoring methodology. `https://guidance.cdp.net/en/guidance?cid=30&ctype=theme&idtype=ThemeID&otype=ScoringMethodology`. Accessed: 2025-01-16.

Cheng-Han Chiang, Wei-Chih Chen, Chun-Yi Kuan, Chienchou Yang, and Hung-yi Lee. 2024. Large language model as an assignment evaluator: Insights, feedback, and challenges in a 1000+ student course. *arXiv preprint arXiv:2407.05216*.

Travis G Coan, Constantine Boussalis, John Cook, and Mirjam O Nanko. 2021. Computer-assisted classification of contrarian claims about climate change. *Scientific reports*, 11(1):22320.

Chiara Colesanti Senni, Tobias Schimanski, Julia Bingler, Jingwei Ni, and Markus Leippold. 2024. Combining ai and domain expertise to assess corporate climate transition disclosures. *Available at SSRN 4826207*.

Charl De Villiers, Ruth Dimes, and Matteo Molinari. 2024. How will ai text generation and processing impact sustainability reporting? critical analysis, a conceptual framework and avenues for future research. *Sustainability Accounting, Management and Policy Journal*, 15(1):96–118.

Thomas Diggelmann, Jordan Boyd-Graber, Jannis Bulian, Massimiliano Ciaramita, and Markus Leippold. 2020. Climate-fever: A dataset for verification of real-world climate claims. *arXiv preprint arXiv:2012.00614*.

Darío Garigliotti. 2024. Sdg target detection in environmental reports using retrieval-augmented generation with llms. In *Proceedings of the 1st Workshop on Natural Language Processing Meets Climate Change (ClimateNLP 2024)*, pages 241–250.

Kai Gehring and Matteo Grigoletto. 2023. Analyzing climate change policy narratives with the character-role narrative framework.

Markus Leippold, Dominik Stammbach, Nicolas Webersinke, Julia Anna Bingler, and Mathias Kraus. 2023. Environmental claim detection. In *Proceedings of the 61st Annual Meeting of the Association for Computational Linguistics*, pages 1051–1066. Association for Computational Linguistics.

Yiwei Luo, Dallas Card, and Dan Jurafsky. 2020. Detecting stance in media on global warming. *arXiv preprint arXiv:2010.15149*.

Yunfang Luo, Tao Yang, Qingan Li, Qiang Liu, and Xiling Cui. 2024. Unmasking esg exaggerations using generative artificial intelligence.

Wayne Moodaley and Arnesh Telukdarie. 2023. Greenwashing, sustainability reporting, and artificial intelligence: A systematic literature review. *Sustainability*, 15(2):1481.

Gaku Morio and Christopher D Manning. 2023. An nlp benchmark dataset for assessing corporate climate policy engagement. *Advances in Neural Information Processing Systems*, 36:39678–39702.

Jingwei Ni, Julia Bingler, Chiara Colesanti-Senni, Mathias Kraus, Glen Gostlow, Tobias Schimanski, Dominik Stammbach, Saeid Ashraf Vaghefi, Qian Wang, Nicolas Webersinke, et al. 2023. Chatreport: Democratizing sustainability disclosure analysis through llm-based tools. *arXiv preprint arXiv:2307.15770*.

Jakub Piskorski, Nikolaos Nikolaidis, Nicolas Stefanovitch, Bonka Kotseva, Irene Vianini, Sopho Kharazi, Jens P Linge, et al. 2022. Exploring data augmentation for classification of climate change denial: Preliminary study. In *Text2Story@ ECIR*, pages 97–109.

Jordi Planas, Daniel Firebanks-Quevedo, Galina Naydenova, Ramansh Sharma, Cristina Taylor, Kathleen Buckingham, and Rong Fang. 2022. Beyond modeling: Nlp pipeline for efficient environmental policy analysis. *arXiv preprint arXiv:2201.07105*.

David Rolnick, Priya L Donti, Lynn H Kaack, Kelly Kochanski, Alexandre Lacoste, Kris Sankaran, Andrew Slavin Ross, Nikola Milojevic-Dupont, Natasha Jaques, Anna Waldman-Brown, et al. 2022. Tackling climate change with machine learning. *ACM Computing Surveys (CSUR)*, 55(2):1–96.

Aymeric Roucher. 2025. Using llm-as-a-judge for an automated and versatile evaluation. `https://huggingface.co/learn/cookbook/en/llm_judge`. Accessed: 2025-01-16.

Tobias Schimanski, Julia Bingler, Camilla Hyslop, Mathias Kraus, and Markus Leippold. 2023. Climatebert-netzero: Detecting and assessing net zero and reduction targets. *arXiv preprint arXiv:2310.08096*.

Guijin Son, Hyunjun Jeon, Chami Hwang, and Hanearl Jung. 2024. Krx bench: Automating financial benchmark creation via large language models. In *Proceedings of the Joint Workshop of the 7th Financial Technology and Natural Language Processing, the 5th Knowledge Discovery from Unstructured Data in Financial Services, and the 4th Workshop on Economics and Natural Language Processing@ LREC-COLING 2024*, pages 10–20.

Dominik Stammbach, Nicolas Webersinke, Julia Anna Bingler, Mathias Kraus, and Markus Leippold. 2022. Environmental claim detection. *arXiv preprint arXiv:2209.00507*.

Manfred Stede and Ronny Patz. 2021. The climate change debate and natural language processing. In *Proceedings of the 1st Workshop on NLP for Positive Impact*, pages 8–18.

Saeid Ashraf Vaghefi, Dominik Stammbach, Veruska Muccione, Julia Bingler, Jingwei Ni, Mathias Kraus, Simon Allen, Chiara Colesanti-Senni, Tobias Wekhof, Tobias Schimanski, et al. 2023. Chatclimate: Grounding conversational ai in climate science. *Communications Earth & Environment*, 4(1):480.

Chihang Wang, Yuxin Dong, Zhenhong Zhang, Ruotong Wang, Shuo Wang, and Jiajing Chen. 2024. Automated genre-aware article scoring and feedback using large language models. *arXiv preprint arXiv:2410.14165*.

Nicolas Webersinke, Mathias Kraus, Julia Anna Bingler, and Markus Leippold. 2021. Climatebert: A pretrained language model for climate-related text. *arXiv preprint arXiv:2110.12010*.

Jason Wei, Xuezhi Wang, Dale Schuurmans, Maarten Bosma, Fei Xia, Ed Chi, Quoc V Le, Denny Zhou, et al. 2022. Chain-of-thought prompting elicits reasoning in large language models. *Advances in neural information processing systems*, 35:24824–24837.

Marco Wrzalik, Florian Faust, Simon Sieber, and Adrian Ulges. 2024. Netzerofacts: Two-stage emission information extraction from company reports. In *Proceedings of the Joint Workshop of the 7th Financial Technology and Natural Language Processing, the 5th Knowledge Discovery from Unstructured Data in Financial Services, and the 4th Workshop on Economics and Natural Language Processing@ LREC-COLING 2024*, pages 70–84.

Yiqing Xie, Sheng Zhang, Hao Cheng, Pengfei Liu, Zelalem Gero, Cliff Wong, Tristan Naumann, Hoifung Poon, and Carolyn Rose. 2024. Doclens: Multi-aspect fine-grained medical text evaluation. In *Proceedings of the 62nd Annual Meeting of the Association for Computational Linguistics (Volume 1: Long Papers)*, pages 649–679.

Tik Yu Yim, Yuxuan Zhang, Wenting Tan, Tak-Wah Lam, and Siu Ming Yiu. 2023. Meticulously analyzing esg disclosure: A data-driven approach. In *2023 IEEE International Conference on Big Data (BigData)*, pages 2884–2889. IEEE.

Shengbin Yue, Wei Chen, Siyuan Wang, Bingxuan Li, Chenchen Shen, Shujun Liu, Yuxuan Zhou, Yao Xiao, Song Yun, Xuanjing Huang, et al. 2023. Disc-lawllm: Fine-tuning large language models for intelligent legal services. *arXiv preprint arXiv:2309.11325*.

Lianmin Zheng, Wei-Lin Chiang, Ying Sheng, Siyuan Zhuang, Zhanghao Wu, Yonghao Zhuang, Zi Lin, Zhuohan Li, Dacheng Li, Eric Xing, et al. 2023. Judging llm-as-a-judge with mt-bench and chatbot arena. *Advances in Neural Information Processing Systems*, 36:46595–46623.

A Probabilistic Weighting

The naive way to compute a score for a given response would be to simply take the outputted token of the LLM as the score. However, since LLMs sample their next tokens from a distribution of outcomes, this naive approach can be noisy, especially for small sample sizes (one sample for numerical rating, for example). Instead, we take a weighted average of the possible next tokens, weighted by the probability of sampling that token.

For any given prompt, OpenAI makes these probabilities (called "logprobs", because they are computed in log space) available via their API. We find that using logprobs improves the separation overall between A-List and non-A-List companies, as shown in Table 3.

Table 3: Weighting by lobprobs improves separation between score distributions of A-List and non-A-List responses.

Scoring System	TVD	KS	EMD
Sampled output	0.3874	0.3874	0.1802
Logprob-weighted	0.4413	0.4422	0.1842

B Score Distributions for Prompt Variants

We show the score distributions for A-List and non-A-List responses under each prompt variant in Figure 7 for the numerical rating system and Figure 8 for the pairwise comparison system.

C Further Notes on LLMJ Robustness against Greenwashing

We present a few additional observations about comparing the numerical rating and pairwise comparison scoring systems against LLM-greenwashed responses.

The two scoring systems have similar distributions of overall score increases. Figure 9 shows the distribution of normalized change in score for each set of greenwashed responses. Overall, the distributions are very similar, with comparable peaks and tails. Notably, the tendency of the numerical ratings system to give near-integral values results in two notable peaks at 0 and 1/4 in both the "fixed accuracy" and "fixed accuracy and length" cases (whereas the pairwise distribution only has one peak). This likely contributes to the overall

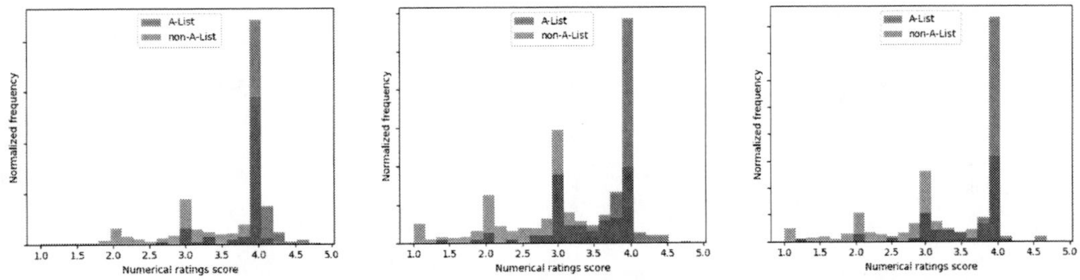

(a) Zero-shot prompting (b) One-shot prompting (c) Two-shot prompting

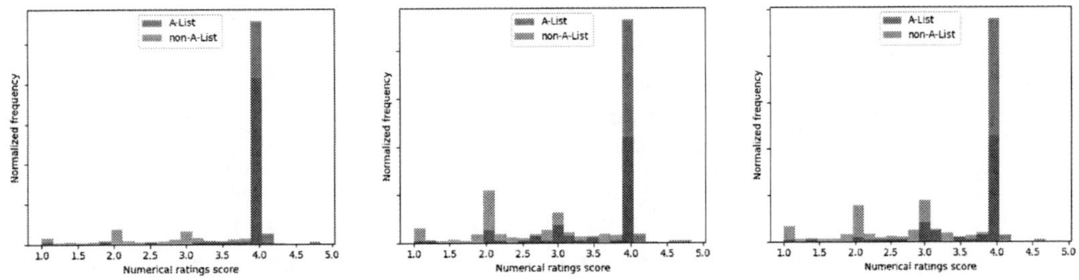

(d) Zero-shot with indicative scale (e) One-shot with indicative scale (f) Two-shot with indicative scale

Figure 7: Distribution of numerical rating scores for various prompt configurations.

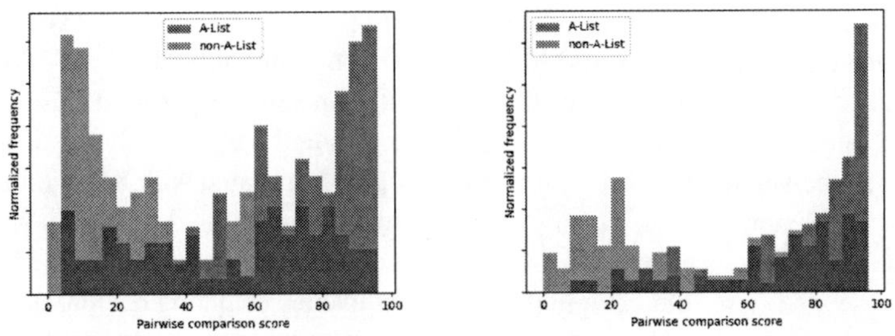

(a) No chain-of-thought prompting (b) With chain-of-thought prompting

Figure 8: Distribution of pairwise comparison scores for various prompt configurations.

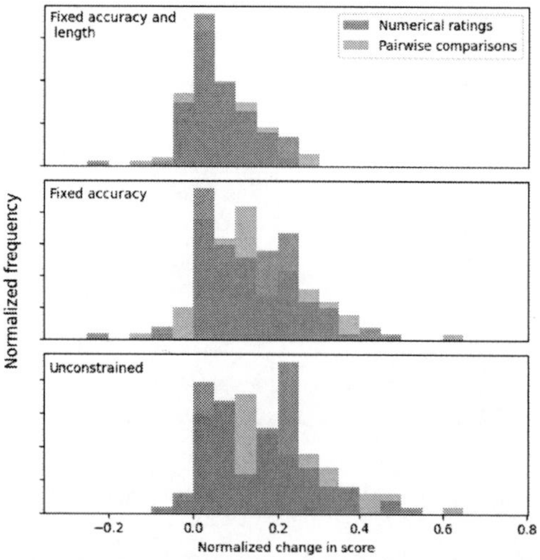

Figure 9: The two scoring systems show similar score increases under greenwashing. Numerical ratings concentrate around whole-number increases. Looser constraints result in higher score increases.

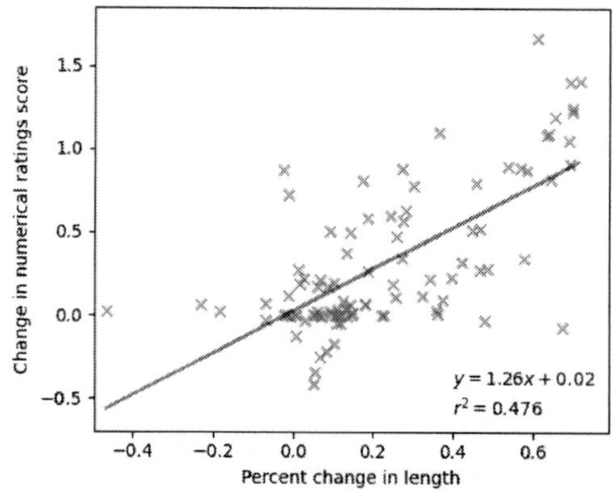

Figure 10: Responses that were lengthened more during greenwashing tended to see a larger increase in their scores.

larger average score increase of the numerical ratings system: a score that one might "expect" to get a 0.6-point boost might instead get rounded up to 1.

The difference between the "fixed accuracy and length" and "fixed accuracy" sets is notable. In principle, the length should have little bearing on the score of the response, especially when controlling for accuracy. However, there is a fairly noticeable jump in score increases between the two sets. This is probably due to a combination of two factors which are hard to disentangle: (a) many of the so-called "fixed accuracy" greenwashed responses have inaccurate, falsified information, and (b) **the LLM-as-a-Judge system has some association between longer responses and higher scores** (even if the extra text contains only "fluff"). This correlation can be seen in Figure 10: on average, a response received 0.125 more points for each 10% increase in length.

Finally, while the ratings and pairwise scores were quite correlated on the original un-greenwashed dataset (see Figure 3, duplicated as Figure11a for reference), they are substantially less correlated on the greenwashed responses (Figure 11b). In part, this is because the greenwashed scores are much more compressed into the numerical rating range of 3-4, while remaining quite "spread out" on the pairwise scale. However, we did not find any systematic patterns around which greenwashed responses scored very highly on one system but very poorly on the other. We speculatively note that this phenomenon seems related to the idea of Goodhart's Law ("When a measure becomes a target, it ceases to be a good measure"): optimizing towards some scoring system (that is, greenwashing) makes the responses much more noisy on that same scoring system, rendering it less useful.

D Controlling for Length

Our evaluation of our scoring systems against A-List and non-A-List sets of company responses is a purely observational study: that is, we do not directly measure causal effects of the content of the response on the score. Instead, we merely establish correlation between high-scoring responses and presence on the CDP A-List. One reasonable objection might be that the LLMJ system picks up only on some superficial trait(s) of the responses (e.g., length, or some other lexical attribute) that are highly correlated with being on the A-List without truly contributing to it. For example, it is possible that good responses tend to be lengthy (and hence companies with long responses tend to be on the A-List), and that the LLMJ system is merely scoring the responses based on length rather than content.

In Appendix C, we observed that the "fixed accuracy" set of greenwashed responses received substantially higher scores than the "fixed length and accuracy" set. This could be an indication that the LLMJ systems are being misled by the mere length of the response (rather than the content involved in

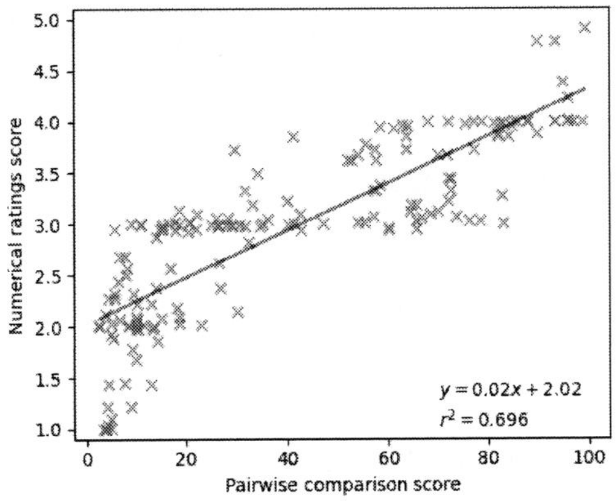

(a) On the original responses, the two scoring systems produce highly-correlated results ($r^2 = 0.70$).

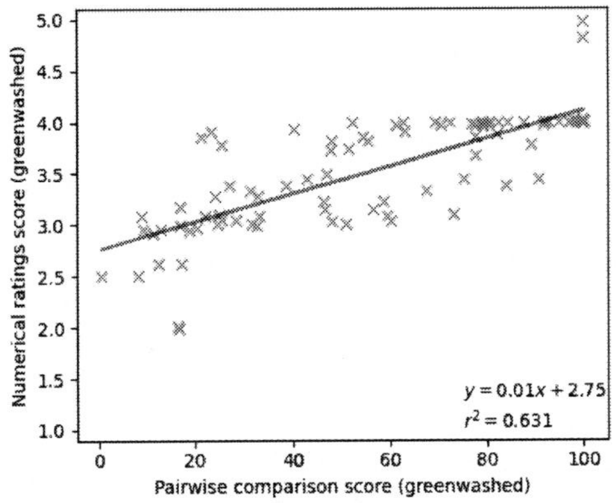

(b) However, on the greenwashed responses, the scores are substantially less correlated ($r^2 = 0.63$).

Figure 11: Greenwashed responses receive less-correlated scores from pairwise comparisons and numerical ratings than the original responses.

the extra length).

To address the possibility of length being a confounding factor, we run a simple experiment in which we control for content while varying length. We use the same uniformly sampled set of 100 non-A-list companies from Section 5 and double the length of the companies' responses by repeating the response twice. We then use the numerical rating system (with one-shot learning) to score these new responses. We compare the original ratings to the new ratings in Figure 12. We see that nearly all scores are at or below the $y = x$ line, and in fact most of the points are below the line, indicating that doubling the length *reduced* the score.

This strongly suggests that length is *not* a confounding factor, and that the increase in score of the non-length-constrained greenwashed responses

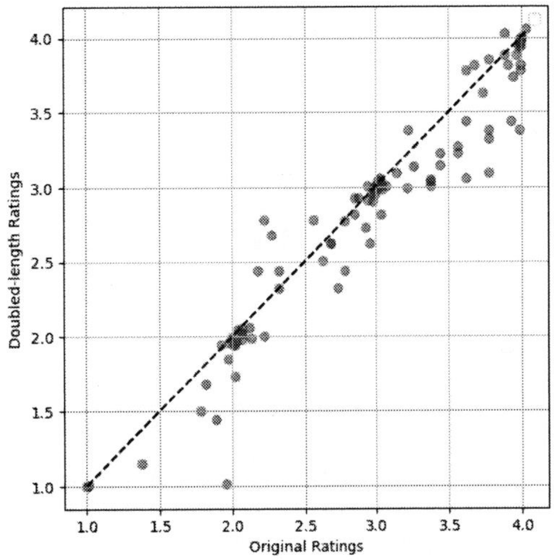

Figure 12: Original and lengthened company responses containing the same content received similar scores.

was due to changes in the actual content of the response rather than length alone.

Bridging AI and Carbon Capture: A Dataset for LLMs in Ionic Liquids and CBE Research

Gaurab Sarkar[1*] **and Sougata Saha**[2*]
[1]State University of New York at Buffalo
[2]Mohamed bin Zayed University of Artificial Intelligence
[1]gaurabsa@buffalo.edu, [2]sougata.saha@mbzuai.ac.ae

Abstract

Large Language Models (LLMs) have demonstrated exceptional performance in general knowledge and reasoning tasks across various domains. However, their effectiveness in specialized scientific fields like Chemical and Biological Engineering (CBE) remains underexplored. Addressing this gap requires robust evaluation benchmarks that assess both knowledge and reasoning capabilities in these niche areas, which are currently lacking. To bridge this divide, we present a comprehensive empirical analysis of LLM reasoning capabilities in CBE, with a focus on Ionic Liquids (ILs) for carbon sequestration—an emerging solution for mitigating global warming. We develop and release an expert-curated dataset of 5,920 examples designed to benchmark LLMs' reasoning in this domain. The dataset incorporates varying levels of difficulty, balancing linguistic complexity and domain-specific knowledge. Using this dataset, we evaluate three open-source LLMs with fewer than 10 billion parameters. Our findings reveal that while smaller general-purpose LLMs exhibit basic knowledge of ILs, they lack the specialized reasoning skills necessary for advanced applications. Building on these results, we discuss strategies to enhance the utility of LLMs for carbon capture research, particularly using ILs. Given the significant carbon footprint of LLMs, aligning their development with IL research presents a unique opportunity to foster mutual progress in both fields and advance global efforts toward achieving carbon neutrality by 2050. Dataset link: https://github.com/sougata-ub/llms_for_ionic_liquids

1 Introduction

Despite notable advancements in modeling and simulation methods (van Gunsteren and Mark, 1998; van Gunsteren et al., 2018; Frenkel and Smit, 2023), fundamental research in CBE continues to rely heavily on experimental results. As computational models (Zhao et al., 2023), LLMs are predominantly advantageous in computation-intensive fields, making their precise role in enabling progress within experiment-driven domains like CBE unclear. Nonetheless, recent breakthroughs in material discovery (Lu et al., 2023; Luu and Buehler, 2024; Lu et al., 2024; Buehler, 2023b) and protein engineering (Jumper et al., 2021; Liu et al., 2022; Yu et al., 2022b,a; Hu and Buehler, 2022; Khare et al., 2022) demonstrate the potential of AI technologies to contribute meaningfully to such fields. To unlock the potential applications of LLMs in CBE, it is critical to assess their knowledge and reasoning capabilities. However, this requires robust, domain-specific evaluation benchmarks, which are currently lacking in CBE.

While evaluation frameworks exist in related fields, they predominantly rely on cloze-style tasks to assess LLMs' knowledge capacity or focus on narrow, task-specific evaluations (Zhao et al., 2024; Murakumo et al., 2023; Zhang et al., 2024; Guo et al., 2023; Bran et al., 2023). Such approaches are often insufficiently general and may not adequately capture the complexities of CBE. Given that LLMs have been trained on a vast corpus of publicly available online data (Villalobos et al., 2022, 2024), studies (Chu et al., 2025) have shown that these models can easily memorize and regurgitate information during cloze-style factual assessments. This limitation provides only a superficial understanding of LLM capabilities across domains. Furthermore, the concept of knowledge extends beyond factual recall to include its application (*p-knowledge*) (Fierro et al., 2024). Therefore, evaluating knowledge capacity alone fails to capture reasoning ability, hindering the practical deployment of LLMs, particularly in fields like CBE, where their utility remains uncertain. To address this gap, we introduce a reasoning evaluation test-bed de-

*Both authors contributed equally to this paper.

Proceedings of the 2nd Workshop on Natural Language Processing Meets Climate Change (ClimateNLP 2025), pages 32–48

signed to more effectively estimate LLMs' applicability in such domains.

Global warming caused by greenhouse gas emissions remains a critical challenge (Wang et al., 2016; Sanz-Pérez et al., 2016), necessitating accelerated research into effective carbon capture solutions (Sheridan et al., 2018). Meeting the ambitious carbon-neutral target of the 2015 Paris Agreement by 2050 (Rhodes, 2016) requires not only reducing carbon emissions but also investing in technologies to remove CO_2 from the atmosphere. Among potential solutions, *Ionic Liquids* (ILs) (Zanco et al., 2021) stand out as promising candidates for CO_2 separation processes due to their non-volatile, non-toxic nature (*"green solvents"*), ease of regeneration, and high CO_2 absorption efficiency. However, experimentation with ILs and achieving industrial scalability are resource-intensive and costly, a challenge that AI technologies like LLMs could help address. In this paper, we take a foundational step toward exploring the role of LLMs in supporting carbon capture research using ILs. Specifically, we assess the potential of general-purpose LLMs in domain-specific scenarios by constructing a test bed of 5,920 expert-curated examples, spanning varying levels of difficulty, to evaluate the factual knowledge and reasoning capabilities of these models in the context of ILs. We benchmark three open-weight LLMs—Llama 3.1-8B (Dubey et al., 2024), Mistral-7B (Jiang et al., 2023), and Gemma-9B (Team et al., 2024)—on this dataset. Given the absence of prior research in this area, our work represents a critical step toward identifying the potential applications of LLMs in IL research. Furthermore, leveraging LLMs for CO_2 capture research offers an opportunity to indirectly address concerns about their environmental impact (Patterson et al., 2021; Strubell et al., 2019; Faiz et al., 2024; Li et al., 2023; Rillig et al., 2023) by aligning their use with climate solutions. Our contributions are as follows:

- **Dataset Creation:** Using ILs for carbon capture as a use case, we construct and publicly share[1] a textual entailment test bed containing 5,920 expert-curated samples designed to evaluate LLM reasoning capabilities in CBE.
- **Benchmarking:** We systematically benchmark three open-weight LLMs—Llama 3.1-8B, Mistral-7B, and Gemma-9B—on the test bed and share the resulting insights.
- **Analysis:** We discuss the implications of our results and the broader potential for LLMs to advance IL research and CO_2 capture technologies.

2 Related Work

2.1 Ionic Liquids for Carbon Capture

COP21 showed that amongst 196 participating countries, China, the United States, and India comprise the top three nations by share of worldwide CO_2 emissions. While the United States has pledged to reach "net-zero" by 2050, the deadlines set by China and India (the two most populous countries) are 2060 and 2070 respectively (Rhodes, 2016; Guiot and Cramer, 2016; Dimitrov, 2016; Robbins, 2016). In an attempt to offset the rising atmospheric carbon dioxide levels, carbon sequestration has emerged as an effective field of research and the timely development of materials and methods is pivotal for the efficient capture of CO_2 (Wang et al., 2016; Sanz-Pérez et al., 2016). Ionic Liquids have presented themselves as an excellent solution for CO_2 capture due to their environmentally friendly nature (Blanchard et al., 1999, 2001; Pérez-Salado Kamps et al., 2003; Anthony et al., 2002; Zeng et al., 2017; Husson-Borg et al., 2003; Aghaie et al., 2018; Ramdin et al., 2012). Thorough experimentation, with ILs, to provide a practical solution is time-conducive and entails high cost (Sheridan et al., 2018; Maginn, 2009). In that regard, various machine learning methods have found use to alleviate dependence on experiments (Cao et al., 2018; Baskin et al., 2022; Dhakal and Shah, 2022; Feng et al., 2022; Paduszynski, 2016).

2.2 LLMs for Scientific Research

Recently, LLMs (Brown, 2020; Chowdhery et al., 2023; Taylor et al., 2022; OpenAI et al., 2024) have gained significant popularity with a wide range of possibilities (Ge et al., 2024; Bubeck et al., 2023; Nadkarni et al., 2021; Beltagy et al., 2019; Schick et al., 2024; Buehler, 2023a; Luu et al., 2023; Mialon et al., 2023; Wei et al., 2023), and the integration of these transformer-based models into the fields of materials science and discovery has yielded tremendous results. Leveraging the abilities of LLMs has been beneficial in various downstream tasks such as protein design and folding (Jumper et al., 2021; Liu et al., 2022; Yu et al., 2022b,a; Hu and Buehler, 2022; Khare et al., 2022), material discovery (Lu et al., 2023;

[1]Dataset available at: https://github.com/sougata-ub/llms_for_ionic_liquids

Luu and Buehler, 2024; Lu et al., 2024; Buehler, 2023b), educational tasks (Lim et al., 2023; Milano et al., 2023; Inguva et al., 2021) and chemistry-related tasks (Castro Nascimento and Pimentel, 2023; White, 2023; Jablonka et al., 2023). The reliability of LLMs is still a massive topic of discussion, and their accuracy is often determined by the size and complexity of the model. Despite their promises, present pitfalls include the issues of hallucinations and fact recall, which warrants a careful validation of the model's output and its eventual ramifications (Hu and Buehler, 2023; Azamfirei et al., 2023; Kandpal et al., 2023; Varshney et al., 2023; Ji et al., 2023; McKenna et al., 2023; Harrer, 2023). Invariably, training and using such networks comes at a huge environmental cost, largely in terms of carbon emissions (Li et al., 2023; Patterson et al., 2021; Strubell et al., 2019; Faiz et al., 2024; Rillig et al., 2023).

The power of LLMs can aid carbon capture by helping researchers with their advances to address the growing problem of global warming and offset the model's carbon footprint to reach the end goal of 'net-zero' carbon emissions.

3 A Practical Test for Knowledge

Although there are several standard definitions of knowledge in Philosophy (Sartwell, 1992; Nozick, 2016; Williamson, 2005; Zagzebski, 2017; Austin, 1961), the most prevalent ones for non-human entities like LLMs are *tb* and *p-knowledge* (Fierro et al., 2024). Most knowledge probing tasks test for *tb-knowledge*, where the model passes the test if it can recall an answer. For example, probing for factual questions like "What is the capital of Germany?" Such tests are weak estimates of knowledge and hold little pragmatic significance, especially in domains like CBE, where the intended use of LLMs is still unclear. LLMs as reasoners can be of better practical use in such domains. Although some methods estimate the model's uncertainty (Huang et al., 2024, 2023; Ye et al., 2024; Geng et al., 2023), they still pertain to *tb-knowledge*. However, a more complete measure of knowledge is *p-knowledge*, which tests a model's capability to use knowledge in practical tasks. For example, sociodemographic prompting (Saha et al., 2025; Pandey et al., 2025; Li et al., 2024b; AlKhamissi et al., 2024; Nadeem et al., 2021; Nangia et al., 2020; Wan et al., 2023; Jha et al., 2023; Li et al., 2024a; Cao et al., 2023; Tanmay et al., 2023; Rao et al., 2023) such as "What would a German find difficult to understand from a text X?" necessitates a model to reason from a group's perspective, which requires prerequisite knowledge. Motivated to create stronger test beds, we set up an *entailment task* to benchmark LLMs' factual capacity in CBE, where the model is provided a claim and a list of propositions and is tasked with determining all propositions that entail the claim or none. Thus, testing the model's reasoning capabilities in a practical setting is warranted.

3.1 Argument Structures

A claim constitutes one or more facts (propositions), where some are evident (explicit) from the text, and some are assumed (implicit) to be known by the reader (enthymemes) (Walton, 1996; Besnard and Hunter, 2008; Walton et al., 2008; Bitzer, 2020). Within a field (such as CBE), the degree of knowledge of the assumed propositions is subjective and varies by person, which impacts the understanding of the claim. For example, the claim "Ionic Liquids are low-melting, non-volatile salts which categorize them within the green solvents category" explicitly informs that (i) Ionic Liquids are low-melting, non-volatile salts. (ii) Ionic Liquids are categorized as green solvents. It also entails that low-melting and non-volatile salts are green solvents, which might be unknown (or partially known) to someone from CBE[2]. The degree of knowledge about the implicit assumption is subjective and varies within the domain[3]. We aim to test this domain-specific knowledge in LLMs via an entailment task.

3.2 The Entailment Task

Hypothesizing that **knowledgeable agents should perform consistently, irrespective of the adversaries**, we create an *entailment task* with the following setups to benchmark LLMs' reasoning capacity, where the model is provided a claim and a list of propositions and tasked to determine all propositions that entail the claim, if applicable.

1. Change the number of adversaries: (i) Keeping the number of entailing propositions constant for a claim, the number of non-entailing propositions should not affect the model's entailment

[2]This is different from general knowledge. For example, understanding the claim also requires knowledge of "low-melting, non-volatile salts" and "green solvents", which is an assumed prerequisite for a domain expert.

[3]We are only interested in domain-specific knowledge. An outsider might not possess such knowledge.

performance. (ii) When provided with only non-entailing options and an additional *"none of the above"* option, a consistent agent should always choose the *"none"* option. A drop in performance indicates a lack of knowledge and supposedly more reliance on linguistic cues for entailment.

2. Introduce linguistic perturbations: A knowledgeable agent should be invariant to paraphrased options. Failure to do so indicates reliance on linguistic cues instead of factual cues for entailment.

2. Apply common sense: Knowledgeable agents should not be derailed by incorrect facts that can be discerned by common sense.

3.3 Dataset Creation

The dataset is created in multiple phases, employing two expert annotators, one with a background in CBE and another from Computer Science and Linguistics (CSL). The CBE expert has domain knowledge of ILs for carbon capture, while the CSL expert is generally unaware of the domain. Figure 1 illustrates the data creation pipeline with an actual example. We detail the pipeline below:

Phase 1 encompassed knowledge creation, where the CBE expert constructed paragraphs capturing the different aspects of carbon capture using ionic liquids. The aspects encompassed the need for carbon capture, ionic liquids, their physical and chemical characteristics, and their advantages. Next, the annotator extracted claims from the paragraphs, which are sentences containing salient knowledge pertaining to ionic liquids for carbon capture, yielding 74 in total.

Phase 2 encompassed identifying the explicit and implicit propositions from each claim and implemented in two stages: (i) **LLM-based annotation:** We prompted Mistral-7B-Instruct-v0.2 (Jiang et al., 2023) to identify the explicit and implicit propositions from a claim. As depicted in Figure 5 (Appendix A), the prompt comprised a short task description and three examples of how to perform the task, followed by the actual claim for annotation. (ii) **Expert evaluations:** The CBE expert extensively evaluated the model response by editing, deleting, or unchanging each model-identified proposition. Additionally, for each claim, the expert added propositions that were missed by the model (if any). Overall, 48 (65%) of the 74 LLM-based annotations were deemed correct by the expert and were unmodified, yielding 164 propositions across 74 claims.

Phase 3 encompassed data standardization. The propositions, being fundamental pieces of knowledge, are universal. Hence, in this phase, we standardized the propositions across all claims. Using sentence transformers (Reimers and Gurevych, 2019), we clustered the propositions by their embedding cosine-similarity[4] and computationally marked propositions belonging to the same cluster as equivalent. The CBE expert evaluated the clustering results, which were accurate in only 28% of cases. The expert annotated and rectified the incorrect cluster assignments, yielding 125 universal propositions across all 74 claims.

Phase 4 involved constructing false variants of the propositions at three difficulty levels: (i) **Low:** Invalid version of a proposition, and can be discerned using common sense reasoning. For example, the proposition *"Ionic liquids can be categorized as conventional or task-specific"* was augmented to *"Ionic liquids can be categorized as conventional or task-specific only while recharging batteries."* (ii) **Medium:** Invalid version of a proposition that might need a mix of common sense and knowledge of science for discerning. For example, *"Ionic liquids can be categorized as conventional or task-specific due to specific environmental conditions and chemical habitability."* (iii) **High:** Determining invalidity requires considerable knowledge about ILs. For example, *"Ionic liquids can be categorized as conventional or task-specific based on molecular weight, isotope atom count, and hydrogen bonding capabilities."* All variants were manually constructed by the CBE expert and evaluated by the CSL expert, who does not know ILs. The CSL expert evaluated 60 random propositions (15 original and 15 from each level of difficulty) by determining if the proposition was correct or assigning a level of difficulty if they thought it was incorrect. Comparing their response with the original labels, the expert attained an F1 score of 67% in discerning factual correctness. For the incorrect propositions, the expert attained F1 scores of 80%, 15%, and 42% for levels 1, 2, and 3, indicating the difficulty of the options for a non-expert.

In **phase 5**, we introduced linguistic variations in the original and all three incorrect variants of each proposition by paraphrasing. We prompted the Llama-3.1-8B instruction-tuned variant (Dubey et al., 2024) using the prompt *"Paraphrase the following text without changing the meaning of the*

[4]We used the 'all-MiniLM-L6-v2' model for computing embeddings.

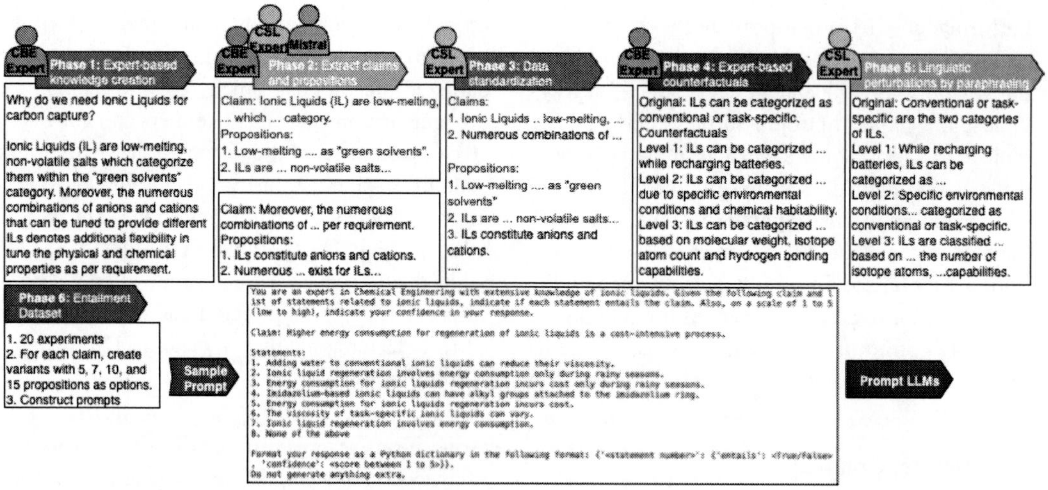

Figure 1: Dataset Creation Pipeline

Group	Description	Id	Experiment	Correct options		Incorrect options		Median F1			Std Dev F1		
				Present	Para-phrased	Difficulty	Para-phrased	Gemma	Llama	Mistral	Gemma	Llama	Mistral
0	Baselines	1	orig+random	Yes	No	Random	No	49.0	**66.0**	55.0	21.7	**16.1**	18.9
		2	para+random		Yes			49.5	**63.0**	57.0	21.8	**14.0**	18.5
1	Only providing incorrect options	3	none+level1	No	No	Level 1	No	9.0	**30.0**	1.5	5.5	6.8	**3.1**
		4	none+level2			Level 2		2.0	**29.0**	0.0	4.9	12.3	**1.0**
		5	none+level3			Level 3		3.0	**21.5**	0.0	4.3	8.6	**0.5**
		6	none+level1-para			Level 1	Yes	3.0	**17.5**	0.0	1.7	7.3	**0.5**
		7	none+level2-para			Level 2		0.0	**17.5**	0.0	2.0	6.4	**0.5**
		8	none+level3-para			Level 3		1.0	**18.5**	0.0	5.3	2.4	**0.0**
2	Difficulty level of incorrect options	9	orig+level1	Yes	No	Level 1	No	35.0	**73.5**	62.5	26.6	**14.6**	19.6
		10	orig+level2			Level 2		32.0	**68.5**	60.5	19.9	**12.3**	18.7
		11	orig+level3			Level 3		29.5	**67.5**	58.0	18.0	**12.1**	16.9
3	Paraphrasing the correct options	12	para+level1	Yes	Yes	Level 1	No	40.5	**66.5**	62.0	24.1	**14.4**	20.1
		13	para+level2			Level 2		34.5	**66.0**	60.0	23.3	**13.5**	18.1
		14	para+level3			Level 3		31.5	**64.0**	60.0	18.8	**13.0**	17.0
4	Paraphrasing the incorrect options	15	orig+level1-para	Yes	No	Level 1	Yes	39.5	**66.0**	57.0	22.1	**11.7**	17.3
		16	orig+level2-para			Level 2		35.5	**64.5**	57.5	17.0	**11.1**	17.1
		17	orig+level3-para			Level 3		34.0	**63.0**	58.5	13.5	**10.5**	16.6
5	Paraphrasing all options	18	para+level1-para	Yes	Yes	Level 1	Yes	33.5	**63.5**	58.0	**10.7**	11.2	16.9
		19	para+level2-para			Level 2		35.5	**64.5**	59.5	15.6	**11.8**	17.6
		20	para+level3-para			Level 3		36.5	**60.5**	58.5	11.9	**10.8**	17.8

Table 1: Definitions of experiments and aggregated model results (median F1 and standard deviation) across experiments with 5, 7, 10, and 15 options. The best scores are highlighted in bold.

text. Text: <text>" and resorted to greedy decoding for paraphrasing.

Using the 74 claims, the 125 original propositions, and their incorrect and paraphrased variants, we constructed the test set for the entailment task in **phase 6**. For each claim, we created variants with 5, 7, 10, and 15 propositions as options. Listed in Table 1, we constructed 20 experiments using different permutations of the original and paraphrased versions of the correct and incorrect propositions, yielding a dataset of 5,920 examples. On average, the claims contain 14 words, the original propositions contain 12, and the incorrect propositions contain 17.

3.4 Experiments

Listed in Table 1, we group the 20 experiments into five groups and test our three hypotheses in Section 3.2. Comprising two experiments, Group 0 serves as the baseline. By only providing incorrect propositions with their difficulty and stylistic variations, the experiments in Group 1 test the model's knowledgeability by measuring the propensity of selecting the "none" option. Group 2 quantifies the effect of varying the difficulty levels of the incorrect options while keeping the original propositions unchanged. Group 3 perturbs the original proposition by paraphrasing and measures the impact of changing the incorrect option difficulty levels. Groups 4 and 5 measure a model's invariance to linguistic variations and the difficulty levels of the incorrect options. For all groups, we experiment with 5, 7, 10, and 15 propositions as options to test for the model's capability of being invariant to additional incorrect options. Except for group 1, the number of correct options varies from 1 to 5. As depicted in Figure 1 (Phase 6), we construct prompts from each example and probe the Llama-3.1-8B-Instruct, Mistral-7B-Instruct-v0.3, and gemma-2-9b-it, setting the temperature to 0. We process each model response and use Llama-3.1-8B-Instruct to rectify ill-formatted outputs. Figures 6 and 7 (Appendix A) illustrate the prompts for the entailment task and correcting the ill-formatted LLM responses.

4 Results and Observations

Table 1 shares the model-wise median F1 score and the standard deviation across all options (5, 7, 10, and 15). Figures 2, 3, and 4 plot the precision and recall scores for all groups of experiments. The baseline results (Group 0) in Table 1 indicate that LLMs are knowledgeable about ionic liquids and carbon capture. Llama performs the best, followed by Mistral and Gemma. However, paraphrasing the original propositions (Id 2) reduces Llama's performance, which contrasts with Mistral and Gemma, where the performance increases. This effect of stylistic perturbations on the model results shows a tendency to rely on linguistic cues.

Effect of the number of incorrect options

We observe a correlation between model performance and the number of incorrect options in Figures 2, 3, and 4. The precision scores for all models drop with more incorrect options, indicating an adverse effect of the number of adversaries on their reasoning capabilities. For Llama and Mistral, the recall scores remain mostly consistent, but drop for Gemma. Nonetheless, as depicted in Table 1, the standard deviation of Llama is the lowest, followed by Mistral and Gemma. For Llama and Mistral, this decline in precision but constant recall scores indicates a propensity to make more predictions as the number of options increases without changing the prediction for the correct propositions. On the contrary, increasing the number of adversaries causes Gemma to change the prediction for the correct propositions, indicating an unreliability of utilizing facts for reasoning.

Effect of the difficulty of incorrect options

Comparing experiment Id 1 with Group 2 and Id 2 with Group 3 in Table 1 and Figure 2, we observe that increasing the difficulty level of the adversarial facts hampers the model performance for Llama and Mistra, which is the opposite for Gemma. The comparisons indicate that the experiments comprising random adversaries (Orig/para+random) are more challenging test beds than the difficulty-controlled adversaries, especially for Llama and Mistral. We hypothesize that since we gradually balance between common sense and domain-specific knowledge across three difficulty levels, higher performance in level 1 can be due to the model's capability of common sense reasoning, which decreases as the difficulty increases, requiring more domain-specific knowledge. However, using random adversaries presents less scope for common-sense reasoning and requires domain-knowledge-based reasoning for entailment resolution. Gemma, on the other hand, is more reliant on syntactic cues than reasoning. Hence, it falters when provided with factually incorrect yet syntactically similar options to the claim. This is also evident from Gemma's decreasing recall scores in Figure 2, compared to Llama and Mistral, which are more consistent.

Effect of only incorrect propositions as options

Compared to the baseline (Group 0) in Table 1, in Group 1, the performance of all models drastically reduces when presented with only incorrect facts and a "none" option to choose from. Mistral and Gemma perform worse than Llama, with median F1 scores < 10 for all experiments and near zero for some. All models perform worse with paraphrased incorrect options. Figure 4 plots the **pr**ecision, **re**call, and **f1** scores for Group 1 experiments. Interestingly, for all three models, sometimes the precision increases with higher options in some experiments. For Gemma, the precision scores increase while the recall decreases with an increase in incorrect choices. On the contrary, for Llama and Mistral, the precision and recall scores increase for some experiments. For Llama, presenting 7 and 10 options yields higher F1 scores for most experiments compared to 5 options. Mistral yields higher F1 scores when prompted with 7 choices compared to other options. We hypothesize that for Llama and Mistral, increasing the choices provides more inter-option reasoning opportunities, resulting in higher F1 scores. We also think the position of the "none" option in the prompt might be a confounding variable, which we leave for future work. Nonetheless, when only presented with incorrect facts and a "none" option, the drastic reduction in

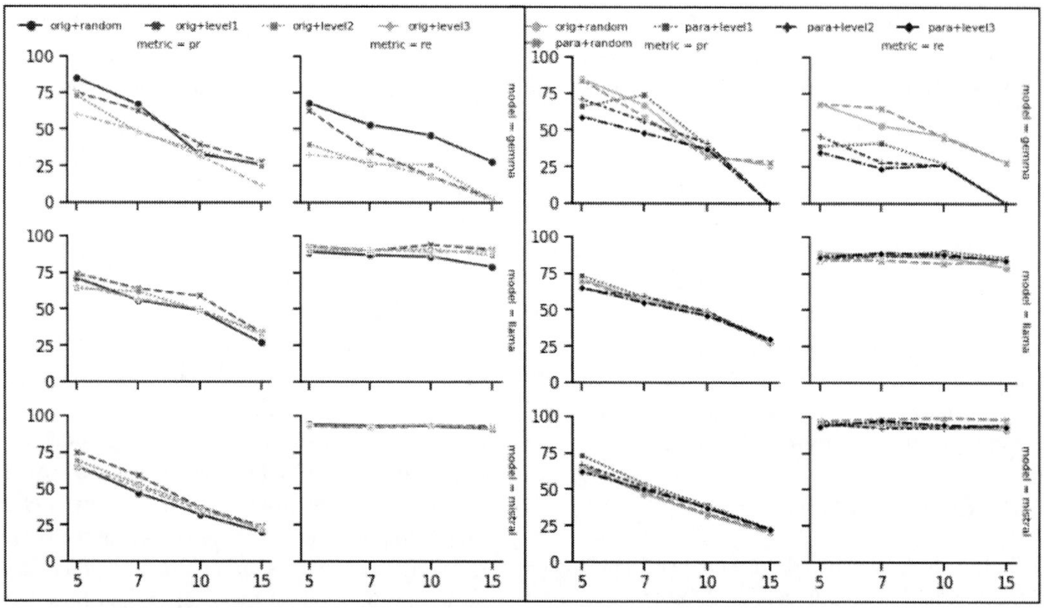

Figure 2: Model-wise precision and recall for experiments in Group 2 (left) and Group 3 (right).

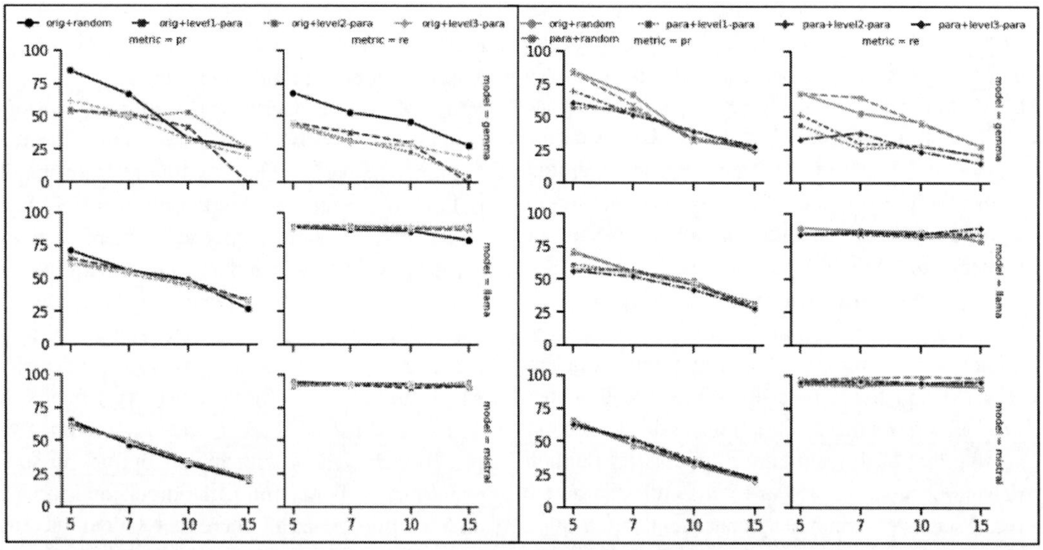

Figure 3: Model-wise precision and recall for experiments in Group 4 (left) and Group 5 (right).

performance for all models indicates that although LLMs contain facts about ionic liquids, they can't reliably utilize and reason with them for complex tasks.

Effect of paraphrasing

Comparing Groups 2 and 3 in Table 1, although paraphrasing the correct options reduces the F1 score across all difficulty levels for Llama and Mistral, paraphrasing the incorrect options in Group 4 has a higher diminishing effect on the model performance than Group 2, which is the opposite for Gemma. We hypothesize that this might be due to Gemma's reliance on linguistic cues for entailment compared to Llama and Mistral, where Gemma relies more on syntactic similarity than semantics.

Comparing Groups 3 and 5, paraphrasing the incorrect options reduces the F1 score across all difficulty levels for Llama and Mistral, which is the opposite for Gemma, except for experiment 15. Comparing Groups 4 and 5, paraphrasing the cor-

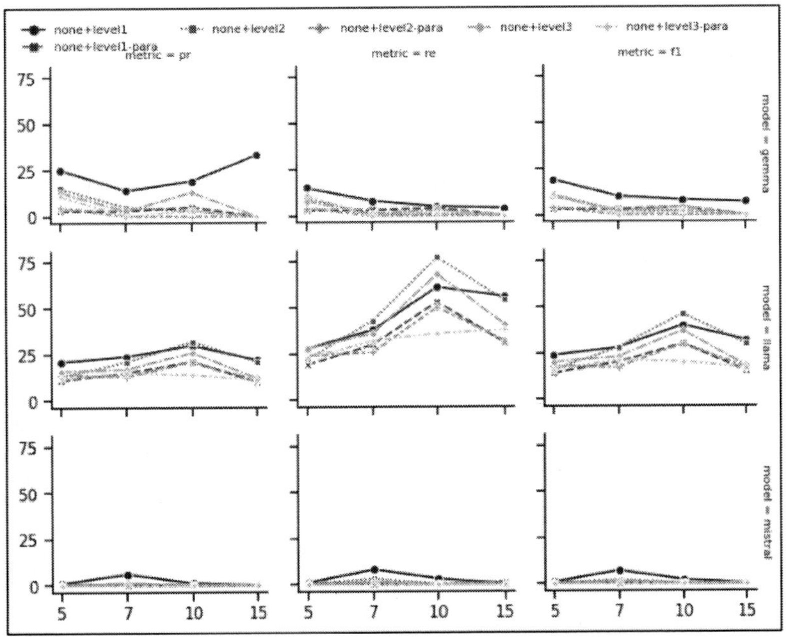

Figure 4: Model-wise precision, recall, and F1 for experiments in comparison suite 5.

rect options reduces the F1 score for Llama across all difficulty levels. On the contrary, the F1 score increases or remains the same for Gemma and Mistral, except for experiment 18. We hypothesize that since the correct and incorrect options share syntactic similarities, they get equally transformed while paraphrasing, causing their paraphrased versions to maintain syntactic similarity, which weaker reasoning models like Gemma exploit. We leave the testing out of this hypothesis as future work.

Overall, Llama performs best across all experiments, followed by Mistral and Gemma. Our results indicate that although LLMs possess knowledge of ionic liquids and carbon capture, their domain-specific reasoning capabilities are limited. The performance drop in Group 1 experiments is drastic for all models and sometimes near zero for Mistral and Gemma, which questions their reasoning capabilities.

5 Discussion

Our experiments indicate that smaller LLMs struggle to coherently reason within the domain-specific constraints and choose non-probable options in the entailment task. This is likely because LLMs are general-purpose and not geared to niche domains such as ILs. We propose that LLMs should be fine-tuned for CBE using curated datasets. Pre-training the models on domain-specific data, fine-tuning using PEFT (Mangrulkar et al., 2022) methods like LoRA (Hu et al., 2021), or in-context learning and efficient methods such as RAG (Lewis et al., 2020; Gao et al., 2024) should help impart the domain-specific knowledge and constraints, which requires collaborative advancements in the intersection of LLMs and CBE. Such domain-specific LLMs can scale IL research by assisting researchers in the bottlenecked areas of data analysis, experiment design, and property predictions. Furthermore, they can serve as educational guides to researchers willing to gain familiarity with the field. This work should be a valuable resource for researchers eager to evaluate LLMs for varied fields and collaboratively help attain the sustainability goals of the UN[5].

6 Conclusion

Global warming remains a pressing challenge, necessitating scalable and interdisciplinary solutions such as carbon capture. To address this need, we propose leveraging LLMs to support research on Ionic Liquids, a promising avenue for carbon capture. As a foundational step, we construct and publicly share an expert-curated dataset designed to evaluate LLMs' knowledge and reasoning capabilities within the specialized domain of Ionic Liquids. Our benchmarking of three open-weight

[5] https://sdgs.un.org/goals

LLMs—Llama, Gemma, and Mistral—reveals that while general-purpose models, particularly Llama, demonstrate a strong grasp of Ionic Liquid-related knowledge, they fall short in domain-specific reasoning tasks. Building on these findings, we outline potential pathways for LLMs to advance Ionic Liquid research, including their use as agents in simulations, reasoners for material discovery and design, and educational tools to help researchers familiarize themselves with the field. Moreover, optimizing LLMs for climate research not only advances carbon capture efforts but also offers a dual benefit by mitigating the models' own carbon footprint. This alignment between AI innovation and environmental goals supports the broader aim of achieving carbon neutrality by 2050.

Limitations

This study has some notable limitations. Firstly, we only evaluate three open-weight models with less than 10B parameters for their knowledge and reasoning ability with ILs. Although extraneous experiments with larger and open-API models indicate a similar trend, they are not quantified and non-generalizable. Secondly, our entailment test set is not an exhaustive resource for IL research. It contains limited facts and only tests reasoning capabilities through entailment. We need more diverse datasets that probe the reasoning capabilities of LLMs from multiple aspects. Thirdly, we do not experiment with fine-tuning the models on our dataset and measure their impact on reasoning, which we intend as future work. Also, our work is limited to two expert evaluators and might benefit from multiple experts. Despite these limitations, our research takes a foundational step in the interdisciplinary field of LLMs for ionic liquid research, which is very nascent.

Ethics Statement

We confirm that all conducted experiments are solely for academic purposes and adhere to ethical standards. The expert evaluators were appropriately compensated for their tasks, following all administrative and regulatory policies. The shared dataset strictly pertains to ionic liquids. It does not contain potentially explicit and sensitive content that might exhibit bias, be hurtful, or offend anyone.

References

Mahsa Aghaie, Nima Rezaei, and Sohrab Zendehboudi. 2018. A systematic review on co2 capture with ionic liquids: Current status and future prospects. *Renewable and sustainable energy reviews*, 96:502–525.

Badr AlKhamissi, Muhammad ElNokrashy, Mai Alkhamissi, and Mona Diab. 2024. Investigating cultural alignment of large language models. In *Proceedings of the 62nd Annual Meeting of the Association for Computational Linguistics (Volume 1: Long Papers)*, pages 12404–12422, Bangkok, Thailand. Association for Computational Linguistics.

Jennifer L Anthony, Edward J Maginn, and Joan F Brennecke. 2002. Solubilities and thermodynamic properties of gases in the ionic liquid 1-n-butyl-3-methylimidazolium hexafluorophosphate. *The Journal of Physical Chemistry B*, 106(29):7315–7320.

John L Austin. 1961. Other minds.

Razvan Azamfirei, Sapna R Kudchadkar, and James Fackler. 2023. Large language models and the perils of their hallucinations. *Critical Care*, 27(1):120.

Igor Baskin, Alon Epshtein, and Yair Ein-Eli. 2022. Benchmarking machine learning methods for modeling physical properties of ionic liquids. *Journal of Molecular Liquids*, 351:118616.

Iz Beltagy, Kyle Lo, and Arman Cohan. 2019. Scibert: A pretrained language model for scientific text. *arXiv preprint arXiv:1903.10676*.

Philippe Besnard and Anthony Hunter. 2008. *Elements of argumentation*, volume 47. MIT press Cambridge.

Lloyd F Bitzer. 2020. Aristotle's enthymeme revisited. In *Landmark Essays on Aristotelian Rhetoric*, pages 179–191. Routledge.

Lynnette A Blanchard, Zhiyong Gu, and Joan F Brennecke. 2001. High-pressure phase behavior of ionic liquid/co2 systems. *The Journal of Physical Chemistry B*, 105(12):2437–2444.

Lynnette A Blanchard, Dan Hancu, Eric J Beckman, and Joan F Brennecke. 1999. Green processing using ionic liquids and co2. *Nature*, 399(6731):28–29.

Andres M Bran, Sam Cox, Oliver Schilter, Carlo Baldassari, Andrew D White, and Philippe Schwaller. 2023. Chemcrow: Augmenting large-language models with chemistry tools. *arXiv preprint arXiv:2304.05376*.

Tom B Brown. 2020. Language models are few-shot learners. *arXiv preprint arXiv:2005.14165*.

Sébastien Bubeck, Varun Chandrasekaran, Ronen Eldan, Johannes Gehrke, Eric Horvitz, Ece Kamar, Peter Lee, Yin Tat Lee, Yuanzhi Li, Scott Lundberg, et al. 2023. Sparks of artificial general intelligence: Early experiments with gpt-4. *arXiv preprint arXiv:2303.12712*.

Markus J Buehler. 2023a. Generative pretrained autoregressive transformer graph neural network applied to the analysis and discovery of novel proteins. *Journal of Applied Physics*, 134(8).

Markus J Buehler. 2023b. Melm, a generative pretrained language modeling framework that solves forward and inverse mechanics problems. *Journal of the Mechanics and Physics of Solids*, 181:105454.

Lingdi Cao, Peng Zhu, Yongsheng Zhao, and Jihong Zhao. 2018. Using machine learning and quantum chemistry descriptors to predict the toxicity of ionic liquids. *Journal of hazardous materials*, 352:17–26.

Yong Cao, Li Zhou, Seolhwa Lee, Laura Cabello, Min Chen, and Daniel Hershcovich. 2023. Assessing cross-cultural alignment between ChatGPT and human societies: An empirical study. In *Proceedings of the First Workshop on Cross-Cultural Considerations in NLP (C3NLP)*, pages 53–67, Dubrovnik, Croatia. Association for Computational Linguistics.

Cayque Monteiro Castro Nascimento and André Silva Pimentel. 2023. Do large language models understand chemistry? a conversation with chatgpt. *Journal of Chemical Information and Modeling*, 63(6):1649–1655.

Aakanksha Chowdhery, Sharan Narang, Jacob Devlin, Maarten Bosma, Gaurav Mishra, Adam Roberts, Paul Barham, Hyung Won Chung, Charles Sutton, Sebastian Gehrmann, et al. 2023. Palm: Scaling language modeling with pathways. *Journal of Machine Learning Research*, 24(240):1–113.

Tianzhe Chu, Yuexiang Zhai, Jihan Yang, Shengbang Tong, Saining Xie, Dale Schuurmans, Quoc V. Le, Sergey Levine, and Yi Ma. 2025. Sft memorizes, rl generalizes: A comparative study of foundation model post-training. *Preprint*, arXiv:2501.17161.

Pratik Dhakal and Jindal K Shah. 2022. A generalized machine learning model for predicting ionic conductivity of ionic liquids. *Molecular Systems Design & Engineering*, 7(10):1344–1353.

Radoslav S Dimitrov. 2016. The paris agreement on climate change: Behind closed doors. *Global environmental politics*, 16(3):1–11.

Abhimanyu Dubey, Abhinav Jauhri, Abhinav Pandey, Abhishek Kadian, Ahmad Al-Dahle, Aiesha Letman, Akhil Mathur, Alan Schelten, Amy Yang, Angela Fan, Anirudh Goyal, Anthony Hartshorn, Aobo Yang, Archi Mitra, Archie Sravankumar, Artem Korenev, Arthur Hinsvark, Arun Rao, Aston Zhang, Aurelien Rodriguez, Austen Gregerson, Ava Spataru, Baptiste Roziere, Bethany Biron, Binh Tang, Bobbie Chern, Charlotte Caucheteux, Chaya Nayak, Chloe Bi, Chris Marra, Chris McConnell, Christian Keller, Christophe Touret, Chunyang Wu, Corinne Wong, Cristian Canton Ferrer, Cyrus Nikolaidis, Damien Allonsius, Daniel Song, Danielle Pintz, Danny Livshits, David Esiobu, Dhruv Choudhary, Dhruv Mahajan, Diego Garcia-Olano, Diego Perino, Dieuwke Hupkes, Egor Lakomkin, Ehab AlBadawy, Elina Lobanova, Emily Dinan, Eric Michael Smith, Filip Radenovic, Frank Zhang, Gabriel Synnaeve, Gabrielle Lee, Georgia Lewis Anderson, Graeme Nail, Gregoire Mialon, Guan Pang, Guillem Cucurell, Hailey Nguyen, Hannah Korevaar, Hu Xu, Hugo Touvron, Iliyan Zarov, Imanol Arrieta Ibarra, Isabel Kloumann, Ishan Misra, Ivan Evtimov, Jade Copet, Jaewon Lee, Jan Geffert, Jana Vranes, Jason Park, Jay Mahadeokar, Jeet Shah, Jelmer van der Linde, Jennifer Billock, Jenny Hong, Jenya Lee, Jeremy Fu, Jianfeng Chi, Jianyu Huang, Jiawen Liu, Jie Wang, Jiecao Yu, Joanna Bitton, Joe Spisak, Jongsoo Park, Joseph Rocca, Joshua Johnstun, Joshua Saxe, Junteng Jia, Kalyan Vasuden Alwala, Kartikeya Upasani, Kate Plawiak, Ke Li, Kenneth Heafield, Kevin Stone, Khalid El-Arini, Krithika Iyer, Kshitiz Malik, Kuenley Chiu, Kunal Bhalla, Lauren Rantala-Yeary, Laurens van der Maaten, Lawrence Chen, Liang Tan, Liz Jenkins, Louis Martin, Lovish Madaan, Lubo Malo, Lukas Blecher, Lukas Landzaat, Luke de Oliveira, Madeline Muzzi, Mahesh Pasupuleti, Mannat Singh, Manohar Paluri, Marcin Kardas, Mathew Oldham, Mathieu Rita, Maya Pavlova, Melanie Kambadur, Mike Lewis, Min Si, Mitesh Kumar Singh, Mona Hassan, Naman Goyal, Narjes Torabi, Nikolay Bashlykov, Nikolay Bogoychev, Niladri Chatterji, Olivier Duchenne, Onur Çelebi, Patrick Alrassy, Pengchuan Zhang, Pengwei Li, Petar Vasic, Peter Weng, Prajjwal Bhargava, Pratik Dubal, Praveen Krishnan, Punit Singh Koura, Puxin Xu, Qing He, Qingxiao Dong, Ragavan Srinivasan, Raj Ganapathy, Ramon Calderer, Ricardo Silveira Cabral, Robert Stojnic, Roberta Raileanu, Rohit Girdhar, Rohit Patel, Romain Sauvestre, Ronnie Polidoro, Roshan Sumbaly, Ross Taylor, Ruan Silva, Rui Hou, Rui Wang, Saghar Hosseini, Sahana Chennabasappa, Sanjay Singh, Sean Bell, Seohyun Sonia Kim, Sergey Edunov, Shaoliang Nie, Sharan Narang, Sharath Raparthy, Sheng Shen, Shengye Wan, Shruti Bhosale, Shun Zhang, Simon Vandenhende, Soumya Batra, Spencer Whitman, Sten Sootla, Stephane Collot, Suchin Gururangan, Sydney Borodinsky, Tamar Herman, Tara Fowler, Tarek Sheasha, Thomas Georgiou, Thomas Scialom, Tobias Speckbacher, Todor Mihaylov, Tong Xiao, Ujjwal Karn, Vedanuj Goswami, Vibhor Gupta, Vignesh Ramanathan, Viktor Kerkez, Vincent Gonguet, Virginie Do, Vish Vogeti, Vladan Petrovic, Weiwei Chu, Wenhan Xiong, Wenyin Fu, Whitney Meers, Xavier Martinet, Xiaodong Wang, Xiaoqing Ellen Tan, Xinfeng Xie, Xuchao Jia, Xuewei Wang, Yaelle Goldschlag, Yashesh Gaur, Yasmine Babaei, Yi Wen, Yiwen Song, Yuchen Zhang, Yue Li, Yuning Mao, Zacharie Delpierre Coudert, Zheng Yan, Zhengxing Chen, Zoe Papakipos, Aaditya Singh, Aaron Grattafiori, Abha Jain, Adam Kelsey, Adam Shajnfeld, Adithya Gangidi, Adolfo Victoria, Ahuva Goldstand, Ajay Menon, Ajay Sharma, Alex Boesenberg, Alex Vaughan, Alexei Baevski, Allie Feinstein, Amanda Kallet, Amit Sangani, Anam Yunus, Andrei Lupu, Andres Alvarado, Andrew Caples, Andrew Gu, Andrew Ho, Andrew Poulton, Andrew Ryan, Ankit Ramchandani, Annie Franco, Aparajita Saraf, Arkabandhu Chowdhury, Ashley Gabriel,

Ashwin Bharambe, Assaf Eisenman, Azadeh Yazdan, Beau James, Ben Maurer, Benjamin Leonhardi, Bernie Huang, Beth Loyd, Beto De Paola, Bhargavi Paranjape, Bing Liu, Bo Wu, Boyu Ni, Braden Hancock, Bram Wasti, Brandon Spence, Brani Stojkovic, Brian Gamido, Britt Montalvo, Carl Parker, Carly Burton, Catalina Mejia, Changhan Wang, Changkyu Kim, Chao Zhou, Chester Hu, Ching-Hsiang Chu, Chris Cai, Chris Tindal, Christoph Feichtenhofer, Damon Civin, Dana Beaty, Daniel Kreymer, Daniel Li, Danny Wyatt, David Adkins, David Xu, Davide Testuggine, Delia David, Devi Parikh, Diana Liskovich, Didem Foss, Dingkang Wang, Duc Le, Dustin Holland, Edward Dowling, Eissa Jamil, Elaine Montgomery, Eleonora Presani, Emily Hahn, Emily Wood, Erik Brinkman, Esteban Arcaute, Evan Dunbar, Evan Smothers, Fei Sun, Felix Kreuk, Feng Tian, Firat Ozgenel, Francesco Caggioni, Francisco Guzmán, Frank Kanayet, Frank Seide, Gabriela Medina Florez, Gabriella Schwarz, Gada Badeer, Georgia Swee, Gil Halpern, Govind Thattai, Grant Herman, Grigory Sizov, Guangyi, Zhang, Guna Lakshminarayanan, Hamid Shojanazeri, Han Zou, Hannah Wang, Hanwen Zha, Haroun Habeeb, Harrison Rudolph, Helen Suk, Henry Aspegren, Hunter Goldman, Ibrahim Damlaj, Igor Molybog, Igor Tufanov, Irina-Elena Veliche, Itai Gat, Jake Weissman, James Geboski, James Kohli, Japhet Asher, Jean-Baptiste Gaya, Jeff Marcus, Jeff Tang, Jennifer Chan, Jenny Zhen, Jeremy Reizenstein, Jeremy Teboul, Jessica Zhong, Jian Jin, Jingyi Yang, Joe Cummings, Jon Carvill, Jon Shepard, Jonathan McPhie, Jonathan Torres, Josh Ginsburg, Junjie Wang, Kai Wu, Kam Hou U, Karan Saxena, Karthik Prasad, Kartikay Khandelwal, Katayoun Zand, Kathy Matosich, Kaushik Veeraraghavan, Kelly Michelena, Keqian Li, Kun Huang, Kunal Chawla, Kushal Lakhotia, Kyle Huang, Lailin Chen, Lakshya Garg, Lavender A, Leandro Silva, Lee Bell, Lei Zhang, Liangpeng Guo, Licheng Yu, Liron Moshkovich, Luca Wehrstedt, Madian Khabsa, Manav Avalani, Manish Bhatt, Maria Tsimpoukelli, Martynas Mankus, Matan Hasson, Matthew Lennie, Matthias Reso, Maxim Groshev, Maxim Naumov, Maya Lathi, Meghan Keneally, Michael L. Seltzer, Michal Valko, Michelle Restrepo, Mihir Patel, Mik Vyatskov, Mikayel Samvelyan, Mike Clark, Mike Macey, Mike Wang, Miquel Jubert Hermoso, Mo Metanat, Mohammad Rastegari, Munish Bansal, Nandhini Santhanam, Natascha Parks, Natasha White, Navyata Bawa, Nayan Singhal, Nick Egebo, Nicolas Usunier, Nikolay Pavlovich Laptev, Ning Dong, Ning Zhang, Norman Cheng, Oleg Chernoguz, Olivia Hart, Omkar Salpekar, Ozlem Kalinli, Parkin Kent, Parth Parekh, Paul Saab, Pavan Balaji, Pedro Rittner, Philip Bontrager, Pierre Roux, Piotr Dollar, Polina Zvyagina, Prashant Ratanchandani, Pritish Yuvraj, Qian Liang, Rachad Alao, Rachel Rodriguez, Rafi Ayub, Raghotham Murthy, Raghu Nayani, Rahul Mitra, Raymond Li, Rebekkah Hogan, Robin Battey, Rocky Wang, Rohan Maheswari, Russ Howes, Ruty Rinott, Sai Jayesh Bondu, Samyak Datta, Sara Chugh, Sara Hunt, Sargun Dhillon, Sasha Sidorov, Satadru Pan, Saurabh Verma, Seiji Yamamoto, Sharadh Ramaswamy, Shaun Lindsay, Shaun Lindsay, Sheng Feng, Shenghao Lin, Shengxin Cindy Zha, Shiva Shankar, Shuqiang Zhang, Shuqiang Zhang, Sinong Wang, Sneha Agarwal, Soji Sajuyigbe, Soumith Chintala, Stephanie Max, Stephen Chen, Steve Kehoe, Steve Satterfield, Sudarshan Govindaprasad, Sumit Gupta, Sungmin Cho, Sunny Virk, Suraj Subramanian, Sy Choudhury, Sydney Goldman, Tal Remez, Tamar Glaser, Tamara Best, Thilo Kohler, Thomas Robinson, Tianhe Li, Tianjun Zhang, Tim Matthews, Timothy Chou, Tzook Shaked, Varun Vontimitta, Victoria Ajayi, Victoria Montanez, Vijai Mohan, Vinay Satish Kumar, Vishal Mangla, Vítor Albiero, Vlad Ionescu, Vlad Poenaru, Vlad Tiberiu Mihailescu, Vladimir Ivanov, Wei Li, Wenchen Wang, Wenwen Jiang, Wes Bouaziz, Will Constable, Xiaocheng Tang, Xiaofang Wang, Xiaojian Wu, Xiaolan Wang, Xide Xia, Xilun Wu, Xinbo Gao, Yanjun Chen, Ye Hu, Ye Jia, Ye Qi, Yenda Li, Yilin Zhang, Ying Zhang, Yossi Adi, Youngjin Nam, Yu, Wang, Yuchen Hao, Yundi Qian, Yuzi He, Zach Rait, Zachary DeVito, Zef Rosnbrick, Zhaoduo Wen, Zhenyu Yang, and Zhiwei Zhao. 2024. The llama 3 herd of models. *Preprint*, arXiv:2407.21783.

Ahmad Faiz, Sotaro Kaneda, Ruhan Wang, Rita Osi, Prateek Sharma, Fan Chen, and Lei Jiang. 2024. Llmcarbon: Modeling the end-to-end carbon footprint of large language models. *arXiv preprint arXiv:2309.14393*.

Haijun Feng, Pingan Zhang, Wen Qin, Weiming Wang, and Huijing Wang. 2022. Estimation of solubility of acid gases in ionic liquids using different machine learning methods. *Journal of Molecular Liquids*, 349:118413.

Constanza Fierro, Ruchira Dhar, Filippos Stamatiou, Nicolas Garneau, and Anders Søgaard. 2024. Defining knowledge: Bridging epistemology and large language models. In *Proceedings of the 2024 Conference on Empirical Methods in Natural Language Processing*, pages 16096–16111, Miami, Florida, USA. Association for Computational Linguistics.

Daan Frenkel and Berend Smit. 2023. *Understanding molecular simulation: from algorithms to applications*. Elsevier.

Yunfan Gao, Yun Xiong, Xinyu Gao, Kangxiang Jia, Jinliu Pan, Yuxi Bi, Yi Dai, Jiawei Sun, Meng Wang, and Haofen Wang. 2024. Retrieval-augmented generation for large language models: A survey. *Preprint*, arXiv:2312.10997.

Yingqiang Ge, Wenyue Hua, Kai Mei, Juntao Tan, Shuyuan Xu, Zelong Li, Yongfeng Zhang, et al. 2024. Openagi: When llm meets domain experts. *Advances in Neural Information Processing Systems*, 36.

Jiahui Geng, Fengyu Cai, Yuxia Wang, Heinz Koeppl, Preslav Nakov, and Iryna Gurevych. 2023. A survey of confidence estimation and calibration in large language models. *arXiv preprint arXiv:2311.08298*.

Joel Guiot and Wolfgang Cramer. 2016. Climate change: The 2015 paris agreement thresholds and mediterranean basin ecosystems. *Science*, 354(6311):465–468.

Taicheng Guo, Bozhao Nan, Zhenwen Liang, Zhichun Guo, Nitesh Chawla, Olaf Wiest, Xiangliang Zhang, et al. 2023. What can large language models do in chemistry? a comprehensive benchmark on eight tasks. *Advances in Neural Information Processing Systems*, 36:59662–59688.

Stefan Harrer. 2023. Attention is not all you need: the complicated case of ethically using large language models in healthcare and medicine. *EBioMedicine*, 90.

Edward J. Hu, Yelong Shen, Phillip Wallis, Zeyuan Allen-Zhu, Yuanzhi Li, Shean Wang, Lu Wang, and Weizhu Chen. 2021. Lora: Low-rank adaptation of large language models. *Preprint*, arXiv:2106.09685.

Yiwen Hu and Markus J Buehler. 2022. End-to-end protein normal mode frequency predictions using language and graph models and application to sonification. *ACS nano*, 16(12):20656–20670.

Yiwen Hu and Markus J Buehler. 2023. Deep language models for interpretative and predictive materials science. *APL Machine Learning*, 1(1).

Hsiu-Yuan Huang, Yutong Yang, Zhaoxi Zhang, Sanwoo Lee, and Yunfang Wu. 2024. A survey of uncertainty estimation in llms: Theory meets practice. *arXiv preprint arXiv:2410.15326*.

Yuheng Huang, Jiayang Song, Zhijie Wang, Shengming Zhao, Huaming Chen, Felix Juefei-Xu, and Lei Ma. 2023. Look before you leap: An exploratory study of uncertainty measurement for large language models. *arXiv preprint arXiv:2307.10236*.

Pascale Husson-Borg, Vladimir Majer, and Margarida F Costa Gomes. 2003. Solubilities of oxygen and carbon dioxide in butyl methyl imidazolium tetrafluoroborate as a function of temperature and at pressures close to atmospheric pressure. *Journal of Chemical & Engineering Data*, 48(3):480–485.

Pavan Inguva, Vijesh J Bhute, Thomas NH Cheng, and Pierre J Walker. 2021. Introducing students to research codes: A short course on solving partial differential equations in python. *Education for Chemical Engineers*, 36:1–11.

Kevin Maik Jablonka, Qianxiang Ai, Alexander Al-Feghali, Shruti Badhwar, Joshua D Bocarsly, Andres M Bran, Stefan Bringuier, L Catherine Brinson, Kamal Choudhary, Defne Circi, et al. 2023. 14 examples of how llms can transform materials science and chemistry: a reflection on a large language model hackathon. *Digital Discovery*, 2(5):1233–1250.

Akshita Jha, Aida Mostafazadeh Davani, Chandan K Reddy, Shachi Dave, Vinodkumar Prabhakaran, and Sunipa Dev. 2023. SeeGULL: A stereotype benchmark with broad geo-cultural coverage leveraging generative models. In *Proceedings of the 61st Annual Meeting of the Association for Computational Linguistics (Volume 1: Long Papers)*, pages 9851–9870, Toronto, Canada. Association for Computational Linguistics.

Ziwei Ji, Nayeon Lee, Rita Frieske, Tiezheng Yu, Dan Su, Yan Xu, Etsuko Ishii, Ye Jin Bang, Andrea Madotto, and Pascale Fung. 2023. Survey of hallucination in natural language generation. *ACM Computing Surveys*, 55(12):1–38.

Albert Q. Jiang, Alexandre Sablayrolles, Arthur Mensch, Chris Bamford, Devendra Singh Chaplot, Diego de las Casas, Florian Bressand, Gianna Lengyel, Guillaume Lample, Lucile Saulnier, Lélio Renard Lavaud, Marie-Anne Lachaux, Pierre Stock, Teven Le Scao, Thibaut Lavril, Thomas Wang, Timothée Lacroix, and William El Sayed. 2023. Mistral 7b. *Preprint*, arXiv:2310.06825.

John Jumper, Richard Evans, Alexander Pritzel, Tim Green, Michael Figurnov, Olaf Ronneberger, Kathryn Tunyasuvunakool, Russ Bates, Augustin Žídek, Anna Potapenko, et al. 2021. Highly accurate protein structure prediction with alphafold. *nature*, 596(7873):583–589.

Nikhil Kandpal, Haikang Deng, Adam Roberts, Eric Wallace, and Colin Raffel. 2023. Large language models struggle to learn long-tail knowledge. *arXiv preprint arXiv:2211.08411*.

Eesha Khare, Constancio Gonzalez-Obeso, David L Kaplan, and Markus J Buehler. 2022. Collagentransformer: end-to-end transformer model to predict thermal stability of collagen triple helices using an nlp approach. *ACS Biomaterials Science & Engineering*, 8(10):4301–4310.

Patrick Lewis, Ethan Perez, Aleksandra Piktus, Fabio Petroni, Vladimir Karpukhin, Naman Goyal, Heinrich Küttler, Mike Lewis, Wen-tau Yih, Tim Rocktäschel, et al. 2020. Retrieval-augmented generation for knowledge-intensive nlp tasks. *Advances in Neural Information Processing Systems*, 33:9459–9474.

Huihan Li, Liwei Jiang, Nouha Dziri, Xiang Ren, and Yejin Choi. 2024a. CULTURE-GEN: Revealing global cultural perception in language models through natural language prompting. In *First Conference on Language Modeling*.

Huihan Li, Liwei Jiang, Jena D Hwang, Hyunwoo Kim, Sebastin Santy, Taylor Sorensen, Bill Yuchen Lin, Nouha Dziri, Xiang Ren, and Yejin Choi. 2024b. Culture-gen: Revealing global cultural perception in language models through natural language prompting. *arXiv preprint arXiv:2404.10199*.

Pengfei Li, Jianyi Yang, Mohammad A Islam, and Shaolei Ren. 2023. Making ai less" thirsty": Uncovering and addressing the secret water footprint of ai models. *arXiv preprint arXiv:2304.03271*.

Weng Marc Lim, Asanka Gunasekara, Jessica Leigh Pallant, Jason Ian Pallant, and Ekaterina Pechenkina. 2023. Generative ai and the future of education: Ragnarök or reformation? a paradoxical perspective from management educators. *The international journal of management education*, 21(2):100790.

Frank YC Liu, Bo Ni, and Markus J Buehler. 2022. Presto: Rapid protein mechanical strength prediction with an end-to-end deep learning model. *Extreme Mechanics Letters*, 55:101803.

Wei Lu, David L Kaplan, and Markus J Buehler. 2024. Generative modeling, design, and analysis of spider silk protein sequences for enhanced mechanical properties. *Advanced Functional Materials*, 34(11):2311324.

Wei Lu, Nic A Lee, and Markus J Buehler. 2023. Modeling and design of heterogeneous hierarchical bioinspired spider web structures using deep learning and additive manufacturing. *Proceedings of the National Academy of Sciences*, 120(31):e2305273120.

Rachel K Luu and Markus J Buehler. 2024. Bioinspiredllm: Conversational large language model for the mechanics of biological and bio-inspired materials. *Advanced Science*, 11(10):2306724.

Rachel K Luu, Marcin Wysokowski, and Markus J Buehler. 2023. Generative discovery of de novo chemical designs using diffusion modeling and transformer deep neural networks with application to deep eutectic solvents. *Applied Physics Letters*, 122(23).

Edward J Maginn. 2009. Molecular simulation of ionic liquids: current status and future opportunities. *Journal of Physics: Condensed Matter*, 21(37):373101.

Sourab Mangrulkar, Sylvain Gugger, Lysandre Debut, Younes Belkada, Sayak Paul, and Benjamin Bossan. 2022. Peft: State-of-the-art parameter-efficient fine-tuning methods. `https://github.com/huggingface/peft`.

Nick McKenna, Tianyi Li, Liang Cheng, Mohammad Javad Hosseini, Mark Johnson, and Mark Steedman. 2023. Sources of hallucination by large language models on inference tasks. *arXiv preprint arXiv:2305.14552*.

Grégoire Mialon, Roberto Dessì, Maria Lomeli, Christoforos Nalmpantis, Ram Pasunuru, Roberta Raileanu, Baptiste Rozière, Timo Schick, Jane Dwivedi-Yu, Asli Celikyilmaz, Edouard Grave, Yann LeCun, and Thomas Scialom. 2023. Augmented language models: a survey. *arXiv preprint arXiv:2302.07842*.

Silvia Milano, Joshua A McGrane, and Sabina Leonelli. 2023. Large language models challenge the future of higher education. *Nature Machine Intelligence*, 5(4):333–334.

Kusuri Murakumo, Naruki Yoshikawa, Kentaro Rikimaru, Shogo Nakamura, Kairi Furui, Takamasa Suzuki, Hiroyuki Yamasaki, Yuki Nishigaya, Yuzo Takagi, and Masahito Ohue. 2023. Llm drug discovery challenge: A contest as a feasibility study on the utilization of large language models in medicinal chemistry. In *AI for Accelerated Materials Design-NeurIPS 2023 Workshop*.

Moin Nadeem, Anna Bethke, and Siva Reddy. 2021. StereoSet: Measuring stereotypical bias in pretrained language models. In *Proceedings of the 59th Annual Meeting of the Association for Computational Linguistics and the 11th International Joint Conference on Natural Language Processing (Volume 1: Long Papers)*, pages 5356–5371, Online. Association for Computational Linguistics.

Rahul Nadkarni, David Wadden, Iz Beltagy, Noah A Smith, Hannaneh Hajishirzi, and Tom Hope. 2021. Scientific language models for biomedical knowledge base completion: an empirical study. *arXiv preprint arXiv:2106.09700*.

Nikita Nangia, Clara Vania, Rasika Bhalerao, and Samuel R. Bowman. 2020. CrowS-pairs: A challenge dataset for measuring social biases in masked language models. In *Proceedings of the 2020 Conference on Empirical Methods in Natural Language Processing (EMNLP)*, pages 1953–1967, Online. Association for Computational Linguistics.

Robert Nozick. 2016. Knowledge and scepticism. In *Readings in Formal Epistemology: Sourcebook*, pages 587–603. Springer.

OpenAI, Josh Achiam, Steven Adler, Sandhini Agarwal, Lama Ahmad, Ilge Akkaya, Florencia Leoni Aleman, Diogo Almeida, Janko Altenschmidt, Sam Altman, Shyamal Anadkat, Red Avila, Igor Babuschkin, Suchir Balaji, Valerie Balcom, Paul Baltescu, Haiming Bao, Mohammad Bavarian, Jeff Belgum, Irwan Bello, Jake Berdine, Gabriel Bernadett-Shapiro, Christopher Berner, Lenny Bogdonoff, Oleg Boiko, Madelaine Boyd, Anna-Luisa Brakman, Greg Brockman, Tim Brooks, Miles Brundage, Kevin Button, Trevor Cai, Rosie Campbell, Andrew Cann, Brittany Carey, Chelsea Carlson, Rory Carmichael, Brooke Chan, Che Chang, Fotis Chantzis, Derek Chen, Sully Chen, Ruby Chen, Jason Chen, Mark Chen, Ben Chess, Chester Cho, Casey Chu, Hyung Won Chung, Dave Cummings, Jeremiah Currier, Yunxing Dai, Cory Decareaux, Thomas Degry, Noah Deutsch, Damien Deville, Arka Dhar, David Dohan, Steve Dowling, Sheila Dunning, Adrien Ecoffet, Atty Eleti, Tyna Eloundou, David Farhi, Liam Fedus, Niko Felix, Simón Posada Fishman, Juston Forte, Isabella Fulford, Leo Gao, Elie Georges, Christian Gibson, Vik Goel, Tarun Gogineni, Gabriel Goh, Rapha Gontijo-Lopes, Jonathan Gordon, Morgan Grafstein, Scott Gray, Ryan Greene, Joshua Gross, Shixiang Shane Gu, Yufei Guo, Chris Hallacy, Jesse Han, Jeff Harris, Yuchen He, Mike Heaton, Johannes Heidecke, Chris Hesse, Alan Hickey, Wade Hickey, Peter Hoeschele, Brandon Houghton, Kenny Hsu, Shengli Hu, Xin Hu, Joost Huizinga, Shantanu Jain, Shawn Jain, Joanne Jang, Angela Jiang, Roger Jiang, Haozhun

Jin, Denny Jin, Shino Jomoto, Billie Jonn, Heewoo Jun, Tomer Kaftan, Łukasz Kaiser, Ali Kamali, Ingmar Kanitscheider, Nitish Shirish Keskar, Tabarak Khan, Logan Kilpatrick, Jong Wook Kim, Christina Kim, Yongjik Kim, Jan Hendrik Kirchner, Jamie Kiros, Matt Knight, Daniel Kokotajlo, Łukasz Kondraciuk, Andrew Kondrich, Aris Konstantinidis, Kyle Kosic, Gretchen Krueger, Vishal Kuo, Michael Lampe, Ikai Lan, Teddy Lee, Jan Leike, Jade Leung, Daniel Levy, Chak Ming Li, Rachel Lim, Molly Lin, Stephanie Lin, Mateusz Litwin, Theresa Lopez, Ryan Lowe, Patricia Lue, Anna Makanju, Kim Malfacini, Sam Manning, Todor Markov, Yaniv Markovski, Bianca Martin, Katie Mayer, Andrew Mayne, Bob McGrew, Scott Mayer McKinney, Christine McLeavey, Paul McMillan, Jake McNeil, David Medina, Aalok Mehta, Jacob Menick, Luke Metz, Andrey Mishchenko, Pamela Mishkin, Vinnie Monaco, Evan Morikawa, Daniel Mossing, Tong Mu, Mira Murati, Oleg Murk, David Mély, Ashvin Nair, Reiichiro Nakano, Rajeev Nayak, Arvind Neelakantan, Richard Ngo, Hyeonwoo Noh, Long Ouyang, Cullen O'Keefe, Jakub Pachocki, Alex Paino, Joe Palermo, Ashley Pantuliano, Giambattista Parascandolo, Joel Parish, Emy Parparita, Alex Passos, Mikhail Pavlov, Andrew Peng, Adam Perelman, Filipe de Avila Belbute Peres, Michael Petrov, Henrique Ponde de Oliveira Pinto, Michael, Pokorny, Michelle Pokrass, Vitchyr H. Pong, Tolly Powell, Alethea Power, Boris Power, Elizabeth Proehl, Raul Puri, Alec Radford, Jack Rae, Aditya Ramesh, Cameron Raymond, Francis Real, Kendra Rimbach, Carl Ross, Bob Rotsted, Henri Roussez, Nick Ryder, Mario Saltarelli, Ted Sanders, Shibani Santurkar, Girish Sastry, Heather Schmidt, David Schnurr, John Schulman, Daniel Selsam, Kyla Sheppard, Toki Sherbakov, Jessica Shieh, Sarah Shoker, Pranav Shyam, Szymon Sidor, Eric Sigler, Maddie Simens, Jordan Sitkin, Katarina Slama, Ian Sohl, Benjamin Sokolowsky, Yang Song, Natalie Staudacher, Felipe Petroski Such, Natalie Summers, Ilya Sutskever, Jie Tang, Nikolas Tezak, Madeleine B. Thompson, Phil Tillet, Amin Tootoonchian, Elizabeth Tseng, Preston Tuggle, Nick Turley, Jerry Tworek, Juan Felipe Cerón Uribe, Andrea Vallone, Arun Vijayvergiya, Chelsea Voss, Carroll Wainwright, Justin Jay Wang, Alvin Wang, Ben Wang, Jonathan Ward, Jason Wei, CJ Weinmann, Akila Welihinda, Peter Welinder, Jiayi Weng, Lilian Weng, Matt Wiethoff, Dave Willner, Clemens Winter, Samuel Wolrich, Hannah Wong, Lauren Workman, Sherwin Wu, Jeff Wu, Michael Wu, Kai Xiao, Tao Xu, Sarah Yoo, Kevin Yu, Qiming Yuan, Wojciech Zaremba, Rowan Zellers, Chong Zhang, Marvin Zhang, Shengjia Zhao, Tianhao Zheng, Juntang Zhuang, William Zhuk, and Barret Zoph. 2024. Gpt-4 technical report. *arXiv preprint arXiv:2303.08774*.

Kamil Paduszynski. 2016. In silico calculation of infinite dilution activity coefficients of molecular solutes in ionic liquids: critical review of current methods and new models based on three machine learning algorithms. *Journal of chemical information and modeling*, 56(8):1420–1437.

Saurabh Kumar Pandey, Harshit Budhiraja, Sougata Saha, and Monojit Choudhury. 2025. CULTURALLY YOURS: A reading assistant for cross-cultural content. In *Proceedings of the 31st International Conference on Computational Linguistics: System Demonstrations*, pages 208–216, Abu Dhabi, UAE. Association for Computational Linguistics.

David Patterson, Joseph Gonzalez, Quoc Le, Chen Liang, Lluis-Miquel Munguia, Daniel Rothchild, David So, Maud Texier, and Jeff Dean. 2021. Carbon emissions and large neural network training. *arXiv preprint arXiv:2104.10350*.

Álvaro Pérez-Salado Kamps, Dirk Tuma, Jianzhong Xia, and Gerd Maurer. 2003. Solubility of co2 in the ionic liquid [bmim][pf6]. *Journal of Chemical & Engineering Data*, 48(3):746–749.

Mahinder Ramdin, Theo W de Loos, and Thijs JH Vlugt. 2012. State-of-the-art of co2 capture with ionic liquids. *Industrial & Engineering Chemistry Research*, 51(24):8149–8177.

Abhinav Sukumar Rao, Aditi Khandelwal, Kumar Tanmay, Utkarsh Agarwal, and Monojit Choudhury. 2023. Ethical reasoning over moral alignment: A case and framework for in-context ethical policies in LLMs. In *Findings of the Association for Computational Linguistics: EMNLP 2023*, pages 13370–13388, Singapore. Association for Computational Linguistics.

Nils Reimers and Iryna Gurevych. 2019. Sentence-bert: Sentence embeddings using siamese bert-networks. In *Proceedings of the 2019 Conference on Empirical Methods in Natural Language Processing*. Association for Computational Linguistics.

Christopher J. Rhodes. 2016. The 2015 paris climate change conference: Cop21. *Science Progress*, 99(1):97–104. PMID: 27120818.

Matthias C Rillig, Marlene Ågerstrand, Mohan Bi, Kenneth A Gould, and Uli Sauerland. 2023. Risks and benefits of large language models for the environment. *Environmental Science & Technology*, 57(9):3464–3466.

Anthony Robbins. 2016. How to understand the results of the climate change summit: Conference of parties21 (cop21) paris 2015. *Journal of public health policy*, 37(2):129–132.

Sougata Saha, Saurabh Kumar Pandey, Harshit Gupta, and Monojit Choudhury. 2025. Reading between the lines: Can llms identify cross-cultural communication gaps? *Preprint*, arXiv:2502.09636.

Eloy S Sanz-Pérez, Christopher R Murdock, Stephanie A Didas, and Christopher W Jones. 2016. Direct capture of co2 from ambient air. *Chemical reviews*, 116(19):11840–11876.

Crispin Sartwell. 1992. Why knowledge is merely true belief. *The Journal of Philosophy*, 89(4):167–180.

Timo Schick, Jane Dwivedi-Yu, Roberto Dessì, Roberta Raileanu, Maria Lomeli, Eric Hambro, Luke Zettlemoyer, Nicola Cancedda, and Thomas Scialom. 2024. Toolformer: Language models can teach themselves to use tools. *Advances in Neural Information Processing Systems*, 36.

Quintin R Sheridan, William F Schneider, and Edward J Maginn. 2018. Role of molecular modeling in the development of co2–reactive ionic liquids. *Chemical reviews*, 118(10):5242–5260.

Emma Strubell, Ananya Ganesh, and Andrew McCallum. 2019. Energy and policy considerations for deep learning in nlp. *arXiv preprint arXiv:1906.02243*.

Kumar Tanmay, Aditi Khandelwal, Utkarsh Agarwal, and Monojit Choudhury. 2023. Probing the moral development of large language models through defining issues test. *Preprint*, arXiv:2309.13356.

Ross Taylor, Marcin Kardas, Guillem Cucurull, Thomas Scialom, Anthony Hartshorn, Elvis Saravia, Andrew Poulton, Viktor Kerkez, and Robert Stojnic. 2022. Galactica: A large language model for science. *arXiv preprint arXiv:2211.09085*.

Gemma Team, Morgane Riviere, Shreya Pathak, Pier Giuseppe Sessa, Cassidy Hardin, Surya Bhupatiraju, Léonard Hussenot, Thomas Mesnard, Bobak Shahriari, Alexandre Ramé, Johan Ferret, Peter Liu, Pouya Tafti, Abe Friesen, Michelle Casbon, Sabela Ramos, Ravin Kumar, Charline Le Lan, Sammy Jerome, Anton Tsitsulin, Nino Vieillard, Piotr Stanczyk, Sertan Girgin, Nikola Momchev, Matt Hoffman, Shantanu Thakoor, Jean-Bastien Grill, Behnam Neyshabur, Olivier Bachem, Alanna Walton, Aliaksei Severyn, Alicia Parrish, Aliya Ahmad, Allen Hutchison, Alvin Abdagic, Amanda Carl, Amy Shen, Andy Brock, Andy Coenen, Anthony Laforge, Antonia Paterson, Ben Bastian, Bilal Piot, Bo Wu, Brandon Royal, Charlie Chen, Chintu Kumar, Chris Perry, Chris Welty, Christopher A. Choquette-Choo, Danila Sinopalnikov, David Weinberger, Dimple Vijaykumar, Dominika Rogozińska, Dustin Herbison, Elisa Bandy, Emma Wang, Eric Noland, Erica Moreira, Evan Senter, Evgenii Eltyshev, Francesco Visin, Gabriel Rasskin, Gary Wei, Glenn Cameron, Gus Martins, Hadi Hashemi, Hanna Klimczak-Plucińska, Harleen Batra, Harsh Dhand, Ivan Nardini, Jacinda Mein, Jack Zhou, James Svensson, Jeff Stanway, Jetha Chan, Jin Peng Zhou, Joana Carrasqueira, Joana Iljazi, Jocelyn Becker, Joe Fernandez, Joost van Amersfoort, Josh Gordon, Josh Lipschultz, Josh Newlan, Ju yeong Ji, Kareem Mohamed, Kartikeya Badola, Kat Black, Katie Millican, Keelin McDonell, Kelvin Nguyen, Kiranbir Sodhia, Kish Greene, Lars Lowe Sjoesund, Lauren Usui, Laurent Sifre, Lena Heuermann, Leticia Lago, Lilly McNealus, Livio Baldini Soares, Logan Kilpatrick, Lucas Dixon, Luciano Martins, Machel Reid, Manvinder Singh, Mark Iverson, Martin Görner, Mat Velloso, Mateo Wirth, Matt Davidow, Matt Miller, Matthew Rahtz, Matthew Watson, Meg Risdal, Mehran Kazemi, Michael Moynihan, Ming Zhang, Minsuk Kahng, Minwoo Park, Mofi Rahman, Mohit Khatwani, Natalie Dao, Nenshad Bardoliwalla, Nesh Devanathan, Neta Dumai, Nilay Chauhan, Oscar Wahltinez, Pankil Botarda, Parker Barnes, Paul Barham, Paul Michel, Pengchong Jin, Petko Georgiev, Phil Culliton, Pradeep Kuppala, Ramona Comanescu, Ramona Merhej, Reena Jana, Reza Ardeshir Rokni, Rishabh Agarwal, Ryan Mullins, Samaneh Saadat, Sara Mc Carthy, Sarah Perrin, Sébastien M. R. Arnold, Sebastian Krause, Shengyang Dai, Shruti Garg, Shruti Sheth, Sue Ronstrom, Susan Chan, Timothy Jordan, Ting Yu, Tom Eccles, Tom Hennigan, Tomas Kocisky, Tulsee Doshi, Vihan Jain, Vikas Yadav, Vilobh Meshram, Vishal Dharmadhikari, Warren Barkley, Wei Wei, Wenming Ye, Woohyun Han, Woosuk Kwon, Xiang Xu, Zhe Shen, Zhitao Gong, Zichuan Wei, Victor Cotruta, Phoebe Kirk, Anand Rao, Minh Giang, Ludovic Peran, Tris Warkentin, Eli Collins, Joelle Barral, Zoubin Ghahramani, Raia Hadsell, D. Sculley, Jeanine Banks, Anca Dragan, Slav Petrov, Oriol Vinyals, Jeff Dean, Demis Hassabis, Koray Kavukcuoglu, Clement Farabet, Elena Buchatskaya, Sebastian Borgeaud, Noah Fiedel, Armand Joulin, Kathleen Kenealy, Robert Dadashi, and Alek Andreev. 2024. Gemma 2: Improving open language models at a practical size. *Preprint*, arXiv:2408.00118.

Wilfred F. van Gunsteren, Xavier Daura, Niels Hansen, Alan E. Mark, Chris Oostenbrink, Sereina Riniker, and Lorna J. Smith. 2018. Validation of molecular simulation: An overview of issues. *Angewandte Chemie International Edition*, 57(4):884–902.

Wilfred F. van Gunsteren and Alan E. Mark. 1998. Validation of molecular dynamics simulation. *The Journal of Chemical Physics*, 108(15):6109–6116.

Neeraj Varshney, Wenlin Yao, Hongming Zhang, Jianshu Chen, and Dong Yu. 2023. A stitch in time saves nine: Detecting and mitigating hallucinations of llms by actively validating low-confidence generation. *arXiv preprint arXiv:2307.03987*.

Pablo Villalobos, Anson Ho, Jaime Sevilla, Tamay Besiroglu, Lennart Heim, and Marius Hobbhahn. 2024. Position: will we run out of data? limits of llm scaling based on human-generated data. In *Proceedings of the 41st International Conference on Machine Learning*, ICML'24. JMLR.org.

Pablo Villalobos, Jaime Sevilla, Lennart Heim, Tamay Besiroglu, Marius Hobbhahn, and Anson Ho. 2022. Will we run out of data? an analysis of the limits of scaling datasets in machine learning. *arXiv preprint arXiv:2211.04325*, 1.

Douglas Walton, Christopher Reed, and Fabrizio Macagno. 2008. *Argumentation schemes*. Cambridge University Press.

Douglas N Walton. 1996. *Argument structure: A pragmatic theory*. University of Toronto Press Toronto.

Yixin Wan, Jieyu Zhao, Aman Chadha, Nanyun Peng, and Kai-Wei Chang. 2023. Are personalized stochastic parrots more dangerous? evaluating persona biases in dialogue systems. In *Findings of the Association for Computational Linguistics: EMNLP 2023*, pages 9677–9705, Singapore. Association for Computational Linguistics.

Shaofei Wang, Xueqin Li, Hong Wu, Zhizhang Tian, Qingping Xin, Guangwei He, Dongdong Peng, Silu Chen, Yan Yin, Zhongyi Jiang, et al. 2016. Advances in high permeability polymer-based membrane materials for co2 separations. *Energy & Environmental Science*, 9(6):1863–1890.

Jason Wei, Xuezhi Wang, Dale Schuurmans, Maarten Bosma, Brian Ichter, Fei Xia, Ed Chi, Quoc Le, and Denny Zhou. 2023. Chain-of-thought prompting elicits reasoning in large language models. *arXiv preprint arXiv:2201.11903*.

Andrew D White. 2023. The future of chemistry is language. *Nature Reviews Chemistry*, 7(7):457–458.

Timothy Williamson. 2005. Knowledge, context, and the agent's. *Contextualism in philosophy*, page 91.

Fanghua Ye, Mingming Yang, Jianhui Pang, Longyue Wang, Derek F Wong, Emine Yilmaz, Shuming Shi, and Zhaopeng Tu. 2024. Benchmarking llms via uncertainty quantification. *arXiv preprint arXiv:2401.12794*.

Chi-Hua Yu, Wei Chen, Yu-Hsuan Chiang, Kai Guo, Zaira Martin Moldes, David L Kaplan, and Markus J Buehler. 2022a. End-to-end deep learning model to predict and design secondary structure content of structural proteins. *ACS biomaterials science & engineering*, 8(3):1156–1165.

Chi-Hua Yu, Eesha Khare, Om Prakash Narayan, Rachael Parker, David L Kaplan, and Markus J Buehler. 2022b. Colgen: An end-to-end deep learning model to predict thermal stability of de novo collagen sequences. *Journal of the mechanical behavior of biomedical materials*, 125:104921.

Linda Zagzebski. 2017. What is knowledge? *The Blackwell guide to epistemology*, pages 92–116.

Stefano E Zanco, José-Francisco Pérez-Calvo, Antonio Gasós, Beatrice Cordiano, Viola Becattini, and Marco Mazzotti. 2021. Postcombustion co2 capture: a comparative techno-economic assessment of three technologies using a solvent, an adsorbent, and a membrane. *ACS Engineering Au*, 1(1):50–72.

Shaojuan Zeng, Xiangping Zhang, Lu Bai, Xiaochun Zhang, Hui Wang, Jianji Wang, Di Bao, Mengdie Li, Xinyan Liu, and Suojiang Zhang. 2017. Ionic-liquid-based co2 capture systems: structure, interaction and process. *Chemical reviews*, 117(14):9625–9673.

Di Zhang, Wei Liu, Qian Tan, Jingdan Chen, Hang Yan, Yuliang Yan, Jiatong Li, Weiran Huang, Xiangyu Yue, Wanli Ouyang, et al. 2024. Chemllm: A chemical large language model. *arXiv preprint arXiv:2402.06852*.

Haochen Zhao, Xiangru Tang, Ziran Yang, Xiao Han, Xuanzhi Feng, Yueqing Fan, Senhao Cheng, Di Jin, Yilun Zhao, Arman Cohan, et al. 2024. Chemsafetybench: Benchmarking llm safety on chemistry domain. *arXiv preprint arXiv:2411.16736*.

Wayne Xin Zhao, Kun Zhou, Junyi Li, Tianyi Tang, Xiaolei Wang, Yupeng Hou, Yingqian Min, Beichen Zhang, Junjie Zhang, Zican Dong, et al. 2023. A survey of large language models. *arXiv preprint arXiv:2303.18223*.

A Appendix

```
<<SYS>>Given a claim identify the implicit and explicit assumptions that needs to be verified to ascertain its validity.<</SYS>>
[INST] A claim makes several implicit and explicit assumptions ranging from the basic level to more complex ones.
Identify the assumptions made by a claim which must be verified to ascertain the overall validity of the claim.
## Example 1
Below is an example of extracting the assumptions from a claim. The assumptions are ordered from the basic level to more complex ones.
Claim: Using supported ionic liquid membranes for separation of ... gas molecule have the same order of decrease in values. [/INST]
Assumptions:
1. Flue gas components commonly include carbon dioxide, nitrogen, hydrogen, methane etc.
2. Different components ... interact differently with the anion of an ionic liquid.
3. The absorption rates of carbon-dioxide, .. follows the decreasing order of ..
[INST] ## Example 2
Below is another example of extracting the assumptions from a claim. The assumptions are ordered from the basic level to more complex.
Claim: Most ionic liquids tend to be costlier than ... and hence effective regeneration of ... ionic liquid batches. [/INST]
Assumptions:
1. Ionic liquids used for carbon dioxide capture can be regenerated.
2. Ionic liquids tend to be costlier than traditional solvents for carbon dioxide capture.
3. Effective regeneration of ionic liquids is important ... batches.
[INST] ## Example 3
Below is another example of extracting assumptions from claim. Note that the claim is a fact. Hence it does not contain any assumptions.
Claim: Ionic liquids can be termed as conventional. [/INST]
Assumptions:
1. No assumptions as the claim is a fact.
[INST] Identify the assumption from the following claim. Order them from the basic level to more complex ones. Respond with none if the claim is a fact.
Claim: Task-specific ionic liquids absorb carbon dioxide. [/INST]
Assumptions:
1. Ionic liquids exist in various task-specific forms.
2. Carbon dioxide is a gas species that can interact with ionic liquids.
....
```

Prompt | 3-shot Examples | Response

Figure 5: Mistral 3-shot prompt to automatically extract and generate the missing assumptions from claims.

```
You are an expert in Chemical Engineering with extensive knowledge of ionic liquids. Given the following claim and list of statements related to ionic liquids, indicate if each statement entails the claim. Also, on a scale of 1 to 5 (low to high), indicate your confidence in your response.

Claim: Higher energy consumption for regeneration of ionic liquids is a cost-intensive process.

Statements:
1. Adding water to conventional ionic liquids can reduce their viscosity.
2. Ionic liquid regeneration involves energy consumption only during rainy seasons.
3. Energy consumption for ionic liquids regeneration incurs cost only during rainy seasons.
4. Imidazolium-based ionic liquids can have alkyl groups attached to the imidazolium ring.
5. Energy consumption for ionic liquids regeneration incurs cost.
6. None of the above

Format your response as a Python dictionary in the following format: {'<statement number>': {'entails': <True/False>, 'confidence': <score between 1 to 5>}}.
Do not generate anything extra.
```
Prompt

```
{'1': {'entails': False, 'confidence': 5}, '2': {'entails': True, 'confidence': 5}, '3': {'entails': True, 'confidence': 5}, '4': {'entails': False, 'confidence': 5}, '5': {'entails': True, 'confidence': 5}, '6': {'entails': False, 'confidence': 5}}
```
Response

Figure 6: Sample prompt for the entailment task.

```
Correct the incorrect input as a Python dictionary in the following format: {{'<number between 1 to {n_statements}>': {{'entails': <Boolean True/False>, 'confidence': <Integer score between 1 to 5>}}}}
Please keep the following points in mind:
1. The size of the output dictionary must be {n_statements} and the dictionary elements must be numbered between 1 to {n_statements}.
2. If there are missing dictionary elements, add them as {{'<number>': {{'entails': False, 'confidence': 1}}}}
3. If the 'entails' key is missing in a dictionary element, add 'entails': False
4. If the 'confidence' key is missing in a dictionary element, add 'confidence': 1
5. If there are additional outside numbers 1 to {n_statements}, remove them.
6. Do not generate the code.

Incorrect input: {'0': {'entails': False}, '2': {'entails': False, 'confidence': 5}, '3': {'entail': False, 'confidence': 5}, '4': {'entails': False, 'confidence': 5}, '5': {'entails': False, 'confidence': 5} 5: {'entails': True, 'confidence': 5},}
```
Prompt

```
{'1': {'entails': False, 'confidence': 1}, '2': {'entails': False, 'confidence': 5}, '3': {'entails': False, 'confidence': 5}, '4': {'entails': False, 'confidence': 5}, '5': {'entails': False, 'confidence': 5}, 6: {'entails': True, 'confidence': 5}}
```
Response

Figure 7: Sample prompt for correcting the LLM response using Llama.

Applying the Character-Role Narrative Framework with LLMs to Investigate Environmental Narratives in Scientific Editorials and Tweets

Francesca Grasso
University of Turin
Dep. of Computer Science
Turin, Italy
fr.grasso@unito.it

Stefano Locci
University of Turin
Dep. of Computer Science
Turin, Italy
stefano.locci@unito.it

Manfred Stede
University of Potsdam
Department of Linguistics
Potsdam, Germany
stede@uni-potsdam.de

Abstract

Communication aiming to persuade an audience uses strategies to frame certain entities in 'character roles' such as hero, villain, victim, or beneficiary, and to build narratives around these ascriptions. The *Character-Role Framework* is an approach to model these narrative strategies, which has been used extensively in the Social Sciences and is just beginning to get attention in Natural Language Processing (NLP). This work extends the framework to scientific editorials and social media texts within the domains of ecology and climate change. We identify characters' roles across expanded categories (human, natural, instrumental) at the entity level, and present two annotated datasets: 1,559 tweets from the *Ecoverse* dataset and 2,150 editorial paragraphs from *Nature & Science*. Using manually annotated test sets, we evaluate four state-of-the-art Large Language Models (LLMs) (GPT-4o, GPT-4, GPT-4-turbo, LLaMA-3.1-8B) for character-role detection and categorization, with GPT-4 achieving the highest agreement with human annotators. We then apply the best-performing model to automatically annotate the full datasets, introducing a novel entity-level resource for character-role analysis in the environmental domain.

1 Introduction

There is a long history in the literature demonstrating how stories are central to how humans understand and communicate about the world, with language playing a key role in constructing and delivering specific messages (Armstrong and Ferguson, 2010; Polkinghorne, 1988). This is particularly important when examining linguistic representations of climate change (CC) and environmental issues (Wolters et al., 2021; Stibbe, 2015, 2021; Jones et al., 2022). Studies from various disciplines, such as ecolinguistics (Fill and Muhlhausler, 2006; Alexander and Stibbe, 2014), political and social sciences (Nerlich et al., 2010; Grundmann and Krishnamurthy, 2010), have demonstrated how environmental and CC narratives are crucial in understanding how individuals and entities like governments and media interpret and relate to ecological issues and the natural world and as a consequence, how they behave towards them (Fløttum and Gjerstad, 2017). The linguistic construction of entities as social actors in these 'stories' can reveal the author's *framing* choices and communicative intent (Hulme, 2015). For instance, "Expanding oil drilling operations will boost economic growth and create thousands of jobs in struggling communities" frames oil drilling positively, emphasizing economic benefits while downplaying environmental concerns. The *Character-Role Framework* – introduced by Gehring and Grigoletto (2023) and drawing on the Narrative Policy Framework (NPF) (Jones and McBeth, 2010; Jones, 2018) – is based on the premise that framing entities in specific roles (hero, villain, victim, beneficiary) is key to understanding a narrative's intent and potential effects. While prior work using this framework has primarily appeared in social and political sciences (Bergstrand and Jasper, 2018; Wolters et al., 2021), its adaptation to Natural Language Processing (NLP) tasks remains limited. Existing studies have either focused on policy narratives (Gehring and Grigoletto, 2023) or explored related tasks like character-role extraction (Stammbach et al., 2022) focusing on a higher-level analyses (e.g. at the paragraph-level). Moreover, Frermann et al. (2023) extended framing analysis to the document-level by integrating narrative media framing with entity roles. In this paper, we extend and adapt the Character-Role Framework to investigate CC and environmental narratives. Specifically, we focus on identifying characters across three categories (human, instrumental, natural) and four roles (hero, villain, victim, beneficiary) at the *entity level*, applying the framework to both scientific editorials and social media. To achieve this, we first created two

Proceedings of the 2nd Workshop on Natural Language Processing Meets Climate Change (ClimateNLP 2025), pages 49–67
July 31, 2025 ©2025 Association for Computational Linguistics

manually annotated test sets: (i) 150 CC-related scientific editorial paragraphs from *Nature* and *Science* (Stede et al. (2023)) and (ii) 300 tweets from the *Ecoverse* dataset (Grasso et al., 2024a), covering a wide range of environmental topics. Characters were identified and roles assigned based on linguistic cues, subsequently categorized as human, instrumental, or natural - a category newly introduced in this work. We evaluated four Large Language Models (LLMs) (GPT-4o, GPT-4, GPT-4-turbo, and Llama-3.1-8B) as additional annotators to measure alignment with human annotations. We subsequently applied the best-performing models to larger datasets, resulting in **1,559** tweets and **2,150** editorial paragraphs annotated for further analysis. Our contributions are fivefold: (i) We offer a novel approach for analyzing CC and environmental texts using the Character-Role Framework across social media and scientific editorials. (ii) We extend the framework by adopting a bottom-up entity-level approach and introducing the "natural" character category. (iii) We release two new annotated datasets for narrative analysis in this domain. (iv) We evaluate LLMs on the entity-level character-role detection and categorization tasks, marking a significant advancement for this framework in NLP[1]. (v) We conduct a qualitative error analysis of model misclassifications and provide preliminary insights into emerging narrative patterns.

2 Theoretical Background

The Character-Role Framework, introduced by Gehring and Grigoletto (2023), builds on foundational work in climate change (CC) narratives, such as those by Fløttum and Gjerstad (2017, 2013)[2]. Their theoretical and methodological framework applies the concept of policy narrative from the Narrative Policy Framework (NPF) (Jones, 2018) to CC discourse, recognizing the role of 'stories' used to communicate and discuss CC issues in shaping opinions and behaviors. This aligns with traditions in ecolinguistics and ecocriticism, which emphasize the importance of studying how language shapes perceptions, behaviors, and actions regarding environmental issues (Stibbe, 2015, 2021; Fill and Muhlhausler, 2006). The NPF adopts a structuralist approach to narrative and posits that they can be generalizable and have an identifiable structure and measurable elements (e.g., characters, setting) (Jones et al., 2022). Among the narrative components, "characters" play a prominent role in that they determine and are determined by the "plot". Key character roles include: (i) *victims*, who are harmed or at risk of harm; (ii) *villains*, responsible for causing harm; (iii) *heroes*, who work to resolve the harm; and (iv) *beneficiaries*, who gain from the events described. Gehring and Grigoletto (2023) further distinguish characters as either human (individuals or entities made up of people) or instrumental (abstract entities like policies or laws). Narratives are categorized as either simple (involving a single character) or complex (involving multiple characters). This framework theory and the assumption on which it is based can be easily extended and applied to other communicative units, such as social media and scientific communication, as each communicative act entails a rather specific (more or less overt) communicative intention. In environmental narratives, any entity can take on the role of a character (Gehring and Grigoletto, 2023): framing institutions, natural entities, or even concrete objects in specific roles can influence perceptions, preferences, and actions towards these entities or events. For example, Kuha (2017) highlights the crucial role of linguistic cues in shaping how language users represent both themselves and other social actors, especially in terms of agency and responsibility.

3 Related Work

Character Roles, Narratives, and Related Tasks The study of environmental narratives has traditionally been rooted in the Humanities, for instance in fields such as Ecolinguistics and Ecocriticism (Alexander and Stibbe, 2014; Stibbe, 2015, 2021). Much research specifically on climate change (CC) narratives has been situated in the Political and Social Sciences, sometimes using the Narrative Policy Framework (NPF) to analyze topics like the political discourse on environment (Peterson, 2021), COVID-19 narratives (Peterson et al., 2021) or economy reports (Goldberg-Miller and Skaggs, 2022). There is only little work in the NLP field on ecology-oriented corpora so far (Grasso et al., 2024a; Bosco et al., 2023), but CC-related topics

[1]The code, the complete set of prompts and the anonymized datasets are available in this repository: `https://github.com/stefanolocci/Character-Role-Narrative-Framework-LLMs`.

[2]We are aware of the huge body of literary-science-oriented research on narratology, in the tradition of Propp and Bakhtin, but for reasons of space we do not make a comparison here but limit ourselves to the more social-science-related view.

have recently gained traction within the NLP community (Stede and Patz, 2021; Grasso et al., 2024b; Stammbach et al., 2024). Recent work has explored character-role extraction in NLP, such as identifying *villain* roles using rule-based approaches and BERT (Klenner et al., 2021) or extracting character roles via zero-shot question-answering (Stammbach et al., 2022). In the context of CC narratives, Gehring and Grigoletto (2023) applied the character-role framework to US policy discourse on Twitter, focusing on economic narratives and a narrow character set at the tweet level. Beyond social-media and short-text analysis, Zhou et al. (2024) used LLMs to analyze CC narratives extracting latent moral messaging from North American and Chinese news. Our work extends this line of research by introducing entity-level annotations across multiple roles and categories, thus offering a broader and more fine-grained analysis of environmental narratives. The character-role task bears similarities to the field of entity-level sentiment detection (Rønningstad et al., 2023), where linguistic indicators like polarity-inducing verbs or modifiers are used to determine whether a certain entity is being portrayed positively or negatively.

LLMs and the CC/Environment Domain In the wider field of applying NLP to the CC and environment domain, notable contributions include Bulian et al. (2023) and Zhu and Tiwari (2023), who propose evaluation frameworks for analyzing LLM responses to CC topics. Koldunov and Jung (2024) developed a prototype tool using LLMs to provide localized climate-related data, while Leippold et al. (2024) created an AI tool for fact-checking CC claims utilizing LLMs. Thulke et al. (2024) introduced a family of domain-specific LLMs designed to synthesize interdisciplinary research on CC. Grasso and Locci (2024) assessed the performance and self-evaluation capabilities of different LLMs in classification tasks within the CC and environmental domain, while Grasso et al. (2025) proposed a novel framework for assessing anthropocentric bias in LLM-generated texts. Fore et al. (2024) experimented with LLMs for CC topics, finding that, while effective with fine-tuning, to ensure accuracy they require safeguards against misinformation.

4 Dataset Creation and Annotation

This section outlines how we applied, adapted, and extended the Character-Role Framework by adopting a novel bottom-up approach. We focus on building two manually annotated datasets for character-role detection within CC and environmental narratives. These datasets will serve as test sets for evaluating the performance of the LLMs before expanding to create the final, larger datasets.

4.1 NatSciEdCC and Ecoverse

We did not aim to restrict the scope of our investigation to a specific (sub)domain or a limited set of possible entities (which we refer to as a top-down approach), as we believe that relying on keywords or predefined lists could limit the diversity of characters discovered in a broader environmental domain. Instead, our goal was to capture a more heterogeneous and comprehensive set of character-roles, even if it might increase the complexity of the task and pose challenges to both human annotation and models' performance. To still ensure domain consistency while maintaining diversity, we used two datasets that cover various subtopics and discussions related to CC and environmental issues:

(i) The *NatSciEdCC* corpus (Stede et al., 2023) consists of 490 plain text files from *Nature* and *Science* editorials related to climate change. The texts are segmented into single paragraphs of varying lengths and annotated with multiple dimensions, including topicality (CC relevance) and frame coding.

(ii) *Ecoverse* (Grasso et al., 2024a) is a dataset of 3,023 tweets covering various environmental topics, including CC. It is annotated for eco-relevance, environmental impact, and the author's stance toward environmental causes (supportive, neutral, or skeptical/opposing). The dataset is openly available under a CC BY-SA 4.0 license.

4.2 Data Cleaning and Dataset Creation

Our goal was to create two datasets that contain rich and diverse narratives, focusing on texts with well-defined narrative elements while minimizing overly vague or noisy content. Given the heterogeneous nature of the language in *Ecoverse* tweets, ranging from formal news sources to informal user posts, and the structured language of scientific editorials, we applied tailored filtering steps to ensure meaningful and balanced content for analysis.

Filtering *Ecoverse* To maximize narrative diversity and reduce noise, we applied the following filtering steps: (i) Tweets unrelated to environmental or climate change (CC) topics were excluded, based on pre-existing eco-relevance annotations. (ii) To minimize the inclusion of overly hashtag-heavy tweets, which tend to lack substantial content, we removed tweets containing more than three hashtags. From the resulting set, we selected 300 tweets for manual annotation. To ensure diversity, we randomly sampled: (i) 180 tweets from environmental publications and news outlets (e.g., National Geographic, New York Times); (ii) 120 tweets from individual users, equally divided into 60 supportive tweets and 60 skeptical/opposing tweets. After combining these selections, we shuffled them to create a diverse dataset for manual annotation.

Filtering *NatSciEdCC* Similarly, for scientific editorials, we applied the following steps to ensure meaningful and balanced content: (i) We excluded extremely short paragraphs (fewer than 24 tokens) to focus on texts with sufficient narrative structure for analysis. (ii) We selected paragraphs with the highest topicality scores related to CC, based on existing annotations. (iii) To capture a broad spectrum of narrative tones, we performed sentiment analysis (Hutto and Gilbert, 2014) on the paragraphs and selected the 50 most positive, 50 most negative paragraphs, and 50 with mid-range sentiment. This ensured a balanced dataset of 150 paragraphs. Given that paragraphs are significantly longer than tweets—often two to four times the length—we determined that a dataset of 150 paragraphs would be sufficient for manual annotation and analysis.

4.3 Dataset Annotation

4.3.1 Character Definition

As Gehring and Grigoletto (2023) note, any entity can be a character in a narrative, making it useful to distinguish broader categories. To adapt our analysis to different text types (social media and editorials) and a wider range of ecological topics beyond climate change policy discussions, we expanded previous definitions of characters. In addition to "human" and "instrumental" characters, we introduced a novel third category—*natural characters*. This decision is informed by ecolinguistics (Stibbe, 2015, 2021; Fill and Muhlhausler, 2006), which shows how natural elements are often personified or attributed with agency in everyday language.

Language use frequently constructs natural entities as sentient or volitional, as seen in expressions in our datasets such as "nature destroys" "the forest heals", "the land is threatened", "rivers are stressed". These strategies also hide the true human agents behind these processes (Kuha, 2017). Recognizing this, the inclusion of natural characters enriches our analysis and opens possibilities for future ecocritical discourse analysis, where the use of such verbs and agency can be further examined.

Thus, our final character categories include:

• *Human Characters*: Individuals or groups (e.g., corporations, governments, organizations) whose actions, inactions, or beliefs significantly influence the narrative.

• *Instrumental Characters*: More abstract entities (e.g., policies, laws, technologies) or human-driven processes (e.g., "urbanization", "deforestation") that play key roles in the narrative and are produced or initiated by human characters.

• *Natural Characters*: Non-human entities (e.g., soil, oceans, animals) and natural phenomena (e.g., "climate change", "pandemics") when they are portrayed as playing an active or passive role in the narrative.

4.3.2 Task Description and Guidelines

The task and guidelines were consistent across both tweets and editorial paragraphs, with slight adjustments made during the annotation process to accommodate the differences in text types. Unlike Gehring and Grigoletto (2023), who annotated at the tweet level, we opted for a finer granularity by annotating at the entity level.

Guidelines Annotators received detailed guidelines[3], which mirrored the prompts later used for LLMs. These were based on the character-role definitions from Gehring and Grigoletto (2023) but tailored to the different text types. Annotators were tasked with identifying prominent characters by assigning them one of four roles: **Hero**, **Villain**, **Victim**, or **Beneficiary**. Linguistic indicators such as polarized or action-driven words — modifiers, verbs — (e.g., "heal", "save") helped determine roles. For editorials, where language can be more subtle, annotators also considered the 'overall sentiment' towards an entity when linguistic cues were not directly adjacent. Annotation proceeded sentence by sentence, and annotators paid attention to role shifts, where a character's role could evolve within a sentence or paragraph. They were instructed to focus on the author's communicative intent and avoid assumptions based on external world knowledge. Only nouns or noun phrases were eligible for labeling, including pronouns like "we",

[3]Guidelines of both character-role and character categorization tasks are provided in Appendix A.3.

which often reflect social actors portraying themselves. This was particularly relevant in tweets, where first-person pronouns are frequently used to express personal ideas. Below are examples of tweets (1)-(2) and editorials (3) with expected annotations, following the guidelines, with villains in burgundy red, heroes in green, victims in orange, and beneficiaries in blue.

(1) *Humboldt penguins face existential threats from climate change and overfishing—but also from habitat theft, as the penguins use guano for nesting while humans covet it for fertilizer.*

(2) *With #climate change impacting agriculture, Genetically Modified crops offer promising solutions, including reducing greenhouse gas emissions. However, the focus on profit over #sustainability risks farmer livelihoods & the #environment.*

(3) *US President Donald Trump is promoting a retrograde energy agenda and has vowed to pull the United States out of the Paris agreement. Still, despite their efforts, Trump's allies have been unsuccessful in stopping the rise of renewable energy companies, while local communities are benefiting from this.*

4.3.3 Annotation Process

The annotation process consisted of two consecutive phases: (i) character-role annotation for the datasets of **300** tweets and **150** editorial paragraphs, and (ii) character categorization for a subset of these annotations. In the second phase, 50 tweets and 50 editorial paragraphs were selected for categorizing entities into three categories: human, instrumental, and natural. Both annotation tasks were carried out using the Label Studio open-source data labeling tool[4], with a NER template and a tailored labeling setup.

Character-Role Annotation The primary character-role annotation task was conducted by two annotators, both part of the same research team. One is an author of this paper. One annotator self-identified as a male social scientist, and the other as a female linguist. To ensure consistency and a shared understanding of the guidelines, an iterative two-step training process was undertaken. Initially, the two annotators performed a pilot annotation on a secondary dataset of 20 tweets and 15 editorial paragraphs (with the same distribution as in the main datasets). After completing the annotation, the annotators compared their results, discussed disagreements and differing interpretations, and worked together to refine the guidelines where necessary. This process was repeated until both annotators were confident in applying the guidelines consistently. Following the pilot phase, the annotators began work on the main datasets, annotating two initial batches of 50 tweets and 20 paragraphs respectively. After these batches, we monitored the Inter-Annotator Agreement (IAA) to measure consistency. Annotators discussed any problematic cases and further refined the guidelines accordingly. Once most issues were addressed, the annotation of the remaining tweets and paragraphs continued without the need for further discussion sessions.

Categorization Task The second annotation task involved assigning one of the three character types - human, instrumental, natural - to previously labeled entities. A third annotator, a student member of our research group who self-identified as non-binary, was provided with task-specific guidelines. This annotator worked on a subset of 50 tweets and 50 paragraphs previously annotated for character/roles, containing respectively 101 and 373 labeled entities. This subset was chosen based on agreement between the two annotators in the first task to ensure higher consistency. The decision to use only a subset was made because, despite the reduced number of texts, each paragraph and tweet contained a significant number of annotated entities. Moreover, this task was deemed relatively objective, so only one annotator was used. Table 8 in Appendix A.1 reports the label distribution for this task. Both character-role annotated datasets and character categorization subsets were then used to instruct the LLMs for automatic character-role detection and categorization, as discussed in Section 5.

4.4 IAA and Datasets Statistics

To measure the agreement between the two annotators, we treated the character-role task as a Named Entity Recognition (NER) task, taking into account two elements: the text spans of each annotated entity within the text unit (either a paragraph or a tweet) and the label assigned to that text span. Agreement was achieved if the two annotators annotated the same entity with either an identical text span or overlapping text spans (e.g., "the President" vs. "President"). We used Precision, Recall, and F1-score to calculate the agreement, as these metrics account for both span overlap and label consistency (Brandsen et al., 2020; Hripcsak and Rothschild, 2005).

[4]https://labelstud.io

- IAA Tweets (300): Precision = 0.80, Recall = 0.73, F1-score = 0.76.
- IAA Editorial Paragraphs (150): Precision: 0.81, Recall: 0.87, F1-score: 0.84

Detailed comments and insights on disagreements are reported in Section 6. Table 1 reports the label distribution among annotators throughout the datasets. Table 7 in Appendix A.1 report datasets statistics.

Label	Tweets			Editorials		
	A1	A2	GPT4	A1	A2	GPT4
Hero	187	216	273	311	285	256
Beneficiary	199	175	177	163	73	55
Villain	112	166	210	285	290	280
Victim	82	87	179	135	143	172
Total labels	580	644	839	894	791	763

Table 1: Label Distribution for Tweets and Editorials test sets for A1, A2 and best model (A1: Annotator 1, A2: Annotator 2).

5 Experiments with LLMs: Methodology

5.1 Motivation and Models Employed

We aimed to evaluate the performance of large language models (LLMs) on the previously unexplored tasks of entity-level character-role detection and character categorization. The main advantage of using LLMs is their ability to perform well with less task-specific training data, as they are pretrained on vast amounts of text. Additionally, LLMs have performed well in similar tasks such as NER (Wang et al., 2023; Villena et al., 2024). However, they are also susceptible to hallucinations, where they generate outputs not grounded in the input data (Mittal et al., 2024). Therefore, we employed careful prompt engineering and iterative testing to optimize their performance in this highly subjective and complex task. We experimented with the following LLMs, covering both closed and open models: **GPT-4o**[5]; **GPT-4-turbo**; **GPT-4** (et al., 2024); **Llama-3.1-8B**[6].

Our methodology for the models' classification experiments proceeded in three phases: (i) an *Exploratory Phase* to refine prompt design and model setup, (ii) an *Exploitation Phase* to assess model performance against human annotators, and (iii) *Classification Phase* to apply the best-performing model and optimal setting across larger datasets of tweets and editorials.

5.2 Exploratory Phase: Prompt Design

The clarity of prompts is crucial for generating accurate outputs in classification tasks (Deldjoo, 2023). Given that small adjustments in prompts or model settings (e.g., temperature) can significantly impact results, we conducted exploratory experiments to refine both the prompts and model setup. **Character-role Task** We tested a zero-shot setting on a small sample of 5-7 textual units for each text type (paragraphs and tweets), iterating through different prompt strategies across models (GPT-4, GPT-4o, GPT-4-turbo, and Llama-3.1-8B-Instruct)[7]. This allowed us to monitor output variations and refine the prompts accordingly. The prompts closely mirrored the guidelines provided to human annotators, and we also experimented with different output formats. By the end of this phase, we finalized the best prompt format for each model. To ensure consistency with human annotations, we used an in-line tag annotation format.

Character Categorization Task Given the simpler nature of the character categorization task, we directly applied a few-shot setting, leveraging insights from the main task's exploratory phase. The output format remained consistent with the character-role detection task, using in-line tagging for the category names.

Some examples of the prompts used are provided in Appendix A.4, while the complete set of prompts, including all versions for all the LLMs, can be found in the GitHub repository linked earlier.

5.3 Experimental Setup

We conducted the experiments on the Paperspace platform[8], utilizing a configuration that includes an NVIDIA P6000 24GB GPU, 30GB of RAM, and a 8-core CPU. We employed the Huggingface Pipeline abstraction[9] to load the Meta-Llama-3.1-8B-Instruct. For the GPT models, we utilized OpenAI's APIs[10]. After conducting a qualitative manual analysis of responses generated at various temperature settings (ranging from 0.1 to 0.9), we

[5] https://openai.com/index/hello-gpt-4o/

[6] https://llama.meta.com/

[7] We also experimented with Llama-2-13B, but it frequently hallucinated, so we excluded it from further testing.

[8] https://www.paperspace.com/

[9] https://huggingface.co/docs/transformers/main_classes/pipelines

[10] https://openai.com/blog/openai-api

determined that a value of 0.2 provided the best balance between coherence and adherence to the prompts.

5.4 Exploitation Phase

After the prompts and parameters optimization, we conducted the main experiment to compare the performance of the LLMs against human annotators.
Character-role task We tested three prompt settings: (i) zero-shot, (ii) one-shot, and (iii) few-shot. Each model's performance was evaluated on the 300-tweet and 150-editorial test sets, measuring agreement with the two human annotators using Precision, Recall, and F1-score. The best F1-scores were achieved by GPT-4 with the few-shot setting for tweets (0.65) and the one-shot setting for editorials (0.70). Full agreement results for all settings and models are presented in Tables 2 and 3.

Tables 1 and 7 provide a comparison of GPT-4 label distribution and statistics against human annotations. Section 6 discusses the results and offers insights into the models' classification errors.
Character Categorization Task We tested the models in a few-shot setting on 50 tweets and 50 paragraphs manually annotated by Annotator 3. The task was to predict the correct character category (human, instrumental, or natural) for each previously labeled entity. The best-performing model was GPT-4o for tweets (F1: 0.88), and GPT-4o and GPT-4-turbo for paragraphs (F1: 0.78). Full performance metrics for all models (Precision, Recall, and F1-score) can be found in Table 4.

5.5 Classification Phase

In the final phase, we applied the best-performing model, GPT-4, in a few-shot configuration for character-role classification to automatically label two larger datasets: the 1,259 eco-related tweets from the *Ecoverse* dataset and an additional 2,000 editorial paragraphs from *NatSciEdCC*. The selection of these 2k paragraphs followed the same criteria used for the creation of the 150-paragraph test set, as detailed in Section 4.2. The aim was to scale the character-role detection process and create two fully annotated datasets. After merging the test sets with these new annotations, we obtained: (i) a **1,559**-tweet dataset and (ii) a **2,150**-editorial paragraph dataset. Finally, we applied the best-performing model for character categorization, GPT-4o, to label all characters in both datasets, assigning them one of three categories: human, instrumental, or natural. Table 5 shows the role distribution among these categories, and Table 6 provides the dataset statistics.

6 Results and Discussion

As shown in Tables 2 and 3, GPT-4 generally outperformed the other models, while Llama-3.1-8B showed the lowest agreement, particularly struggling with longer paragraphs, where it tended to hallucinate. GPT-4o delivered strong results for editorial paragraphs. For tweets, the few-shot setting yielded the best results (F1: 0.60 for Annotator 1 and F1: 0.65 for Annotator 2), while editorials saw the highest score in the one-shot setting (F1: 0.70). GPT-4o also showed high precision in zero-shot settings for editorials. Interestingly, in GPT-4o experiments, the one-shot setup performed slightly better the few-shot one in both editorial and tweet tasks (F1: 0.70 vs. F1: 0.66 for editorials, F1: 0.59 vs. F1: 0.52 for tweets). This may indicate that providing fewer examples helps the model generalize better, avoiding overfitting and inconsistencies. Models consistently aligned more with Annotator 2 for tweets, with minimal differences in editorials. Interestingly, GPT-4o aligned more closely with Annotator 1 in editorials. Overall, models performed better in recognizing character-roles in editorials, likely due to their structured and homogeneous nature.

6.1 Disagreements Analysis

To better understand the areas of disagreement, we have to distinguish two levels: (i) disagreement on text spans, where one annotator recognized a character and the other did not (i.e., presence/absence), and (ii) disagreement on labels for the same text span, where both annotators agreed on the entity but assigned different roles. For instance, in tweets, more than half of the disagreements between GPT-4 and Annotator 2 (522 instances) resulted from mismatched labels, while fewer disagreements (452) stemmed from mismatched text spans. This indicates that while role assignment may be subjective, the model effectively identified prominent characters. Importantly, we observed that disagreements between human annotators and between humans and models often arose from similar challenges. **Reliance on world knowledge**: Both models and annotators sometimes relied on external world knowledge rather than strictly following the author's intent. For example, GPT-4 and Annotator 1 labeled "fossil fuel companies"

Model	Setting	Annotator1			Annotator2		
		Precision	Recall	F1-score	Precision	Recall	F1-score
GPT-4	one-shot	0.52	0.64	0.57	**0.57**	0.68	0.62
	zero-shot	0.48	0.61	0.54	0.54	0.66	0.59
	few-shot	0.52	**0.72**	**0.60**	**0.57**	**0.76**	**0.65**
GPT-4o	one-shot	0.48	0.62	0.54	0.53	0.67	0.59
	zero-shot	0.44	0.58	0.50	0.50	0.65	0.56
	few-shot	0.47	0.59	0.52	0.53	0.65	0.58
GPT-4-Turbo	one-shot	**0.53**	0.55	0.54	**0.57**	0.57	0.57
	zero-shot	0.46	0.34	0.39	0.51	0.37	0.43
	few-shot	0.52	0.63	0.57	0.55	0.65	0.60
Llama3.1	one-shot	0.41	0.44	0.42	0.45	0.46	0.45
	zero-shot	0.37	0.27	0.31	0.44	0.32	0.37
	few-shot	0.37	0.65	0.47	0.40	0.68	0.51

Table 2: Models Performance on Tweets test set in terms of IAA with human annotators.

Model	Setting	Annotator1			Annotator2		
		Precision	Recall	F1-score	Precision	Recall	F1-score
GPT-4	one-shot	0.73	0.54	0.62	0.85	**0.59**	**0.70**
	zero-shot	0.75	0.49	0.59	0.88	0.55	0.68
	few-shot	0.72	0.61	0.66	0.78	**0.59**	0.67
GPT-4o	one-shot	0.71	0.55	0.62	0.86	0.58	0.69
	zero-shot	**0.79**	0.42	0.55	**0.89**	0.53	0.66
	few-shot	0.73	0.61	**0.67**	0.82	0.55	0.66
GPT-4-Turbo	one-shot	0.66	0.51	0.58	0.80	0.49	0.61
	zero-shot	0.69	0.37	0.48	0.85	0.38	0.53
	few-shot	0.66	**0.67**	0.66	0.72	0.62	0.67
Llama3.1	one-shot	0.61	0.52	0.56	0.68	0.33	0.44
	zero-shot	0.64	0.25	0.36	0.76	0.42	0.54
	few-shot	0.56	0.64	0.60	0.65	0.57	0.61

Table 3: Models Performance on Paragraphs test set in terms of IAA with human annotators.

Model	Precision	Recall	F1-score
Tweets 50 test set			
GPT-4	**0.86**	0.89	0.87
GPT-4o	**0.86**	**0.90**	**0.88**
GPT-4-Turbo	0.77	0.79	0.78
LLama3.1	0.63	0.75	0.68
Editorials 50 test set			
GPT-4	0.76	0.76	0.76
GPT-4o	0.76	**0.79**	**0.78**
GPT-4-Turbo	**0.77**	**0.79**	**0.78**
LLama3.1	0.65	0.65	0.65

Table 4: Model performance on the categorization task for the 50-tweet and 50-editorial paragraph datasets (few-shot setting).

as villains, likely due to their general environmental impact, and "poor countries" as victims, possibly influenced by the linguistic cue "poor", even when the narrative did not explicitly frame them as such. **Subtle framing of characters**: Some entities played subtle yet important roles, making it difficult to decide whether to annotate them. Models, especially in tweets, tended to over-annotate compared to humans (as seen in Tables 1 and 7), though this over-annotation followed consistent patterns. For instance, in a tweet calling Croatia a "remarkable biodiversity hot spot", the model labeled Croatia as a beneficiary, or in "we can develop #green habits", it labeled "green habits" as a hero. However, model and human alignment was much closer in editorials, with GPT-4's label distribution closely matching Annotator 1 and 2. **Role ambiguity**: Some entities played multiple roles within the same text. For example, "endangered species" in a tweet could be labeled as victims of habitat destruction but also as beneficiaries of conservation efforts. Finally, **calculation strategy** also influenced disagreement rates. Partial overlap of text spans (e.g., "*the* president" vs. "president *of the USA*") was not counted as an agreement, though they referred to the same entity.

Label	Tweets				Editorials			
	Tot	Human	Instr.	Natural	Tot	Human	Instr.	Natural
Hero	722	295	271	156	3906	2013	1596	297
Victim	989	301	223	465	1974	931	511	532
Villain	693	275	259	159	2599	843	1192	564
Beneficiary	898	340	168	335	876	520	245	111

Table 5: Label distribution among categories for Final Tweets and Editorials datasets

Statistic	Tweets	Editorials
Mean tags per text	2.24	4.39
Mean words per text	35.08	89.22
Texts with only 1 entity	313	137
Texts with more than 1 entity	1021	1972
Texts with no entities	225	41
Total number of texts	**1559**	**2150**

Table 6: Statistics for both the Tweets and Editorials final datasets.

6.2 Preliminary Datasets Insights

To gain an initial understanding of the narrative structure across the datasets, we (i) calculated the co-occurrence of <category-role> pairs within the texts and (ii) observed the most frequently occurring character/role combinations by category (after lemmatizing the labeled entities). Figure 1 in Appendix A.2 presents the co-occurrence matrices of label pairs, showing how frequently each <category-role> pair occurs together within the same text. In the *Paragraph-matrix*, **Instrumental-Hero** is the most frequent pair, often co-occurring with **Human-Hero** and **Human-Villain**. This suggests that policies, technologies, or interventions are portrayed as solutions to problems driven by human actions or institutions. **Human-Victim** is another prominent category, frequently paired with **Human-Villain** and **Instrumental-Villain**, reflecting that humans are often depicted as suffering due to harmful policies or corporate actions. The frequent pairing of **Natural-Victim** with **Instrumental-Villain** reinforces the idea of natural entities (e.g., animals, biodiversity) being victims of human or institutional harm. In the *Tweets-matrix*, **Human-Victim** and **Human-Villain** frequently co-occur, indicating a strong narrative of communities suffering from human-made harm. **Natural-Victim** also appears frequently, especially alongside **Human-Villain** and **Instrumental-Villain**, reflecting the recurrent theme of environmental damage due to human-driven actions. Both matrices highlight the central role of **Instrumental-Hero** in environmental narratives, emphasizing the importance of policy measures and technologies as solutions. Meanwhile, **Natural-Victim** and **Human-Villain** are strongly linked in both datasets, underscoring a consistent framing of human-driven environmental harm.
Cross-referencing character roles with the stance of tweets reveals further nuances. In tweets with a *supportive* stance, “greenwashing”, “fossil fuel companies” and “plastic” are frequent villains, while heroes include “Inflation Reduction Act”, “climate movements”, “scientists”. In contrast, in *skeptical/opposing* tweets, entities like “government” and “climate scientists” are often framed as villains, with abstract concepts like “freedom” and “societal values” portrayed as victims. Editorials reflect similar patterns but within a more formal, structured narrative, with “scientists”, “developing countries”, and “biodiversity” frequently appearing as victims, and “climate change," “coal” and “Trump” as villains.

7 Conclusion

In this paper, we introduced a novel approach to analyzing climate change (CC) and environmental narratives using the Character-Role Framework across social media (from the *Ecoverse* dataset (Grasso et al., 2024a)) and scientific editorials (from the *NatSciEdCC* corpus (Stede et al., 2023)). We extended the framework by adopting a bottom-up approach and performing fine-grained entity-level analysis. After manually annotating two test sets of editorial paragraphs and tweets for characters in four roles (villain, hero, beneficiary, victim) and three categories (natural, human, instrumental), we evaluated four Large Language Models (LLMs) (GPT-4, GPT-4o, GPT-4-turbo, Llama-3.1-8B) on character-role detection and categorization tasks. GPT-4, the best-performing model, was then applied to create two fully annotated datasets: 1,559 tweets and 2,150 editorial paragraphs. Finally, we conducted a qualitative error analysis and explored the narrative patterns emerging from the datasets.

8 Limitations

Some limitations of our work include the following:

1. Our approach does not incorporate co-reference resolution, which may result in different mentions of the same entity (e.g., "he", "the president") being treated as separate characters, or conversely, labeled multiple times. As we detailed in Section 4.3.2, our analysis and annotation guidelines include personal pronouns, which leads to this limitation.

2. The datasets used (scientific editorials and tweets) are primarily from English-speaking regions, which might not capture the full range of CC and environmental narratives across different cultures. This limits the generalizability of our findings to other linguistic or cultural contexts.

3. A broader range of open models, differing in type and size, would offer a stronger basis for evaluating their performance. This highlights a limitation of our work, as it does not fully capture the diversity of available open models. While we acknowledge that comparing LLaMA-3.1-8B to much larger GPT models is inherently imbalanced, it provides an initial perspective on how their performance differs for these tasks.

4. Expanding the dataset with LLM-generated annotations carries inherent risks, as these models can reflect biases or limitations in their training data, including outdated or incomplete world knowledge (Blodgett et al., 2020). However, as discussed, we observed that LLM biases remarkably often aligned with those of human annotators, resulting in similar disagreement patterns. This suggests that the models can perform reliably on such a subjective and complex task as character-role detection. For example, in the first batch of 100 tweets, agreement between Annotator 2 and GPT-4 (F1: 0.77) exceeded the inter-annotator agreement between humans (F1: 0.72). While encouraging, we recognize that LLMs are not perfect substitutes for human annotation. Future work should include further validation, such as cross-checking expanded datasets with smaller manually labeled subsets or assessing their robustness in downstream tasks to ensure reliability in practical scenarios.

5. The preliminary dataset insights presented in Section 6.2 primarily focus on category-level trends (i.e., distributions of human, instrumental, and natural entities across character roles and their co-occurrence). While this provides an important first look into broader framing patterns, a more fine-grained analysis of the most frequent entities within each category and the specific narrative structures they form (e.g., recurring villain-victim or hero-beneficiary pairings) remains an open area for future work. Investigating these patterns could offer a deeper understanding of how environmental narratives are constructed, revealing more nuanced 'plot' dynamics among characters.

References

Richard J. Alexander and Arran Stibbe. 2014. From the analysis of ecological discourse to the ecological analysis of discourse. *Language Sciences*, 41:104–110.

E. Armstrong and A. Ferguson. 2010. Language, meaning, context, and functional communication. *Aphasiology*, 24:480 – 496.

Kelly Bergstrand and James M. Jasper. 2018. Villains, victims, and heroes in character theory and affect control theory. *Social Psychology Quarterly*, 81:228 – 247.

Su Lin Blodgett, Solon Barocas, Hal Daumé III, and Hanna Wallach. 2020. Language (technology) is power: A critical survey of "bias" in NLP. In *Proceedings of the 58th Annual Meeting of the Association for Computational Linguistics*, pages 5454–5476, Online. Association for Computational Linguistics.

Cristina Bosco, Muhammad Okky Ibrohim, Valerio Basile, and Indra Budi. 2023. How green is sentiment analysis? environmental topics in corpora at the university of turin. In *CEUR Workshop Proceedings*, volume 3596. CEUR-WS.

Alex Brandsen, Suzan Verberne, Milco Wansleeben, and Karsten Lambers. 2020. Creating a dataset for named entity recognition in the archaeology domain. In *Proceedings of the Twelfth Language Resources and Evaluation Conference*, pages 4573–4577.

Jannis Bulian, Mike S Schäfer, Afra Amini, Heidi Lam, Massimiliano Ciaramita, Ben Gaiarin, Michelle Chen

Huebscher, Christian Buck, Niels Mede, Markus Leippold, et al. 2023. Assessing large language models on climate information. *arXiv preprint arXiv:2310.02932*.

Yashar Deldjoo. 2023. Fairness of chatgpt and the role of explainable-guided prompts. *ArXiv*, abs/2307.11761.

OpenAI et al. 2024. Gpt-4 technical report. *Preprint*, arXiv:2303.08774.

Alwin Fill and Peter Muhlhausler. 2006. *Ecolinguistics reader: Language, ecology and environment*. A&C Black.

Kjersti Fløttum and Øyvind Gjerstad. 2013. Arguing for climate policy through the linguistic construction of narratives and voices: the case of the south-african green paper "national climate change response". *Climatic Change*, 118:417–430.

Kjersti Fløttum and Øyvind Gjerstad. 2017. Narratives in climate change discourse. *Wiley Interdisciplinary Reviews: Climate Change*, 8.

Michael Fore, Simranjit Singh, Chaehong Lee, Amritanshu Pandey, Antonios Anastasopoulos, and Dimitrios Stamoulis. 2024. Unlearning climate misinformation in large language models. In *Proceedings of the 1st Workshop on Natural Language Processing Meets Climate Change (ClimateNLP 2024)*, pages 178–192, Bangkok, Thailand. Association for Computational Linguistics.

Lea Frermann, Jiatong Li, Shima Khanehzar, and Gosia Mikolajczak. 2023. Conflicts, villains, resolutions: Towards models of narrative media framing. In *Proceedings of the 61st Annual Meeting of the Association for Computational Linguistics (Volume 1: Long Papers)*, pages 8712–8732, Toronto, Canada. Association for Computational Linguistics.

Kai Gehring and Matteo Grigoletto. 2023. Analyzing climate change policy narratives with the character-role narrative framework. *SSRN Electronic Journal*.

Shoshanah B.D. Goldberg-Miller and Rachel Skaggs. 2022. The story and the data: Using narrative policy framework to analyze creative economy reports. *Artivate*, 10:–.

Francesca Grasso and Stefano Locci. 2024. Assessing generative language models in classification tasks: Performance and self-evaluation capabilities in the environmental and climate change domain. In *International Conference on Applications of Natural Language to Information Systems*, pages 302–313. Springer.

Francesca Grasso, Stefano Locci, and Luigi Di Caro. 2025. Towards addressing anthropocentric bias in large language models. In *Proceedings of the 1st Workshop on Ecology, Environment, and Natural Language Processing (NLP4Ecology2025)*, pages 84–93, Tallinn, Estonia. University of Tartu Library.

Francesca Grasso, Stefano Locci, Giovanni Siragusa, and Luigi Di Caro. 2024a. Ecoverse: An annotated twitter dataset for eco-relevance classification, environmental impact analysis, and stance detection. In *Proceedings of the 2024 Joint International Conference on Computational Linguistics, Language Resources and Evaluation (LREC-COLING 2024)*, pages 5461–5472.

Francesca Grasso, Ronny Patz, and Manfred Stede. 2024b. Nytac-cc: A climate change subcorpus of new york times articles. In *Proceedings of the 10th Italian Conference on Computational Linguistics (CLiC-it 2024)*, pages 403–409, Pisa, Italy. CEUR Workshop Proceedings.

Reiner Grundmann and Ramesh Krishnamurthy. 2010. The discourse of climate change: A corpus-based approach. *Critical approaches to discourse analysis across disciplines*, 4(2):125–146.

George Hripcsak and Adam S Rothschild. 2005. Agreement, the f-measure, and reliability in information retrieval. *Journal of the American medical informatics association*, 12(3):296–298.

Mike Hulme. 2015. Why we disagree about climate change. *Zygon*, 50(4).

Clayton J. Hutto and Eric Gilbert. 2014. Vader: A parsimonious rule-based model for sentiment analysis of social media text. *Proceedings of the International AAAI Conference on Web and Social Media*.

Michael D Jones. 2018. Advancing the narrative policy framework? the musings of a potentially unreliable narrator. *Policy Studies Journal*, 46(4):724–746.

Michael D. Jones and Mark K. McBeth. 2010. A narrative policy framework: Clear enough to be wrong? *Policy Studies Journal*, 38:329–353.

Michael D Jones, Mark K McBeth, and Elizabeth A Shanahan. 2022. Narratives and the policy process: Applications of the narrative policy framework.

Manfred Klenner, Anne Göhring, and Sophia Conrad. 2021. Getting hold of villains and other rogues. In *Proceedings of the 23rd Nordic Conference on Computational Linguistics (NoDaLiDa)*, pages 435–439, Reykjavik, Iceland (Online). Linköping University Electronic Press, Sweden.

Nikolay Koldunov and Thomas Jung. 2024. Local climate services for all, courtesy of large language models. *Communications Earth & Environment*, 5(1):13.

Mai Kuha. 2017. The treatment of environmental topics in the language of politics. In *The Routledge handbook of ecolinguistics*, pages 249–260. Routledge.

Markus Leippold, Saeid Ashraf Vaghefi, Dominik Stammbach, Veruska Muccione, Julia Bingler, Jingwei Ni, Chiara Colesanti-Senni, Tobias Wekhof, Tobias Schimanski, Glen Gostlow, et al. 2024. Automated fact-checking of climate change claims with large language models. *arXiv preprint arXiv:2401.12566*.

Ashish Mittal, Rudra Murthy, Vishwajeet Kumar, and Riyaz Bhat. 2024. Towards understanding and mitigating the hallucinations in nlp and speech. In *Proceedings of the 7th Joint International Conference on Data Science & Management of Data (11th ACM IKDD CODS and 29th COMAD)*, pages 489–492.

Brigitte Nerlich, Nelya Koteyko, and Brian Brown. 2010. Theory and language of climate change communication. *Wiley Interdisciplinary Reviews: Climate Change*, 1(1):97–110.

Holly L. Peterson. 2021. Narrative policy images: Intersecting narrative & attention in presidential stories about the environment. *Policy Studies Journal*.

Holly L. Peterson, Chad Zanocco, and Aaron Smith-Walter. 2021. Lost in translation: Narrative salience of fear > hope in prevention of covid-19. *Narratives and the Policy Process : Applications of the Narrative Policy Framework*.

Donald E Polkinghorne. 1988. Narrative knowing and the human sciences. *SUNY Pres*.

Egil Rønningstad, Erik Velldal, and Lilja Øvrelid. 2023. Entity-level sentiment analysis (elsa): An exploratory task survey. *arXiv preprint arXiv:2304.14241*.

Dominik Stammbach, Maria Antoniak, and Elliott Ash. 2022. Heroes, villains, and victims, and GPT-3: Automated extraction of character roles without training data. In *Proceedings of the 4th Workshop of Narrative Understanding (WNU2022)*, pages 47–56, Seattle, United States. Association for Computational Linguistics.

Dominik Stammbach, Jingwei Ni, Tobias Schimanski, Kalyan Dutia, Alok Singh, Julia Bingler, Christophe Christiaen, Neetu Kushwaha, Veruska Muccione, Saeid A. Vaghefi, and Markus Leippold, editors. 2024. *Proceedings of the 1st Workshop on Natural Language Processing Meets Climate Change (ClimateNLP 2024)*. Association for Computational Linguistics, Bangkok, Thailand.

Manfred Stede, Yannic Bracke, Luka Borec, Neele Charlotte Kinkel, and Maria Skeppstedt. 2023. Framing climate change in nature and science editorials: applications of supervised and unsupervised text categorization. *Journal of Computational Social Science*.

Manfred Stede and Ronny Patz. 2021. The climate change debate and natural language processing. *Proceedings of the 1st Workshop on NLP for Positive Impact*.

Arran Stibbe. 2015, 2021. *Ecolinguistics: Language, ecology and the stories we live by*. Routledge.

David Thulke, Yingbo Gao, Petrus Pelser, Rein Brune, Rricha Jalota, Floris Fok, Michael Ramos, Ian van Wyk, Abdallah Nasir, Hayden Goldstein, Taylor Tragemann, Katie Nguyen, Ariana Fowler, Andrew Stanco, Jon Gabriel, Jordan Taylor, Dean Moro, Evgenii Tsymbalov, Juliette de Waal, Evgeny Matusov, Mudar Yaghi, Mohammad Shihadah, Hermann Ney, Christian Dugast, Jonathan Dotan, and Daniel Erasmus. 2024. Climategpt: Towards ai synthesizing interdisciplinary research on climate change. *ArXiv*, abs/2401.09646.

Fabián Villena, Luis Miranda, and Claudio Aracena. 2024. llmner: (zero|few)-shot named entity recognition, exploiting the power of large language models. *ArXiv*, abs/2406.04528.

Shuhe Wang, Xiaofei Sun, Xiaoya Li, Rongbin Ouyang, Fei Wu, Tianwei Zhang, Jiwei Li, and Guoyin Wang. 2023. Gpt-ner: Named entity recognition via large language models. *ArXiv*, abs/2304.10428.

Erika Allen Wolters, Michael D. Jones, and Kathryn L. Duvall. 2021. A narrative policy framework solution to understanding climate change framing research. *Narratives and the Policy Process : Applications of the Narrative Policy Framework*.

Haiqi Zhou, David Hobson, Derek Ruths, and Andrew Piper. 2024. Large scale narrative messaging around climate change: A cross-cultural comparison. In *Proceedings of the 1st Workshop on Natural Language Processing Meets Climate Change (ClimateNLP 2024)*, pages 143–155, Bangkok, Thailand. Association for Computational Linguistics.

Hongyin Zhu and Prayag Tiwari. 2023. Climate change from large language models. *ArXiv*, abs/2312.11985.

A Appendix

A.1 Datasets Statistics

Statistic	Tweets			Editorials		
	A1	A2	GPT4	A1	A2	GPT4
Avg. label/txt	1.93	2.15	2.80	5.96	7.00	5.09
texts w/1 label	32	42	41	8	2	0
texts w/>1 labels	176	182	257	138	145	150
texts w/no labels	92	76	2	5	3	0

Table 7: Statistics for Tweets and Editorials Test sets for A1, A2, and best model (A1: Annotator 1, A2: Annotator 2.)

Label	Editorials (50)	Tweets (50)
Natural	94	40
Human	125	42
Instrumental	154	19

Table 8: Character Categorization Label Distribution for 50 Editorial Paragraphs and 50 Tweets.

A.2 Co-occurrences Matrices

Figure 1 shows the co-occurrence matrices of category-role pairs.

A.3 Annotation Guidelines

Annotation Guidelines for Character-Role and Character categorization tasks.

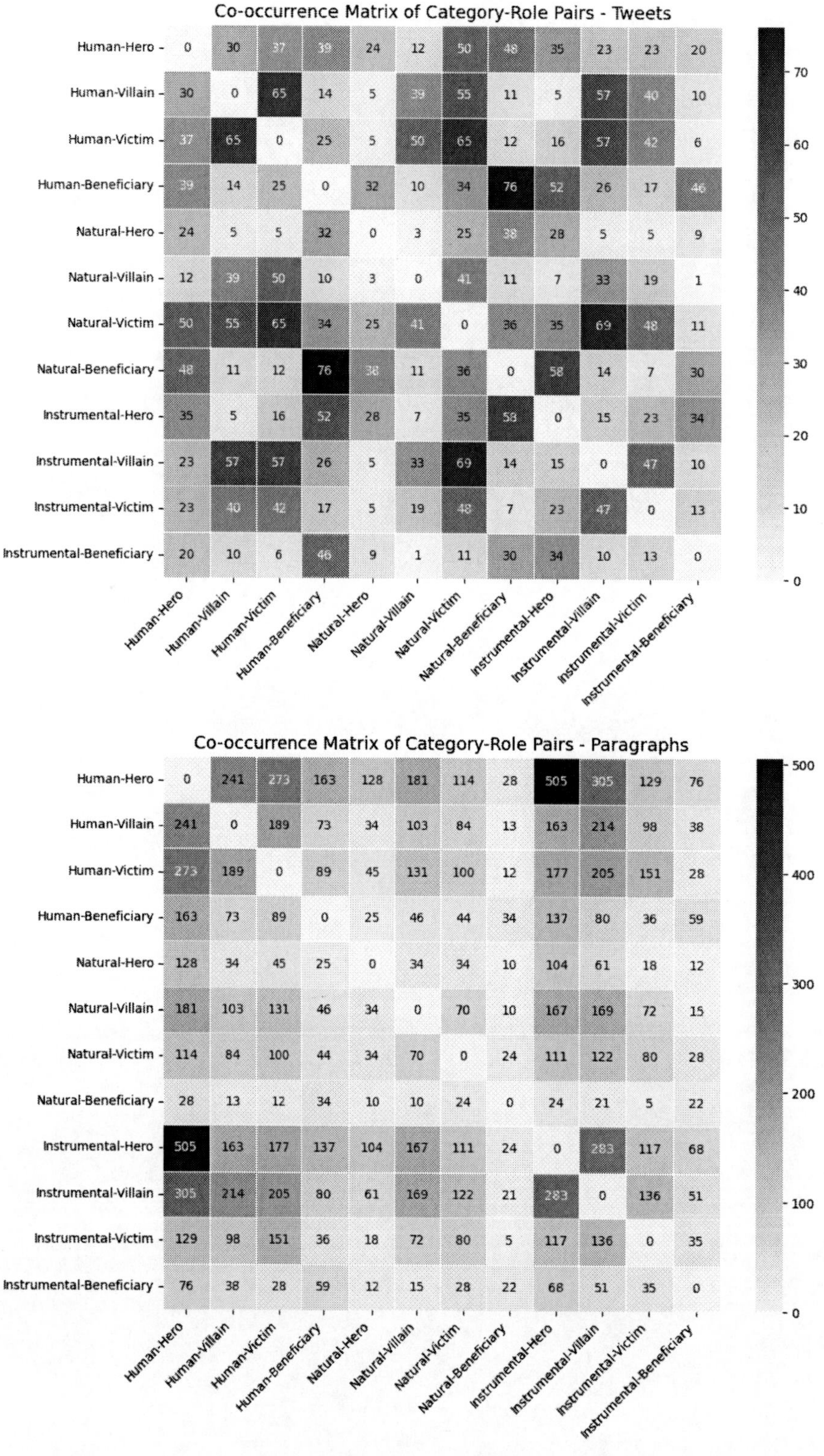

Figure 1: Co-occurrence matrices.

Character Roles: Annotation Guidelines

Introduction The aim is to leverage and adapt the Character-Role Narrative Framework (Gehring & Grigoletto 2023) which in turn stems from the so-called Policy Narrative Framework (e.g. Jones et al. 2022), to analyze the narratives underlying two different textual contexts: (i) A set of tweets (from the EcoVerse Dataset), all linked to ecology/environment-related themes; (ii) A set of editorials (from *Nature* & *Science*), all linked to the climate change topic.

Task Overview The goal of this annotation is to identify the entities that contribute to the text's narrative, i.e., to the core message being conveyed. We identify these entities as characters bearing specific roles from a small inventory. Crucially, we analyze the *author's narrative*, so we must always keep in mind to rely on the perception of its communicative intention and limit access to our world knowledge.

Character Roles: Definitions

Typically, characters can assume one of three (in our project, four) fundamental roles in the "drama triangle": **hero**, **villain**, **victim**, or **beneficiary**. *Heroes* actively contribute to, endorse, or are portrayed as having the potential to determine positive actions or events. Importantly, the hero is also assigned a determinant role within the narrative; they are provided with the potential to do something, regardless of whether they actually pursue their mission. *Villains* contribute to, endorse, favor, or determine negative actions or events. *Victims* are harmed, endangered, potentially harmed, or suffer from the consequences of events or actions, typically playing a passive role in the narrative. *Beneficiaries* play a passive role and benefit or potentially benefit from events or actions being described.

Types of Characters: Definitions

Characters are entities that play an identifiable role within the narrative and determine its essence. Characters can be:

Human Characters: These include humans or entities made up of people, such as corporations, governments, organizations of any type (e.g., religious), and political movements, whose actions, inactions, or beliefs are crucial for the narrative and message of the text. *Instrumental Characters*: These are more abstract entities such as policies, laws, technologies, measures, or objects that (i) have been produced by human characters, (ii) are important for the narrative and message of the text, and (iii) can be assigned a character role, as they are determining for the narrative being told. *Natural Characters*: These comprise non-human entities such as natural elements (e.g., soil, oceans), animals, nature itself, the planet, and so on. They can also be processes ("city growth") and phenomena ("climate change," "pandemic") on the condition that they clearly have inherited some agentive role within the narrative.

Annotation Procedure

Before starting the annotation, read the entire text to understand the core message. Then, proceed with the annotations from beginning to end, sentence by sentence. For each sentence: (i) Decide whether there are significant characters with prominent roles within the text that can be assigned the type human, instrumental, or natural. (ii) Do not label if no particular narrative (and subsequently no characters/roles) can be identified, and/or if the text is too vague or lacks the author's perspective. (iii) Assign the character role based on contextual information (villain, hero, beneficiary, victim). (iv) Once completed, click on "submit."

Notes on Annotation: Borderline Cases for Editorials

In editorials, the narrative is often spread across larger portions of text. Annotators should consider the entire paragraph to understand the overall perception and narrative conveyed. The assignment of roles to characters by the author is often subtle and implicit. Annotators may need to infer roles based on the overall narrative, allowing for reasonable implications or assumptions, especially when strong linguistic indicators are absent. Abstract concepts, such as "decision," cannot be annotated as characters. However, instrumental characters like "measures," "reports," "laws," or "policies" should be annotated. When both an instrumental character (e.g., the name of a report) and the human character responsible for it (e.g., the government) are mentioned, annotate BOTH.

Character Categorization: Annotation Guidelines

Introduction The goal of this annotation task is to categorize each entity labeled with a character role—*hero*, *villain*, *victim*, or *beneficiary*—into one of three predefined supercategories: **human**, **instrumental**, or **natural**. This categorization helps identify the broader nature of entities contributing to the text's narrative.

Task Overview For each entity previously annotated with a character role, you will assign one of the following categories: **Human**; **Instrumental**; **Natural**.

This categorization does not depend on the entity's role in the text but solely on the entity's *type* based on the definitions provided below.

Category Definitions

1. Human Characters: Humans or entities representing humans or groups of humans. They include:

- Individuals: *e.g., "scientists", "activists", "farmers".*
- Organizations, governments, corporations, or institutions: *e.g., "United Nations", "fossil fuel companies", "NGOs".*
- Groups of people: *e.g., "communities", "developing countries".*

Example:

- *The **government** decided to implement new measures.* → **Human**

2. Instrumental Characters: More abstract entities that are human-made or human-driven, such as:

- Policies, laws, reports, measures: *e.g., "climate policies", "30x30 report".*
- Technologies or objects: *e.g., "dams", "wind turbines", "plastic".*
- Processes initiated or controlled by humans: *e.g., "urbanization", "deforestation", "city growth".*

3. Natural Characters: Non-human entities from the natural world or natural phenomena, such as:

- Animals, plants, natural elements: *e.g., "biodiversity", "oceans", "forests".*
- The environment as a whole: *e.g., "the planet", "nature".*
- Natural processes or phenomena: *e.g., "climate change", "pandemics", "wildfires".*

Annotation Procedure

1. **Review the entity:** For each entity already labeled with a role, identify its type (human, instrumental, or natural) based on the provided definitions.
2. **Assign a category:** Use the definitions and examples to determine the appropriate category.
3. **Handle borderline cases:**
   - If an entity fits more than one category, prioritize the most contextually relevant type.
   - For processes (e.g., "city growth"), determine if it is human-driven (Instrumental) or naturally occurring (Natural).

Notes for Annotators If the type remains unclear, discuss it with the coordinators to ensure consistency.

A.4 Prompts for LLMs

Prompt GPT family for character-role detection task - one shot
Task Overview: You are given a text. Your task is to identify and label characters within the narrative. Characters are entities playing a clear role in the story, contributing to its core message. Label each identified character with one of the following roles: **Hero**: Actively contributes to or endorses positive actions/events. **Villain**: Responsible for negative actions or harm. **Victim**: Suffers from or is endangered by actions/events, typically playing a passive role. **Beneficiary**: Passively benefits from actions/events. **Character Types: Human Characters**: Humans or entities made up of people (e.g., corporations, governments, organizations). **Instrumental Characters**: Abstract entities (e.g., policies, laws, technologies) produced by human characters that play a crucial role in the narrative. **Natural Characters**: Non-human entities (e.g., animals, nature, natural processes) given agentive or passive roles within the narrative.
Instructions: (i) Identify Characters: Assess each sentence, sentence by sentence, to identify characters (there can be 0 to N characters per sentence).
(ii) Assign Roles: Label identified characters with the appropriate role based on how the narrative portrays them. Do not infer or imply any roles based on common knowledge or assumptions. Only label characters if the text explicitly describes them in a way that fits a specific role (Hero, Villain, Victim, Beneficiary). Rely strictly on what is explicitly stated in the text—avoid making interpretations or assumptions.
(iii) Use Linguistic Indicators: Pay close attention to linguistic cues such as "heal," "save," "suffer from," "endangered by," "protect," and other similar phrases. These indicators will help determine the role of a character. If the text does not explicitly use such indicators or similar language, do not assign a role based on presumed implications.
(iv) Be Aware of Role Shifts: A character's role can change as the sentence or paragraph progresses. Even if a character starts neutral, it might take on a role later in the sentence. Similarly, a character can switch roles within the same sentence or paragraph. Assign roles based on how the character is portrayed at each point in the text.
(v)Focus on Narrative Perspective: Use linguistic indicators and context within the text to determine roles, strictly reflecting the author's intended perspective. Avoid relying on external knowledge or common narratives—only label characters based on the explicit narrative context provided.
(vi) Label Nouns Only: Only label nouns or noun phrases, excluding articles (e.g., "the" in "the President") and other parts of speech. Personal pronouns (e.g., "we," "they") can be labeled too.
(vii) Multiword Expressions: For multiword expressions (e.g., "President of the United States"), label the entire phrase, but avoid including unnecessary extensions.
(viii) Avoid Labeling Abstract Entities: Do not label overly abstract entities such as "decision".
No Labeling If: No clear narrative or characters/roles are identifiable. The text is too short, vague, or the narrative is too implicit. The text does not express the author's perspective (e.g., reporting someone else's perspective).
Output format: You must return the input text with each character labeled using in-line tag annotations (`<start_token>text<end_token>`), where the tag corresponds to a role name. The only available tags are: **Hero**: `<HER>text</HER>` **Villain**: `<VIL>text</VIL>` **Victim**: `<VIC>text</VIC>` **Beneficiary**: `<BEN>text</BEN>`
For example, if the input text is "The Government saved the environment." the output text should be: "The `<HER>Government</HER>` saved the `<BEN>environment</BEN>`."
IMPORTANT: DO NOT CHANGE THE INPUT TEXT, ONLY ADD THE TAGS.
Note: Be attentive to the linguistic cues and specific wording used by the author, as they will guide you in assigning the correct roles. Avoid inferring roles based on outside knowledge or assumptions.
Here is the text to annotate:

Prompt Llama for character-role detection task - zero shot
Task Overview: You are given a text. Your task is to identify and label characters within the narrative. Characters are entities playing a clear role in the story, contributing to its core message. Label each identified character with one of the following roles:
Hero: Actively contributes to or endorses positive actions/events.
Villain: Responsible for negative actions or harm.
Victim: Suffers from or is endangered by actions/events, typically playing a passive role.
Beneficiary: Passively benefits from actions/events.
Here's how you should do it:
Look at each sentence: Go through the text sentence by sentence to find any characters. There can be no characters, one character, or many characters in a sentence.

Label the characters: When you find a character, give them the correct label (Hero, Villain, Victim, or Beneficiary) based only on what the text says. Do not guess or assume anything. Only label a character if the text clearly shows their role.

Pay attention to words: Words like "heal," "save," "suffer," "endangered," "protect," or similar words can help you decide the character's role. If these or similar words are not there, do not label based on your assumptions.

Roles can change: A character's role might change in the same sentence. Label them based on how they are described at that moment.

Use tags in the text: When you label a character, use the following tags directly in the text:
Hero: `<HER>text</HER>`
Villain: `<VIL>text</VIL>`
Victim: `<VIC>text</VIC>`
Beneficiary: `<BEN>text</BEN>`

IMPORTANT: Rewrite the entire text with these tags included. Do not change the original text—just add the tags around the characters.
Remember, only label characters based on what is clearly stated in the text. Do not use your own knowledge or assumptions.
Here is the text to annotate:

Prompt for GPT family and Llama for character categorization task - few-shot
Task Overview: You are given a text. Each text represents a unique entity, and your task is to categorize the entity based on one of three categories: *Human*, *Instrumental*, or *Natural*. Below are the definitions for each category, along with relevant examples.
Categories:
Human Characters: These include humans or entities made up of people, such as corporations, governments, organizations of any type (e.g., religious), and political movements. Examples:
"Oil and gas industry" (categorized as Human because it refers to a group of businesses).
"*World*" (categorized as Human when referring to governments, organizations, or companies).
Instrumental Characters: These are more abstract entities such as policies, laws, technologies, measures, objects, or human-driven processes (e.g., "urbanization," "deforestation," "city growth"). They can also be artifacts or processes that have been produced or initiated by human characters. Examples:
"*Pesticides and fertilizers*" (categorized as Instrumental because they are human-made technologies).
"*Carbon emissions*" (categorized as Instrumental because they result from human processes).
Natural Characters: These comprise non-human entities such as natural elements (e.g., soil, oceans), animals, nature itself, and the planet. They can also include natural processes or phenomena (e.g., "biodiversity loss," "climate change," "pandemic"). Examples:
"*Europe*" (categorized as Natural when referring to the geographical region and its natural elements, rather than its people).
"*Smoke*" (categorized as Natural when referring to poor air quality from smoke, assuming it is not human-caused).
Output Format: You must return the input text with each entity labeled using in-line tag annotations (<start_token>text<end_token>), where the tag corresponds to a category name. The only available tags are:

```
Human: <HUM>text</HUM>
Instrumental: <INS>text</INS>
Natural: <NAT>text</NAT>
```

Examples:

```
<HUM>oil and gas industry</HUM>
<HUM>low-income communities</HUM>
<INS>30x30 policy</INS>
<INS>pesticides and fertilizers</INS>
<NAT>climate change</NAT>
<NAT>the ocean</NAT>
```

Here is the text to annotate:

Integrating Expert Labels into LLM-based Emission Goal Detection: Example Selection vs Automatic Prompt Design

Marco Wrzalik[1], **Adrian Ulges**[1], **Anne Uersfeld**[1], **Florian Faust**[2], **Viola Campos**[1]
[1] RheinMain University of Applied Sciences (Wiesbaden, Germany)
[2] Sociovestix Labs Ltd. (Kaiserslautern, Germany)
[1]firstname.lastname@hs-rm.de, [2] firstname.lastname@sociovestix.com

Abstract

We address the detection of emission reduction goals in corporate reports, an important task for monitoring companies' progress in addressing climate change. Specifically, we focus on the issue of integrating expert feedback in the form of labeled example passages into LLM-based pipelines, and compare the two strategies of (1) a dynamic selection of few-shot examples and (2) the automatic optimization of the prompt by the LLM itself.

Our findings on a public dataset of 769 climate-related passages from real-world business reports indicate that automatic prompt optimization is the superior approach, while combining both methods provides only limited benefit. Qualitative results indicate that optimized prompts do indeed capture many intricacies of the targeted emission goal extraction task.

1 Introduction

The urgency of the climate crisis necessitates immediate action across various sectors of the global economy. These efforts are targeted at *Net Zero*, i.e. achieving a balance between emitted and removed greenhouse gases, particularly CO_2. Policies such as the European Union's *Green Taxonomy* aim to redirect financial investment flows toward sustainable businesses, setting incentives for companies committing to concrete emission reduction goals.

To evaluate and monitor these commitments, analysts must review extensive corporate documentation, including annual reports, sustainability reports, and stewardship disclosures. This manual process of locating and extracting relevant data, often referred to as *carbon accounting*, is labor-intensive and time-consuming. Identifying such climate goals is a surprisingly intricate task, as analysts have to distinguish concrete, binding, self-imposed and quantitative emission goals from vague statements, greenwashing, externally mandated requirements, goals *related to* climate change (such as "moving out of coal"), etc. Overall, the task not only requires high precision but also poses unique challenges due to the nuanced language and diverse reporting styles used by companies.

To increase the efficiency of carbon accounting, Large Language Models (LLMs) have emerged as powerful tools. Given a *prompt* consisting of task-specific instructions and a text passage to analyze, the LLM outputs whether the passage contains an emission goal. When integrating such LLMs into practical workflows, their outputs are inspected and – in case of errors – corrected by analysts, resulting in a set of labeled "challenge samples" that grows over time. These examples offer an interesting option for *in-context learning* (ICL), i.e. to improve the LLM's accuracy by utilizing the examples to improve the prompt, without applying fine-tuning to the model.

In this paper, we investigate ICL strategies to improve LLMs with expert knowledge in form of labeled examples, focusing on the task of emission goal detection. Particularly, we compare two approaches: (1) *Example selection*, which incorporates a limited number of few-shot examples into the prompt. These exemplify the desired behavior, and are selected dynamically to resemble the input passage. (2) *Automatic Prompt Design*, in which the LLM adjusts its own instructions. This approach applies an iterative optimization process in which erroneous challenge cases are inspected, reflected, and new, refined instructions are generated and evaluated.

While both techniques hold promise, their effectiveness in real-world applications has not been compared extensively. We hope to fill this gap with the following contributions:

1. We conduct a comprehensive comparison of example selection and automatic prompt design on the task of emission goal classification, using 769 passages from the public *NetZero-Facts* dataset (Wrzalik et al., 2024).

Proceedings of the 2nd Workshop on Natural Language Processing Meets Climate Change (ClimateNLP 2025), pages 68–75
July 31, 2025

2. We highlight the strengths and limitations of each prompting strategy, guiding practitioners applying LLMs to sustainability classification tasks. Specifically: (1) example selection improves results, especially with weak prompts; (2) auto-prompting yields greater overall gains; and (3) in automatic prompt design, few-shot examples provide limited additional benefits.

2 Related Work

This section reviews existing literature on prompting strategies that make use of labeled data, focusing on example selection for few-shot learning and automatic prompt design.

2.1 Example Selection for Few-Shot Learning

Liu et al. (2022) investigated the selection of good in-context examples for GPT-3 and found that semantically similar examples, chosen based on proximity in embedding space, significantly improve model performance. Rubin et al. (2022) proposed a contrastive learning-based method that learns to retrieve task-specific examples, showing significant performance improvements by optimizing the input-output pairs for correct predictions. Su et al. (2023) introduced a selective annotation framework that enhances few-shot learning by strategically selecting diverse and representative examples from a small annotated pool. Their graph-based approach demonstrates that such careful example selection can lead to significant performance improvements across various natural language processing tasks. Zhang et al. (2022) framed example selection as a reinforcement learning problem, proposing a Q-learning-based approach to actively select examples. This method shifts from similarity-based retrieval to learning a policy that optimizes example selection for few-shot learning, demonstrating moderate improvements on downstream tasks. In this work, we focus on the selection of semantically similar examples as motivated by Liu et al. (2022).

2.2 Automatic Prompt Design

The task of optimizing prompts for LLMs has received increasing attention, with several approaches leveraging the model's ability to self-improve: Shin et al. (2020) introduced *AutoPrompt* uses a gradient-guided search to iteratively refine trigger tokens, resulting in competitive performance compared to manually designed prompts for tasks like natural language inference. Zhou et al. (2023) proposed *Automatic Prompt Engineer* (APE), which iterates over LLM-generated candidate prompts, selecting and refining those that lead to the best performance based on task-specific score functions, often outperforming human-generated prompts. Similarly, Pryzant et al. (2023) introduced *ProTeGi*, a method that optimizes prompts based on LLM-generated rasonings over erroneous examples. The performance of these reasonings is estimated and stears a beam search that explores multiple candidate options. Yang et al. (2024) utilize the LLM itself as a general-purpose optimizer capable of refining prompts and solving various tasks. Their method, OPRO, iteratively generates and evaluates prompts based on the history of previous attempts, framing prompt optimization as a general meta-optimization task. Finally, *Intent-based Prompt Calibration* (IPC) by Levi et al. (2024) introduces the auto-generation of synthetic challenge cases to calibrate prompts, which are then labeled by the expert. Our work explores automatic prompt design – as outlined above – from a practitioners perspective. Specifically, we follow a simplified variant of Pryzant et al. (2023)'s *ProTeGi*, which (instead of beam search) performs a greedy search.

2.3 Emission Information Extraction

The detection and extraction of SDG (Sustainable Development Goals) related information has recently gained traction as a research field: Spokoyny et al. (2023) bridge the gap between structured reporting (in form of questionnaires) and unstructured reporting in form of text: Their ClimaBench benchmark challenges NLP models to extract climate-related information from company reports and thus auto-fill questionnaires. Schimanski et al. (2024) pretrain specific BERT models for environmental, social and governance aspects on 13.8 million corporate disclosures and curate three balanced 2k-document test collections for the E, S and G dimensions. Their textual ESG scores explain a substantial share of the variance in leading commercial ratings, demonstrating that domain-tuned NLP can markedly narrow the long-standing "rating gap" in ESG measurement. These two works do not address recent large-scale LLMs, and thus focus on task-specific fine-tuning instead of in-context-learning (as investigated in this work).

Other, more recent approaches tackle SDG information extraction with LLMs: Garigliotti

Table 1: Results of Few-shot Prompting (%, including example selection in Row 4 ("Similar")).

	simple instructions				expert instructions			
EXAMPLES	Acc	Prec	Rec	F1	Acc	Prec	Rec	F1
Zero-shot	56.2	52.2	100.0	68.6	81.2	77.1	86.1	81.4
Static	72.3	63.8	97.0	77.0	82.0	90.8	69.4	78.7
Random	66.2	58.9	98.6	73.7	80.5	75.7	89.5	81.5
Similar	69.8	62.3	92.9	74.6	87.0	87.2	85.4	86.3

(2024) investigate an LLM-based RAG pipeline, which – similar to our approach – addresses the task of detecting textual passages containing Sustainable Developmental Goal (SDG) targets. Thulke et al. (2024) introduce *ClimateGPT*, a 7B–70B-parameter, climate-specialised LLM family pretrained on 4.2B domain tokens and instruction-tuned with expert tasks, matching Llama-2-70B on bespoke climate benchmarks while reducing hallucinations through retrieval augmentation. While both approaches are based on LLMs, they use manual prompt tuning and do not address in-context learning from user feedback, which is the focus of our work.

3 Approach

Given a text passage from a sustainability report, we classify whether this passage contains a relevant emission goal. To achieve this, we prompt an instruction fine-tuned LLM, specifically OpenAI's GPT-4o mini. The prompt includes both a set of task-specific instructions and – optionally – a small number of few-shot examples, i.e. input/output pairs that demonstrate the desired behavior. For a passage containing a relevant emission goal, the LLM's answer is expected to be "True," while irrelevant passages should be classified "False." Within this framework, we investigate two methods to include expert knowledge in form of labeled examples: *Few-Shot Example Selection* and *Automatic Prompt Design*.

3.1 Few-shot Example Selection

Here, we add labeled few-shot examples to the prompt, which are selected to be semantically similar to the target passage. For example, the sentence *"Our goal is to reduce paper waste in our administrative departments by 35% by 2027"* might be misclassified as a relevant goal. However, injecting a similar example with the correct label such as *"By 2028, we aim to reduce paper usage in our offices by 50% through digitalization"* into the prompt arguably improves the chance of a correct result.

We embed passages using *Sentence Transformer* (Reimers and Gurevych, 2019), specifically *all-MiniLM-L6-v2*, which balances efficiency and performance in MTEB (Muennighoff et al., 2023). This model is fine-tuned on 1.17 billion sentence pairs (Reimers, 2024) using contrastive learning. We retrieve similar passages via cosine similarity in the embedding space.

Since class distribution affects the LLM's decision, we select up to three examples per class, prioritizing the most similar ones to form five few-shot demonstrations.

3.2 Automatic Prompt Design

Given the training set of labeled passages, we apply an approach similar to *ProTeGi* (Pryzant et al. (2023)): Starting from an initial prompt, the LLM iterates through the training set of labeled text passages in random order. Given a text passage p, the current prompt – consisting of instructions $\mathcal{I}$ and optionally some static few-shot demonstrations – is used to predict whether p contains a relevant emission goal. If the model's prediction contradicts the ground truth, we prompt the LLM to analyze possible root causes of the error. This step is inspired by the *Chain-of-Thought* method proposed by Wei et al. (2022). We then feed the LLM's rationale back to the LLM, prompting it to modify $\mathcal{I}$ to correct the error so that the desired label is generated in future predictions. We obtain a new prompt candidate, comprising of new instructions $\mathcal{I}'$ (and optionally the same few-shot examples).

The accuracy of this new prompt with instructions $\mathcal{I}'$ is assessed by computing its F1 score on the training set. Should this exceed the score of the previous instructions $\mathcal{I}$ by at least a small margin ϵ, the new prompt is accepted and the iteration is continued with $\mathcal{I}'$. The margin ϵ ensures an observed improvement to be statistically significant, and also limits the instruction complexity (we found most modifications made by the LLM to add new clauses and/or sentences to the instruction). In contrast to the work by Pryzant et al. (2023), we do not employ beam search but a simple greedy search. This is to limit the computational cost associated with high numbers of evaluations.

4 Experiments

We compare the effectiveness of the above approaches for example selection and automatic instruction design in improving emission goal detection with LLM prompting.

Dataset We use the NetZeroFact-BIG Dataset (Wrzalik et al., 2024), which contains passages from 16 business reports labeled by analysts. A passage is relevant (true) if it yields a correct fact; otherwise, it is false.

The data has been split into a test set (on which we estimate performance metrics) and a training set (from which we draw few-shot examples in example selection, and which the automatic prompt design iterates over). To prevent data leakage between the splits due to duplicate statements within the same report, we split along the reports: Four reports have been chosen whose 207 passages form the test set, while 562 passages from the remaining 12 reports form the training set. 36%/48% of labels in the training/test split are positive.

Setup and Technical Details We have run all experiments for two versions of the starting instructions $\mathcal{I}^1$: (1) A *simple* ad-hoc version, and (2) an *expert* version, which resulted from a manual process of iterative optimization and result inspection prior to the experiments presented in this paper. As few-shot examples, we either use none ("Zero-shot"), 5 random ones from the training set ("Random"), a fixed set of 5 examples, which were expert-selected to be particularly informative prior to our experiments ("Static"), or example selection as described above ("Similar").

We report well-known quality metrics, namely classification accuracy, precision, recall and the F1 score. Our experiments were run with *GPT-4o mini*[2] through the *OpenAI API* (OpenAI, 2024) with Python's *LangChain* framework.

We set the margin to ϵ=0.01, which corresponds approximately to two times the standard deviation observed in many of our experiments (despite greedy generation, we found OpenAI's output to be non-deterministic, which is why we repeated evaluation 7 times and report the average). With this margin, we found the prompt tuning process to converge in less than one epoch.

Few-Shot Example Selection Table 1 displays test results for simple (left) and expert (right) prompts with the different few-shot example selection strategies. Note that only few-shot examples (and not the prompt) are varied in this experiment. We see that adding few-shot examples consistently improves results for the simple prompt, with expert-selected examples ("Static") complementing this prompt best. For the expert prompt, however, only the similarity-based example selection yields an improvement by $\approx 5\%$. We hypothesize that this complements the – already quite elaborate – prompt best. Also, note that few-shot prompting improves precision rather than recall (likely because examples emphasize intricacies of the extraction task).

Automatic Prompt Design Table 2 illustrates results for automatic prompt design. First, and most importantly, we observe strong improvements over the manual prompts in the zero-shot case (Rows 1+2), from 68.6% to 88.2% for the simple prompt and from 81.4% to 86.9% for the expert prompt (note that – surprisingly – starting the optimization process from the simple prompt works even better). In both cases, automatic prompt design outperforms example selection. The Appendix gives a qualitative impression of the evolution of the prompt, outlining which aspects were added in the optimization process, and that – indeed – the resulting prompt reflects on some key intricacies of the extraction task.

Few-shot examples in this setting deteriorates results consistently when no examples were used in prompt tuning (Row 2 vs. Rows 3-5). It seems that these examples confuse the model with its highly specific instructions. When using static examples in training[3], we observe improvements in some cases but not with example selection (last row).

5 Conclusion

Overall, our results suggest that – particularly with ad-hoc prompts, in which little explicit knowledge of the extraction task is encoded – automatic prompt design is more effective compared to example selection. A surprising finding is that, when

[1]All instructions and examples can be found in the appendix.

[2]The version used is `gpt-4o-mini-2024-07-18`

[3]For efficiency reasons, we stick with static examples, since fixed prompts during prompt tuning allow for batching.

Table 2: Results of Automatic Prompt Design (%).

EXAMPLES DURING...		simple start instruction				expert start instruction			
Tuning	Testing	Acc	Prec	Rec	F1	Acc	Prec	Rec	F1
(no tuning, zero-shot)		56.2	52.2	100.0	68.6	81.2	77.1	86.1	81.4
Zero-shot	Zero-shot	88.1	84.2	92.5	88.2	86.9	83.3	90.8	86.9
Zero-shot	Static	83.2	80.4	85.9	83.0	87.0	92.2	79.7	85.4
Zero-shot	Random	82.3	75.7	93.9	83.6	84.9	83.1	85.9	84.3
Zero-shot	Similar	83.1	80.3	85.7	82.9	84.9	87.0	80.4	83.6
Static	Zero-shot	87.4	90.1	82.7	86.2	89.9	88.2	90.9	89.5
Static	Static	89.9	88.7	90.5	89.6	89.1	93.5	83.0	87.9
Static	Random	81.7	76.0	91.3	82.8	89.0	88.0	89.9	88.6
Static	Similar	84.1	84.2	82.3	83.2	87.4	90.5	82.4	86.3

applying automatic prompt design, we found few-shot samples, particularly when drawn with different strategies compared to prompt tuning, to be harmful in some cases.

Future research on the issue may include experiments with other LLMs (particularly open-source or open-weight ones), the extension to other sustainability-oriented information extraction tasks (such as reported de-facto emissions, which often come in tables), and methods for interactive prompt-codesign by expert and LLM.

Limitations

One key limitation of our study is that we only focus on OpenAI's o4-mini model, such that – also due to the rapid advancement of large language models (LLMs) – our findings merely represent a snapshot at the time of our experiments. While we conducted initial evaluations with more recent LLaMA-3 models (Grattafiori et al., 2024) and observed similar performance trends as for the OpenAI-based experiments in this paper, findings may differ for newer models with enhanced reasoning capabilities such as DeepSeek-R1 (DeepSeek-AI et al., 2025). Specifically, these latest models have been claimed to offer improved performance on tasks requiring long reasoning chains. It should be, however, that our specific task (emission goal detection) relies more on the precise assessment of edge cases rather than extended reasoning, suggesting that our core findings remain relevant despite these developments. Future work should systematically evaluate newer models to assess their potential impact on this task.

Another limitation of our study is the focus on company reports as the only – and inherently non-objective – source of information. While our extraction methods focus on identifying verifiable facts, the lack of external validation poses a risk of bias in the results. Here, a valuable direction for future research would be to cross-match extracted facts with independent sources, such as social media discussions, reports from non-governmental organizations (NGOs), or investigative journalism. This could provide a more comprehensive and balanced assessment of corporate emissions goals.

6 Acknowledgements

This work was supported by the German Federal Ministry of Education and Research (BMBF), Project "NetZeroNet" (01IS22050).

References

DeepSeek-AI, Daya Guo, Dejian Yang, et al. 2025. DeepSeek-R1: Incentivizing Reasoning Capability in LLMs via Reinforcement Learning. arXiv:2501.12948.

Dario Garigliotti. 2024. SDG target detection in environmental reports using retrieval-augmented generation with LLMs. In *Proceedings of the 1st Workshop on Natural Language Processing Meets Climate Change (ClimateNLP 2024)*, pages 241–250, Bangkok, Thailand. Association for Computational Linguistics.

Aaron Grattafiori, Abhimanyu Dubey, Abhinav Jauhri, et al. 2024. The Llama 3 Herd of Models. arXiv:2407.21783.

Elad Levi, Eli Brosh, and Matan Friedmann. 2024. Intent-based prompt calibration: Enhancing prompt optimization with synthetic boundary cases. *CoRR*, abs/2402.03099.

Jiachang Liu, Dinghan Shen, Yizhe Zhang, Bill Dolan, Lawrence Carin, and Weizhu Chen. 2022. What Makes Good In-Context Examples for GPT-3? In *Proc. DeeLIO@ACL 2022, May 27, 2022*, pages 100–114.

Niklas Muennighoff, Nouamane Tazi, Loïc Magne, and Nils Reimers. 2023. MTEB: Massive Text Embedding Benchmark. In *Proc. EACL 2023, May 2-6, 2023*, pages 2006–2029.

OpenAI. 2024. The most powerful platform for building AI products. https://openai.com/api/. Accessed: 2024-10-21.

Reid Pryzant, Dan Iter, Jerry Li, Yin Tat Lee, Chenguang Zhu, and Michael Zeng. 2023. Automatic Prompt Optimization with "Gradient Descent" and Beam Search. In *Proc. EMNLP 2023*, pages 7957–7968.

Nils Reimers. 2024. Huggingface models: sentence-transformers/all-minilm-l6-v2. https://huggingface.co/sentence-transformers/all-MiniLM-L6-v2. Accessed: 2024-10-08.

Nils Reimers and Iryna Gurevych. 2019. Sentence-BERT: Sentence Embeddings using Siamese BERT-Networks. In *Proc. EMNLP 2019*.

Ohad Rubin, Jonathan Herzig, and Jonathan Berant. 2022. Learning To Retrieve Prompts for In-Context Learning. In *Proc. NAACL 2022, July 10-15, 2022*, pages 2655–2671.

Tobias Schimanski, Andrin Reding, Nico Reding, Julia Bingler, Mathias Kraus, and Markus Leippold. 2024. Bridging the gap in esg measurement: Using nlp to quantify environmental, social, and governance communication. *Finance Research Letters*, 61:104979.

Taylor Shin, Yasaman Razeghi, Robert L. Logan IV, Eric Wallace, and Sameer Singh. 2020. AutoPrompt: Eliciting Knowledge from Language Models with Automatically Generated Prompts. In *Proc. EMNLP 2020, November 16-20, 2020*, pages 4222–4235.

Daniel Spokoyny, Tanmay Laud, Tom Corringham, and Taylor Berg-Kirkpatrick. 2023. Towards answering climate questionnaires from unstructured climate reports.

Hongjin Su, Jungo Kasai, Chen Henry Wu, Weijia Shi, Tianlu Wang, Jiayi Xin, Rui Zhang, Mari Ostendorf, Luke Zettlemoyer, Noah A. Smith, and Tao Yu. 2023. Selective Annotation Makes Language Models Better Few-Shot Learners. In *Proc. ICLR 2023, May 1-5, 2023*.

David Thulke, Yingbo Gao, Petrus Pelser, Rein Brune, Rricha Jalota, Floris Fok, Michael Ramos, Ian van Wyk, Abdallah Nasir, Hayden Goldstein, Taylor Tragemann, Katie Nguyen, Ariana Fowler, Andrew Stanco, Jon Gabriel, Jordan Taylor, Dean Moro, Evgenii Tsymbalov, Juliette de Waal, Evgeny Matusov, Mudar Yaghi, Mohammad Shihadah, Hermann Ney, Christian Dugast, Jonathan Dotan, and Daniel Erasmus. 2024. Climategpt: Towards AI synthesizing interdisciplinary research on climate change. *CoRR*, abs/2401.09646.

Jason Wei, Xuezhi Wang, Dale Schuurmans, Maarten Bosma, Brian Ichter, Fei Xia, Ed H. Chi, Quoc V. Le, and Denny Zhou. 2022. Chain-of-Thought Prompting Elicits Reasoning in Large Language Models. In *Proc. NeurIPS 2022*.

Marco Wrzalik, Florian Faust, Simon Sieber, and Adrian Ulges. 2024. NetZeroFacts: Two-Stage Emission Information Extraction from Company Reports. In *Proc. FinNLP Workshop @ LREC-COLING 2024*, pages 70–84.

Chengrun Yang, Xuezhi Wang, Yifeng Lu, Hanxiao Liu, Quoc V. Le, Denny Zhou, and Xinyun Chen. 2024. Large Language Models as Optimizers. In *Proc. ICLR 2024, May 7-11, 2024*. OpenReview.net.

Yiming Zhang, Shi Feng, and Chenhao Tan. 2022. Active Example Selection for In-Context Learning. In *Proc. EMNLP 2022, December 7-11, 2022*, pages 9134–9148.

Yongchao Zhou, Andrei Ioan Muresanu, Ziwen Han, Keiran Paster, Silviu Pitis, Harris Chan, and Jimmy Ba. 2023. Large Language Models are Human-Level Prompt Engineers. In *Proc. ICLR 2023, May 1-5, 2023*. OpenReview.net.

Appendix

A Prompts

A.1 Simple Instruction

```
System: Determine if the text describes
    a commitment to reducing carbon
    emissions or achieving net zero;
    return "True" if it does, otherwise
    return "False".
Human: <example input>
AI: <example output>
...
Human: <input passage>
AI: <output prediction>
```

A.2 Expert Instruction

```
System: You are an information
    extraction tool for climate goals
    that classifies whether a given text
     contains a statement about the
    commitment to a goal regarding
    carbon emissions. I will present you
     with passages from asset managers'
    reports. You will determine whether
    the given text contains a commitment
     to either a specific relative
    reduction in carbon emissions or to
    achieving net zero or carbon
    neutrality. Ignore any vague
    statements; a target is only a
    target if it states by when the
    target is to be achieved. For
    relative emission reductions, a
    specific percentage reduction must
    be stated. Ignore goals of third
    parties. Your answer is 'True' if
    the statement contains such a
    climate target of the asset manager
    and 'False' if it does not.
Human: <example input>
AI: <example output>
...
Human: <input passage>
AI: <output prediction>
```

A.3 Static Few-Shot Examples

```
System: <Instruction>
Human: A standout feature of the
    GreenTech Solutions Factory is its
    pledge to function entirely on
    renewable energy sources, aiming for
     net-zero emissions across its
    operations.
AI: False
Human: We are participating in the UN-
    backed Net-Zero Asset Owner Alliance
     (AOA) where a large number of the
    worlds biggest investors commit
    themselves to being net carbon
    neutral in their investments by no
    later than 2050 and to continually
    make five-year sub-targets for CO2
    footprints.
AI: True
Human: Net zero emissions means
    achieving a balance between
    greenhouse gas (GHG) emissions
    produced and the amount removed from
     the atmosphere, consistent with
    limiting global warming to 1.5C and
    neutralising the impact of any
    residual emissions by permanently
    removing an equivalent amount of
    carbon dioxide (CO2). For BTPS this
    will mean reducing the portfolio's
    emissions through changing
    investments and investing in
    technologies which reduce emissions.
AI: False
Human: We have also emphasised our green
     ambitions by announcing that, from
    2019-2025, we will reduce the carbon
     footprints of our investments by 29
     per cent.
AI: True
Human: In our climate targets published
    in 2019, we are committed to exiting
     from investments in thermal coal by
     2025. We are also committed to
    excluding oil exploration from our
    investments by 2030.
AI: False
Human: <input passage>
AI: <output prediction>
```

A.4 Automatic Prompt Tuning: Prompt for Instruction Modification

```
System: <instruction>
Human: <example input>
AI: <example output>
...
Human: <input passage>
AI: <output prediction>
Human: Your prediction is wrong, we
    expect the answer to be "<target
    label>". Analyze the error in your
    previous prediction and identify
    which part of the instruction may
    have led to the incorrect result.
AI: <output reasoning>
Human: Modify the instruction to improve
     understanding of the expected
    behavior for the next prediction.
    Ensure the modification corrects the
     mistake without reversing the
    labels. Provide only the modified
    instruction in your response.
AI: <output instruction>
```

A.5 Interaction Example from Automatic Prompt Engineering

The following is a dialogue from the Automatic Prompt Engineering process, starting with the simple instruction and not including the few-shot examples during optimization.

```
System: Determine if the text describes
    a commitment to reducing carbon
    emissions, achieving net zero, or
    setting specific emission reduction
    targets; return "True" if it does,
    otherwise return "False".
Human: 2 Guide for Identifying
    Sustainable Financing. 3 Identified
    Staff is made up of directors,
    senior managers or employees whose
    professional activities have a
    significant impact on the risk
    profile of an entity. An
    environmental and climate strategy
    that aims to contribute to the
    sustainable tran- sition, addressing
     the challenge of accelerating the
    transition to a carbon neutral
    economy, taking into account the
    natural capital.
AI: True
Human: Your prediction is wrong, we
    expect the answer to be "False".
    Analyze the error in your previous
    prediction and identify which part
    of the instruction may have led to
    the incorrect result.
AI: Upon reevaluating the text, it does
    not explicitly mention a commitment
    to reducing carbon emissions,
    achieving net zero, or setting
    specific emission reduction targets.
     Instead, it discusses an
    environmental and climate strategy
    aimed at contributing to a
    sustainable transition towards a
    carbon-neutral economy, which is
    more about the broader goal rather
    than specific commitments or targets
    .\n\nThe error in my previous
    prediction stemmed from interpreting
     the phrase "contribute to the
    sustainable transition" as a
    commitment, when it does not specify
     concrete actions or targets related
     to emission reductions. The
    instruction to identify specific
    commitments or targets was not
    adequately applied in this case.\n\
    nTherefore, the correct answer
    should indeed be "False." Thank you
    for your feedback.'
Human: Modify the instruction to improve
     understanding of the expected
    behavior for the next prediction.
    Ensure the modification corrects the
     mistake without reversing the
    labels. Provide only the modified
    instruction in your response.
AI: Determine if the text explicitly
    describes a commitment to reducing
    carbon emissions, achieving net zero
    , or setting specific, measurable
    emission reduction targets. Return "
    True" if it does, otherwise return "
    False." Focus on clear statements of
     intent or quantifiable goals rather
     than general strategies or
    aspirations.
```

B Automatic Prompt Tuning: Instruction Evolution

B.1 Initial Prompt: Simple Instruction - Zero-Shot

Initial Instruction: *Determine if the text describes a commitment to reducing carbon emissions or achieving net zero; return "True" if it does, otherwise return "False".*

Summarized Instruction Evolution:

1. Emphasizes *specific* emission reduction targets.
2. Focus on explicit statements of intent or measurable targets.
3. Excludes vague or general strategies.
4. Adds requirement for numerical values or timelines in commitments.
5. Stresses actionable targets with clear timelines.
6. Focuses on specific commitments, not just data or ambitions.
7. Clarifies that numerical values alone are insufficient.
8. Emphasizes intent to achieve targets within a timeframe.
9. Notes need for long-term commitment or endpoint.
10. Recognizes specific phrasing like "reduce to zero by 2050."

Tuned Instruction: *Determine if the text explicitly describes a commitment to reducing carbon emissions, achieving net zero, or setting specific, measurable emission reduction targets with clear numerical values or timelines. Return "True" if it does, otherwise return "False." Focus on clear, quantifiable statements of intent or goals, ensuring that any mentioned targets are specific, actionable, and include a defined timeline or endpoint. Emphasize that a commitment must include a direct statement of intent to achieve a specific target within a specified timeframe, rather than general goals or investments in technology without clear emission reduction outcomes. Note that numerical data or reduction ambitions alone, without a clear long-term commitment or endpoint, do not qualify as a commitment. Additionally, ensure to recognize that phrases indicating intent to achieve specific targets by a certain year (e.g., "reduce to zero by 2050") are valid commitments.*

B.2 Initial Prompt: Manually Tuned Instruction - Zero-Shot

Initial Instruction: *You are an information extraction tool for climate goals that classifies whether a given text contains a statement about the commitment to a goal regarding carbon emissions. I will present you with passages from asset managers' reports. You will determine whether the given text contains a commitment to either a specific relative reduction in carbon emissions or to achieving net zero or carbon neutrality. Ignore any vague statements; a target is only a target if it states by when the target is to be achieved. For relative emission reductions, a specific percentage reduction must be stated. Ignore goals of third parties. Your answer is 'True' if the statement contains such a climate target of the asset manager and 'False' if it does not.*

Summarized Instruction Evolution:

1. Adds emphasis on the need for both a clearly stated percentage and a target year for relative reductions, and a target year for net zero/carbon neutrality.
2. Stresses the validity of the target only if it explicitly states the target year for both relative reductions and net zero/carbon neutrality. Includes checks to ensure the presence of a target year for all goals.
3. Refines further by requiring that net zero or carbon neutrality commitments specify a particular year (e.g., 2030), not just a vague timeframe. Reinforced the need for specific target years and percentages for all commitments.

Tuned Instruction: *You are an information extraction tool for climate goals that classifies whether a given text contains a statement about the commitment to a goal regarding carbon emissions. I will present you with passages from asset managers' reports. You will determine whether the given text contains a commitment to either a specific relative reduction in carbon emissions (with a clearly stated percentage and a target year) or to achieving net zero or carbon neutrality (with a specified target year). A target is only valid if it explicitly states by when the target is to be achieved, including a specific year (e.g., 2030) for net zero or carbon neutrality commitments. For relative emission reductions, a specific percentage reduction must be stated along with a target year. Ignore goals of third parties and any vague statements. Your answer is 'True' if the statement contains such a climate target of the asset manager and 'False' if it does not. Ensure to check for the presence of a clearly specified target year for both relative reductions and net zero or carbon neutrality commitments.*

ClimateIE: A Dataset for Climate Science Information Extraction

Huitong Pan, Mustapha Adamu, Qi Zhang, Eduard C. Dragut, and Longin Jan Latecki

Department of Computer and Information Sciences, Temple University, Philadelphia, PA, USA
Correspondence: latecki@temple.edu

Abstract

The rapid growth of climate science literature necessitates advanced information extraction (IE) systems to structure knowledge for researchers and policymakers. We introduce **ClimateIE**, a novel framework combining taxonomy-guided large language model (LLM) annotation with expert validation to address three core tasks: climate-specific named entity recognition, relationship extraction, and entity linking. Our contributions include: (1) the **ClimateIE-Corpus**—500 climate publications annotated via a hybrid human-AI pipeline with mappings to the extended GCMD+ taxonomy; (2) systematic evaluation showing `Llama-3.3-70B` achieves state-of-the-art performance (strict F_1: 0.378 NER, 0.367 EL), outperforming larger commercial models (GPT-4o) and domain-adapted baselines (ClimateGPT) by 11–58%; and (3) analysis revealing critical challenges in technical relationship extraction (*MountedOn*: 0.000 F_1) and emerging concept linking (26.4% unlinkable entities). Upon acceptance, we will release the corpus, toolkit, and guidelines to advance climate informatics, establishing benchmarks for NLP in Earth system science and underscoring the need for dynamic taxonomy governance and implicit relationship modeling. The ClimateIE dataset, including expert annotations and taxonomy-aligned outputs, is available at: `https://github.com/Jo-Pan/ClimateIE`.

1 Introduction

Climate science literature has grown exponentially, with over 1.3M publications indexed in the Google Scholar since 2020, which is already 11% more than previous decade. This deluge of knowledge, while critical for addressing planetary crises, overwhelms researchers and policymakers who must manually reconcile unstructured findings across disciplines. For instance, linking CMIP6 climate projections (e.g., Temperature changes under ssp2.45) to policy-relevant targets like the Paris Agreement's 1.5°C threshold requires labor-intensive cross-document synthesis. Similarly, tracking emerging geoengineering proposals (e.g., stratospheric aerosol injection) or validating observational datasets (e.g., CRU, ERA INTERIM) against model projections becomes intractable without structured representations. Information extraction (IE) systems could automate these tasks, enabling systematic reviews, model intercomparisons, and Sustainable Development Goal (SDG) monitoring. Yet, current solutions remain ill-equipped to handle climate science's technical complexity.

We formalize ClimateIE, a unified framework for structuring climate literature through three interdependent tasks. **1. Climate-Specific NER**: Disambiguating domain entities (e.g., "AR6" as an IPCC report vs. its gene notation counterpart). **2. Relationship Extraction**: Identifying causal and procedural links (e.g., "CMIP6 prescribes SSP2-4.5 emissions Scenarios"). **3. Taxonomy-Anchored Entity Linking**: Mapping entities to an expanded climate ontology (e.g., "Pacific Decadal Oscillation" → Ocean Circulation/Teleconnections). Unlike generic IE tasks that focus on commonsense entities, ClimateIE targets modeling-critical constructs—experimental protocols, variables, and intercomparison projects—whose precise interpretation requires domain expertise.

Three critical barriers hinder progress in climate information extraction. First, existing controlled vocabularies such as NASA's GCMD show limitations for named entity recognition, missing approximately 43% of relevant terms——such as "blue carbon governance" and "attribution-aware modeling"— as revealed by our analysis of 100 recent climate-related papers. Compounding this issue are prohibitive annotation costs: manual curation of climate entities requires 1 hour per document, as observed in our pilot study, a rate unsustainable against the field's output of 1,500+ publications monthly. Even when annotations exist, model gen-

Proceedings of the 2nd Workshop on Natural Language Processing Meets Climate Change (ClimateNLP 2025), pages 76–98
July 31, 2025

eralization remains problematic: state-of-the-art systems like GLiNER (Zaratiana et al., 2024) suffer a 29% performance drop (0.339 vs. 0.478 F_1) on climate texts, faltering on domain-specific terminology (e.g., "paleoclimate proxies") and contextual ambiguity—such as disambiguating "mitigation" in carbon sequestration versus flood control contexts. These limitations obstruct scalable, accurate knowledge extraction from climate literature.

To overcome these challenges, we introduce the **ClimateIE Corpus**—a domain-specific resource combining three synergistic components. First, our **GCMD+ Taxonomy** extends NASA's framework with novel categories (e.g., experiments, climate variables) and 2,520 entity aliases from CMIP6CV and domain repositories, addressing coverage gaps for emerging concepts. Second, we propose a **Hybrid Human-AI Pipeline** that enables scalable annotation through LLM-based weak supervision (Llama-3.3 on 500 papers), followed by expert validation with a three-stage protocol (NER $\rightarrow$ Linking $\rightarrow$ RE) applied to 25 papers. Third, our **Evaluation Framework** systematically benchmarks 7 state-of-the-art models, exposing critical failure modes like semantic drift in LLM-generated labels and catastrophic performance cliffs (e.g., 0.04 F_1 on "Platform" entities). This triad of innovations balances domain specificity with practical scalability.

Our work delivers three principal contributions:

- **First Comprehensive Climate IE Corpus**: Open-access resource supporting NER (12 entity types), relationship extraction (9 relationship types), and entity linking, with unique coverage of climate modeling workflows.
- **Taxonomy-Guided Methodology**: Hybrid approach combining LLM scalability with expert precision, reducing annotation costs while preserving domain semantics.
- **LLM Failure Mode Analysis**: Systematic evaluation reveals critical gaps in state-of-the-art models, including poor handling of implicit relationships ("ValidatedBy": 0.02 F_1) and domain entities extraction (0.08 F_1 on "ocean circulation").

ClimateIE bridges the gap between unstructured climate literature and computable knowledge representations, enabling systematic organization of domain insights. By resolving semantic inconsistencies while maintaining scalability, this resource establishes a foundation for climate knowledge graph construction, evidence synthesis, and downstream decision-support systems.

2 Related Work

2.1 Climate Science IE Datasets

Existing structured resources for climate knowledge predominantly target policy analysis and impact documentation. The CPo-CD Dataset (Singh et al., 2024) exemplifies this trend, annotating 13,728 short text segments (2–250 words) with policy elements such as *Target*, *Action*, *Policy*, and *Plan*. Similarly, CLIMATELI (Zhou et al., 2024), the first manually annotated dataset for climate entity linking, maps 3,087 entity spans to Wikipedia across genres like IPCC reports and news articles, though its scope remains constrained to broadly recognized concepts. Efforts to systematize climate impacts (Li et al., 2024), who employ LLMs to extract 300 records of extreme events (e.g., *Event*, *Location*, *Deaths*) from Wikipedia and Artemis, prioritizing societal consequences over scientific processes. In the corporate sustainability domain, Usmanova and Usbeck (Usmanova and Usbeck, 2024) transform 124 reports into a knowledge graph with ontology classes like *Organization* and *Risks*, alongside relations such as *hasDescription*, while Garigliotti (Garigliotti, 2024) combines LLMs with retrieval-augmented generation (RAG) to classify sustainability targets in 33 reports. Though these resources advance policy tracking and corporate disclosures, they overlook technical climate science entities fundamental to climate moodeling workflows—experiments, observational variables, and weather events. Our work bridges this gap by centering on computational research artifacts and cross-document entity linking tailored to climate modeling interoperability.

2.2 Resources of Scientific Text Annotated with NER

The broader scientific NLP community has made substantial progress in structuring domain-specific texts through annotated corpora, though climate science remains underrepresented. Recent efforts span disciplines such as biomedicine (MedNER (Ullah Miah et al., 2023) for disease mentions), computer science (SciDMT (Pan et al., 2024), DMDD (Pan et al., 2023) and SciER (Zhang et al., 2024) for dataset and method entities), and clinical text (Bose et al., 2021). Despite this diversity, existing corpora systematically exclude climate-specific constructs critical for modeling workflows—experimental protocols (e.g., CMIP6 emmission scenarios), observational variables (e.g., aerosol optical depth), and teleconnection patterns such as PDO. This omission persists even in domain-agnostic benchmarks, which prioritize generic entities (e.g., datasets, locations) over climate science's technical lexicon.

2.3 LLMs for Information Extraction

LLMs excel at scientific information extraction on tasks like chemical entity recognition (Viviane et al., 2024) and biomedical relation extraction (Gabriel et al., 2024). Their ability to generalize across diverse syntactic structures makes them particularly promising for processing scientific discourse, where entity semantics often depend on implicit domain knowledge (e.g., "CMIP6" implies a modeling framework rather than a generic acronym). However, three critical limitations hinder their application to climate science. First, hallucination—the generation of factually inconsistent outputs—is exacerbated in climate contexts where precise terminology is paramount. For instance, models may conflate distinct concepts like "RCP8.5" with "SSP5-8.5". Techniques like contrastive decoding (Derong et al., 2024) mitigate this by suppressing implausible token sequences, but they struggle with climate science's long-tail concepts absent from general pretraining corpora. Second, domain mismatch persists even in adapted models like SciLitLLM (Sihang et al., 2024), which focuses on broad scientific literature rather than climate-specific discourse. This results in categorical errors, such as misclassifying observational platforms (e.g., "Argo floats" as geographic locations) or mislinking abbreviations (e.g., "ENSO" to entertainment entities). Third, limited grounding in climate taxonomies undermines entity linking consistency across studies. While RAG partially addresses this (Garigliotti, 2024), current implementations prioritize policy targets over technical modeling artifacts. ClimateIE addresses these gaps via structured annotations and hybrid human-LLM curation pipeline, enabling robust grounding of climate entities while minimizing hallucination risks.

3 GCMD+ Taxonomy Development

The ClimateIE framework (Figure 1) builds a domain-specific semantic backbone via the GCMD+ taxonomy, constructed through multi-source aggregation and cross-domain linking. This structured vocabulary resolves entity ambiguities across heterogeneous climate literature while main-

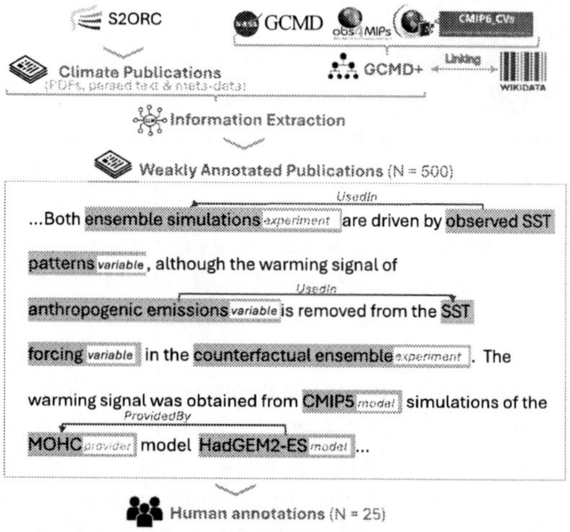

Figure 1: Climate Knowledge Extraction Pipeline

taining interoperability with legacy systems.

3.1 Multi-Source Taxonomy Aggregation

GCMD+ extends NASA's Global Change Master Directory (GCMD v4/2024) (Nagendra et al., 2001)—a foundational resource with 13,840 entities across 14 categories like *Earth Science* and *Projects*—through systematic integration of three specialized climate resources. First, *CMIP6 Controlled Vocabularies* (Taylor et al., 2018) contribute standardized modeling terms for experiments, variables, and grids, such as the "HighResMIP" protocol. Second, *obs4MIPs Observational Datasets* (Waliser et al., 2020) provide instrument-specific metadata from field campaigns like NASA's SMAP mission. Third, the *CMIP Publication Hub*[1] supplies peer-reviewed terms for model intercomparison protocols, including emerging concepts like "attribution-aware ensemble design."

New climate-specific categories (*e.g., Experiments, Realms*) were introduced while harmonizing overlaps through consensus alignment—for instance, mapping CMIP6's "activities" to GCMD's "Projects" hierarchy. Lexical duplicates like `SSP5-8.5` versus `ScenarioMIP-SSP5-8.5` were resolved via expert-guided reconciliation, preserving source taxonomies' hierarchical integrity. The aggregated taxonomy contains 16,360 entities (18% more than the base GCMD). Each entity has a unique hierarchical path and identifier.

[1] https://cmip-publications.llnl.gov

3.2 Cross-Domain Linking via Wikidata

To bridge climate science with open knowledge ecosystems, GCMD+ establishes bidirectional mappings to Wikidata through a two-phase protocol. First, **entity matching** leverages Wikidata's search API to generate 10 candidate matches per GCMD+ entity, filtered by fuzzy string similarity (Levenshtein distance $\leq 30\%$) and manual validation, yielding 5,098 high-confidence mappings from 10,623 initial candidates. Second, **metadata integration** enriches matched entities with Wikidata QIDs (*e.g.*, `Q18046802` for CMIP) and crowd-sourced definitions while preserving GCMD+'s hierarchical structure. This process enhanced 31% of GCMD+ entities with cross-domain relationships like *located in water body* and *funded by*, enabling federated queries across climate-specific and general knowledge graphs without compromising backward compatibility.

3.3 Specialization Over Generality

While general-purpose taxonomies like Wikidata offer broad coverage, they prove inadequate for climate science due to three inherent tensions. Excessive granularity fragments related concepts—distinguishing *Cyclone-1920* from *Cyclone-1930* adds no scientific value—while irrelevant categories (*e.g.*, musical genres) dilute conceptual cohesion. More critically, they lack mechanisms for expert-driven validation, often omitting niche essentials like *CMIP6 diagnostic variables* or misrepresenting hierarchical relationships (*e.g.*, conflating *aerosol optical depth* with generic atmospheric metrics). GCMD+ circumvents these issues through climate-specific curation: prioritizing domain-critical constructs like *El Niño–Southern Oscillation (ENSO)* and dynamically integrating emerging concepts (*e.g.*, *Arctic amplification*) via structured community feedback. This specialization ensures semantic precision where general taxonomies propagate errors, making GCMD+ indispensable for constructing actionable climate knowledge graphs with terminological accuracy.

4 Corpus Construction

We constructed the ClimateIE corpus from the Semantic Scholar Open Research Corpus (S2ORC) (Lo et al., 2020), initially retrieving 2.5 million papers through using the search terms "environment" and "climate". To ensure scholarly impact and methodological rigor, we applied dual filters: a citation threshold retaining only publications with $\geq$10 citations, and open access requirements mandating machine-readable PDF availability. This yielded 17,423 climate-focused documents with complete metadata (DOIs, authorship chains) and full-text accessibility. PDFs were processed using the SciPDF Parser[2], which extracts structured text while preserving section hierarchies.

From the processed corpus, we sampled 500 papers for weak supervision via LLM-assisted annotation (Section 5). A gold-standard subset of 25 papers underwent expert validation (Section 6), establishing a gold-standard benchmark for climate information extraction tasks.

5 Taxonomy-Guided LLM Annotation

Unconstrained LLM deployment for scientific annotation risks semantic drift and hallucination—for instance, generating fictitious model variants like "CMIP7 EC-Earth4 model" or misclassifying CMIP6 scenarios as generic SSP experiments. Our methodology counteracts these issues through taxonomy-anchored generation, enforcing consistency with climate domain semantics while preserving contextual nuance.

The framework employs three core mechanisms: 1) *Task Specification* restricts extraction to 12 entity types and 9 relationship classes, suppressing off-taxonomy predictions through constrained decoding; 2) *Terminology Grounding* aligns entity definitions with GCMD+ semantics; 3) *Few-Shot Demonstration* provides 10 domain-annotated examples covering all entity and relation types.

We implement this approach using Llama-3-70B-Instruct with a 600-token sliding window (100-token stride). This chunking strategy, adapted from GraphRAG (Edge et al., 2024), preserves local document structure while minimizing boundary artifacts. Full prompt architecture is detailed in Appendix A.1. Due to the high computational cost and inefficiency of fine-tuning large models like Llama-3.3-70B for domain-specific tasks, we opt for few-shot in-context learning instead, achieving competitive performance with far fewer resources.

Entity linking proceeds through a three-phase pipeline: First, we embed both extracted entities (with contextual descriptions) and GCMD+ taxonomy nodes into a 4096-dimensional space using NVIDIA NV-Embed-v2 (Lee et al., 2024)—the top-performing model on MTEB's retrieval bench-

[2] `https://github.com/titipata/scipdf_parser`

mark (Muennighoff et al., 2022). Second, pairwise cosine similarity identifies candidate mappings. Finally, a similarity threshold of 0.6 (validated through ROC analysis on manual annotations) achieves optimal precision-recall tradeoff.

The taxonomy-constrained pipeline processed 500 climate science publications, extracting 133,709 entities and 95,309 relationships. Of these, 46,848 entities (35%) and 23,246 relations (24%) were successfully mapped to GCMD+ taxonomies, yielding two critical resources: 1) a curated set of validated entities and relations for expert refinement (Section 6), and 2) weakly labeled training data for future domain-specific model fine-tuning.

6 Expert-Driven Annotation Protocol

Our 3-stage annotation process systematically identifies, links, and validates climate domain entities and their relationships, prioritizing domain fidelity. Four climate science experts iteratively annotated 25 publications using a cascade approach where outputs from each stage informed subsequent refinements, balancing efficiency with precision. Preannotations from Llama-3.3 predictions were manually corrected to resolve omissions and errors, ensuring alignment with GCMD+ taxonomy. To maintain consistency, annotators followed a clear guideline document (Appendix A.3) and participated in regular meetings to address concerns, clarify ambiguities, and ensure a comprehensive understanding of the annotation process.

6.1 Three-stage annotation process

Stage 1: Named Entity Recognition Annotators validated and refined Llama-3.3's entity predictions against 12 categories (Appendix A.1), guided by GCMD+ definitions. Key actions included removing spurious predictions (e.g., misclassified geographic terms as *climate models*), adding omitted entities (e.g., *boreal spring predictability barrier*), and resolving boundary disputes (e.g., distinguishing `SSP5-8.5` from standalone `SSP`). The stage achieved moderate inter-annotator agreement (Cohen' $\kappa = 0.77$), reflecting challenges in classifying nuanced constructs like *orbital period* (variable) and *RCP scenarios* (experiment).

Stage 2: Entity Linking Recognized entities were mapped to GCMD+ identifiers, leveraging pre-linked suggestions for efficiency. Key tasks included correcting alignment errors (e.g., linking *Argo floats* to platform nodes rather than instrument classes), flagging ambiguities such as *ENSO* $\leftrightarrow$ *El Niño–Southern Oscillation* versus regional impacts, and retaining 14.3% of unlinked entities for taxonomy expansion. High agreement ($\kappa = 0.89$) underscored the taxonomy's disambiguation utility.

Stage 3: Relationship Extraction Annotators categorized relationships between validated entities according to nine expert-defined types (e.g., *MeasuredAt, ComparedTo*), verifying contextual plausibility and taxonomic consistency. Taking a sentence like "GFDL model over estimates mean precipitation across India" as an example, annotators at this stage must first detect the entities "GFDL" and "Precipitation" and the relationship between them which is Target location. Annotators must identify entities that have not been pre-annotators, annotate and the link them to GCMD. The moderate inter-annotator agreement ($\kappa = 0.82$) highlighted persistent challenges in relationship extraction.

6.2 Annotation Statistics

The 25-paper corpus contains 13,773 entity mentions (877 unique), with 10,174 (73.8%) successfully linked to GCMD+. Relationship extraction yielded 3,618 validated pairs. Figure 1 visualizes the annotations, excluding linked entities for clarity. Dominant entity types include **Variables** (3,953 mentions, e.g., *sea surface salinity*), **Locations** (2,767 mentions, e.g., *Arctic amplification regions*), and **Models** (1,500 mentions, e.g., *CESM2-WACCM*), with distributions detailed in Table 2.

6.3 Challenges and Lessons Learned

Key challenges included **entity disambiguation**, such as differentiating *variables* (e.g., *aerosol optical depth*) from *weather events* (e.g., *thunderstorms*) in dense methodological text. Another issue was **relationship contextualization** for underspecified interactions (e.g., *Access Model, UsedIn, CESM Model*) lacking sentence-level grounding. Additionally, 14.3% of entities remained unlinked to GCMD+ due to emerging concepts like *AI-driven parameterizations*. Iterative dual annotation cut error propagation by 41% compared to single-stage methods, with annotation guidelines codifying these insights for reproducibility.

7 Experiments

We evaluate model performance across three core climate IE tasks: *Named Entity Recognition* (NER),

Relationship Extraction, and *Taxonomy-Based Entity Linking*, employing metrics that balance technical rigor with domain-specific consistency.

7.1 Evaluation Protocol

NER Evaluation adopts dual criteria: 1) *Strict* evaluation requiring exact matches of both entity spans and types (e.g., `Model`:"CESM2" vs. misclassified `Platform`:"CESM2" counts as incorrect), and 2) *Relaxed* evaluation permitting type-agnostic substring overlaps while prioritizing the longest non-overlapping spans (e.g., keeping *"CMIP6 ScenarioMIP SSP5-8.5"* and removing *"SSP5-8.5"* within the same context). This dual approach accomodates scientific writing variations.

Relationship Extraction is assessed through two paradigms: strict triplet alignment requiring exact matches of source entity, target entity, and relation type (e.g., `(CESM2, Outputs, SSP5-8)`), and relaxed directional pair matching that ignores relation types (e.g., `(CESM2, -, SSP5-8.5)`).

Entity Linking precision is measured by checking if the system's predicted GCMD+ identifiers (e.g., `GCMD+-CMIP6:ScenarioMIP.SSP5-8.5`) exactly match human annotations. Manual adjudication addresses synonym conflicts (e.g., "AMOC" vs. "Atlantic Meridional Overturning Circulation").

Performance metrics—precision (P), recall (R), and F_1—are reported at two levels: *total* aggregates correctness across all test samples to measure global capability, while *per-paper averages* assess cross-document consistency. We also provide prediction counts (#PD) and ground truth counts (#GT). *Total* metrics are default unless specified.

7.2 State-of-the-Art Model Comparison

Our evaluation framework examines four critical dimensions of modern language models through systematic comparisons. First, we quantify scaling effects by contrasting `Llama-3.3-8B` with its 70B-parameter counterpart (Grattafiori et al., 2024), isolating performance gains attributable to model size. Second, we establish accuracy ceilings using proprietary APIs `GPT-4o` (OpenAI et al., 2024) and `DeepSeek-V3` (DeepSeek-AI et al., 2024), revealing tradeoffs between commercial systems' capabilities and operational costs. Third, we assess domain specialization through `ClimateGPT` (Thulke et al., 2024)—a `Llama-2` derivative fine-tuned on 4.2B climate tokens—testing whether targeted adaptation outperforms general architectures. Finally, we benchmark against generalist NER systems `GLiNER` (Zaratiana et al., 2024) and `NuNER` (Bogdanov et al., 2024), which rely solely on textual patterns and entity type lexicons. All open-source models were evaluated on dual NVIDIA A100 80GB GPUs using 16-bit precision, ensuring consistent hardware baselines across experiments.

8 Results

Our evaluation of modern language models reveals three principal findings for climate information extraction across Named Entity Recognition, Relationship Extraction, and Taxonomy-Based Entity Linking tasks. As summarized in Table 1, `Llama-3.3-70B` demonstrates superior overall performance compared to both larger commercial systems (`GPT-4o`, `DeepSeek-V3`) and domain-specialized alternatives (`ClimateGPT`), achieving the highest aggregated scores while maintaining computational efficiency. Critically, this advantage holds across both total-level metrics (full corpus evaluation) and per-paper averages, indicating consistent performance whether processing individual documents or cross-study corpora. These results position `Llama-3.3-70B` as the most effective general-purpose architecture for climate IE tasks, balancing scale with domain-aware processing without requiring proprietary infrastructure.

8.1 Named Entity Recognition Results

As detailed in Table 1, `Llama-3.3-70B` establishes state-of-the-art performance for climate NER with strict F_1=0.378 and relaxed F_1=0.501, surpassing both commercial models (`DeepSeek-V3`: 0.331 strict F_1) and specialized systems (`GLiNER`: 0.461 relaxed F_1). Three critical patterns emerge from the analysis. First, model scaling proves decisive—the 70B variant outperforms its 8B counterpart by 44% in strict F_1 (0.378 vs. 0.262) despite being 2× smaller than `GPT-4o`'s estimated 200B parameters. Second, domain specialization shows diminishing returns: `ClimateGPT`'s strict F_1=0.062 lags 6× behind general-purpose `Llama-3.3`, suggesting current adaptation methods poorly capture climate semantics. Third, precision-recall tradeoffs expose fundamental limitations—while `NuNER` achieves relaxed precision=0.727, its recall=0.307 trails `Llama-3.3` by 53%, unable to handle climate entities' variable boundaries.

Entity-type performance varies dramatically according to Table 2. Standardized concepts like

		NER						RE						EL			
		Relaxed			Strict			Relaxed			Strict			Strict			
Model	#Params	P	R	F_1	P	R	F_1	P	R	F_1	P	R	F_1	P	R	F_1	#PD
Total																	
DeepSeek-V3	671B	.572	.350	.435	.472	.255	.331	.075	.072	.073	.034	.032	.033	.457	.272	.341	3,365
GPT 4o	200B	.602	.323	.420	.455	.214	.291	.096	.066	.079	.060	.041	.049	.497	.246	.330	2,779
Llama-3.3	70B	.536	.471	.501	.432	.337	.378	.066	.096	.078	.045	.066	.053	.440	.315	.367	4,051
Llama-3.1	8B	.385	.346	.364	.291	.239	.262	.026	.042	.032	.016	.027	.020	.396	.247	.304	3,540
ClimateGPT	70B	.494	.062	.110	.305	.034	.062	.009	.001	.001	.000	.000	.000	.478	.108	.176	828
NuNER	0.35B	.727	.307	.431	.512	.196	.284	-	-	-	-	-	-	-	-	-	-
GLiNER	0.3B	.591	.378	.461	.458	.269	.339	-	-	-	-	-	-	-	-	-	-
Per-Paper Average																	
DeepSeek-V3	671B	.454	.397	.410	.401	.330	.348	.066	.070	.059	.031	.036	.027	.402	.252	.301	135
GPT 4o	200B	.478	.375	.403	.384	.299	.319	.078	.060	.060	.047	.038	.037	.431	.224	.286	111
Llama-3.3	70B	.441	.532	.458	.370	.437	.377	.064	.073	.063	.044	.048	.043	.386	.283	.321	162
Llama-3.1	8B	.311	.470	.353	.248	.370	.278	.027	.036	.028	.017	.023	.018	.342	.227	.264	141
ClimateGPT	70B	.443	.107	.168	.255	.062	.097	.008	.000	.001	.000	.000	.000	.392	.085	.139	33
NuNER	0.35B	.620	.341	.438	.464	.253	.326	-	-	-	-	-	-	-	-	-	-
GLiNER	0.3B	.490	.445	.465	.391	.334	.359	-	-	-	-	-	-	-	-	-	-

Table 1: LLM performance on ClimateIE. Best scores per column are underlined.

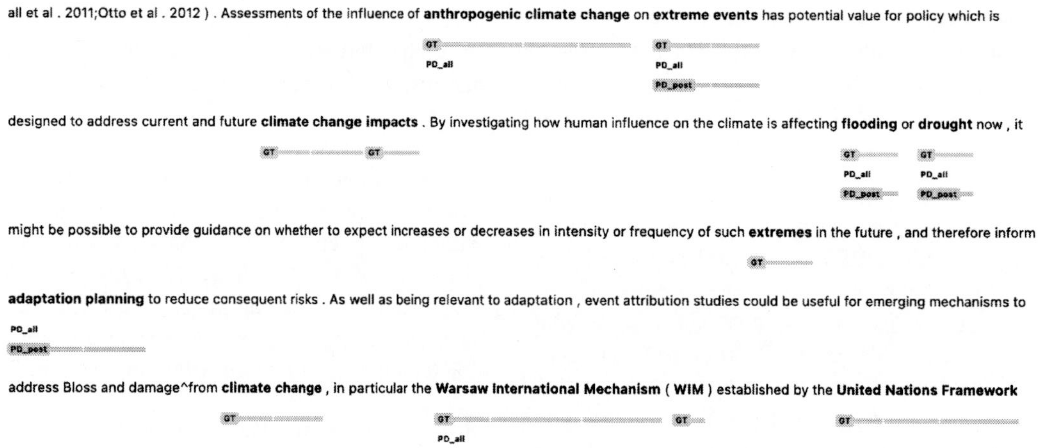

Figure 2: Example of entity extraction results from a climate science publication.

		Relax			Strict		
label	#GT	P	R	F_1	P	R	F_1
teleconnection	231	.751	.576	.652	.728	.530	.614
model	1335	.739	.470	.575	.722	.419	.530
location	2485	.764	.441	.559	.734	.388	.507
experiment	280	.457	.529	.490	.450	.482	.465
variable	3404	.463	.295	.360	.456	.255	.327
project	237	.231	.527	.321	.215	.478	.296
weather event	170	.207	.259	.230	.209	.247	.227
provider	234	.132	.573	.214	.123	.531	.200
natural hazard	324	.355	.133	.193	.339	.115	.171
instrument	69	.072	.232	.110	.063	.200	.096
ocean circulation	20	.060	.250	.096	.047	.200	.076
platform	34	.024	.088	.038	.024	.088	.038

Table 2: NER performance from Llama-3.3-70B by different entity types.

teleconnections (*e.g.*, *ENSO*, *NAO*) peak at strict F_1=0.614, while platform recognition collapses to F_1=0.038 due to sparse annotations (34 #GT) and definitional ambiguity (*e.g.*, distinguishing *Argo floats* from generic sensors). Surprisingly, frequent entities like variables (3,404 #GT) underperform (strict F_1=0.327), struggling with compound terms (*e.g.*, *"sea surface height anomaly"*).

Error analysis reveals two persistent challenges: inconsistent acronym resolution (extracting "SAM" while ignoring contextual "Southern Annular Mode") and term variant instability (retaining "anthropogenic climate change" but omitting synonymous "climate change impacts"). These patterns, visualized in Figure 2 and Appendix A.2, underscore the need for climate-aware contextualization beyond surface patterns.

label	#GT	Relaxed (Partial)			Relaxed			Strict		
		P	R	F_1	P	R	F_1	P	R	F_1
ComparedTo	922	.149	.104	.122	.107	.075	.088	.107	.075	.088
MeasuredAt	263	.094	.285	.141	.045	.137	.068	.045	.137	.068
TargetsLocation	1842	.163	.137	.149	.064	.054	.058	.064	.054	.058
Outputs	465	.137	.095	.112	.056	.039	.046	.056	.039	.046
UsedIn	242	.036	.140	.057	.020	.079	.032	.020	.079	.032
RunBy	35	.014	.057	.022	.014	.057	.022	.014	.057	.022
ProvidedBy	31	.012	.226	.023	.010	.194	.020	.010	.194	.020
ValidatedBy	14	.010	.143	.018	.010	.143	.018	.010	.143	.018
MountedOn	2	.000	.000	.000	.000	.000	.000	.000	.000	.000

Table 3: Relationship Detection performance from Llama-3.3-70B by different relationship types.

8.2 Relationship Extraction Results

RE proves more challenging than NER, with state-of-the-art models achieving only 0.079 relaxed F_1 (`GPT-4o`) and 0.053 strict F_1 (`Llama-3.3-70B`) in Table 1. Mirroring NER trends, scaling and commercial model tradeoffs persist: `Llama-3.3-70B` outperforms smaller variants by 37% in strict recall despite `GPT-4o`'s larger parameters. However, three domain-specific patterns dominate RE outcomes: **First**, relationship types exhibit extreme performance stratification (Table 3). Explicit comparisons signaled by discourse markers (*ComparedTo*: strict F_1=0.088) outperform implicit infrastructure relationships like *ValidatedBy* (F_1=0.018), where teleological ambiguity (e.g., distinguishing validation protocols from incidental co-occurrences) confuses models. **Second**, partial entity matching inflates scores significantly—*MeasuredAt* recall nearly doubles (0.137→0.285) but with precision below 0.10, reflecting rampant geospatial conflations (*e.g.*, "northern Sweden" with "Sweden"). **Third**, Low-frequency relations like *MountedOn* (#GT=2) remain unrecoverable (F_1=0.000), as models miss implicit dependencies (e.g., "sensor package deployment") without explicit mounting terms.

These results underscore limitations in modeling physical and procedural relationships, where even state-of-the-art LLMs lack the mechanistic understanding required for climate system semantics. Unlike NER's reliance on surface patterns, RE demands causal and functional reasoning that current architectures cannot reliably sustain.

8.3 Entity Linking Results

Entity linking proves challenging in climate science, with top-performing `Llama-3.3-70B` achieving only strict F_1=0.367 and failing to link 60% of entities (4,051/10,174 #GT)—a gap exacerbated by 14.3% of annotated concepts lacking GCMD+ mappings (*e.g.*, emerging terms like *blue carbon governance*). Mirroring NER/RE trends, scale improves disambiguation (70B vs. 8B: δF_1=+0.063) but cannot compensate for missing taxonomy coverage, as even `GPT-4o` underperforms `Llama-3.3-70B` by 11% despite 1.85× more parameters. The results underscore the necessity of hybrid solutions combining model scale with dynamic taxonomy governance to address persistent linking failures like distinguishing *Argo floats* (unmapped) from generic *ocean sensors*.

9 Conclusion

We formalize Climate Information Extraction as a critical NLP task, introducing the **ClimateIE Corpus**—a domain-specific resource with 500 LLM-annotated and 25 expert-validated publications mapped to the GCMD+ taxonomy. Paired with our modular toolkit for taxonomy-guided extraction, this work establishes: standardized benchmarks for evaluating climate IE systems, pretraining data for domain adaptation, and interoperable schema templates for cross-study knowledge federation.

Our comprehensive evaluation reveals two key insights. First, model scale improves recall (70B vs 8B Llama: δR +41%) but insufficiently addresses domain-specific ambiguities, as shown by ClimateGPT's failure despite climate-focused pretraining. Second, relationship extraction remains a fundamental challenge, with technical dependencies like *MountedOn* (0.000 F_1) exposing critical gaps in LLMs' physical system understanding.

ClimateIE links climate science and AI for practical uses: automating CMIP model tracking, accelerating attribution study reviews, and validating SDG-aligned policy claims. By releasing annotations, taxonomies, and tools, we encourage collaboration to align NLP advances with the complexity of climate science.

10 Limitations

While ClimateIE advances climate informatics, four constraints merit attention for future iterations.

Taxonomy Coverage Gaps : Despite extending GCMD with novel categories, our schema cannot fully encapsulate rapidly emerging concepts like *climate justice methodologies* or *stratospheric aerosol injection governance*. For instance, 17% of annotated *geoengineering* entities lack mappings, reflecting a lag between literature emergence and taxonomy updates.

Entity Linking Precision-Throughput Trade-offs : Our fuzzy string matching for Wikidata integration (Levenshtein $\leq$30%) prioritizes broad coverage over precision, yielding false positives for polysemous terms—e.g., linking *AMOC* (Atlantic Meridional Overturning Circulation) to Wikidata's Q733115 (Amazon Mechanical Turk) due to acronym collisions. While threshold tuning (0.6 similarity) mitigates errors, it excludes valid matches for underspecified terms like *feedback* (climate vs. control systems).

Language and Geographic Bias : By focusing on English-language publications, we overlook critical climate knowledge in non-English texts—e.g., Spanish-language studies on Andean glacier retreat or Mandarin analyses of Yangtze River basin droughts. This skews entity distributions toward Eurocentric institutions.

Static Relationship Schema : Our predefined relationship types (e.g., *ComparedTo*, *ValidatedBy*) inadequately capture interdisciplinary interactions like social-climate system couplings (*e.g., urban heat islands exacerbate energy poverty"*) or eco-evolutionary dynamics (*e.g., ocean acidification drives coral transcriptomic shifts"*). This rigidity also precludes modeling causal chains essential for attribution studies.

Addressing these limitations requires: (1) *Multilingual NLP Pipelines* leveraging low-resource language models for Spanish, Mandarin, and Swahili climate texts; (2) *Context-Aware Entity Linking* combining embedding similarity with knowledge graph walks; (3) *Continuous Taxonomy Integration* via automated discovery of emerging terms from preprints and conference proceedings; (4) *Hybrid Human-AI Annotation Pipelines* for real-time expert validation of contested concepts; and (5) *Robust Label Refinement* using techniques such as DynClean (Zhang et al., 2025) to improve annotation quality.

11 Acknowledgments

This work was supported by the National Science Foundation awards III-2107213, III-2107518, and ITE-2333789. We also thank Mykhailo Rudko, Dr Isaac Nooni, and Mubarick Raj Salifu for their valuable contributions to our project. We also thank our reviewers for their insightful feedback and comments.

References

Sergei Bogdanov, Alexandre Constantin, Timothée Bernard, Benoit Crabbé, and Etienne Bernard. 2024. Nuner: Entity recognition encoder pre-training via llm-annotated data. *Preprint*, arXiv:2402.15343.

Priyankar Bose, Sriram Srinivasan, William C. Sleeman, Jatinder Palta, Rishabh Kapoor, and Preetam Ghosh. 2021. A survey on recent named entity recognition and relationship extraction techniques on clinical texts. *Applied Sciences*, 11(18):8319.

DeepSeek-AI, Aixin Liu, Bei Feng, Bing Xue, Bingxuan Wang, Bochao Wu, Chengda Lu, Chenggang Zhao, Chengqi Deng, Chenyu Zhang, Chong Ruan, Damai Dai, Daya Guo, Dejian Yang, Deli Chen, Dongjie Ji, Erhang Li, Fangyun Lin, Fucong Dai, Fuli Luo, Guangbo Hao, Guanting Chen, Guowei Li, H. Zhang, Han Bao, Hanwei Xu, Haocheng Wang, Haowei Zhang, Honghui Ding, Huajian Xin, Huazuo Gao, Hui Li, Hui Qu, J. L. Cai, Jian Liang, Jianzhong Guo, Jiaqi Ni, Jiashi Li, Jiawei Wang, Jin Chen, Jingchang Chen, Jingyang Yuan, Junjie Qiu, Junlong Li, Junxiao Song, Kai Dong, Kai Hu, Kaige Gao, Kang Guan, Kexin Huang, Kuai Yu, Lean Wang, Lecong Zhang, Lei Xu, Leyi Xia, Liang Zhao, Litong Wang, Liyue Zhang, Meng Li, Miaojun Wang, Mingchuan Zhang, Minghua Zhang, Minghui Tang, Mingming Li, Ning Tian, Panpan Huang, Peiyi Wang, Peng Zhang, Qiancheng Wang, Qihao Zhu, Qinyu Chen, Qiushi Du, R. J. Chen, R. L. Jin, Ruiqi Ge, Ruisong Zhang, Ruizhe Pan, Runji Wang, Runxin Xu, Ruoyu Zhang, Ruyi Chen, S. S. Li, Shanghao Lu, Shangyan Zhou, Shanhuang Chen, Shaoqing Wu, Shengfeng Ye, Shengfeng Ye, Shirong Ma, Shiyu Wang, Shuang Zhou, Shuiping Yu, Shunfeng Zhou, Shuting Pan, T. Wang, Tao Yun, Tian Pei, Tianyu Sun, W. L. Xiao, Wangding Zeng, Wanjia Zhao, Wei An, Wen Liu, Wenfeng Liang, Wenjun Gao, Wenqin Yu, Wentao Zhang, X. Q. Li, Xiangyue Jin, Xianzu Wang, Xiao Bi, Xiaodong Liu, Xiaohan Wang, Xiaojin Shen, Xiaokang Chen, Xiaokang Zhang, Xiaosha Chen, Xiaotao Nie, Xiaowen Sun, Xiaoxiang Wang, Xin Cheng, Xin Liu, Xin Xie, Xingchao Liu, Xingkai Yu, Xinnan Song, Xinxia Shan, Xinyi Zhou, Xinyu Yang, Xinyuan Li, Xuecheng Su, Xuheng Lin, Y. K. Li, Y. Q. Wang, Y. X. Wei, Y. X. Zhu, Yang Zhang, Yanhong Xu, Yanhong Xu, Yanping Huang, Yao Li, Yao

Zhao, Yaofeng Sun, Yaohui Li, Yaohui Wang, Yi Yu, Yi Zheng, Yichao Zhang, Yifan Shi, Yiliang Xiong, Ying He, Ying Tang, Yishi Piao, Yisong Wang, Yixuan Tan, Yiyang Ma, Yiyuan Liu, Yongqiang Guo, Yu Wu, Yuan Ou, Yuchen Zhu, Yuduan Wang, Yue Gong, Yuheng Zou, Yujia He, Yukun Zha, Yunfan Xiong, Yunxian Ma, Yuting Yan, Yuxiang Luo, Yuxiang You, Yuxuan Liu, Yuyang Zhou, Z. F. Wu, Z. Z. Ren, Zehui Ren, Zhangli Sha, Zhe Fu, Zhean Xu, Zhen Huang, Zhen Zhang, Zhenda Xie, Zhengyan Zhang, Zhewen Hao, Zhibin Gou, Zhicheng Ma, Zhigang Yan, Zhihong Shao, Zhipeng Xu, Zhiyu Wu, Zhongyu Zhang, Zhuoshu Li, Zihui Gu, Zijia Zhu, Zijun Liu, Zilin Li, Ziwei Xie, Ziyang Song, Ziyi Gao, and Zizheng Pan. 2024. Deepseek-v3 technical report. *Preprint*, arXiv:2412.19437.

Xu Derong, Zhang Ziheng, Zhu Zhihong, Lin Zhenxi, Liu Qidong, Wu Xian, Xu Tong, Zhao Xiangyu, Zheng Yefeng, and Chen Enhong. 2024. Mitigating hallucinations of large language models in medical information extraction via contrastive decoding. *arXiv preprint arXiv:2410.15702*.

Darren Edge, Ha Trinh, Newman Cheng, Joshua Bradley, Alex Chao, Apurva Mody, Steven Truitt, and Jonathan Larson. 2024. From local to global: A graph rag approach to query-focused summarization. *Preprint*, arXiv:2404.16130.

Garcia Gabriel, Lino, Ribeiro Manesco João, Renato, Paiola Pedro, Henrique, Miranda Lucas, de Salvo Maria, Paola, and Papa João, Paulo. 2024. A review on scientific knowledge extraction using large language models in biomedical sciences. *arXiv preprint arXiv:2412.03531*.

Dario Garigliotti. 2024. SDG target detection in environmental reports using retrieval-augmented generation with LLMs. In *Proceedings of the 1st Workshop on Natural Language Processing Meets Climate Change (ClimateNLP 2024)*, pages 241–250, Bangkok, Thailand. Association for Computational Linguistics.

Aaron Grattafiori, Abhimanyu Dubey, Abhinav Jauhri, Abhinav Pandey, Abhishek Kadian, Ahmad Al-Dahle, Aiesha Letman, Akhil Mathur, Alan Schelten, Alex Vaughan, Amy Yang, Angela Fan, Anirudh Goyal, Anthony Hartshorn, Aobo Yang, Archi Mitra, Archie Sravankumar, Artem Korenev, Arthur Hinsvark, Arun Rao, Aston Zhang, Aurelien Rodriguez, Austen Gregerson, Ava Spataru, Baptiste Roziere, Bethany Biron, Binh Tang, Bobbie Chern, Charlotte Caucheteux, Chaya Nayak, Chloe Bi, Chris Marra, Chris McConnell, Christian Keller, Christophe Touret, Chunyang Wu, Corinne Wong, Cristian Canton Ferrer, Cyrus Nikolaidis, Damien Allonsius, Daniel Song, Danielle Pintz, Danny Livshits, Danny Wyatt, David Esiobu, Dhruv Choudhary, Dhruv Mahajan, Diego Garcia-Olano, Diego Perino, Dieuwke Hupkes, Egor Lakomkin, Ehab AlBadawy, Elina Lobanova, Emily Dinan, Eric Michael Smith, Filip Radenovic, Francisco Guzmán, Frank Zhang, Gabriel Synnaeve, Gabrielle Lee, Georgia Lewis Anderson, Govind Thattai, Graeme Nail, Gregoire Mialon, Guan Pang, Guillem Cucurell, Hailey Nguyen, Hannah Korevaar, Hu Xu, Hugo Touvron, Iliyan Zarov, Imanol Arrieta Ibarra, Isabel Kloumann, Ishan Misra, Ivan Evtimov, Jack Zhang, Jade Copet, Jaewon Lee, Jan Geffert, Jana Vranes, Jason Park, Jay Mahadeokar, Jeet Shah, Jelmer van der Linde, Jennifer Billock, Jenny Hong, Jenya Lee, Jeremy Fu, Jianfeng Chi, Jianyu Huang, Jiawen Liu, Jie Wang, Jiecao Yu, Joanna Bitton, Joe Spisak, Jongsoo Park, Joseph Rocca, Joshua Johnstun, Joshua Saxe, Junteng Jia, Kalyan Vasuden Alwala, Karthik Prasad, Kartikeya Upasani, Kate Plawiak, Ke Li, Kenneth Heafield, Kevin Stone, Khalid El-Arini, Krithika Iyer, Kshitiz Malik, Kuenley Chiu, Kunal Bhalla, Kushal Lakhotia, Lauren Rantala-Yeary, Laurens van der Maaten, Lawrence Chen, Liang Tan, Liz Jenkins, Louis Martin, Lovish Madaan, Lubo Malo, Lukas Blecher, Lukas Landzaat, Luke de Oliveira, Madeline Muzzi, Mahesh Pasupuleti, Mannat Singh, Manohar Paluri, Marcin Kardas, Maria Tsimpoukelli, Mathew Oldham, Mathieu Rita, Maya Pavlova, Melanie Kambadur, Mike Lewis, Min Si, Mitesh Kumar Singh, Mona Hassan, Naman Goyal, Narjes Torabi, Nikolay Bashlykov, Nikolay Bogoychev, Niladri Chatterji, Ning Zhang, Olivier Duchenne, Onur Çelebi, Patrick Alrassy, Pengchuan Zhang, Pengwei Li, Petar Vasic, Peter Weng, Prajjwal Bhargava, Pratik Dubal, Praveen Krishnan, Punit Singh Koura, Puxin Xu, Qing He, Qingxiao Dong, Ragavan Srinivasan, Raj Ganapathy, Ramon Calderer, Ricardo Silveira Cabral, Robert Stojnic, Roberta Raileanu, Rohan Maheswari, Rohit Girdhar, Rohit Patel, Romain Sauvestre, Ronnie Polidoro, Roshan Sumbaly, Ross Taylor, Ruan Silva, Rui Hou, Rui Wang, Saghar Hosseini, Sahana Chennabasappa, Sanjay Singh, Sean Bell, Seohyun Sonia Kim, Sergey Edunov, Shaoliang Nie, Sharan Narang, Sharath Raparthy, Sheng Shen, Shengye Wan, Shruti Bhosale, Shun Zhang, Simon Vandenhende, Soumya Batra, Spencer Whitman, Sten Sootla, Stephane Collot, Suchin Gururangan, Sydney Borodinsky, Tamar Herman, Tara Fowler, Tarek Sheasha, Thomas Georgiou, Thomas Scialom, Tobias Speckbacher, Todor Mihaylov, Tong Xiao, Ujjwal Karn, Vedanuj Goswami, Vibhor Gupta, Vignesh Ramanathan, Viktor Kerkez, Vincent Gonguet, Virginie Do, Vish Vogeti, Vítor Albiero, Vladan Petrovic, Weiwei Chu, Wenhan Xiong, Wenyin Fu, Whitney Meers, Xavier Martinet, Xiaodong Wang, Xiaofang Wang, Xiaoqing Ellen Tan, Xide Xia, Xinfeng Xie, Xuchao Jia, Xuewei Wang, Yaelle Goldschlag, Yashesh Gaur, Yasmine Babaei, Yi Wen, Yiwen Song, Yuchen Zhang, Yue Li, Yuning Mao, Zacharie Delpierre Coudert, Zheng Yan, Zhengxing Chen, Zoe Papakipos, Aaditya Singh, Aayushi Srivastava, Abha Jain, Adam Kelsey, Adam Shajnfeld, Adithya Gangidi, Adolfo Victoria, Ahuva Goldstand, Ajay Menon, Ajay Sharma, Alex Boesenberg, Alexei Baevski, Allie Feinstein, Amanda Kallet, Amit Sangani, Amos Teo, Anam Yunus, Andrei Lupu, Andres Alvarado, Andrew Caples, Andrew Gu, Andrew Ho, Andrew Poulton, Andrew Ryan, Ankit Ramchandani, Annie Dong, Annie Franco, Anuj Goyal, Aparajita Saraf, Arkabandhu Chowdhury, Ashley Gabriel, Ashwin Bharambe, Assaf Eisenman, Azadeh Yazdan, Beau James, Ben Maurer, Benjamin Leonhardi, Bernie Huang, Beth Loyd, Beto De Paola, Bhargavi Paranjape, Bing Liu, Bo Wu, Boyu Ni, Braden Hancock, Bram Wasti, Brandon Spence, Brani Stojkovic, Brian Gamido, Britt Montalvo, Carl Parker, Carly Burton, Catalina Mejia, Ce Liu, Changhan Wang, Changkyu Kim, Chao Zhou, Chester Hu, Ching-Hsiang Chu, Chris Cai, Chris Tindal, Christoph Feichtenhofer, Cynthia Gao, Damon Civin, Dana Beaty, Daniel Kreymer, Daniel Li, David Adkins, David Xu, Davide Testuggine, Delia David, Devi Parikh, Diana Liskovich, Didem Foss, Dingkang Wang, Duc Le, Dustin Holland, Edward Dowling, Eissa Jamil, Elaine Montgomery, Eleonora Presani, Emily Hahn, Emily Wood, Eric-Tuan Le, Erik Brinkman, Esteban Arcaute, Evan Dunbar, Evan Smothers, Fei Sun, Felix Kreuk, Feng Tian, Filippos Kokkinos, Firat Ozgenel, Francesco Caggioni, Frank Kanayet, Frank Seide, Gabriela Medina Florez, Gabriella Schwarz, Gada Badeer, Georgia Swee, Gil Halpern, Grant Herman, Grigory Sizov, Guangyi, Zhang, Guna Lakshminarayanan, Hakan Inan, Hamid Shojanazeri, Han Zou, Hannah Wang, Hanwen Zha, Haroun Habeeb, Harrison Rudolph, Helen Suk, Henry Aspegren, Hunter Goldman, Hongyuan Zhan, Ibrahim Damlaj, Igor Molybog, Igor Tufanov, Ilias Leontiadis, Irina-Elena Veliche, Itai Gat, Jake Weissman, James Geboski, James Kohli, Janice Lam, Japhet Asher, Jean-Baptiste Gaya, Jeff Marcus, Jeff Tang, Jennifer Chan, Jenny Zhen, Jeremy Reizenstein, Jeremy Teboul, Jessica Zhong, Jian Jin, Jingyi Yang, Joe Cummings, Jon Carvill, Jon Shepard, Jonathan McPhie, Jonathan Torres, Josh Ginsburg, Junjie Wang, Kai Wu, Kam Hou U, Karan Saxena, Kartikay Khandelwal, Katayoun Zand, Kathy Matosich, Kaushik Veeraraghavan, Kelly Michelena, Keqian Li, Kiran Jagadeesh, Kun Huang, Kunal Chawla, Kyle Huang, Lailin Chen, Lakshya Garg, Lavender A, Leandro Silva, Lee Bell, Lei Zhang, Liangpeng Guo, Licheng Yu, Liron Moshkovich, Luca Wehrstedt, Madian Khabsa, Manav Avalani, Manish Bhatt, Martynas Mankus, Matan Hasson, Matthew Lennie, Matthias Reso, Maxim Groshev, Maxim Naumov, Maya Lathi, Meghan Keneally, Miao Liu, Michael L. Seltzer, Michal Valko, Michelle Restrepo, Mihir Patel, Mik Vyatskov, Mikayel Samvelyan, Mike Clark, Mike Macey, Mike Wang, Miquel Jubert Hermoso, Mo Metanat, Mohammad Rastegari, Munish Bansal, Nandhini Santhanam, Natascha Parks, Natasha White, Navyata Bawa, Nayan Singhal, Nick Egebo, Nicolas Usunier, Nikhil Mehta, Nikolay Pavlovich Laptev, Ning Dong, Norman Cheng, Oleg Chernoguz, Olivia Hart, Omkar Salpekar, Ozlem Kalinli, Parkin Kent, Parth Parekh, Paul Saab, Pavan Balaji, Pedro Rittner, Philip Bontrager, Pierre Roux, Piotr Dollar, Polina Zvyagina, Prashant Ratanchandani, Pritish Yuvraj, Qian Liang, Rachad Alao, Rachel Rodriguez, Rafi Ayub, Raghotham Murthy, Raghu Nayani, Rahul Mitra, Rangaprabhu Parthasarathy, Raymond Li, Rebekkah Hogan, Robin Battey, Rocky Wang, Russ Howes, Ruty Rinott, Sachin Mehta, Sachin Siby, Sai Jayesh Bondu, Samyak Datta, Sara Chugh, Sara Hunt, Sargun Dhillon, Sasha Sidorov, Satadru Pan, Saurabh Mahajan, Saurabh Verma, Seiji Yamamoto, Sharadh Ramaswamy, Shaun Lindsay, Shaun Lindsay, Sheng Feng, Shenghao Lin, Shengxin Cindy Zha, Shishir Patil, Shiva Shankar, Shuqiang Zhang, Shuqiang Zhang, Sinong Wang, Sneha Agarwal, Soji Sajuyigbe, Soumith Chintala, Stephanie Max, Stephen Chen, Steve Kehoe, Steve Satterfield, Sudarshan Govindaprasad, Sumit Gupta, Summer Deng, Sungmin Cho, Sunny Virk, Suraj Subramanian, Sy Choudhury, Sydney Goldman, Tal Remez, Tamar Glaser, Tamara Best, Thilo Koehler, Thomas Robinson, Tianhe Li, Tianjun Zhang, Tim Matthews, Timothy Chou, Tzook Shaked, Varun Vontimitta, Victoria Ajayi, Victoria Montanez, Vijai Mohan, Vinay Satish Kumar, Vishal Mangla, Vlad Ionescu, Vlad Poenaru, Vlad Tiberiu Mihailescu, Vladimir Ivanov, Wei Li, Wenchen Wang, Wenwen Jiang, Wes Bouaziz, Will Constable, Xiaocheng Tang, Xiaojian Wu, Xiaolan Wang, Xilun Wu, Xinbo Gao, Yaniv Kleinman, Yanjun Chen, Ye Hu, Ye Jia, Ye Qi, Yenda Li, Yilin Zhang, Ying Zhang, Yossi Adi, Youngjin Nam, Yu, Wang, Yu Zhao, Yuchen Hao, Yundi Qian, Yunlu Li, Yuzi He, Zach Rait, Zachary DeVito, Zef Rosnbrick, Zhaoduo Wen, Zhenyu Yang, Zhiwei Zhao, and Zhiyu Ma. 2024. The llama 3 herd of models. *Preprint*, arXiv:2407.21783.

Chankyu Lee, Rajarshi Roy, Mengyao Xu, Jonathan Raiman, Mohammad Shoeybi, Bryan Catanzaro, and Wei Ping. 2024. Nv-embed: Improved techniques for training llms as generalist embedding models. *arXiv preprint arXiv:2405.17428*.

Ni Li, Shorouq Zahra, Mariana Brito, Clare Flynn, Olof Görnerup, Koffi Worou, Murathan Kurfali, Chanjuan Meng, Wim Thiery, Jakob Zscheischler, Gabriele Messori, and Joakim Nivre. 2024. Using LLMs to build a database of climate extreme impacts. In *Proceedings of the 1st Workshop on Natural Language Processing Meets Climate Change (ClimateNLP 2024)*, pages 93–110, Bangkok, Thailand. Association for Computational Linguistics.

Kyle Lo, Lucy Lu Wang, Mark Neumann, Rodney Kinney, and Daniel Weld. 2020. S2ORC: The semantic scholar open research corpus. In *Proceedings of the 58th Annual Meeting of the Association for Computational Linguistics*, pages 4969–4983, Online. Association for Computational Linguistics.

Niklas Muennighoff, Nouamane Tazi, Loïc Magne, and Nils Reimers. 2022. Mteb: Massive text embedding benchmark. *arXiv preprint arXiv:2210.07316*.

Kishan Nagendra, Omran A. Bukhres, Srinivasan Sikkupparbathyam, Marcelo Areal, Zina Ben-Miled, Lola M. Olsen, Chris Gokey, David Kendig, Tom Northcutt, Rosy Cordova, and Gene Major. 2001. Nasa global change master directory: an implementation of asynchronous management protocol in a heterogeneous distributed environment. *Proceedings 3rd International Symposium on Distributed Objects and Applications*, pages 136–145.

OpenAI, :, Aaron Hurst, Adam Lerer, Adam P. Goucher, Adam Perelman, Aditya Ramesh, Aidan Clark, AJ Ostrow, Akila Welihinda, Alan Hayes, Alec Radford, Aleksander Mądry, Alex Baker-Whitcomb, Alex Beutel, Alex Borzunov, Alex Carney, Alex Chow, Alex Kirillov, Alex Nichol, Alex Paino, Alex Renzin, Alex Tachard Passos, Alexander Kirillov, Alexi Christakis, Alexis Conneau, Ali Kamali, Allan Jabri, Allison Moyer, Allison Tam, Amadou Crookes, Amin Tootoochian, Amin Tootoonchian, Ananya Kumar, Andrea Vallone, Andrej Karpathy, Andrew Braunstein, Andrew Cann, Andrew Codispoti, Andrew Galu, Andrew Kondrich, Andrew Tulloch, Andrey Mishchenko, Angela Baek, Angela Jiang, Antoine Pelisse, Antonia Woodford, Anuj Gosalia, Arka Dhar, Ashley Pantuliano, Avi Nayak, Avital Oliver, Barret Zoph, Behrooz Ghorbani, Ben Leimberger, Ben Rossen, Ben Sokolowsky, Ben Wang, Benjamin Zweig, Beth Hoover, Blake Samic, Bob McGrew, Bobby Spero, Bogo Giertler, Bowen Cheng, Brad Lightcap, Brandon Walkin, Brendan Quinn, Brian Guarraci, Brian Hsu, Bright Kellogg, Brydon Eastman, Camillo Lugaresi, Carroll Wainwright, Cary Bassin, Cary Hudson, Casey Chu, Chad Nelson, Chak Li, Chan Jun Shern, Channing Conger, Charlotte Barette, Chelsea Voss, Chen Ding, Cheng Lu, Chong Zhang, Chris Beaumont, Chris Hallacy, Chris Koch, Christian Gibson, Christina Kim, Christine Choi, Christine McLeavey, Christopher Hesse, Claudia Fischer, Clemens Winter, Coley Czarnecki, Colin Jarvis, Colin Wei, Constantin Koumouzelis, Dane Sherburn, Daniel Kappler, Daniel Levin, Daniel Levy, David Carr, David Farhi, David Mely, David Robinson, David Sasaki, Denny Jin, Dev Valladares, Dimitris Tsipras, Doug Li, Duc Phong Nguyen, Duncan Findlay, Edede Oiwoh, Edmund Wong, Ehsan Asdar, Elizabeth Proehl, Elizabeth Yang, Eric Antonow, Eric Kramer, Eric Peterson, Eric Sigler, Eric Wallace, Eugene Brevdo, Evan Mays, Farzad Khorasani, Felipe Petroski Such, Filippo Raso, Francis Zhang, Fred von Lohmann, Freddie Sulit, Gabriel Goh, Gene Oden, Geoff Salmon, Giulio Starace, Greg Brockman, Hadi Salman, Haiming Bao, Haitang Hu, Hannah Wong, Haoyu Wang, Heather Schmidt, Heather Whitney, Heewoo Jun, Hendrik Kirchner, Henrique Ponde de Oliveira Pinto, Hongyu Ren, Huiwen Chang, Hyung Won Chung, Ian Kivlichan, Ian O'Connell, Ian O'Connell, Ian Osband, Ian Silber, Ian Sohl, Ibrahim Okuyucu, Ikai Lan, Ilya Kostrikov, Ilya Sutskever, Ingmar Kanitscheider, Ishaan Gulrajani, Jacob Coxon, Jacob Menick, Jakub Pachocki, James Aung, James Betker, James Crooks, James Lennon, Jamie Kiros, Jan Leike, Jane Park, Jason Kwon, Jason Phang, Jason Teplitz, Jason Wei, Jason Wolfe, Jay Chen, Jeff Harris, Jenia Varavva, Jessica Gan Lee, Jessica Shieh, Ji Lin, Jiahui Yu, Jiayi Weng, Jie Tang, Jieqi Yu, Joanne Jang, Joaquin Quinonero Candela, Joe Beutler, Joe Landers, Joel Parish, Johannes Heidecke, John Schulman, Jonathan Lachman, Jonathan McKay, Jonathan Uesato, Jonathan Ward, Jong Wook Kim, Joost Huizinga, Jordan Sitkin, Jos Kraaijeveld, Josh Gross, Josh Kaplan, Josh Snyder, Joshua Achiam, Joy Jiao, Joyce Lee, Juntang Zhuang, Justyn Harriman, Kai Fricke, Kai Hayashi, Karan Singhal, Katy Shi, Kavin Karthik, Kayla Wood, Kendra Rimbach, Kenny Hsu, Kenny Nguyen, Keren Gu-Lemberg, Kevin Button, Kevin Liu, Kiel Howe, Krithika Muthukumar, Kyle Luther, Lama Ahmad, Larry Kai, Lauren Itow, Lauren Workman, Leher Pathak, Leo Chen, Li Jing, Lia Guy, Liam Fedus, Liang Zhou, Lien Mamitsuka, Lilian Weng, Lindsay McCallum, Lindsey Held, Long Ouyang, Louis Feuvrier, Lu Zhang, Lukas Kondraciuk, Lukasz Kaiser, Luke Hewitt, Luke Metz, Lyric Doshi, Mada Aflak, Maddie Simens, Madelaine Boyd, Madeleine Thompson, Marat Dukhan, Mark Chen, Mark Gray, Mark Hudnall, Marvin Zhang, Marwan Aljubeh, Mateusz Litwin, Matthew Zeng, Max Johnson, Maya Shetty, Mayank Gupta, Meghan Shah, Mehmet Yatbaz, Meng Jia Yang, Mengchao Zhong, Mia Glaese, Mianna Chen, Michael Janner, Michael Lampe, Michael Petrov, Michael Wu, Michele Wang, Michelle Fradin, Michelle Pokrass, Miguel Castro, Miguel Oom Temudo de Castro, Mikhail Pavlov, Miles Brundage, Miles Wang, Minal Khan, Mira Murati, Mo Bavarian, Molly Lin, Murat Yesildal, Nacho Soto, Natalia Gimelshein, Natalie Cone, Natalie Staudacher, Natalie Summers, Natan LaFontaine, Neil Chowdhury, Nick Ryder, Nick Stathas, Nick Turley, Nik Tezak, Niko Felix, Nithanth Kudige, Nitish Keskar, Noah Deutsch, Noel Bundick, Nora Puckett, Ofir Nachum, Ola Okelola, Oleg Boiko, Oleg Murk, Oliver Jaffe, Olivia Watkins, Olivier Godement, Owen Campbell-Moore, Patrick Chao, Paul McMillan, Pavel Belov, Peng Su, Peter Bak, Peter Bakkum, Peter Deng, Peter Dolan, Peter Hoeschele, Peter Welinder, Phil Tillet, Philip Pronin, Philippe Tillet, Prafulla Dhariwal, Qiming Yuan, Rachel Dias, Rachel Lim, Rahul Arora, Rajan Troll, Randall Lin, Rapha Gontijo Lopes, Raul Puri, Reah Miyara, Reimar Leike, Renaud Gaubert, Reza Zamani, Ricky Wang, Rob Donnelly, Rob Honsby, Rocky Smith, Rohan Sahai, Rohit Ramchandani, Romain Huet, Rory Carmichael, Rowan Zellers, Roy Chen, Ruby Chen, Ruslan Nigmatullin, Ryan Cheu, Saachi Jain, Sam Altman, Sam Schoenholz, Sam Toizer, Samuel Miserendino, Sandhini Agarwal, Sara Culver, Scott Ethersmith, Scott Gray, Sean Grove, Sean Metzger, Shamez Hermani, Shantanu Jain, Shengjia Zhao, Sherwin Wu, Shino Jomoto, Shirong Wu, Shuaiqi, Xia, Sonia Phene, Spencer Papay, Srinivas Narayanan, Steve Coffey, Steve Lee, Stewart Hall, Suchir Balaji, Tal Broda, Tal Stramer, Tao Xu, Tarun Gogineni, Taya Christianson, Ted Sanders, Tejal Patwardhan, Thomas Cunninghman, Thomas Degry, Thomas Dimson, Thomas Raoux, Thomas Shadwell, Tianhao Zheng, Todd Underwood, Todor Markov, Toki Sherbakov, Tom Rubin, Tom Stasi, Tomer Kaftan, Tristan Heywood, Troy Peterson, Tyce Walters, Tyna Eloundou, Valerie Qi, Veit Moeller, Vinnie Monaco, Vishal Kuo, Vlad Fomenko, Wayne Chang, Weiyi Zheng, Wenda Zhou, Wesam Manassra, Will Sheu, Wojciech Zaremba, Yash Patil, Yilei Qian, Yongjik Kim, Youlong Cheng, Yu Zhang, Yuchen He, Yuchen Zhang, Yujia Jin, Yunxing Dai, and Yury Malkov. 2024. Gpt-4o system card. *Preprint*, arXiv:2410.21276.

Huitong Pan, Qi Zhang, Cornelia Caragea, Eduard Constantin Dragut, and Longin Jan Latecki. 2024. Scidmt: A large-scale corpus for detecting scientific mentions. In *International Conference on Language Resources and Evaluation*.

Huitong Pan, Qi Zhang, Eduard Constantin Dragut, Cornelia Caragea, and Longin Jan Latecki. 2023. Dmdd: A large-scale dataset for dataset mentions detection. *Transactions of the Association for Computational Linguistics*, 11:1132–1146.

Li Sihang, Huang Jin, Zhuang Jiaxi, Shi Yaorui, Cai Xiaochen, Xu Mingjun, Wang Xiang, Zhang Linfeng, Ke Guolin, and Cai Hengxing. 2024. Scilitllm: How to adapt llms for scientific literature understanding. *arXiv preprint arXiv:2408.15545*.

Prashant Singh, Erik Lehmann, and Mark Tyrrell. 2024. Climate policy transformer: Utilizing NLP to track the coherence of climate policy documents in the context of the Paris agreement. In *Proceedings of the 1st Workshop on Natural Language Processing Meets Climate Change (ClimateNLP 2024)*, pages 1–11, Bangkok, Thailand. Association for Computational Linguistics.

Karl E Taylor, Martin Juckes, V Balaji, Luca Cinquini, Sébastien Denvil, Paul J Durack, Mark Elkington, Eric Guilyardi, Slava Kharin, Michael Lautenschlager, et al. 2018. Cmip6 global attributes, drs, filenames, directory structure, and cv's. *PCMDI Document*.

David Thulke, Yingbo Gao, Petrus Pelser, Rein Brune, Rricha Jalota, Floris Fok, Michael Ramos, Ian van Wyk, Abdallah Nasir, Hayden Goldstein, Taylor Tragemann, Katie Nguyen, Ariana Fowler, Andrew Stanco, Jon Gabriel, Jordan Taylor, Dean Moro, Evgenii Tsymbalov, Juliette de Waal, Evgeny Matusov, Mudar Yaghi, Mohammad Shihadah, Hermann Ney, Christian Dugast, Jonathan Dotan, and Daniel Erasmus. 2024. Climategpt: Towards ai synthesizing interdisciplinary research on climate change. *Preprint*, arXiv:2401.09646.

M. Saef Ullah Miah, Junaida Sulaiman, Talha Bin Sarwar, Saima Sharleen Islam, Mizanur Rahman, and Md. Samiul Haque. 2023. Medical named entity recognition (medner): A deep learning model for recognizing medical entities (drug, disease) from scientific texts. In *IEEE EUROCON 2023 - 20th International Conference on Smart Technologies*, pages 158–162.

Aida Usmanova and Ricardo Usbeck. 2024. Structuring sustainability reports for environmental standards with LLMs guided by ontology. In *Proceedings of the 1st Workshop on Natural Language Processing Meets Climate Change (ClimateNLP 2024)*, pages 168–177, Bangkok, Thailand. Association for Computational Linguistics.

da Silva Viviane, Torres, Rademaker Alexandre, Lionti Krystelle, Giro Ronaldo, Lima Geisa, Fiorini Sandro, Archanjo Marcelo, Carvalho Breno, W., Neumann Rodrigo, Souza Anaximandro, Souza João, Pedro, Valnisio Gabriela, de, Paz Carmen, Nilda, Cerqueira Renato, and Steiner Mathias. 2024. Automated, llm enabled extraction of synthesis details for reticular materials from scientific literature. *arXiv preprint arXiv:2411.03484*.

D. Waliser, P. J. Gleckler, R. Ferraro, K. E. Taylor, S. Ames, J. Biard, M. G. Bosilovich, O. Brown, H. Chepfer, L. Cinquini, P. J. Durack, V. Eyring, P.-P. Mathieu, T. Lee, S. Pinnock, G. L. Potter, M. Rixen, R. Saunders, J. Schulz, J.-N. Thépaut, and M. Tuma. 2020. Observations for model intercomparison project (obs4mips): status for cmip6. *Geoscientific Model Development*, 13(7):2945–2958.

Urchade Zaratiana, Nadi Tomeh, Pierre Holat, and Thierry Charnois. 2024. GLiNER: Generalist model for named entity recognition using bidirectional transformer. In *Proceedings of the 2024 Conference of the North American Chapter of the Association for Computational Linguistics: Human Language Technologies (Volume 1: Long Papers)*, pages 5364–5376, Mexico City, Mexico. Association for Computational Linguistics.

Qi Zhang, Zhijia Chen, Huitong Pan, Cornelia Caragea, Longin Jan Latecki, and Eduard Constantin Dragut. 2024. Scier: An entity and relation extraction dataset for datasets, methods, and tasks in scientific documents. In *Conference on Empirical Methods in Natural Language Processing*.

Qi Zhang, Huitong Pan, Zhijia Chen, Longin Jan Latecki, Cornelia Caragea, and Eduard Dragut. 2025. Dynclean: Training dynamics-based label cleaning for distantly-supervised named entity recognition. *Preprint*, arXiv:2504.04616.

Shijia Zhou, Siyao Peng, and Barbara Plank. 2024. CLIMATELI: Evaluating entity linking on climate change data. In *Proceedings of the 1st Workshop on Natural Language Processing Meets Climate Change (ClimateNLP 2024)*, pages 215–222, Bangkok, Thailand. Association for Computational Linguistics.

A Appendix

A.1 Prompt

Table 4 shows the prompt being used for Climate Science Entity and Relationship Extraction from the climate science literature.

A.2 Entity extraction prediction

We employ regular expressions to align predicted entity names with the input text, enabling precise boundary matching. Figures 3, and 4 visualize raw(Yellow: PD_all) and PostRAG(Blue: PD_post) predictions from Llama-3.3-70B, showcasing examples from evaluation documents.

-Goal-
Given a text document with a preliminary list of potential entities, verify, and identify all entities of the specified types within the text. Note that the initial list may contain missing or incorrect entities. Additionally, determine and label the relationships among the verified entities.

-Entity Types-
A project refers to the scientific program, field campaign, or project from which the data were collected.
A location is a place on Earth, a location within Earth, a vertical location, or a location outside of the Earth.
A model is a sophisticated computer simulation that integrate physical, chemical, biological, and dynamical processes to represent and predict Earth's climate system.
An experiment is a structured simulation designed to test specific hypotheses, investigate climate processes, or assess the impact of various forcings on the climate system.
A platform refers to a system, theory, or phenomenon that accounts for its known or inferred properties and may be used for further study of its characteristics.
A instrument is a device used to measure, observe, or calculate.
A provider is an organization, an academic institution or a commercial company.
A variable is a quantity or a characteristic that can be measured or observed in climate experiments.
A weather event is a meteorological occurrence that impacts Earth's atmosphere and surface over short timescales.
A natural hazard is a phenomenon with the potential to cause significant harm to life, property, and the environment.
A teleconnection is a large-scale pattern of climate variability that links weather and climate phenomena across vast distances.
An ocean circulation is the large-scale movement of water masses in Earth's oceans, driven by wind, density differences, and the Coriolis effect, which regulates Earth's climate.

-Relationship Types and Definitions-
ComparedTo: The source entity is compared to the target entity. Outputs: A climate model, experiment, or project (source entity) outputs data (target entity).
RunBy: Experiments or scenarios (source entity) are run by a climate model (target entity).
ProvidedBy: A dataset, instrument, or model (source entity) is created or managed by an organization (target entity).
ValidatedBy: The accuracy or reliability of model simulations (source entity) is confirmed by datasets or analyses (target entity).
UsedIn: An entity, such as a model, simulation tool, experiment, or instrument (source entity), is utilized within a project (target entity).
MeasuredAt: A variable or parameter (source entity) is quantified or recorded at a geographic location (target entity).
MountedOn: An instrument or measurement device (source entity) is physically attached or installed on a platform (target entity).
TargetsLocation: An experiment, project, model, weather event, natural hazard, teleconnection, or ocean circulation (source entity) is designed to study, simulate, or focus on a specific geographic location (target entity).

-Steps-
1. Identify all entities. For each identified entity, extract the following information:
- entity name: Name of the entity
- entity type: One of the following types: [project, location, model, experiment, platform, instrument, provider, variable]
Format each entity as ("entity"<|><entity name><|><entity type><|><entity description>)
2. From the entities identified from step 1, identify all pairs of (source entity, target entity) that are *clearly related* to each other. For each pair of related entities, extract the following information:
- source entity: name of the source entity
- target entity: name of the target entity
- relationship type: One of the following relationship types: ComparedTo, Outputs, RunBy, ProvidedBy, ValidatedBy, UsedIn, MeasuredAt, MountedOn, TargetsLocation
Format each relationship as ("relationship"<|><source entity><|><target entity><|><relationship type>)
3. Return output in English as a single list of all the entities and relationships identified in steps 1 and 2. Use **** as the list delimiter. Do not output any code or steps for solving the question.
4. When finished, output <|COMPLETE|>

########################
-Examples-
{formatted examples}
########################
-Real Data-
########################
Text: {input text}
Potential Entities: {potential entities}
########################
Output:

Table 4: Prompt Template for Climate Science Entity and Relationship Extraction

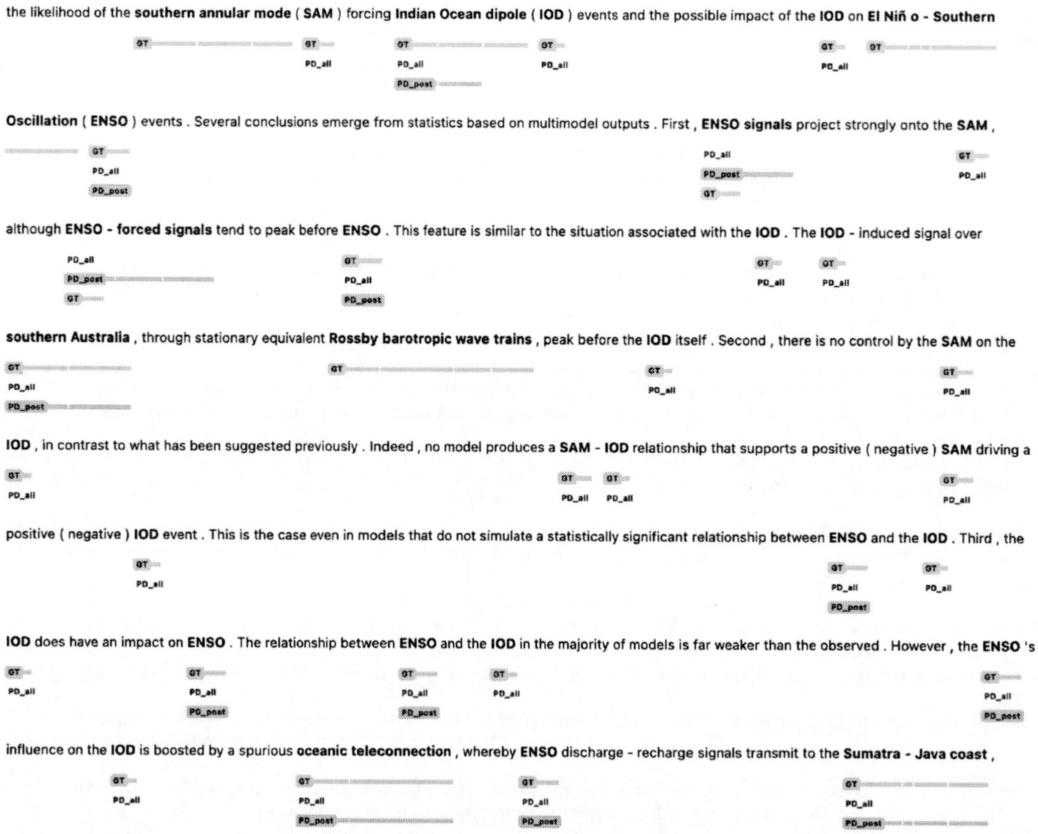

Figure 3: Example 2 of entity extraction results from a climate science publication.

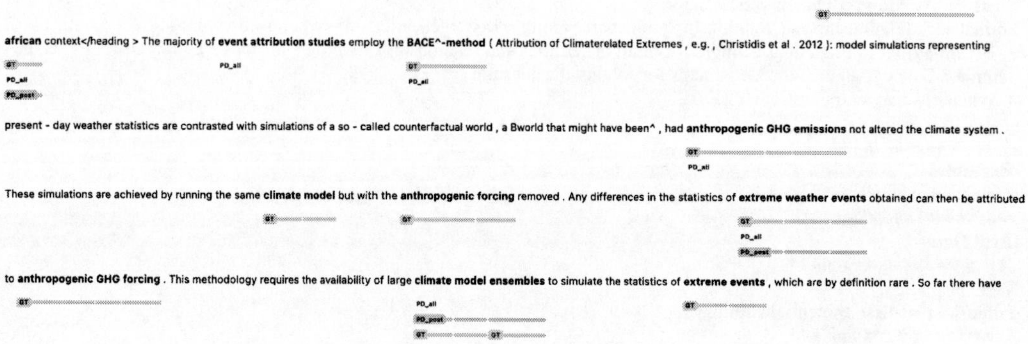

Figure 4: Example 3 of entity extraction results from a climate science publication.

Given the following metadata about an entity in a climate science ontology, which may include the entity's name, ontology path, and a definition (which may be missing), please develop an edited definition suitable for a named entity recognition (NER) task in climate science literature. The definition should be concise, clear, and limited to 150 tokens. Ensure it is precise and emphasizes the entity's unique aspects, avoiding overly general descriptions that could apply to multiple entities. Do not explain; only provide the edited definition. Metadata: {} Edited Definition:

Table 5: Prompt Template for Refining Definitions

A.3 Annotation Guidelines

Annotation guidelines are attached at the end.

Annotation Guideline

STAGE ONE: Named Entity Recognition

1. Introduction

Purpose of the Manual:
This manual provides detailed instructions for annotating climate-related text or terms extracted from scientific literature. It aims to ensure consistency and accuracy in labelling climate entities, data, and models.

Intended Audience:
The guidelines are designed for annotators, including researchers, climate analysts, scientists, and students, who are familiar with climate science terminology and concepts.

Scope of Annotations:
The annotations focus on specific climate entities, including but not limited to:

- **Earth Systems:** Land, ocean, atmosphere, and biosphere entities.
- **Climate Data:** Specific datasets and measurements.
- **Climate Models:** Global and regional climate models.

2. Definitions and Examples of Key Climate Entities

2.1 Earth Systems

Land:
Refers to a specific region or unit of land that can be described and modeled geographically within the framework of a climate model. **Examples:**

- Continents/Regions: Africa, Ethiopia, United Kingdom (UK), high/mid-latitudes, tropics (tropical regions).
- Land Features: Groundwater, river flow, runoff, streamflow, land cover, land use.
- Specific Landmarks: Amazon Rainforest, Himalayas, United States Midwest (Corn Belt), Antarctica.

Atmosphere:
Refers to the layer of gases surrounding the Earth, which plays a vital role in shaping climate and weather patterns and can be modeled geographically within the framework of a climate model.
Examples:

- Atmospheric Layers: Troposphere, mesosphere.
- Climate Phenomena: Temperature, precipitation, wind, evapotranspiration, clouds.
- Weather Systems: Hadley Cells, Ferrel Cells, Trade Winds, Jet Streams, Monsoons, Intertropical Convergence Zone (ITCZ), El Niño-Southern Oscillation (ENSO), Tornadoes, Thunderstorms.

Oceans:
Refers to the large bodies of saltwater that cover about 71% of the Earth's surface and can be modeled geographically within the framework of a climate model. **Examples:**

- Oceans/Seas: Pacific Ocean, Indian Ocean, Atlantic Ocean.
- Oceanic Features: Gulf Stream, Kuroshio Current, Thermohaline Circulation.
- Climate-Related Ocean Phenomena: Ocean acidification, marine heatwaves, coral reefs, upwelling zones, sea ice, continental shelves.

2.2 Climate Data

Refers to detailed, quantitative measurements or simulations of variables that describe various components of the Earth's climate system. **Examples:**

- Datasets: CRU (Climate Research Unit), GPCC (Global Precipitation Climatology Centre), ERA5 (ECMWF Reanalysis 5th Generation).
- Climate Indices: HadCRUT, MERRA-2, GSMP3.

2.3 Climate Models

Refers to computational models used to simulate the Earth's climate system. **Examples:**

2.4 Global Climate Models (GCMs): CCSM4, CNRM-CM5, HadGEM2-ES.

2.5 Regional Climate Models (RCMs): MICRO, ACCESS-ESM1.5.

3. Key Tags or Labels

Guidelines for Tagging:

- Ensure the correct spelling and usage of tags. For example, use "Variables" consistently, not "Variable>" or other variations.
- Review definitions carefully and apply tags or values strictly based on the provided examples and their accurate definitions.
- If uncertain about the definition of an entity, verify its classification (e.g., variable, teleconnection) before tagging.

Tag	Definition and examples
Variable	represents a specific measurable element or attribute of the climate system that is studied or monitored (e.g., cloud cover, temperature (i.e., surface air, ocean, or groundwater), precipitation, wind speed, vapor pressure, geopotential height, humidity (relative, specific) etc.
Project	refers to a coordinated effort or initiative aimed at investigating specific aspects of climate. Projects often involve multiple stakeholders and produce datasets, models, or assessments (e.g., Coupled Model Intercomparison Project Phase 6 (CMIP6))
Location	refers to the geographic region or coordinates being studied or monitored. This can be global, regional, or local. Examples includes West Africa, Central Africa, East Africa, or Southern Africa; tropics or polar regions; high or mid latitudes regions, specific sites (such as the Amazon, Congo Rainforest or Sahara Desert etc).
Model	refers to computational tool used to simulate and predict climate processes and interactions in the Earth system (e.g., HadGEM3, WRF etc)
Provider	refers to the organization or agency responsible for creating, maintaining, or distributing climate data or tools (e.g., NASA (e.g., GISS for climate models, MERRA datasets); ECMWF (e.g., ERA5 reanalysis datasets); NOAA (e.g., NCEP datasets and climate services).
Instrument	refers to the device or tool used to measure climate variables. Instruments can be ground-based, airborne, or spaceborne. Examples includes Radiosondes (balloons for atmospheric measurements); Satellites (e.g., MODIS, GOES, or Sentinel); Rain gauges and anemometers for ground-level data.
Event	An event is an occurrence or phenomenon in the Earth's system that varies in temporal scale, ranging from short-term weather events lasting minutes to days to long-term climate events spanning decades or more. Examples include remote teleconnection such as ENSO, IOD, etc, droughts, floods, etc
Weather event	Weather events are meteorological occurrences that impact Earth's atmosphere and surface over short timescales (hours to days). Common Weather Events; Rainfall (e.g., Drizzle, showers, or steady rain), Snowfall (e.g., Light , or heavy); Thunderstorms (e.g., storms with lightning, thunder, heavy rain, and hail), Wind Events (e.g., breezes, gusts, and strong winds), Cloud Cover (e.g., Clear skies, partly cloudy, overcast), Temperature Changes (Heatwaves or cold snaps), Fog and Mist, Frost, Dew etc.

Natural Hazard	Natural hazards are phenomena with the potential to cause significant harm to life, property, and the environment. Teleconnection refers to large-scale patterns of climate variability that link weather and climate phenomena across vast geographic areas, influencing atmospheric conditions over long distances. Typical examples of hazards can be broadly classified into geophysical (e.g., earthquakes, volcanic eruptions, tsunamis, landslides), meteorological (e.g., cyclones or hurricanes or typhons, tornadoes, heatwaves), hydrological (e.g., floods, flash floods, drought, avalanches), biological (pandemics, plagues, animal borne diseases), and climatological (e.g., wildfires, frost, cold wave) categories.
Ocean circulation	Ocean circulation is the large-scale movement of water masses in the Earth's oceans, driven by wind, density differences, and the Coriolis effect, regulating Earth's climate. Key examples of ocean circulation, categorized into surface currents (Gulf Stream, Kuroshio Current, California Current, Canary Current, Equatorial Currents), deep ocean currents (North Atlantic Deep Water (NADW), Antarctic Bottom Water (AABW), Mediterranean Outflow Water, Indian Ocean Overturning), Global Ocean Circulation Systems (the Global Conveyor Belt, the Atlantic Meridional Overturning Circulation (AMOC).
Teleconnection	Teleconnection is a large-scale patterns of climate variability that link weather and climate phenomena across vast distances. Examples includes El Niño-Southern Oscillation (ENSO; (El Niño or La Niña), North Atlantic Oscillation (NAO), Arctic Oscillation (AO), Pacific Decadal Oscillation (PDO), Indian Ocean Dipole (IOD), Madden-Julian Oscillation (MJO), Atlantic Multi-Decadal Oscillation (AMO), Southern Annular Mode (SAM), Rossby Waves, Walker Circulation, Monsoonal Systems (i.e., Asian Monsoon and West African Monsoon)

4. Example

Example: *"This annotation manual aims to provide consistent methods for annotating climate data. Our primary focus is 09bdb7d909ed6615760571a6aa14051133179aee.xmi"*

<u>**Task one**</u>: see the scientific literature with serial number above.

Role of the annotator: The annotator is expected is to read each sentence carefully. Then, you are required to perform these tasks concurrently.

1. Verify specific pre-annotated climate entries of interest in line 22: (E.g., "clouds", "precipitation", "ENSO") and other scientific terms such as "mid-latitude continents". (see details below for more information).
2. Delete pre-annotated test that involves a "process" or "methods", "tools", frameworks, "instrument of measurements", "units of measurement", "temporal, threshold or range of values" (e.g., convective parameterisation, diurnal, monsoon (see details below for more information).
3. Annotate missing but relevant "un-annotated" text of interest (E.g., Westerly Winds) (see details below on how to annotate).

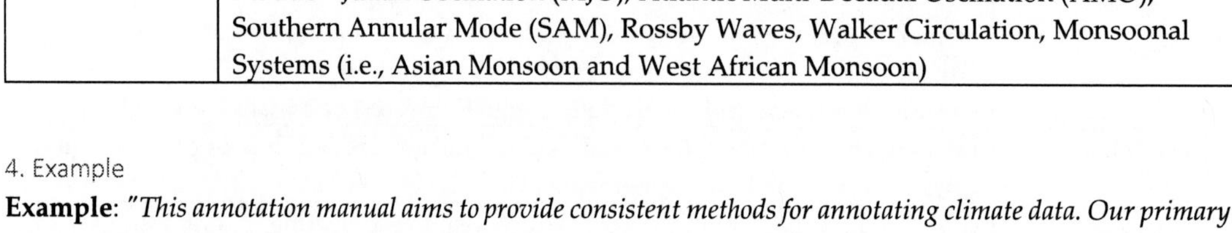
variable

28 The strength of the westerly winds, and therefore the Ekman transport, varies with latitude-the maximum northward surface transport occurs at about 50° S and decreases south of that.

29 Water must be drawn up from below in order to balance the difference between the larger northward transport at 50° S, say, compared with the smaller northward transport at 60° S.

variable location

30 The broad ring of upwelling shown in figure 2a starts close to the Antarctic continent and extends all the way to roughly 50° S.

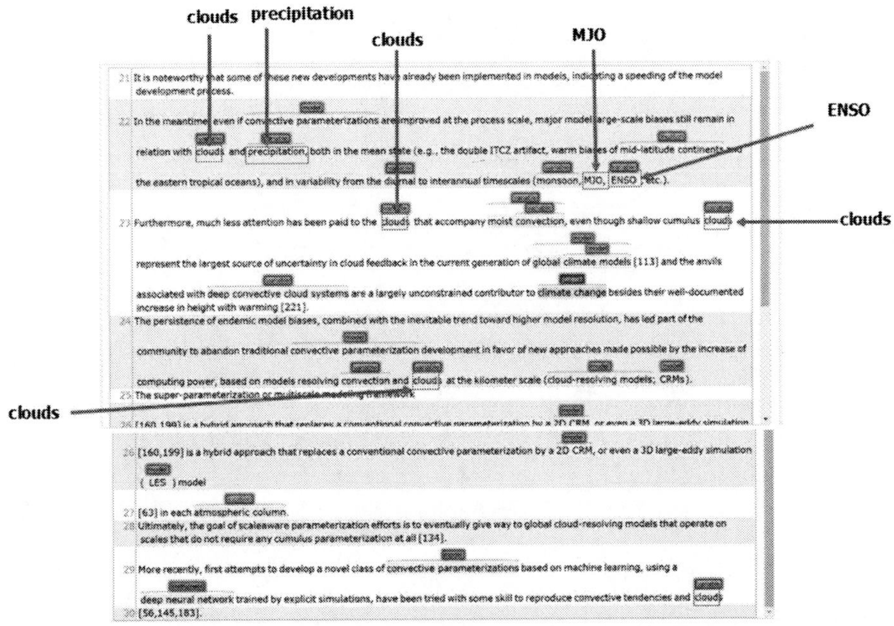

Other Scientific Terms: You may find other climate variables such as temperature, wind speed or wind, sea surface temperature or SST; rainfall, cyclones, aerosols, etc

Delete wrongly pre-annotated climate entities. These may include but not limited to methods, materials, processes, units of measurements, threshold, or range of values, etc

Units of Measurement: (e.g., Celsius for temperature, mm for rainfall, km/h for wind speed).

Thresholds and Ranges: Values or thresholds or ranges. E.g., 10°C for temperature or mm for precipitation."

Standardization: standardizing annotations across climate entities. For example, temperature (delete prefix "minimum or min", "maximum or max", "nighttime", "daytime" for temperature annotations to ensure consistency (e.g. minimum temperature to temperature).

Other Scientific Terms: Phrases that are a scientific term but do not fall into any of the above classes E.g. diurnal, interannual,

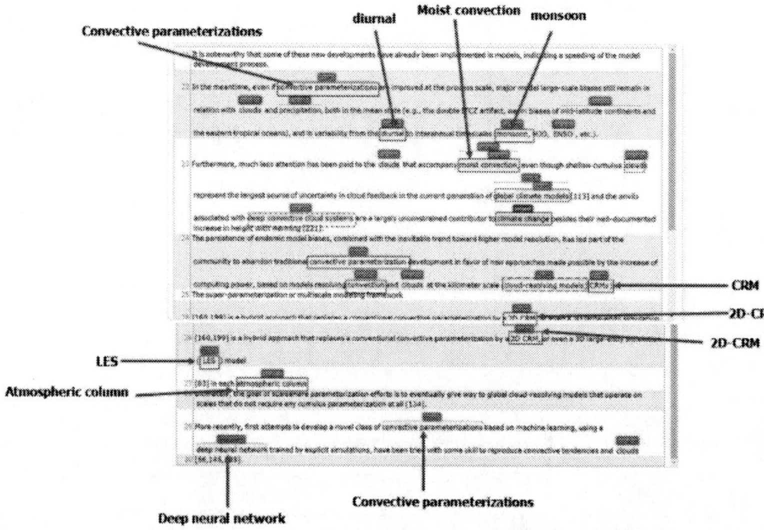

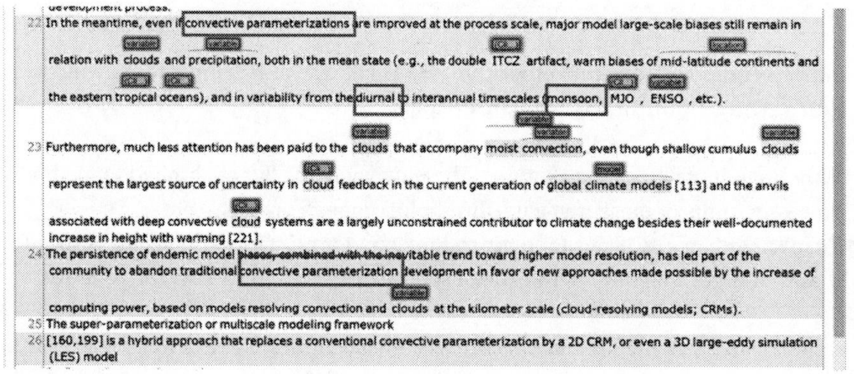

STAGE TWO: Entity Linking

1. Tag Selection Guidelines

- **Allowed Tags**: Only the following values should be selected as tags. Do not type any tags manually; only select from the provided list: project, location, model, experiment, platform, instrument, provider, variable, weather event, natural hazard, teleconnection, ocean circulation
- **Spelling and Formatting**:
  - Ensure all tags are in **lowercase**.
  - Do not use uppercase letters or modify the spellings in any way.
  - If you encounter any foreign or unrecognized tags, do not use them.

2. Annotation Setup

- Open **two tables** simultaneously:
  1. **Annotation Table**: The document or interface where you are performing the annotations.
  2. **Knowledge Base Table**: A reference table or database containing entity identifiers and their corresponding information.

- Use the knowledge base to search for and verify the correct identifiers for each entity. Make sure to check if the definitions and the path match the semantic meaning.

3. Task Description

- **Objective**: Link each entity in the text to its corresponding identifier in the knowledge base.
- **Steps**:
    1. Identify the entity in the text.
    2. Double check the tag from the allowed list (e.g., location, variable, etc.).
    3. Search the knowledge base to find the correct identifier for the entity.
    4. Link the entity to its identifier in the annotation table.

4. Quality Assurance

- Double-check the spelling and formatting of tags.
- Ensure that all entities are linked to the correct identifiers in the knowledge base.
- If an entity cannot be found in the knowledge base, flag it for review rather than making an assumption.

STAGE THREE: Relationship

1. Relationship Types and Definitions

Below are the relationship types to be annotated, along with their definitions and examples. Ensure that you correctly identify the **source entity** and **target entity** for each relationship.

1. **ComparedTo**
    - **Definition**: The source entity is compared to the target entity.
    - **Example**: A climate model, experiment, or project (source entity) outputs data (target entity).
    - **Template**: [Source Entity] ComparedTo [Target Entity]
2. **RunBy**
    - **Definition**: Experiments or scenarios (source entity) are run by a climate model (target entity).
    - **Example**: An experiment (source entity) is executed by a climate model (target entity).
    - **Template**: [Source Entity] RunBy [Target Entity]
3. **ProvidedBy**
    - **Definition**: A dataset, instrument, or model (source entity) is created or managed by an organization (target entity).
    - **Example**: A dataset (source entity) is provided by a research organization (target entity).
    - **Template**: [Source Entity] ProvidedBy [Target Entity]
4. **ValidatedBy**
    - **Definition:** The accuracy or reliability of model simulations (source entity) is confirmed by datasets or analyses (target entity).
    - **Example**: A climate model simulation (source entity) is validated by observational data (target entity).
    - **Template:** [Source Entity] ValidatedBy [Target Entity]
5. **UsedIn**
    - **Definition**: An entity, such as a model, simulation tool, experiment, or instrument (source entity), is utilized within a project (target entity).
    - **Example**: A climate model (source entity) is used in a research project (target entity).
    - **Template**: [Source Entity] UsedIn [Target Entity]
6. **MeasuredAt**

- **Definition**: A variable or parameter (source entity) is quantified or recorded at a geographic location (target entity).
- **Example**: Temperature data (source entity) is measured at a specific weather station (target entity).
- **Template**: [Source Entity] MeasuredAt [Target Entity]

7. **MountedOn**
   - **Definition**: An instrument or measurement device (source entity) is physically attached or installed on a platform (target entity).
   - **Example**: A weather sensor (source entity) is mounted on a satellite (target entity).
   - **Template**: [Source Entity] MountedOn [Target Entity]
8. **TargetsLocation**
   - **Definition**: An experiment, project, model, weather event, natural hazard, teleconnection, or ocean circulation (source entity) is designed to study, simulate, or focus on a specific geographic location (target entity).
   - **Example**: A climate model (source entity) targets the Amazon Rainforest (target entity).
   - **Template**: [Source Entity] TargetsLocation [Target Entity]

2. Annotation Instructions

1. **Identify Entities**:
   - Clearly identify the **source entity** and **target entity** in the text.
   - Ensure that both entities are correctly tagged (e.g., model, location, variable, etc.) before annotating the relationship.
2. **Select Relationship Type**:
   - Choose the most appropriate relationship type from the list above based on the context.
   - Refer to the definitions and examples to ensure accuracy.
3. **Annotate the Relationship**:
   - Use the provided templates to annotate the relationship between the source and target entities.
   - Double-check that the relationship type aligns with the context of the text.
4. **Verify Consistency**:
   - Ensure that the relationship annotation is consistent with the definitions and examples provided.
   - If unsure, consult the knowledge base or flag the relationship for review.

Biodiversity ambition analysis with Large Language Models

Stefan Troost, Roos Immerzeel, Christoph Krüger
Netherlands Environmental Assessment Agency
{Stefan.Troost, Roos.Immerzeel, Christoph.Kruger}@pbl.nl

Abstract

The Kunming-Montreal Global Biodiversity Framework (GBF) has 23 action-oriented global targets for urgent action over the decade to 2030. Parties committing themselves to the targets set by the GBF are required to share their national targets and biodiversity plans. In a case study on the GBF target to reduce pollution risks, we analyze the commitments of 110 different Parties, in 6 different languages. Obtaining satisfactory results for this target, we argue that using Generative AI can be very helpful under certain conditions, and it is a relatively small step to scale up such an analysis for other GBF targets.

1 Introduction

The Global Biodiversity Framework (GBF), adopted at COP15 of the Convention on Biological Diversity (CBD) in 2022, represents a landmark international agreement focused on addressing the unprecedented decline in species and ecosystem health worldwide (Convention on Biological Diversity, 2022a). The framework consists of 4 overarching goals for 2050 and 23 supporting targets for 2030. Additionally, the framework is supported by a monitoring framework (Convention on Biological Diversity, 2022b). The GBF is a global framework, and its implementation depends on the Parties of the CBD. Parties must submit a new or updated National Biodiversity Strategy and Action Plan (NBSAP) and/or submit national targets to the Online Reporting Tool (ORT), which indicates the ambition a Party has regarding its contribution to implementation of the GBF. In a later stage National Reports will be submitted, reporting on the implementation of these ambitions. At COP17, in 2026, progress in implementation of the GBF will be reviewed (Convention on Biological Diversity, 2024; Convention on Biological Diversity, 2025).

Still, several challenges impede effective and efficient assessment of progress towards the GBF's goals and targets. One of the key challenges that occur is the absence of a comprehensive analysis methodology to establish baseline conditions, evaluate country-level commitments, and identify additional measures needed to reach the global goals. This analytical deficit is further complicated by the large amount of data that needs to be assessed, as the GBF consists of 23 targets that will be translated to national policy by the 196 CBD Parties. National policy that is created up until this point has different formats which can be difficult to compare, not solely due to differences between NBSAPs and ORT data; the way in which national commitments are structured within those formats also looks different per country. Language differences add another layer of complexity, as the national commitments can be uploaded in any of the six official UN languages. Finally, analyses of country commitments are complicated by the risk of inconsistent interpretation when humans review these documents.

To address these challenges, we propose a novel approach leveraging a multi-lingual Retrieval-Augmented Generation (RAG) framework. This methodology enables automated analysis of country commitments at scale. The system can process multilingual documents, standardize terminologies, and generate comparable assessment metrics across different national contexts. Similar RAG frameworks have been used previously to assess SDG claims (Garigliotti, 2024) and sustainability reports (Bronzini et al., 2024). However, to the authors knowledge, it has

Proceedings of the 2nd Workshop on Natural Language Processing Meets Climate Change (ClimateNLP 2025), pages 99–112
July 31, 2025 ©2025 Association for Computational Linguistics

not been utilized to assess the GBF. We use GBF target 7 on pollution reduction as a case study. The aim is to create an aggregate view of the country commitments for the GBF target. This target refers to multiple sources of pollution to be reduced by 2030. The fact that Parties have uploaded National Targets that are aligned with this GBF target, does not necessarily mean that all sources of pollution are addressed, with similar ambition levels. We establish a framework of creating classes to identify differences and similarities in focus and ambition levels by analyzing national targets that are uploaded by Parties in the ORT. This way, we get a better understanding what pollution sources are prioritized by Parties when translating the GBF to the national level.

2 Biodiversity ambitions

2.1 National level biodiversity ambitions

The GBF goals and targets are created aiming to reduce threats to biodiversity, to meet people's needs through sustainable use and benefit-sharing, and on tools and solutions for implementation (Convention on Biological Diversity, 2022a). To make it easier for CBD Parties to upload their national targets under the GBF according to the requested reporting template, the ORT was created. This tool should also make it easier for the CBD's Secretariat, and other stakeholders, to analyze national ambitions and implementation.

2.2 Prioritization in target translation

A previous analysis by Kok et al. (2024) concludes that big differences can be observed between Parties regarding the type and amount of information uploaded to the ORT. Based on an analysis of 61 Parties and 6 GBF targets, it was found that not many national targets are specific and quantified. Regarding GBF target 7, on reducing pollution to levels not harmful to biodiversity (see Appendix A for the entire target text), around one third of the national commitments included some type of quantification, which was more than most of the other analyzed targets. However, GBF target 7 refers to all pollution sources that are harmful to biodiversity and Parties are not obligated to create national targets for all sources individually. This results in differences in prioritization within national targets as some Parties focus on the concept of pollution more generally and others focus on specific pollution sources.

This paper aims to gain an insight into the number of Parties that specify types of pollution in their national target creation, in the degree to which these targets are quantified, and in the ambition levels Parties show regarding cutting back on these pollution sources.

3 Using a Large Language Model as an assistant

Large Language Models (LLMs) are known for their versatility. Many applications have emerged since OpenAI released ChatGPT in 2022. In the scientific research area, one of the more popular ideas is to use LLMs for document analysis. However, when using LLMs for this purpose, the most important drawback is the temporary character of the data the LLM was trained with: the specific document may not have been used for training of the LLM. Mainly for this reason, the RAG framework (Lewis et al., 2020) has become popular for document analysis.

3.1 The RAG framework

The RAG framework that was used in this study (ChatPBL, 2024) was set up as a research project, with a focus on evaluation of the complete pipeline on several custom-made document-question-answer test sets.

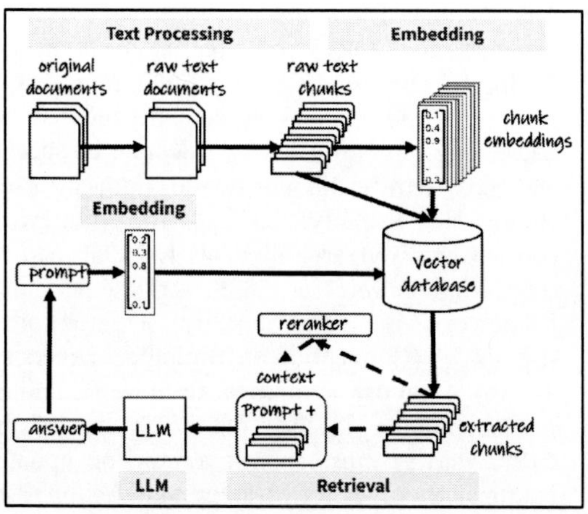

Figure 1: RAG question-answer pipeline

This means that most of the steps in the RAG pipeline (see Figure 1) have been parameterized: swapping between different choices for the components in the pipeline, whether it is for

splitting text into text chunks, embedding of the resulting text chunks, retrieval of relevant chunks from the vector database or the LLM itself, is simply a matter of changing parameters. We refer to Appendix E for the parameter settings used for this analysis.
Figure 1 shows the core question-answer functionality of the application. Our RAG framework also provides a "review" module that allows to ask multiple documents predefined questions sequentially, using this core functionality. All the answers and input that was used are stored for reproducibility purposes. The module has an option to summarize the answers to each of the questions into a synthesis. It is the review module that was used for this case study.

4 Experiments

4.1 Data processing

The ORT data was downloaded from https://ort.cbd.int/#0.4/0/0. For each submitted national target, the country must indicate at least one GBF target to which the national target is aligned. As this research is focused on GBF target 7, we only used the national targets that Parties aligned to this GBF target. The data was downloaded on January 29, 2025. At that time, 110 Parties submitted national targets aligned with GBF target 7 (see Appendix B).
After choosing the relevant columns (Government, National target title, Description, Main policy measures and Aspects of the goal or target are covered, see Figure 2), the texts were then merged together with their column titles and line breaks in between. If the country did not fill in the column 'nan' was added instead. After the merging we obtained a text file for each Party that could be used for further analysis.

The text of GBF target 7 mentions three sources of pollution in particular:

1. Excess nutrients
2. Pesticides and other hazardous chemicals
3. Plastics

For the first two types of pollution, GBF Target 7 specifically states a reduction of *"at least half"* by 2030. This is not the case for pollution caused by plastics, however for comparison purposes we

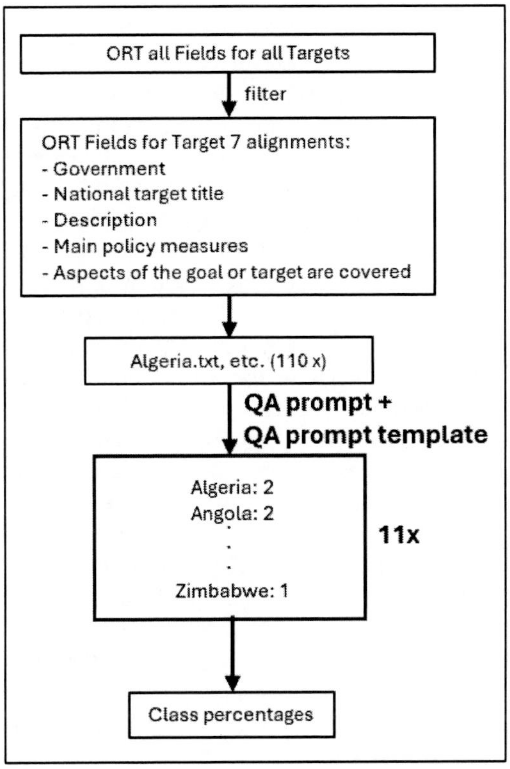

Figure 2: ORT data processing

defined the following classes for each of the three types of pollution:

- Class 1: the type of pollution is not mentioned in the country commitment
- Class 2: the type of pollution is mentioned, but there is no quantified reduction target
- Class 3: the type of pollution is mentioned and quantified, but the target is below the GBF target of at least 50% reduction
- Class 4: the type of pollution is mentioned and quantified, and the target is at or above the GBF target of at least 50% reduction

4.2 Prompting

For each pollution type, prompting the commitments was set up in 2 ways:

- A Question-Answer (QA) prompt was used to ask the Parties' documents (e.g. Algeria.txt) whether the specific pollution type was mentioned and if quantified targets were set for the type mentioned.
- This QA prompt was accompanied by a QA prompt template that instructs he LLM with the task to assign the commitment to one of the four distinguished classes

4.3 Results

Experiments were executed with several different prompts and prompt templates until satisfactory results were obtained on a sample of Parties (the first 20 alphabetically). An analysis of the Parties' commitments learned that the texts had a length short enough for our LLM, gpt-4o, to include them whole in the context. The commitment texts in the sample were manually reviewed by 2 reviewers and assigned to 1 of the 4 classes (see Appendix D). There were some cases where the manual reviews differed, which were discussed by both reviewers. In general, a consensus was quickly reached on these cases, but it is interesting to note that there were also cases where the language used was particularly difficult to interpret. For example, the following commitment description was found in the commitment text of Bangladesh and difficult to assign to a class: *"By 2026, highly hazardous pesticides and chemicals will be identified. By 2030, identified highly hazardous pesticides and chemicals will be phased out."* Depending on how one would read the last part, the phasing out can be interpreted to have ended in 2030 or to start in 2030.

After reaching consensuses for all cases where the manual reviews differed, the manual class assignments were compared with the RAG results. In 4 out of 60 reviews, the RAG results differed from the manual results, see Appendix D. Interestingly, in 3 out of those 4 cases, the manual reviewers also did not agree on a classification.

It was believed by the authors that the sample results were good enough to use the prompts on the complete set of commitments, keeping in mind that we are looking for the overall picture and not so much at the level of individual Parties. To test robustness of the results, the system was run 11 times, with the same settings. simple majority vote was taken for each country, meaning that the most occurring class was taken as the final class assignment.

When the prompts were applied to the full dataset, this resulted in the scores below:

	Target		
	excess nutrients	**pesticides and other hazardous chemicals**	**plastics**
Class 1	45.5%	39.1%	49.1%
Class 2	31.8%	37.3%	35.5%
Class 3	1.8%	1.8%	2.7%
Class 4	20.9%	21.8%	12.7%

Table 1: Class assignments for all Parties

From the table, we can observe that:

- The majority of the commitments either don't mention the pollution type, or don't quantify a target when they do mention them. This goes for all pollution types in this case-study. It should be emphasized however that commitments could have been given by Parties, using other terms and/or for other types of pollution which are out of scope of this case study (e.g. solid waste, light and noise pollution).
- Looking at the three distinguished types of pollution, pollution from pesticides and other hazardous chemicals scores best in terms of ambition levels. In 24 cases (21.8%), commitments to reducing this type of pollution were at or above the 50% reduction GBF target. This is slightly better than the score for excess nutrients and a lot better than the score for plastics, but still a relatively small number.

5 Conclusions

We have learned from this case-study that:

- Formulating satisfactory prompts is not straightforward: several attempts were necessary to produce the prompts in Appendix C.
- Interpretation of the wording in the texts can be difficult, not only for the RAG system but also when reviewing manually. Domain knowledge is important.
- Although RAG classes differed from the manual classes in some cases in the sample of 20 countries, we feel that the method is robust enough for our goal to look at the overall picture regarding Parties' pollution reduction commitments.
- Since we now have established satisfactory prompts, it is very easy to update the analysis

as soon as more Parties upload their commitments.

6 Further research

This analysis is based on a limited set of data and for illustrative purposes. It will be updated once all Parties have submitted commitments to the ORT. Available NBSAPs will also be analyzed in a similar way. Furthermore, this methodology will be used to analyze other GBF targets. Apart from analyzing if Parties include quantifications in their national target-setting, a similar method will be used to look deeper into the policy strategies Parties use to eventually implement their commitments. Together, this research helps in identifying implementation gaps and facilitating more informed policy adjustments at both national and international levels in a way that is feasible and standardized.

Limitations

Language-related challenges persist as biodiversity commitments contain specialized terminology that varies across languages and regulatory contexts. Even advanced multilingual models may struggle with nuanced ecological terms or region-specific biodiversity concepts, potentially missing critical details in commitments.

The non-deterministic nature of LLMs presents another challenge. This variability can complicate efforts to establish standardized assessment metrics and could potentially lead to fluctuating evaluations of national progress. In this case study we see that results between two consecutive runs can differ; however, the overall conclusions are not affected. We've shown the results of 11 runs in this paper, applying a mechanism of majority voting, in a way like it is used in Random Forest algorithm[1] for each Party.

This analysis focuses on two main aspects of national commitments related to the GBF's pollution target: if specific pollution types are mentioned and if commitments related to those types are quantified. The class assignment exercise as described in this paper therefore doesn't show the depth of policy behind the commitments. A manual analysis of the data showed that in some cases goals are not quantified but are backed up by specific measures to cut back on pollution, while there are also cases where highly ambitious quantifications are given which are not supported by any measures (yet).

Another limitation is that this analysis depends on self-assessment of Parties of what national targets are linked to GBF target 7. It could be that national targets aligned with other GBF targets are also directly relevant to this pollution target, without this connection made in the ORT.

These limitations necessitate human oversight in any LLM-based biodiversity analysis system, with domain experts validating model outputs and methodology to ensure accurate representation of global conservation efforts.

[1] https://en.wikipedia.org/wiki/Random_forest

References

Marco Bronzini et al. 2024. Glitter or gold? Deriving structured insights from sustainability reports via large language models.

Convention on Biological Diversity. 2022a. Kunming-Montreal Global Biodiversity Framework. (CBD/COP/15/L.25).

Convention on Biological Diversity. 2022b. Monitoring framework for the Kunming-Montreal Global Biodiversity Framework. (CBD/COP/DEC/15/5).

Convention on Biological Diversity. 2024. Mechanisms for planning, monitoring, reporting and review, including the global review of collective progress in the implementation of the Kunming-Montreal Global Biodiversity Framework to be conducted at the seventeenth and nineteenth meetings of the Conference of the Parties. (CBD/COP/16/L.33).

Convention on Biological Diversity. 2025. Governments agree on the way forward to mobilise the resources needed to protect biodiversity for people and planet [Press release].

ChatPBL. 2024. Chat with your docs!

Dario Garigliotti. 2024. SDG target detection in environmental reports using Retrieval-augmented Generation with LLMs. In *Proceedings of the 1st Workshop on Natural Language Processing Meets Climate Change (ClimateNLP 2024)*, pages 241–250, Bangkok, Thailand. Association for Computational Linguistics.

Marcel Kok, Alexandra Marques, Michel Bakkenes, Jonathan Doelman, Jelle Hilbers, Roos Immerzeel, Christoph Krüger, Aafke Schipper. 2024. A prospective evaluation of the ambition of the Kunming-Montreal Global Biodiversity Framework: *a contribution to the CBD review process', PBL Netherlands Environmental Assessment Agency,* The Hague, The Netherlands.

Patrick Lewis, Ethan Perez, Aleksandra Piktus, Fabio Petroni, Vladimir Karpukhin, Naman Goyal, Heinrich Küttler, Mike Lewis, Wen-tau Yih, Tim Rocktäschel, Sebastian Riedel, Douwe Kiela. 2020. Retrieval-Augmented Generation for Knowledge-Intensive NLP Tasks, *Computation and Language*, arXiv:2005.11401. Version 4

Appendix A: GBF TARGET 7 full text

"Reduce pollution risks and the negative impact of pollution from all sources by 2030, to levels that are not harmful to biodiversity and ecosystem functions and services, considering cumulative effects, including: (a) by reducing excess nutrients lost to the environment by at least half, including through more efficient nutrient cycling and use; (b) by reducing the overall risk from pesticides and highly hazardous chemicals by at least half, including through integrated pest management, based on science, taking into account food security and livelihoods; and (c) by preventing, reducing, and working towards eliminating plastic pollution."

Appendix B: ORT data

The following 110 Parties uploaded their commitments to the ORT by January 29, 2025, related to GBF target 7, in alphabetical order:

Algeria, Angola, Australia, Austria, Azerbaijan, Bangladesh, Benin, Bhutan, Bolivia, Botswana, Burkina Faso, Burundi, Cambodia, Cameroon, Canada, Cape Verde, Central African Republic, Chad, Chile, China, Comoros, Cook Islands, Costa Rica, Cote d'Ivoire, Croatia, Cuba, Cyprus, Democratic Republic of Congo, Djibouti, Dominican Republic, Egypt, Equatorial Guinea, Eritrea, Eswatini, Ethiopia, European Union (27), Fiji, Finland, France, Gabon, Ghana, Honduras, Hungary, India, Indonesia, Iran, Iraq, Ireland, Israel, Japan, Jordan, Kazakhstan, Kenya, Laos, Lebanon, Lesotho, Liberia, Luxembourg, Madagascar, Malawi, Malaysia, Mali, Malta, Marshall Islands, Mexico, Moldova, Mongolia, Morocco, Mozambique, Nauru, Nepal, New Zealand, Niger, Nigeria, Pakistan, Palau, Panama, Papua New Guinea, Paraguay, Peru, Philippines, Qatar, Russia, Rwanda, Samoa, Saudi Arabia, Senegal, Sierra Leone, Slovenia, Somalia, South Africa, South Korea, South Sudan, Spain, Sudan, Suriname, Sweden, Tanzania, Togo, Tunisia, Turkey, Uganda, United Arab Emirates, United Kingdom, Uruguay, Uzbekistan, Vanuatu, Venezuela, Yemen, Zimbabwe

The majority of Parties use English as language for the commitment texts. The other two widely used languages are Spanish and French. Only 4% of the texts are in the other languages as depicted in Figure 3 below.

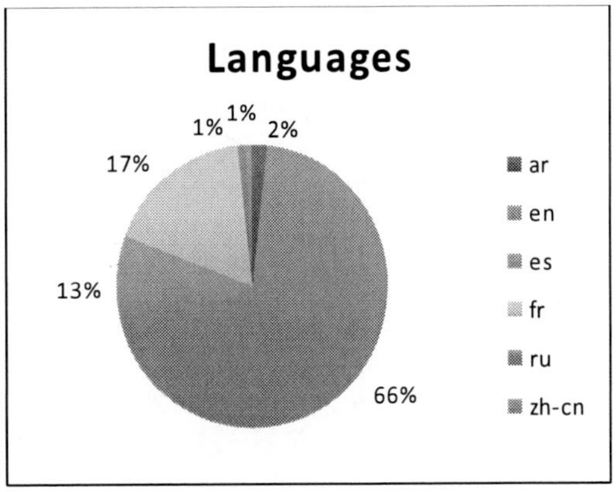

Figure 3: languages used, ar = Arabic, en = English, es = Spanish, fr = French, ru = Russian, zh-cn = Chinese

Appendix C: Prompts

1. Question-Answer (QA) prompt for pollution from excess nutrients: *"Does the text mention any targets specifically for reducing excess nutrients pollution? Excess nutrients pollution is defined as nutrients lost to the environment. Some examples of nutrients are nitrogen, phosphorus and fertilizer. If yes, is the target quantified? A quantified target is defined by a reduction in terms of a percentage, in terms of an absolute value or in terms of "reduce by half" or "phase out""*

2. Question-Answer (QA) prompt for pollution from pesticides and other hazardous chemicals: *"Does the text mention any targets specifically for reducing pesticides and highly hazardous chemicals pollution? If yes, is the target quantified? A quantified target is defined by a reduction in terms of a percentage, in terms of an absolute value or in terms of "reduce by half" or "phase out""*

3. Question-Answer (QA) prompt for pollution from plastics: *"Does the text mention any targets specifically for reducing plastics pollution? Plastics pollution is defined as the use of plastic, single use plastic, or the amount of plastic ending up in the environment. If yes, is the target quantified? A*

quantified target is defined by a reduction in terms of a percentage, in terms of an absolute value or in terms of "reduce by half" or "phase out""

4. The QA prompt template:

   "You are an AI assistant for a document analysis system. Analyze the retrieved document context and return a response based on the User Query below.
   Context: {context}

   User Query: {question}

   Assign each country to one and only one of the following classes:

   class 1: the context doesn't mention pollution from <pollution>
   class 2: the context mentions pollution from <pollution> but has no quantified target to reduce that type of pollution
   class 3: the context mentions pollution from <pollution> but the quantified target is lower than 50% reduction
   class 4: the context mentions pollution from <pollution> AND also the quantified target is at least 50% reduction.

   Just return output in the format country: class number

   Examples:
   - if country X mentions pollution from <pollution> but has no quantified target to reduce that type of pollution, the output would be X: 2
   - if country Y mentions pollution from <pollution> but the quantified target is lower than 50% reduction, the output would be Y: 3"
   Above, <pollution> is either "excess nutrients", "pesticides and chemicals", or "plastics".

Appendix D: ORT commitment reviews

Manually assigned classes of the sample of 20 Parties' commitments, compared with the RAG system results (majority vote of 11 runs for excess nutrients pollution (Table 2), pesticides and other hazardous chemicals (Table 3), and plastics (Table 4)

C1 = pollution type not mentioned
C2 = pollution type mentioned, but no quantified reduction target
C3 = pollution type mentioned, but target below GBF target of at least 50% reduction
C4 = pollution type mentioned, and target at or above GBF target of at least 50% reduction

Rev1 = class assigned by reviewer 1
Rev2 = class assigned by reviewer 2
Rev12 = final consensus of manual review
RAG = class assigned by RAG system

	Rev1	Rev2	Rev12	RAG
Algeria	C2	C2	C2	C2
Angola	C1	C1	C1	C1
Australia	C1	C1	C1	C1
Austria	C4	C4	C4	C4
Azerbaijan	C4	C4	C4	C4
Bangladesh	C2	C2	C2	C2
Benin	C1	C1	C1	**C2**
Bhutan	C1	C1	C1	C1
Bolivia	C1	C1	C1	C1
Botswana	C2	C2	C2	C2
Burkina Faso	C1	C1	C1	C1
Burundi	C2	C2	C2	C2
Cambodia	C1	C1	C1	C1
Cameroon	C2	C2	C2	C2
Canada	C4	C4	C4	C4
Cape Verde	C1	C1	C1	C1
Central African Republic	C1	C1	C1	C1
Chad	C2	C2	C2	C2
Chile	C2	C2	C2	C2
China	C2	C2	C2	C2

Table 2: excess nutrients pollution reviews

	Rev1	Rev2	Rev12	RAG
Algeria	C2	C2	C2	C2
Angola	C1	C1	C1	C1
Australia	C1	C1	C1	C1
Austria	C4	C4	C4	C4
Azerbaijan	C2	C1	C2	C2
Bangladesh	C4	C3	C4	**C2**
Benin	C2	C2	C2	C2
Bhutan	C1	C1	C1	C1
Bolivia	C1	C1	C1	C1
Botswana	C2	C2	C2	C2
Burkina Faso	C1	C1	C1	C1
Burundi	C2	C2	C2	C2
Cambodia	C1	C1	C1	C1
Cameroon	C2	C2	C2	C2
Canada	C4	C4	C4	C4
Cape Verde	C2	C1	C2	C2
Central African Republic	C1	C1	C1	C1
Chad	C1	C1	C1	C1
Chile	C1	C1	C1	C1
China	C4	C3	C3	**C2**

Table 3: pesticides and other hazardous chemicals pollution reviews

	Rev1	Rev2	Rev12	RAG
Algeria	C2	C2	C2	C2
Angola	C1	C2	C2	C2
Australia	C2	C2	C2	C2
Austria	C1	C1	C1	C1
Azerbaijan	C4	C4	C4	C4
Bangladesh	C3	C3	C3	C3
Benin	C1	C1	C1	C1
Bhutan	C1	C1	C1	C1
Bolivia	C1	C1	C1	C1
Botswana	C4	C2	C2	**C4**
Burkina Faso	C1	C1	C1	C1
Burundi	C2	C2	C2	C2
Cambodia	C1	C1	C1	C1
Cameroon	C2	C2	C2	C2
Canada	C2	C2	C2	C2
Cape Verde	C1	C1	C1	C1
Central African Republic	C1	C1	C1	C1
Chad	C1	C1	C1	C1
Chile	C1	C1	C1	C1
China	C2	C2	C2	C2

Table 4: plastics pollution reviews

Appendix E: ChatPBL parameter settings

Text chunking:
NLTKTextSplitter, chunk size of 126000 characters
Chunk embedding:
OpenAI "text-embedding-ada-002"
Retrieval:
Vectorstore retrieval based on similarity search
LLM:
OpenAI "gpt-4o", model version 2024-08-06

Appendix F: Examples country files

Example Ireland:

Government: Ireland

National target title: By 2024, the Environmental Impact Assessment (EIA) (Agriculture) Regulations will be reviewed

Description: nan

Main policy measures: DAFM will review the EIA (Agriculture) Regulations

Aspects of the goal or target are covered: nan

Government: Ireland

National target title: By 2030, address key issues in relation to the Management of Deer in Ireland

Description: nan

Main policy measures: NPWS will continue to work with DAFM and all relevant stakeholders to develop recommendations with the aim of improving the effectiveness of managing wild deer in Ireland.

Aspects of the goal or target are covered: nan

Government: Ireland

National target title: By 2025, Ireland takes enhanced measures to safeguard against the risk of fraud and other indirect effects of its renewable transport fuels policy and targets for the use of biofuels, considering the potential high ILUC-risk and detrimental impact to global biodiversity.

Description: nan

Main policy measures: nan

Aspects of the goal or target are covered: nan

Government: Ireland

National target title: By 2030, the objectives of the NBAP, where relevant, are aligned with and integrated, within the statutory landuse plans of the Regional Assemblies and Planning Authorities and within LBAPs

Description: nan

Main policy measures: All Regional Spatial and Economic Strategies, City and County Development Plans, Local Area Plans and Local Biodiversity Action Plans shall be aligned with the objectives of the National

Aspects of the goal or target are covered: nan

Government: Ireland

National target title: By 2030, shared responsibility for the conservation of biodiversity acted on

Description: nan

Main policy measures: All Public Authorities and private sector bodies move towards no net loss of biodiversity through strategies, planning, mitigation measures, appropriate offsetting and/or investment in Blue-Green infrastructure

Aspects of the goal or target are covered: nan

Government: Ireland

National target title: By 2024, the Industrial Development Agency (IDA) has delivered on the biodiversity measures in its 2021-2024 strategy Driving Recovery and Sustainable Growth

Description: nan

Main policy measures: DETE will work with IDA Ireland to develop biodiversity measures across their property programme, in line with the commitment to biodiversity measures outlined in IDA's 2021-2024 strategy, Driving Recovery and Sustainable Growth

Aspects of the goal or target are covered: nan

Government: Ireland

National target title: By 2024, OPW is working to enhance biodiversity at National Historic Property sites

Description: nan

Main policy measures: OPW will conduct biodiversity audits at multiple sites, implement enhancements and recommendations, and share the data gathered

Aspects of the goal or target are covered: nan

Government: Ireland

National target title: By 2027, implementation of the National Restoration Plan is monitored

Description: nan

Main policy measures: DHLGH and all stakeholders across Government, will monitor implementation of the National Restoration Plan.

Aspects of the goal or target are covered: nan

Government: Ireland

National target title: By 2027, implementation of a National Restoration Plan has begun

Description: nan

Main policy measures: DHLGH and all stakeholders across Government, will put in place restoration measures as described in the National Restoration Plan, within the appropriate timeframes.

Aspects of the goal or target are covered: nan

Government: Ireland

National target title: By 2026, a National Restoration Plan is published

Description: nan

Main policy measures: NPWS and DAFM and other relevant stakeholders will work to align existing indicators and/or establish new ones for monitoring restoration of ecosystems. DHLGH, in collaboration with DAFM, OPW and DECC, and other relevant bodies, will identify synergies between nature restoration and climate change mitigation/adaptation and disaster prevention, and prioritise these measures. DHLGH, in collaboration with DAFM, OPW and DECC will engage with stakeholders and the public during the development of a National Restoration Plan. DHLGH, in collaboration with DAFM, OPW and DECC, will develop a National Restoration Plan .

Aspects of the goal or target are covered: nan

Government: Ireland

National target title: By 2026, Ireland has actively enabled and contributed to the ongoing achievement of OSPAR's North-East Atlantic Environment Strategy 2030 (NEAES)

Description: nan

Main policy measures: DHLGH will continue to work nationally, internationally with OSPAR contracting parties, and with external organisations and bodies to support and ensure effective delivery of the 12 strategic objectives and 54 operational objectives set out in OSPAR's North-East Atlantic Environment Strategy 2030

Aspects of the goal or target are covered: nan

Government: Ireland

National target title: By 2030, address key issues in relation to fire management and emerging wildfire issues in Ireland

Description: nan

Main policy measures: DAFM, NPWS and NFDM will continue to work with all relevant stakeholders to develop a national fire management strategy.

Aspects of the goal or target are covered: nan

Government: Ireland

National target title: By 2024, Enhanced implementation of the Habitats and Birds Directives

Description: nan

Main policy measures: NPWS will complete the selection and notification of sites for the protection of Annex habitats and species listed on the EU Habitats and Birds Directives. NPWS will publish detailed site-specific conservation objectives, along with the approach used, for all existing SACs and SPAs .

Aspects of the goal or target are covered: nan

Government: Ireland

National target title: By 2024, the Management of National Parks are underpinned by Management Plans

Description: nan

Main policy measures: Approve Management Plans for National Parks by 2024 in line with the NPWS Strategic Action Plan

Aspects of the goal or target are covered: nan

Government: Ireland

National target title: By 2027, the revised legislation arising from the NPWS review of the Wildlife Acts is in place

Description: nan

Main policy measures: NPWS will complete a review of Wildlife legislation; NPWS to publish legislation to provide a legal basis for National Parks.

Aspects of the goal or target are covered: nan

Government: Ireland

National target title: By 2025, the Strategic Action Plan resulting from the review of the NPWS is implemented

Description: nan

Main policy measures: NPWS will implement the Strategic Action Plan resulting from the NPWS Review

Aspects of the goal or target are covered: nan

Example China:

Government: China

National target title: 7. 生态空间保护

Description: 优化国土空间开发和保护格局，将生物多样性保护作为国土空间规划的重要内容。严守生态保护红线，加强生态保护红线人为活动管控，开展动态监测及保护成效评估，强化生态环境监督。加强对生物多样性保护优先区域的保护监督，筑牢重点生态功能区格局，完善重点生态功能区配套政策。优化海洋生态安全格局，完善围填海管控和岸线开发管控制度，严守大陆自然岸线保有率目标底线要求。充分衔接国土空间规划分区和用途管制等要求，完善全域覆盖的生态环境分区管控体系，建立差别化的生态环境准入清单。依托生态空间相关监督平台加强重要生态空间动态监测、评估和预警。将生物多样性影响评价纳入大型工程建设、资源开发利用等项目的管理要求，强化事前事中事后全过程监管。到2030 年，重要生态空间得到有效保护，自然生态系统的原真性和完整性得以保持，重要生态系统退化及栖息地丧失得到基本遏制。

Main policy measures: nan

Aspects of the goal or target are covered: nan

Government: China

National target title: 1. 生物多样性政策法规体系

Description: 加快生物多样性保护法治建设，加快出台国家公园法，持续推进野生动植物及其栖息地保护、生物安全、生物资源可持续利用、生物遗传资源获取与惠益分享、生态保护红线、自然保护地以及森林、草原、湿地、河湖、海洋等领域法律法规的制定修订工作，研究起草生物多样性相关传统知识保护条例，完善外来入侵物种名录和管理制度。完善生物多样性保护政策及制度体系，健全生态保护补偿制度，健全野生动物种群调控和致害补偿制度，完善生态环境损害赔

偿制度，完善打击野生动植物非法贸易制度，推行草原森林河流湖泊海湾休养生息，继续实施长江十年禁渔，健全耕地休耕轮作制度。鼓励各地因地制宜出台相应的生物多样性保护地方性法规政策。到2030 年，生物多样性保护及可持续利用相关政策法规全面建立。

Main policy measures: nan

Aspects of the goal or target are covered: nan

AI and Climate Change Discourse: What Opinions Do Large Language Models Present?

Marcelo Sartori Locatelli, Pedro Robles Dutenhefner, Arthur Buzelin, Pedro Alzamora, Yan Aquino, Pedro Bento, Samira Malaquias, Victoria Estanislau, Caio Santana, Lucas Dayrell, Marisa Affonso Vasconcelos, Wagner Meira Jr., Virgilio Almeida
Universidade Federal de Minas Gerais, Belo Horizonte, Brazil
{locatellimarcelo, arthurbuzelin, pedro.loures, yanaquino, pedro.bento, samiramalaquias, victoria.estanislau, caiosantana, lucasdayrell, marisavasconcelos, meira, virgilio}@dcc.ufmg.br, pedroroblesduten@ufmg.br

Abstract

Large Language Models (LLMs) are increasingly used in applications that shape public discourse, yet little is known about whether they reflect distinct opinions on global issues like climate change. This study compares climate change-related responses from multiple LLMs with human opinions collected through the People's Climate Vote 2024 survey (UNDP – United Nations Development Programme and Oxford, 2024). We compare country and LLM's answer probability distributions and apply Exploratory Factor Analysis (EFA) to identify latent opinion dimensions. Our findings reveal that while LLM responses do not exhibit significant biases toward specific demographic groups, they encompass a wide range of opinions, sometimes diverging markedly from the majority human perspective.

1 Introduction

Climate change is one of the most pressing global challenges of our time, shaping policy decisions, influencing public behavior, and driving scientific inquiry (Lahsen and Ribot, 2022). Public opinion plays a critical role in guiding both governmental decisions and societal responses, making its assessment indispensable for understanding the support and resistance that can influence policy effectiveness and climate action. In this context, surveys serve as fundamental tools, providing valuable insights into the diverse perspectives shaping climate discourse and enabling policymakers to craft more responsive and effective climate strategies (Shi et al., 2015). With major summits like G20 and COP30 approaching in 2025, where sustainability and climate change will be central topics (Wonneberger et al., 2020; Lochner et al., 2024), gauging public sentiment is crucial to inform discussions, anticipate challenges, and align policies with public expectations.

Recently, as artificial intelligence technologies advance, LLMs become key players in public opinion formation and information dissemination, as their integration to major search engines – such as Google and Bing – continues to expand (Costello et al., 2024). AI-generated responses frequently precede traditional search results, many of which are also algorithmically curated (Dai et al., 2023). By shaping public discourse, reflecting societal perspectives, and anticipating emerging trends (Yakura et al., 2024; Faruk, 2024), LLMs play a crucial role in how information is accessed and interpreted. Given their widespread reach, critically examining the biases they introduce and reinforce is essential.

Understanding how these models portray critical topics is not merely a technical concern, but a critical factor in assessing their impact on public perception and societal narratives (Wan et al., 2023; Motoki et al., 2024). Researchers caution that, due to LLMs being predominantly trained on data from Western and high-income countries, these models may inherently amplify the perspectives of these regions while also reflecting and perpetuating biases related to race and gender. This can lead to an oversimplification of complex societal issues (Atari et al., 2023; Cheng et al., 2023).

Therefore, assessing the alignment of LLMs in climate-related contexts is crucial. Evaluating their tendencies and biases helps determine their influence on climate narratives and broader societal and political discourses. Comparing their outputs with human opinions across different countries can provide valuable insights into how these models engage with climate discourse (Lee et al., 2024a).

In this study, we aim to examine the perspectives that large language models adopt when generating climate change-related responses. In particular, we assess which opinions their outputs reflect. Since different LLMs are trained on diverse datasets, rely on different algorithms, and are subject to distinct biases (Feng et al., 2023), discrepancies in the information they provide are expected. To address

Proceedings of the 2nd Workshop on Natural Language Processing Meets Climate Change (ClimateNLP 2025), pages 113–125

these concerns, we define the following research questions:

RQ1: To what extent do LLM responses align with different countries and geopolitical groups in climate change surveys?

RQ2: How does prompting LLMs to adopt a given citizenship influence their alignment with human responses?

RQ3: How do LLMs respond to climate-related questions, and what factors influence these responses?

We use responses from the People's Climate Vote 2024 survey (UNDP – United Nations Development Programme and Oxford, 2024), covering 77 countries, as a benchmark to evaluate eight LLMs, including both open-source and proprietary models from diverse companies and regions. The survey consists of closed-ended questions with predefined choices, which we present to the LLMs, instructing them to select the corresponding alternatives. This approach enables us to analyze token probability distributions and measure how closely their outputs align with human responses. Figure 1 illustrates the process of obtaining and evaluating LLM and human responses, highlighting the comparison and analysis framework.

Our findings reveal that, while LLMs do not exhibit systematic biases toward specific geopolitical or demographic groups, their responses often diverge significantly from majority human opinions. In particular, we found that LLMs generally express greater concern about climate change, especially regarding future risks and long-term policy commitments, than the average human respondent. However, their alignment with human perspectives on immediate climate actions varies, with some models displaying notable discrepancies. Additionally, prompting LLMs to adopt a national identity sometimes reduces divergence, but the effect is inconsistent across countries and models. These results highlight the distinct role that LLMs play in shaping climate discourse and underscore the need for careful evaluation of their potential biases and influence on public narratives.

2 Related Work

Understanding the opinions held by large language models (LLMs) has become a key area of study. Santurkar et al. (2023) proposes a framework to evaluate LLM alignment with public opinion, finding significant misalignment with U.S. views, especially in models fine-tuned with human feedback. Similarly, Durmus et al. (2024) compares LLM-generated survey responses with data from the World Values and the PEW Surveys, revealing stronger alignment with opinions in Western and South American countries. They also note that LLMs tend to assign disproportionately high probabilities to single responses, in contrast to the more diverse distributions seen in human responses.

Numerous studies have examined LLM biases across critical topics, like gender (Kotek et al., 2023), cultural perspectives (Naous et al., 2024), standardized tests (Locatelli et al., 2024), and political alignment (Motoki et al., 2024). Recent research has focused on how LLMs simulate public opinion on climate change, with studies like Wan et al. (2023) highlighting misrepresentation of demographic diversity and potential harms such as identity essentialization. Jansen et al. (2023) and Demszky et al. (2023) emphasize that LLMs are not yet reliable substitutes for human survey respondents, often misrepresenting demographic diversity. Additionally, Lee et al. (2024b) investigates social desirability response bias (SDR) in LLMs, finding limited bias with models maintaining consistent responses across varying demographic prompts.

Regarding climate change, Lee et al. (2024a) finds that GPT-based models reflect liberal, higher-income, and highly educated views, but struggle to represent beliefs of non-Hispanic Black Americans. Expanding beyond the U.S., Qu and Wang (2024) identifies regional disparities and biases based on demographic factors and ideological stances.

Our work extends on prior research by analyzing a broader set of LLMs and expanding the geographical scope of climate change simulations. We assess how these models align with human opinions and uncover which point of view they are propagating.

3 Survey Dataset

The survey used in this study is the Peoples' Climate Vote 2024 (UNDP – United Nations Development Programme and Oxford, 2024), the world's largest standalone public opinion survey on climate change. This edition introduced 15 questions organized into three main themes: (1) the direct effects of climate change on daily life, (2) how climate change is being addressed in the participant's coun-

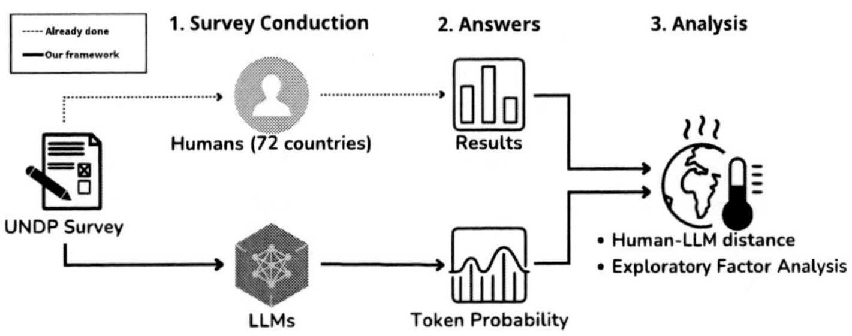

Figure 1: Diagram summarizing the proposed methodology for obtaining and evaluating responses.

try, and (3) preferences for future policy actions[1].

Administered by GeoPoll using Computer Assisted Telephone Interviewing (CATI) and Random Digit Dialing (RDD) methodologies, the survey was conducted in 87 languages, enabling the participation of a broad spectrum of demographic groups. Sample sizes per country typically ranged from 900 to 1,500 respondents, yielding in a total of 73,765 completed interviews from 1.9 million calls across 77 different countries. However, responses from only 72 countries and a global summary were available in the survey dataset.

The dataset provides a structured representation of survey responses, including the distribution of human responses for each alternative across all questions. Each entry contains the full question text, multiple-choice options, and respondents' demographic attributes, like age and education level.

4 Methodology

We evaluate LLM responses by submitting each survey question and its predefined answer choices, prompting the model to select a single-letter response. This allows us to extract log probabilities for each option, which we normalize into a probability distribution for comparison with human responses. Our analysis includes three key components: (i) measuring distributional distance using Jensen-Shannon divergence to compare model outputs with public opinion across countries; (ii) conducting Exploratory Factor Analysis to identify underlying factors influencing responses; and (iii) performing sentence embedding analysis to examine whether LLMs favor answer choices semantically closer to the question in the embedding space.

[1]https://peoplesclimate.vote/about

4.1 Selection of Large Language Models

To ensure a representative analysis, we selected both open-source and proprietary LLMs from diverse companies and countries to assess their alignment with human opinions across different regions. We included GPT-4o as a state-of-the-art LLM, DeepSeek and Qwen as Asian models, LLaMA, Phi, and Grok as U.S.A. representatives, and Mistral as a European counterpart. Open-source models were executed in local machines, while proprietary models were accessed via API.

4.2 Prompts for Multiple-Choice Questions

The prompting strategies in this study simulate real-world scenarios. We employed a zero-shot approach, allowing models to leverage their natural language and contextual understanding to handle unfamiliar questions. Each prompt consists of an instruction explicitly requesting the model to respond with a single letter corresponding to the selected answer, followed by the question and its predefined answer choices. All prompts were written in English, matching the language used in the survey. The prompt used in this study can be found in the Appendix C.

Consistent with current literature (Argyle et al., 2023), we set the models temperature to 0.7 to balance deterministic responses with moderate variability. Additionally, we imposed a strict token limit of 1 to ensure that only a single token—the model's answer – was generated. This setup enabled us to extract log probabilities or logits for the predicted token directly.

To obtain the probability distribution of the model's responses, we first extract the logits for all tokens in the vocabulary, which represent the unnormalized scores assigned to each token. From the logits, we select only those corresponding to the predefined answer choices ("A", "B", "C", "D", etc.). Additionally, we apply a *strip* process to remove leading and trailing whitespace from tokens, ensuring that variations like " A" and "A" are treated identically.

To convert the selected logits into probabilities, we apply the softmax function, which normalizes the values into a probability distribution where all probabilities range between 0 and 1 and sum to 1. The softmax function is defined as:

$$P(y_i) = \frac{e^{z_i}}{\sum_{j=1}^{N} e^{z_j}}$$

where z_i represents the logit for answer choice i, and N is the total number of answer choices.

4.3 Measuring Distances Between Human And LLM Responses

Following prior literature (Locatelli et al., 2024), we use Jensen-Shannon distance as the primary metric to quantify the differences between human responses and those generated by LLMs. Applying a base-2 logarithm, this metric is bounded within the interval $[0, 1]$, enhancing its interpretability.

Let Q be a set of size N representing a collection of survey questions, and let A_q denote the set of possible answers for each question q. Since we extract the log probabilities for the tokens corresponding to the answer choices, we define the following for LLMs:

$$P_m(a|q), \forall a \in A_q, q \in Q, m \in M$$

where m refers to a specific model in our study, and $P_m(a|q)$ is the probability of model m answering question q with alternative a.

Analogously, for human responses, we also have the probability for each possible answer, which is available in the survey data. Thus, we define:

$$P_H(a|q), \forall a \in A_q, q \in Q$$

where $P_H(a|q)$ represents the probability of humans answering question q with alternative a, and this probability distribution is available for each country in the survey.

The distance between human responses and a model m is then calculated as the mean of the Jensen-Shannon distances across all questions:

$$Distance(m, H) = \frac{1}{N} \sum_{q=1}^{N} JS(P_m(A_q|q), P_H(A_q|q))$$

A larger distance indicates a greater divergence between the model and human distributions, while smaller values suggest stronger alignment.

4.4 Evaluating Question-Level Contributions to Global Alignment

To assess the structure of alignment between LLM-generated responses and human opinions at a more granular level, we applied the DISTATIS method (Abdi et al., 2005) to the distance matrix derived from each individual survey question. This approach allows us to combine multiple distance matrices into a shared structure, assigning a weight to each question based on its contribution to the global similarity pattern. Higher weights indicate that a question's distance matrix not only aligns more closely with the overall trend, but also contributes more significantly to shaping the shared structure. In contrast, lower weights suggest that a question's distance relationships deviate more from the common pattern, exerting a smaller influence on the global alignment. We leverage this analysis to evaluate the extent to which each individual question influenced the overall alignment between LLM-generated responses and human opinions, allowing us to identify which questions deviate from the global behavior.

4.5 Exploring Latent Factors in Climate Change Opinions

To understand the underlying structure of climate change opinions, we employ Exploratory Factor Analysis (EFA), a widely used statistical technique in social sciences (Teo, 2014). EFA identifies latent factors that explain patterns of correlation among observed variables, assuming that responses to individual items are influenced by these underlying dimensions. By analyzing response patterns, EFA reveals the structure of opinions in a dataset, reducing complexity while preserving key relationships.

In the context of a climate change survey, these latent factors group countries with similar response distributions on related questions. By interpreting these factors, we gain insights into the values of citizens of different countries and the opinions that large language models might generate.

For the EFA, we first construct a matrix based on our observations. Since most survey questions have ordinal answers, we assigned a value to each alternative ranging from 1 to $|A_q|$, where $|A_q|$ denotes the number of alternatives for question q. Next, we calculated the weighted average score for each

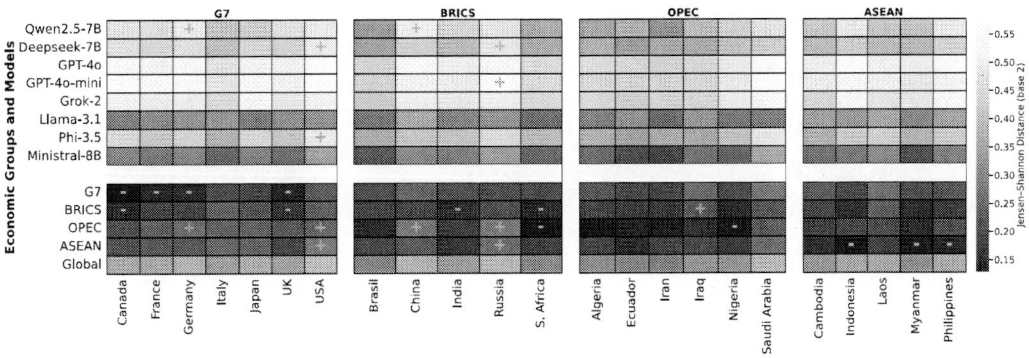

Figure 2: Distance between each country and group/LLM. Distances marked with "-" are lower than the mean distance for that row, while those marked with "+" are significantly higher at significance level 0.05. Note that the country responses are removed from its group's distribution when calculating the distance between the two. A full version including all countries is available in Appendix B.

observation (country or LLM) across question alternatives. This results in an 81x14 matrix, where each row represents a country or LLM, and each column corresponds to the average score for a given question. To simplify the process, we exclude one question whose alternatives were not ordinal. Applying EFA to this matrix reveals latent factors that capture climate-related opinion patterns, providing a more interpretable representation of potential alignments or divergences between LLMs and human respondents across regions.

We use the Factor software (Lorenzo-Seva and Ferrando, 2006), applying unweighted least squares as the optimization method and *promin* rotation to maximize factor simplicity. To determine the optimal number of factors, we use parallel analysis, comparing eigenvalues from our observations with those from Monte-Carlo simulated random data (Allen, 2017).

5 Results

In this section, we present the results of our proposed methodology, addressing the research questions (RQs) posed earlier. The results are organized as follows: Section 5.1 analyzes the distances between probabilities distributions of LLM and human responses, Section 5.2 investigates the effect of conditioning the LLM to be more similar to specific countries, Section 5.3 delves into the characteristics of individual questions, and, finally, Section 5.4 explores the alignment between LLM-generates opinions and human values.

5.1 Assessing LLM Alignment with Regional and Geopolitical Groups

To assess how closely LLM responses align with different human populations, we analyze the distances between models, geopolitical groups, and individual countries. Figure 2 presents these distances, where the probability distribution for each geopolitical organization group is obtained by averaging the distributions of all countries within that group, excluding the country of interest. This grouping approach enhances visualization and interpretability. For instance, when calculating the distance between G7 and the United States, U.S. responses are excluded from the G7 distribution, allowing us to assess whether LLMs align more closely with specific regions or geopolitical groups.

We selected the G7[2], BRICS[3], OPEC[4], and ASEAN[5] as representative geopolitical groups, given their diverse economic and political perspectives. These groups provide a broader context for evaluating alignment patterns. The distributions for each group were obtained by averaging the answers from each of its members.

When comparing LLMs responses to human responses, we find no clear evidence of alignment with any specific group. If LLMs strongly aligned with a group, we would expect significantly lower distances compared to the average for countries in that group. Instead, the distances remain relatively high, suggesting that LLMs do not show a sys-

[2]Canada, France, Germany, Italy, Japan, the United Kingdom and the United States.

[3]Brazil, Russia, India, China and South Africa.

[4]Algeria, Ecuador, Iran, Iraq, Nigeria and Saudi Arabia.

[5]Cambodia, Indonesia, Laos, Myanmar and Philippines.

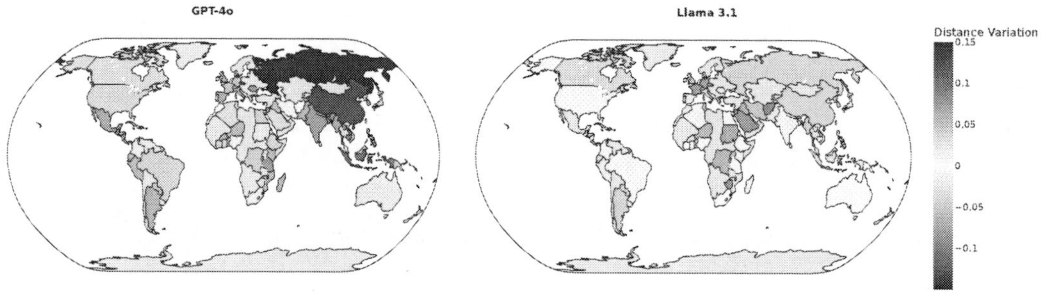

Figure 3: Variation in the distance between each surveyed country and GPT-4o/LLaMa 3.1 8B when prompting the model to respond as a citizen of that country. Positive values (red) indicate greater divergence, while negative values (green) suggest improved alignment with human responses.

tematic preference for any geopolitical or regional group.

This is further corroborated by the fact that the standard deviation in similarity between LLM and human responses across individual survey questions is relatively high ($\sigma > 0.12$ for the distance between LLMs and their closest country responses), indicating inconsistency in model predictions when compared to human populations. Additionally, the minimum distance between an LLM and its closest country is higher than the distance between two geographically or culturally similar countries (median for LLMs = 0.34, median for countries = 0.13).

These findings challenge the assumption that LLMs may be biased towards certain populations, such as Western or developed countries. In the context of climate change, our analysis provide no strong evidence of such biases. Instead, the results suggest LLM-generated response distributions do not closely resemble human distributions in general. Nevertheless, some models generate responses that significantly diverge from those of the populations they might be expected to represent. For example, the responses generated by Chinese LLM Qwen2.5 differ notably from those provided by Chinese citizens.

5.2 LLMs as Virtual Citizens: Can LLMs Adapt to Country-Specific Beliefs?

Since large language models do not inherently produce responses that align with the answer distributions of any specific country, we explored whether prompting techniques could encourage more human-like responses. To test this, we instruct the LLM to act as that countries' (country X) citizen (see Appendix C).

We then measure the impact of this intervention by comparing the distance between the model's new responses and those of country X. Figure 3 shows the change in distance before and after applying this prompt, referred to as *distance variation*. This variation is computed as:

$$\Delta_{\text{dist}}(m, H_X) = \text{dist}(H_X, m_X) - \text{dist}(H_X, m),$$

where H_X represents the human response distribution for country X, m denotes the default LLM response distribution, and m_X corresponds the LLM's response when prompted to act as a citizen of country X.

A positive $\Delta_{\text{dist}}(m, H_X)$ means the customized prompt increased the distance to human responses, whereas a negative value suggests better alignment. This analysis is limited to GPT-4o and LLaMa 3.1 8B Instruct for brevity.

Our findings reveal that, for both models – particularly GPT-4o – assigning a national identity for the LLM to mimic often reduced the distance to the target country. However, in certain cases (e.g. Pakistan), the intervention failed to bring the model's responses closer to human distributions. In some instances, it even increased the divergence, suggesting that the effectiveness of this approach varies depending on the model and the country.

Moreover, LLaMa 3.1 8B failed to reduce its distance to several African, Middle Eastern, and South Asian countries. This may derive from biases in the model's training data, as well as its relatively small number of parameters. The representation of multilingual content in the training corpus, estimated at around 8% (Grattafiori et al., 2024), could have contributed to weaker alignment with regional human responses. Additionally, its reduced model capacity may limit its ability to capture complex cultural nuances.

These results suggest that prompting LLMs to mimic a nation's citizen can sometimes improve

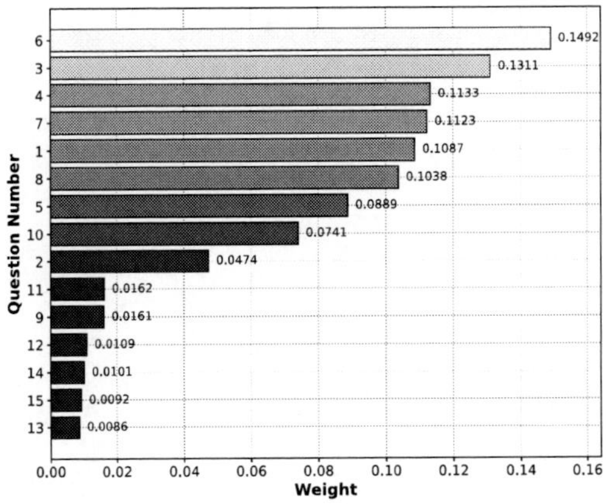

Figure 4: Weights assigned to each survey question using the DISTATIS method.

Model	F1	F2	F3
Deepseek 7b-chat	5.06	3.99	5.19
GPT-4o	4.84	5.63	4.36
GPT-4o-mini	4.91	4.18	4.05
Grok-2	4.77	7.73	5.87
LLaMa-3.1 8B-Instruct	6.12	7.10	8.53
Phi-3.5-mini-Instruct	4.99	4.82	6.03
Ministral 8B-Instruct	5.38	6.84	6.36
Qwen2.5 7B-Instruct	4.88	6.42	4.44

Table 1: Factor scores for each tested LLM. Cells highlighted in red represent values in the top 10%, while those in green represent the bottom 10%, including the countries. Due to the scale we adopt for the answers, a lower value on a factor indicates that the model is more concerned with that aspect of climate change.

alignment with human responses, but the effect is inconsistent across models and regions. Differences in model architecture, training data, and parameter count likely contribute to variations, while increased divergence in certain countries highlights the risks of misrepresentation. This underscores the importance of careful evaluation when using LLMs to simulate national public opinion.

5.3 Question-Level Contributions to LLM-Human Alignment

The results reveal substantial variation in how different survey questions contribute to the overall similarity structure. As shown in Figure 4, question 6, that addresses governmental effectiveness in climate action and question 3, related to concerns for future generations exhibit the highest contributions, suggesting that LLM responses on these topics align more consistently with the overall distance between humans and LLMs.

In contrast, question 15, related to international cooperation and question 13, about educational efforts present the lowest contributions, which suggest that regarding these topics they present a different answer pattern from the one presented by the global similarity pattern. This can be confirmed by looking at the distance matrix of these two questions and noting that it describes a much smaller distance between models and countries, specifically, the tested models seem to be express opinion much closer to the developing responses than on other questions.

5.4 Exploring the Opinions of LLMs on Climate Change

In the previous sections, we found that LLM answer distributions, even when prompted to simulate responses as citizens of specific countries, had very inconsistent alignment with those of human groups. This suggests that the models do not exhibit a strong bias towards any national perspective on climate change issues. However, this analysis alone does not reveal the underlying opinions the models may be expressing. To address this, we now turn to an Exploratory Factor Analysis (EFA) to better understand the models' perspectives on climate change.

Three factors were identified as significant in our analysis (Kaiser-Meyer-Olkin (KMO) test = 0.788, 69.4% explained variance), suggesting that our data is suitable for factor analysis. The full factor loadings are available in Appendix A. By examining the associations between factors and individual survey questions, we found that each factor aligned with one of the main themes of the survey presented in Section 3. Since these themes emerged from question groupings, we defined the factor labels *a posteriori* as:

F1: Future Actions: Concerns about long-term climate policies and commitments.

F2: Present Actions: Focus on immediate efforts and measures to address climate change.

F3: Climate Change and Daily Life: The perceived impact of climate change on everyday life and personal experiences.

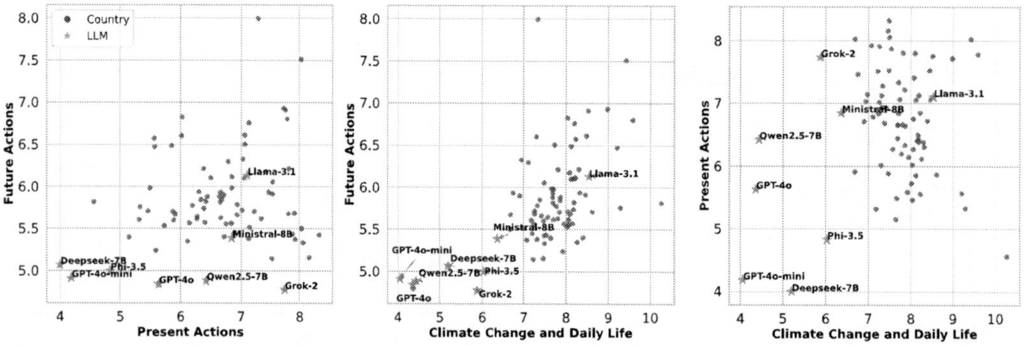

Figure 5: Scatter plots comparing factor scores for countries and LLMs. (a) Future Actions vs. Present Actions: LLMs consistently express concern for future actions, but show variation on present actions. (b) Future Actions vs. Climate Change and Daily Life: LLMs exhibit diverse positions, with LLaMa showing lower climate anxiety than most countries. (c) Present Actions vs. Climate Change and Daily Life: LLM responses differ significantly from human populations, highlighting their distinct perspectives.

Using the weighted sum method (DiStefano et al., 2019), we calculated factor scores for each country and LLM, reflecting their responses to each question. Table 1 presents the scores for LLMs analyzed. A lower score on a given factor suggests that the model is more likely to provide favourable responses to questions related to that factor.

Among the models examined, the GPT-4 family stood out as the most likely to acknowledge the impact of climate change and the importance of government actions across all factors, followed by Phi. In general, we found that LLMs expressed more concern about climate change's effects on daily lives (F3) and future actions (F1) than the average human from most countries. This was not the case for present actions (F2), where LLMs factor scores, except for Phi and GPT-4, aligned more closely with human responses. Notably, Mistral and LLaMA showed the most divergent responses: both models tended to provide more negative assessments regarding present and future actions, but differed on their stance towards F3–LLaMA being more negative than most countries, and Mistral more positive, aligning with other LLMs.

Having analyzed the performance of the models relative to each other, we now compare their responses to human answers. Figure 5 show the positions of the LLMs relative to the countries on these factors. Most models are clear outliers in relation to the factor values, positioning themselves relatively far from the countries' distributions. Even the models that are not clear outliers – LLaMa and Mistral – appear on the border of the cloud of countries, suggesting that the opinions they generate may differ significantly from those of most countries.

In practice, this highlights how unusual the answer distribution from LLMs are when compared to humans, especially when considering the combination of factors. Although some of the concerns of the large language models, in the form of factor scores, individually may approach the opinions of some countries, when assessing all three factors, we notice that the generated response distributions are inconsistent with existing countries.

6 Conclusion

As large language models gain widespread use, understanding the nature of the opinions they generate is crucial, particularly in sensitive areas like climate change. Our analysis of responses from eight LLMs compared with human answers from the People's Climate Vote 2024 survey, reveals that LLMs generally express greater concern about climate change than average human, with their responses differing significantly from human groups.

Furthermore, the higher levels of concern observed in LLM responses may be linked to various stages of model training, though the lack of transparency in training data complicates the identification of specific causes. Future research could explore the impact of these factors on LLM-generated opinions.

It is still unclear whether LLMs should mimic the public opinion or the expert opinion on a given topic. In this study, we focus solely on the first, finding that there is currently little alignment between model generated and people's response on climate change. Nevertheless, future work should explore the latter, as it can be argued that this technology should be used to gently steer people's opinions towards the scientific consent on pressing world problems.

7 Implications

As the use of LLMs as substitutes for human participants in surveys becomes increasingly debated (Jansen et al., 2023), it is crucial to be aware of the limitations these models have when representing diverse groups. As our analysis shows, the answers distributions generated by these models are considerably different from those of humans, and mitigation techniques such as prompting the model to adopt the role of a specific demographic group can only go so far, potentially without risking representational harms.

Another point to consider is that even between LLMs, their answer distributions may vary greatly, and, in some cases, this can lead them to express different views on specific issues. For instance, the degree to which each model values **F2:Present Actions** is significantly different, with LLama-3.1 and Grok-2 showing much higher scores when compared to GPT-4o-mini and Deepseek-7B.

As an user, it is hard to know which kind of bias or point of view an LLM may display a priori and one may be influenced without even realizing. With the trend in decreasing information in LLM model cards, especially in sections related to bias and limitations (Liang et al., 2024), and the sheer number of different models, it is hard to know what kind of information one may receive when interacting with a LLM-powered application. Large language model providers should be encouraged to provide accurate and transparent documentation that can inform the end users of the expected outputs of their products.

Limitations

In our study, we aim to represent a diverse range of cultures by examining the countries available in the Peoples' Climate Vote 2024 survey. However, this focus on countries means we do not account for within-country demographic variations. LLM responses may align closely with specific age, education, gender, religion or other demographic groups, which we leave for future work to explore. For the model selection, we analyze eight widely-used models from diverse companies and countries of origin. However, other state-of-art models, such as Deepseek-V3 and Claude-3, or models tailored for specific languages, could provide valuable insights. Additionally, versions of models used in the study with more parameters, such as LLaMa 3.1 405B Instruct, may offer further improvements. Finally, while we assess model opinions using controlled prompts and survey questions, our findings may not fully reflect the responses these models would generate in real-world applications.

Acknowledgements

This work was partially funded by CNPq, CAPES, FAPEMIG, and IAIA - INCT on AI.

References

H. Abdi, A.J. O'Toole, D. Valentin, and B. Edelman. 2005. Distatis: The analysis of multiple distance matrices. In *2005 IEEE Computer Society Conference on Computer Vision and Pattern Recognition (CVPR'05) - Workshops*, pages 42–42.

Mike Allen. 2017. *The SAGE encyclopedia of communication research methods*. SAGE publications.

Lisa P. Argyle, Ethan C. Busby, Nancy Fulda, Joshua R. Gubler, Christopher Rytting, and David Wingate. 2023. Out of one, many: Using language models to simulate human samples. *Political Analysis*, 31(3):337–351.

Mohammad Atari, Mona J Xue, Peter S Park, Damián Blasi, and Joseph Henrich. 2023. Which humans?

Myra Cheng, Esin Durmus, and Dan Jurafsky. 2023. Marked personas: Using natural language prompts to measure stereotypes in language models. In *Proceedings of the 61st Annual Meeting of the Association for Computational Linguistics (Volume 1: Long Papers)*, pages 1504–1532.

Thomas H. Costello, Gordon Pennycook, and David G. Rand. 2024. Durably reducing conspiracy beliefs through dialogues with ai. *Science*, 385(6714):eadq1814.

Sunhao Dai, Yuqi Zhou, Liang Pang, Weihao Liu, Xiaolin Hu, Yong Liu, Xiao Zhang, and Jun Xu. 2023. Llms may dominate information access: Neural retrievers are biased towards llm-generated texts. *arXiv preprint arXiv:2310.20501*.

Dorottya Demszky, Diyi Yang, David S Yeager, Christopher J Bryan, Margarett Clapper, Susannah Chandhok, Johannes C Eichstaedt, Cameron Hecht, Jeremy Jamieson, Meghann Johnson, et al. 2023. Using large language models in psychology. *Nature Reviews Psychology*, 2(11):688–701.

Christine DiStefano, Min Zhu, and Diana Mindrila. 2019. Understanding and using factor scores: Considerations for the applied researcher. *Practical assessment, research, and evaluation*, 14(1):20.

Esin Durmus, Karina Nguyen, Thomas Liao, Nicholas Schiefer, Amanda Askell, Anton Bakhtin, Carol Chen, Zac Hatfield-Dodds, Danny Hernandez, Nicholas Joseph, Liane Lovitt, Sam McCandlish, Orowa Sikder, Alex Tamkin, Janel Thamkul, Jared Kaplan, Jack Clark, and Deep Ganguli. 2024. Towards measuring the representation of subjective global opinions in language models. In *First Conference on Language Modeling*.

Tanjim Bin Faruk. 2024. Evaluating the performance of large language models in scientific claim detection and classification. *arXiv preprint arXiv:2412.16486*.

Shangbin Feng, Chan Young Park, Yuhan Liu, and Yulia Tsvetkov. 2023. From pretraining data to language models to downstream tasks: Tracking the trails of political biases leading to unfair NLP models. In *Proceedings of the 61st Annual Meeting of the Association for Computational Linguistics (Volume 1: Long Papers)*, pages 11737–11762, Toronto, Canada. Association for Computational Linguistics.

Aaron Grattafiori, Abhimanyu Dubey, Abhinav Jauhri, Abhinav Pandey, Abhishek Kadian, Ahmad Al-Dahle, Aiesha Letman, Akhil Mathur, Alan Schelten, Alex Vaughan, Amy Yang, Angela Fan, et al. 2024. The llama 3 herd of models. *Preprint*, arXiv:2407.21783.

Bernard J. Jansen, Soon gyo Jung, and Joni Salminen. 2023. Employing large language models in survey research. *Natural Language Processing Journal*, 4:100020.

Hadas Kotek, Rikker Dockum, and David Sun. 2023. Gender bias and stereotypes in large language models. In *Proceedings of the ACM collective intelligence conference*, pages 12–24.

Myanna Lahsen and Jesse Ribot. 2022. Politics of attributing extreme events and disasters to climate change. *Wiley Interdisciplinary Reviews: Climate Change*, 13(1):e750.

Sanguk Lee, Tai-Quan Peng, Matthew H. Goldberg, Seth A. Rosenthal, John E. Kotcher, Edward W. Maibach, and Anthony Leiserowitz. 2024a. Can large language models estimate public opinion about global warming? an empirical assessment of algorithmic fidelity and bias. *PLOS Climate*, 3(8):1–14.

Sanguk Lee, Kai-Qi Yang, Tai-Quan Peng, Ruth Heo, and Hui Liu. 2024b. Exploring social desirability response bias in large language models: Evidence from gpt-4 simulations. *arXiv preprint arXiv:2410.15442*.

Weixin Liang, Nazneen Rajani, Xinyu Yang, Ezinwanne Ozoani, Eric Wu, Yiqun Chen, Daniel Scott Smith, and James Zou. 2024. Systematic analysis of 32,111 ai model cards characterizes documentation practice in ai. *Nature Machine Intelligence*, 6(7):744–753.

Marcelo Sartori Locatelli, Matheus Prado Miranda, Igor Joaquim da Silva Costa, Matheus Torres Prates, Victor Thomé, Mateus Zaparoli Monteiro, Tomas Lacerda, Adriana Pagano, Eduardo Rios Neto, Wagner Meira Jr., and Virgilio Almeida. 2024. Examining the behavior of llm architectures within the framework of standardized national exams in brazil. *Proceedings of the AAAI/ACM Conference on AI, Ethics, and Society*, 7(1):879–890.

Jakob H Lochner, Annika Stechemesser, and Leonie Wenz. 2024. Climate summits and protests have a strong impact on climate change media coverage in germany. *Communications Earth & Environment*, 5(1):279.

Urbano Lorenzo-Seva and Pere J Ferrando. 2006. Factor: A computer program to fit the exploratory factor analysis model. *Behavior research methods*, 38(1):88–91.

Fabio Motoki, Valdemar Pinho Neto, and Victor Rodrigues. 2024. More human than human: measuring chatgpt political bias. *Public Choice*, 198(1):3–23.

Tarek Naous, Michael J Ryan, Alan Ritter, and Wei Xu. 2024. Having beer after prayer? measuring cultural bias in large language models. In *Proceedings of the 62nd Annual Meeting of the Association for Computational Linguistics (Volume 1: Long Papers)*, pages 16366–16393, Bangkok, Thailand. Association for Computational Linguistics.

Yao Qu and Jue Wang. 2024. Performance and biases of large language models in public opinion simulation. *Humanities and Social Sciences Communications*, 11(1):1095.

Nils Reimers and Iryna Gurevych. 2019. Sentence-bert: Sentence embeddings using siamese bert-networks. In *Proceedings of the 2019 Conference on Empirical Methods in Natural Language Processing*. Association for Computational Linguistics.

Shibani Santurkar, Esin Durmus, Faisal Ladhak, Cinoo Lee, Percy Liang, and Tatsunori Hashimoto. 2023. Whose opinions do language models reflect? In *Proceedings of the 40th International Conference on Machine Learning*, volume 202 of *Proceedings of Machine Learning Research*, pages 29971–30004. PMLR.

Jing Shi, Vivianne HM Visschers, and Michael Siegrist. 2015. Public perception of climate change: The importance of knowledge and cultural worldviews. *Risk Analysis*, 35(12):2183–2201.

Timothy Teo. 2014. *Handbook of quantitative methods for educational research*. Springer Science & Business Media.

UNDP – United Nations Development Programme and University of Oxford. 2024. Peoples' climate vote 2024. *UNDP (United Nations Development Programme)*.

Yixin Wan, George Pu, Jiao Sun, Aparna Garimella, Kai-Wei Chang, and Nanyun Peng. 2023. "kelly is a warm person, joseph is a role model": Gender biases in LLM-generated reference letters. In *Findings of the Association for Computational Linguistics: EMNLP 2023*, pages 3730–3748, Singapore. Association for Computational Linguistics.

Anke Wonneberger, Marjolein H. C. Meijers, and Andreas R. T. Schuck. 2020. Shifting public engagement: How media coverage of climate change conferences affects climate change audience segments. *Public Understanding of Science*, 29(2):176–193.

Hiromu Yakura, Ezequiel Lopez-Lopez, Levin Brinkmann, Ignacio Serna, Prateek Gupta, and Iyad Rahwan. 2024. Empirical evidence of large language model's influence on human spoken communication. *arXiv preprint arXiv:2409.01754*.

A Factor Loadings for Survey Questions

Table 2 shows the factor loadings for each question. The absolute value of the factor loading indicates how related that question is to the factor. For example, the question "Should your country strengthen or weaken its commitments to address climate change?" is highly associated with the factor **(F1) Future Actions**.

B Complete Jensen-Shannon Distances

Figure 6 shows the mean Jensen-Shannon distance between all surveyed countries and the studied LLMs. Note that the addition of the extra countries adds little information: they are generally further away from the LLMs when compared to the economical/geopolitical groups.

C System Prompt

The following system prompt was used to standardize the responses generated by the LLMs:

> "You will receive a question. You **MUST** respond with only one letter. The possible answers will be presented as follows: A: answer, B: answer, C: answer, etc. You should respond **ONLY** with the letter corresponding to the correct alternative according to you. Do not provide explanations, additional text, or repeat the answer—just the letter."

This prompt ensured that all models produced structured and comparable outputs, facilitating a consistent evaluation of their alignment with human responses.

We append the following instruction to the prompt in order to conduct the analysis proposed in section 5.2, where country X stands for any country we wish the LLM to mimic:

> "You must answer the following question as if you were a typical citizen of **{country X}**, reflecting the general opinions, beliefs, and cultural perspectives of people from that nation."

D Question-Level Contributions Distances

In this section, we present the distances between each country and its group/LLM for Question 6: *"How well is your country addressing climate change?"*. This analysis helps to understand how each country perceives its own efforts in addressing climate change relative to others. The results are shown in Figure 7.

E Semantic Proximity in LLM Responses

Large Language Models rely on internal text representations to generate responses. This raises the question of whether their answer choices are influenced by semantic proximity in the embedding space. To explore this, we analyze if LLMs tend to favor answer choices closer to the questions in a pre-trained sentence embedding space.

For this analysis, we use a pre-trained sentence embedding model to encode both survey questions and answer choices into a shared embedding space. Specifically, we adopt the SentenceTransformer (Reimers and Gurevych, 2019), a bidirectional, encoder-only transformer model. Each question is encoded as a single vector, and each answer choice is separately encoded into the same space.

To assess whether LLMs are more likely to select answer choices semantically closer to the question in embedding space, we computed the correlation between the distance of each answer choice to the question and its selection probability. Figure 8 presents these correlations for the studied models.

The results indicate a clear negative correlation across all LLMs, with values ranging from approximately between -0.30 to -0.55. This suggests that the closer an answer choice is to the question in embedding space, the more likely the model is to select it. While the strength of this effect varies across models, the consistent trend implies that semantic proximity plays a significant role in shaping LLM predictions.

ID	Question Text	(F1) Future Actions	(F2) Present Actions	(F3) Climate Change and Daily Life
1	How often do you think about climate change?	0.097	0.333	0.732
2	Compared with last year, are you more or less worried about climate change?	-0.018	-0.123	0.806
3	How worried are you about the effects of climate change on the next generation?	-0.053	-0.049	0.840
4	Thinking about extreme weather events - such as, droughts, flooding, storms, and extreme heat or cold - was your community's experience this year...	-0.032	-0.027	0.743
5	How much has climate change affected any big decisions for your family, such as where to live or work, or what to buy?	-0.024	0.423	0.548
6	How well is your country addressing climate change?	0.020	0.830	0.010
7	How well are big businesses addressing climate change?	-0.052	0.944	-0.028
8	In your country, who do you think has had the most impact addressing climate change?	N/A	N/A	N/A
9	Should your country strengthen or weaken its commitments to address climate change?	0.758	0.088	0.050
10	How quickly should your country replace coal, oil, and gas with renewable energy, such as power from the wind or sun?	0.281	0.130	0.335
11	How much should your country protect and restore nature, for example, by planting trees or protecting wildlife?	0.835	-0.121	-0.116
12	When it comes to protecting people at risk from extreme weather events, such as storms or extreme heat, should your country provide...	0.932	-0.057	-0.094
13	Should countries work together on climate change even if they disagree on other issues, such as trade or security?	0.537	-0.045	0.197
14	Should rich countries give more or less help to poorer countries to address climate change?	0.824	0.076	0.037
15	Should schools in your country do more or less to teach about climate change?	0.838	0.011	0.010

Table 2: Factor loadings for each survey question. Factors with an absolute value greater than 0.3 are highlighted for easier interpretation. Question 8 was not included in the EFA as its answers are not ordinal, resulting in no factor loadings.

This finding has important implications for how LLMs respond to climate-related survey questions. Survey design typically aims to capture nuanced opinions, but an overreliance on semantic proximity may introduce biases in response selection. If LLMs prioritize answers that semantically closer to the question, they may systematically favor certain perspectives rather than reflecting a broader range of human responses. Moreover, the linguistic style of the question, such as word choice and phrasing, could reinforce these biases, influencing the model's response selection. In climate discourse, for example, questions often contrast immediate versus long-term actions or individual versus governmental responsibility, leading models to disproportionately select semantically aligned answers.

The variation in correlation strength across models also suggests that architecture and training data influence how semantic similarity impacts response selection. Models with stronger correlations might be more susceptible to this effect, limiting their ability to represent a balanced spectrum of climate opinions. This highlights the need to understand the internal biases of LLMs, particularly when using them to simulate public sentiment or inform policy decisions.

Overall, these results provide an initial insight into how sentence embeddings influence LLM decision-making, potentially introducing systematic patterns in response selection. While these findings shed light on the role of semantic alignment in model outputs, further research is needed to deepen this analysis and develop strategies to mitigate such biases. This is particularly crucial to ensure that LLM-generated responses in climate-related surveys and discussion are are scientifically grounded and not unduly influenced by embedding or linguistic biases.

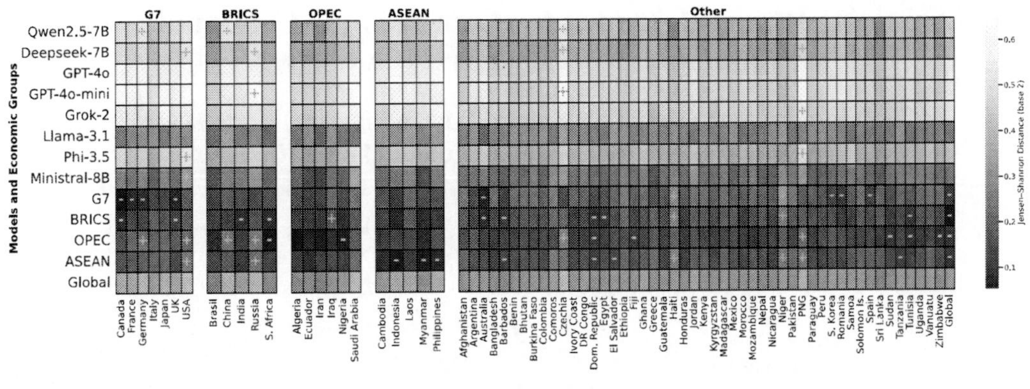

Figure 6: Distance between each country and group/LLM. Distances marked with "-" are significantly lower than the row, while those marked with "+" are significantly higher at the 0.05 significance level. Note that the country responses are removed from their group's distribution when calculating the distance.

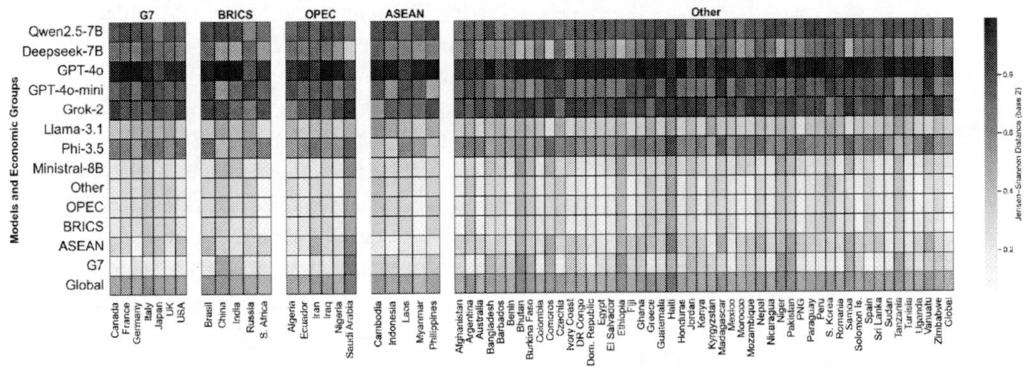

Figure 7: Distance between each country and group/LLM for Question 6.

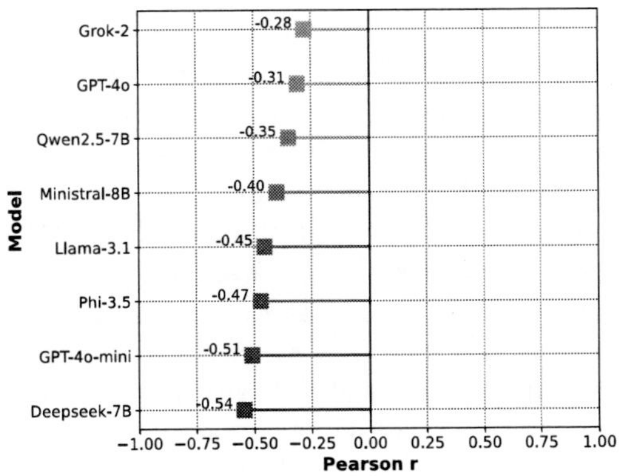

Figure 8: Correlation between the distance of answer choices from the question in embedding space and their selection probability.

Evaluating Retrieval Augmented Generation to Communicate UK Climate Change Information

Arjun Biswas[1], **Hatim Chahout**[2], **Tristan Pigram**[2], **Hang Dong**[1]
Hywel T.P. Williams[1],**Fai Fung**[2], **Hailun Xie**[1]
[1]University of Exeter, Exeter, United Kingdom
[2]Met Office, Exeter, United Kingdom
{ab1574, h.t.p.williams, h.dong2, h.x.xie}@exeter.ac.uk,
{hatim.chahout, fai.fung, tristan.pigram}@metoffice.gov.uk

Abstract

There is a huge demand for information about climate change across all sectors as societies seek to mitigate and adapt to its impacts. However, the volume and complexity of climate information, which takes many formats including numerical, text, and tabular data, can make good information hard to access. Here we use Large Language Models (LLMs) and Retrieval Augmented Generation (RAG) to create an AI agent that provides accurate and complete information from the United Kingdom Climate Projections 2018 (UKCP18) data archive. To overcome the problematic hallucinations associated with LLMs, four phases of experiments were performed to optimize different components of our RAG framework, combining various recent retrieval strategies. Performance was evaluated using three statistical metrics (faithfulness, relevance, coverage) as well as human evaluation by subject matter experts. Results show that the best model significantly outperforms a generic LLM (GPT-3.5) and has high-quality outputs with positive ratings by human experts. The UKCP Chatbot developed here will enable access at scale to the UKCP18 climate archives, offering an important case study of using RAG-based LLM systems to communicate climate information.

1 Introduction

Climate services are data, information, and knowledge provided to support decision-making about climate change (Global Framework for Climate Services, 2025). In the UK, the national government funded the UK Met Office to produce the UK Climate Projections (UKCP) (Lowe et al., 2018), a compilation of high-quality climate models, outputs, and analyses that help organizations prepare and adapt to climate change. Similar efforts are underway in other countries (e.g. Climate Change in Australia (CSIRO and Bureau of Meteorology, 2015) and CH2018 in Switzerland (Fischer et al., 2022)). The audiences for such climate services can be very large and diverse; for example, the UKCP data portal has over 11,000 registered users and is widely used in national government policy (Department for Environment, Food and Rural Affairs, 2024) and environmental regulations (Environment Agency, 2024), as well as business adaptation planning (Anglian Water, 2020) and best practice guidelines for local governments preparing for climate change (ADEPT, 2019). One major challenge is tailoring such services to specific and local user contexts. There are too few human experts to serve the complex climate information needs of such a large and diverse set of users. Generative AI tools offer a potential solution, allowing a user to extract bespoke climate information tailored to their own local context, through simple natural language interfaces. However, it is very important that such tools provide high-quality information; poor quality or incorrect information could cause harm by worsening climate-related decision-making.

In this study, we present the development of an LLM-based climate service that is intended to help deliver UKCP climate information. The UKCP archive contains a wide variety of complex scientific content (Met Office, 2025). A helpdesk is provided and human experts assist the UKCP user community in navigating the complex UKCP archive, offering user guidance and scientific documentation to improve access and utilization. Our tool is conceptualized as an automated support tool that can respond to typical UKCP helpdesk queries with accurate and trustworthy information. If deployed, this will reduce pressure on human experts and allow a greater number of UKCP users to be served. Here we describe our development of this tool in the form of a chatbot that uses Retrieval Augmented Generation (RAG). We evaluate a number of different information chunking, retrieval, ranking, and query expansion strategies,

Proceedings of the 2nd Workshop on Natural Language Processing Meets Climate Change (ClimateNLP 2025), pages 126–141
July 31, 2025 ©2025 Association for Computational Linguistics

creating and testing 14 different RAG pipelines. Performance is evaluated using a range of automated metrics (including a novel *coverage* metric) and human evaluation of outputs by subject matter experts (SME) in climate science. Results show that our RAG-based chatbot communicates accurate and relevant information from the UKCP archive, avoiding hallucinations or deviation from the content in the UKCP archive, and outperforming a non-specialized LLM-based chatbot. Overall positive ratings by human experts are achieved for our best RAG system (S2BH-CHR-MQG5).

2 Background and Related Work

Climate science and projections about future climate change are typically presented as complex datasets, scientific reports, articles, and other technical content. Currently, human climate scientists are needed to interpret this information for non-experts (Intergovernmental Panel On Climate Change (IPCC), 2023). While generative AI might help increase access to climate information at scale, effective decision-making around climate entails accurate translation of complex concepts and provision of trustworthy information. AI tools for communicating climate science must prioritize output accuracy and scientific quality.

Generally, LLMs are prone to hallucinations while having strong generative capability – providing responses that appear grammatically correct, fluent, and authentic, but actually deviate from source inputs (faithfulness) and/or fail on factual accuracy (factualness), offering outdated or incorrect information (Ji et al., 2023; Xu et al., 2024). Answers may also be incomplete, generic, or vague. This has led to methods that provide LLMs with additional domain-specific information to improve performance in applications requiring precise answers (Wu et al., 2023; Peng et al., 2023). Two popular approaches include domain-specific training of LLMs and RAG. Below we summarise studies that use these approaches in the domain of climate change.

Earlier studies adapt encoder-only, discriminative LLMs like BERT (Devlin et al., 2019) for climate communication tasks. ClimateBERT (Webersinke, 2022) was trained on approximately 2 million paragraphs of climate-related information, including reports, scientific paper abstracts, and news articles. The training process included general pre-training, followed by domain-specific training on climate information and then downstream training for specific tasks like classification, sentiment analysis, and fact-checking. Another example is ClimateBERT-NetZero (Schimanski et al., 2023) which fine-tuned BERT to detect whether a text contains a net zero or reduction target, and thus support subsequent data analyses.

Until most recently, generative LLMs are applied to convert climate information. ClimateGPT (Thulke et al., 2024) is a foundation model trained on a large corpus of climate-related texts. Training of ClimateGPT involved pre-training and instruction fine-tuning. As reported (Thulke et al., 2024), the pre-trained model outputs are domain-specific but suffer from hallucinations and cannot provide detailed information. ClimateGPT was then expanded by integrating a simple hierarchical RAG system, leading to improved performance. Also, training LLMs is energy-consuming and cannot easily adapt to new information, e.g., for climate projection.

Retrieval Augmented Generation (RAG) (Lewis et al., 2020) is a process of incorporating information from external databases to increase the answer accuracy of LLMs in domain-specific applications. It works by extracting relevant information (retrieval), processing the retrieved information with other external sources to create a structured prompt (augmentation), and summarising the combined information using an LLM (generation). The study (Fore et al., 2024) shows that RAG helps to improve the factual metrics of answers using in-context learning, which effectively mitigates conflicting information from the training set, for question answering with climate-related claims.

ChatClimate used a RAG-based system to communicate climate information (Vaghefi et al., 2023). This RAG system extracted the top-n pieces of relevant information for a given query from the IPCC Report. More recently, ChatNetZero (Hsu et al., 2024) is a RAG-based chatbot targeting the net zero domain. Our RAG-based system further explores a variety of chunking, retrieval and query rewriting strategies to enhance the RAG process. We focus on the dynamic, future climate projection data, instead of the current climate reports.

Robust evaluation of answer quality and information retrieval strategies is vital to ensure RAG systems' correctness and trustworthiness, as they are highly sensitive to noisy or irrelevant con-

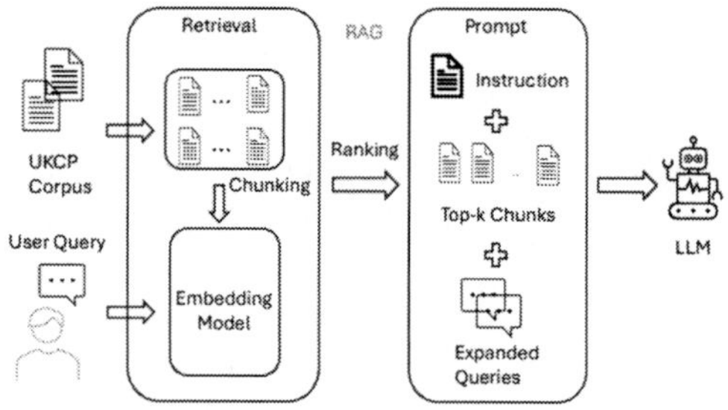

Figure 1: Overall framework of the UKCP Chatbot.

texts (Shi et al., 2023). A notable RAG-focused evaluation tool is RAGAS (Es et al., 2024), but it does not capture all dimensions of trustworthiness and accuracy. In the climate-focused applications above, human evaluators are used to judge answer correctness (Thulke et al., 2024), or automated evaluation is performed using benchmark datasets (Webersinke, 2022). These approaches become difficult in specialized domains, such as climate change and projection, where non-experts may not spot mistakes and there are no domain-specific benchmarks. In this work, we provide a phased automated evaluation with new datasets and human evaluation design with climate experts.

3 Methodology

We develop a conversational question-answering system to provide relevant, accurate, and trustworthy information related to the UKCP archive (hereafter referred to as "UKCP Chatbot"). Answers must be based only on the available UKCP archive; in other words, answers should only use UKCP data and not "general knowledge" or other external information. This makes the task a complex testbed for the faithfulness and hallucination of an LLM-based RAG system.

Our structured RAG system integrates different document chunking strategies, retrieval methods, and query expansion into well-defined retrieval pipelines. Multiple proposed pipelines are evaluated to optimise information retrieval, answer relevance and accuracy. A hybrid evaluation approach was used, incorporating both automated and human assessments. To demonstrate the advantage of the RAG approach, we also compared a general-purpose LLM, GPT-3.5-0125[1], used outside the RAG framework and based only on its general knowledge. The same GPT-3.5 LLM was used within several of our RAG system components so gives a fair comparison. User surveys were conducted to understand subjective perceptions of the UKCP Chatbot.

3.1 System Overview

The overall functionality of the chatbot is shown in Figure 1. The user enters a query q as input. The RAG system then parses the query and extracts the most relevant data chunks from the UKCP corpus. The extracted information is encapsulated into a prompt to a LLM to summarise the information and generate an answer. GPT-3.5 was used for its efficiency and cost-effectiveness. Conversation history is recorded to better understand the context of user queries. The components of the RAG system are optimized by comparison of several alternatives in each case; these choices are described below. The system is developed using a JavaScript front-end interface (see Figure 5 in Appendix F) and Python for back-end data manipulation.

3.2 Data Preparation

The UKCP archive contains diverse UK-focused climate data and information for a wide audience including scientific researchers, policymakers, industry professionals, and members of the public. The archive provides climate projections to the year 2100 based on model projections of future climate conditions for a number of greenhouse gas emission scenarios. Information is presented as published literature, observations, and climate model data. Here we focus on documents available from the UKCP archive, which include scientific reports, fact sheets, technical and guidance documentation, stakeholder engagement materials, and case study reports.

The corpus consists of 85 documents in raw PDF format in complex layouts. From this corpus, four segmented datasets were created by using different chunking approaches: *fixed-length*, *paragraph*, *section*, and *summary* methods. Details of data extraction, document segmentation, data cleaning, and data representation are in Appendix A.

3.3 RAG Framework

To develop the optimal RAG pipeline, we divide the RAG methodology into four components: document segmentation (chunking), chunk retrieval,

[1]https://platform.openai.com/docs/models/gpt-3.5-turbo

Table 1: Design options for RAG system components tested by phased evaluation.

Phase	Component Evaluated	Model ID	Component Variant
1	Chunking	F5	Fixed-Length (5-chunk context)
		F10	Fixed-Length (10-chunk context)
		P5	Paragraph (5-chunk context)
		P10	Paragraph (10-chunk context)
2	Retrieval	H20	Hierarchical (20 documents)
		S2B	Small-to-big
		S2BH15	Hierarchical (15 documents) & Small-to-big
		S2BH20	Hierarchical (20 documents) & Small-to-big
3	Ranking	S2BH-LST	Lost-in-the-middle
		S2BH-CHR	Cohere
		S2BH-DVR	Diversity
		S2BH-REC	Reciprocal
4	Query Expansion	S2BH-CHR-MQG3	Multiple Query Generation (3 queries)
		S2BH-CHR-MQG5	Multiple Query Generation (5 queries)

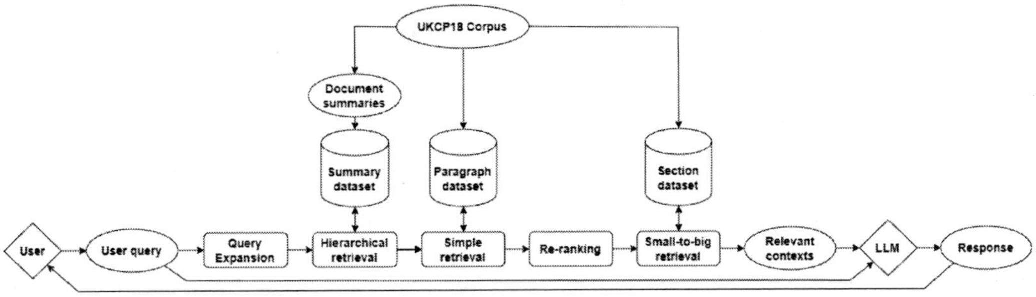

Figure 2: Overview of the RAG framework for model S2BH-CHR-MQG3.

chunk re-ranking, and query expansion. A total of 14 RAG pipelines were evaluated across the four components (see Table 1) and a locally optimal solution for each component was identified.

Since the four components work together in a functional RAG pipeline, component evaluation was performed sequentially in four experimental phases that each identified the best option for one component. This is based on the assumption that the components are independent of each other. The best component option found in each phase was adopted for subsequent phases of testing. This heuristic approach greatly reduces the number of test combinations (as high as 128 considering all possible combinations). Evaluation during this process used automated metrics that are described below; outcomes are presented in the Results section. The final RAG solution chosen for the UKCP Chatbot is visualized schematically in Figure 2 and incorporates the best components selected by this process, with an additional layer of human evaluation/testing (see Section 3.5).

3.3.1 Document segmentation (chunking)

This first phase considers the best low-level chunking strategy and how many chunks are needed in the prompt. It uses a simple retrieval approach based on cosine similarity between vector embeddings of the query and context chunks. We evaluated RAG pipelines that use the top-5 or top-10 most-similar chunks (following ClimateGPT (Thulke et al., 2024) and ChatClimate (Vaghefi et al., 2023)) chosen by two different chunking strategies (fixed-length F, or paragraph-based P). Fixed length and paragraph has the length of 1024 tokens. Later we consider larger-sized section chunks and summary chunks to improve final retrieval, but we do not use them in the initial retrieval stage.

3.3.2 Enhanced Information Retrieval

We introduce two retrieval strategies, *small-to-big* and *hierarchical*, and a third strategy that combines them, to enhance the simple retrieval of top-k relevant paragraph chunks above. These three retrieval algorithms combine multiple chunking strategies (including section and summary chunks) to enhance the final outputs of information retrieval. These enhanced retrieval methods provide more information to answer a user question by extracting longer document sections based on the smaller chunks found by simple retrieval (small-to-big) and by pre-selection of relevant documents prior to simple retrieval (hierarchical). These methods localize the relevant documents and sections to reduce inclusion of irrelevant information.

Small-to-big contextual expansion. Here small paragraph chunks are enhanced with bigger section chunks to increase the amount of relevant content found during information retrieval. First, 10 paragraph chunks are identified (using the P10 model, which provides the most relevant information) and then the document sections containing those paragraphs are also retrieved. The top-5 sections (fitting within the 16k context-length limit of GPT-3.5) most similar to the user query are then used to create the final prompt for question-answering. Expanding from paragraphs to sections increases the relevant/specific information extracted from the corpus and thereby enables better answers to be generated.

Hierarchical filtering. Here a pre-filter is applied to consider only the top-k most relevant documents for initial retrieval of paragraph chunks, creating a two-stage (or hierarchical) retrieval process. We set $k = 20$ to include a large number of documents and allow a more diverse set of chunks to be retrieved. Relevant documents are identified by first creating a summary of each document and then using cosine similarity between the embeddings of the user query and each document summary. The P10 model for simple retrieval is then applied to all paragraph chunks from the top-k relevant documents. This approach can prevent the spurious inclusion of paragraphs from irrelevant documents.

The two approaches above are then combined, leading to pipelines using hierarchical filtering (with 15 or 20 documents retained) followed by small-to-big retrieval. This helps extract the relevant sections from the most relevant documents.

3.3.3 Chunk ranking

Ranking (cf. re-ranking) is prioritization amongst the matching chunks selected by a retrieval method; it applies a rule or strategy to re-order the selected chunks and decide which ones will be included in the prompt. Here we tested four re-ranking strategies which we applied to the combined, hierarchical & small-to-big model, which was chosen as the candidate model for this phase based on the automated evaluation results.

Lost-in-the-middle. Language models can struggle to parse information in a long prompt, most often missing relevant information placed in the middle of a long input sequence (Liu et al., 2024). This re-ranking strategy places the most relevant chunks at the beginning and the end of the prompt, moving the least relevant chunks to the middle, following (deepset, 2025a). Unlike many other rankers, it does not use the query and simply re-orders the list of retrieved chunks.

Cohere is a platform that provides relevance-based re-ranking language models (Shi and Reimers, 2024) trained on query-passage pairs in documents. Here we used the Cohere "rerank-english-v3.0" model[2] , which was fine-tuned to retrieve the most relevant passage for a given query.

Diversity ranking ranks a list of chunks based on the relevance to the query and the diversity of the information in each chunk. The greedy algorithm initially chooses the most similar chunk to the query and then iteratively adds chunks that are, on average, least similar to previously added chunks, until all chunks are ranked (deepset, 2025b). Following the implementation in deepset (2025b), we use a sentence BERT model (Reimers and Gurevych, 2019), here "all-MiniLM-L6-v2"[3], to embed the query and the chunks for ranking.

Reciprocal ranking (Rackauckas, 2024) is used with multiple query generation (see below). For each query, inverse rank scores are calculated for all retrieved chunks:

$$\texttt{reciprocal\_score} = \frac{1}{\texttt{rank} + k} \tag{1}$$

where `rank` is the similarity-based rank of the chunk and k is a smoothing factor. The final ranking is calculated using the mean value of all reciprocal scores for each chunk.

3.3.4 Query Expansion

Retrieval responses are highly dependent on the exact phrasing of the query, so this phase seeks to diversify phrasing to give a more consistent retrieval of information (Rackauckas, 2024). An LLM (GPT-3.5) is utilised to generate multiple versions of the original query, keeping the meaning but varying how it is written. Each version is then used for information retrieval and the combined responses are used collectively to generate an answer to the original query. Here we tested

[2] https://huggingface.co/Cohere/rerank-english-v3.0

[3] https://huggingface.co/sentence-transformers/all-MiniLM-L6-v2

RAG pipelines using 3 or 5 versions of the original query. All chunks retrieved were collated and ranked together using a re-ranker (Cohere) and the original query to determine the top-k (here k=10) chunks used in the prompt. Here we use the S2BH-CHR model as the candidate based on the evaluation of the re-ranking stage of the pipelines. The prompt for generating multiple queries is presented in Appendix B (prompt-1).

3.3.5 Prompt construction

The complete prompt for the proposed RAG framework comprises a detailed *system instruction* and a *user prompt*. The system instruction includes the context for the chatbot (the UKCP archive), the task (question-answering), detailed constraints to ensure that answers are generated only from provided chunks from the UKCP corpus, and the steps to create an answer. The user prompt includes a structured format of the chunks, the query, and an answer mark ("ANSWER:") to prompt the system to generate an answer. The full prompt is presented in Appendix B (prompt-2).

3.4 Evaluation

A combination of automated and human evaluation was used to assess the quality of the UKCP Chatbot.[4] Automated evaluation metrics were used to compare 14 RAG pipeline variants and ChatGPT (GPT-3.5, chosen as a strong baseline example of a general-purpose LLM). Human experts then evaluated four RAG pipelines identified by automated evaluation, to determine the best pipeline overall and characterise user perceptions of the system.

3.4.1 Evaluation data

Two datasets are used for the automated evaluation of outputs from the UKCP Chatbot: (1) A dataset of 250 **synthetic QCA triplets**; and (2) An anonymised dataset of 50 **authentic QA pairs** from the UKCP helpdesk. Details about the dataset creation are available in the Appendix C.

3.4.2 Automated Evaluation Metrics

Automated evaluation here aims to assign metrics to RAG pipeline responses to assess three important characteristics that good answers must contain for our use case: (i) answer relevance; (ii) answer faithfulness; and (iii) answer coverage. Answer relevance measures how well the response aligns with the intent of the user query. Answer faithfulness measures the extent to which the response is based on the source information (or conversely, how much it uses other unsupported content). For ***Relevance*** and ***Faithfulness*** scores, we use metrics provided by RAGAS (Es et al., 2024). As accurate answers to scientific questions (as here in the climate domain) often require a high level of specificity and detail, here we propose a new ***Coverage*** metric, which calculates the proportion of all the named entities, keywords, and numerical values from the context chunks that are given in a generated answer. Details of the above metrics are described in Appendix.D. For the ChatGPT coverage score we compare the answers to the groundtruth context.

We also compare RAG-system scores to those from a baseline LLM (ChatGPT/GPT-3.5). Since relevance and faithfulness are defined using context chunks from a RAG system, the only metric that can be directly compared to a non-RAG LLM is coverage. We compute two metrics: (i) ***ChatGPT mean coverage score*** and (ii) ***Proportion of answers with coverage*** $>$ ***ChatGPT***: the percentage of answers by each RAG pipeline that have a higher coverage score than ChatGPT.

[4]Evaluation data, user testing survey, and implementation of the RAG pipeline are available at `https://github.com/arjun8009/UKCP-Repo-pub`.

3.5 Human Evaluation

Four selected RAG pipelines, chosen as the top-performing pipeline from each of the four phases of automated evaluation, were tested by subject matter experts. An initial screening was conducted by climate experts in the author group to choose the best two of these pipelines for further testing. Interactive evaluation was then performed by a panel of experts (n=10) recruited from UK Met Office staff. Details are given in Appendix E.

Two-stage survey design. Panelists received a survey in two stages, with access to the live chatbot given in the second stage.

In the first stage, two preliminary questions assessed participant background: (Q1a) Duration of professional experience in climate science; and (Q1b) Self-assessed familiarity with UKCP18 data. Next, participants evaluated the quality of answers provided by the chatbot. Four question-answer pairs were chosen from the 50 authentic QA pairs dataset, that could be easily reviewed without extensive additional knowledge. Answers given by the two selected pipelines were provided

Table 2: Automated evaluation metrics calculated for all RAG pipelines

Model Variation	Phase	Faithfulness Mean Score (Sources)	Relevance Mean Score	Coverage Mean Score	Answers with Coverage>ChatGPT	ChatGPT Mean Coverage
F10	1	0.90	0.93	0.34	79.00%	0.26
P10	1	0.89	0.94	0.34	78.00%	0.26
F5	1	0.89	0.93	0.32	72.00%	0.26
P5	1	0.86	0.93	0.32	77.00%	0.26
H20	2	0.81	0.93	0.35	78.00%	0.26
S2BH15	2	0.91	0.94	0.36	85.00%	0.26
S2BH20	2	0.92	0.94	0.36	83.00%	0.26
S2B	2	0.92	0.94	0.34	81.00%	0.26
S2BH-CHR	3	0.93	0.96	0.38	87.00%	0.26
S2BH-DVR	3	0.91	0.94	0.34	84.00%	0.26
S2BH-LST	3	0.90	0.93	0.36	82.00%	0.26
S2BH-REC	3	0.92	0.95	0.35	86.00%	0.26
S2BH-CHR-MQG3	4	0.91	0.90	0.32	81.00%	0.26
S2BH-CHR-MQG5	4	0.92	0.90	0.36	77.00%	0.26

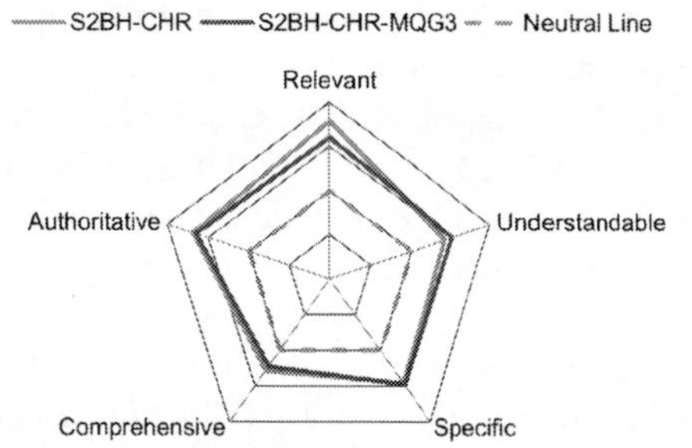

(a) Average human ratings of answer quality (n=10 participants). Scaled from strong disagree (inner) to strong agree (outer).

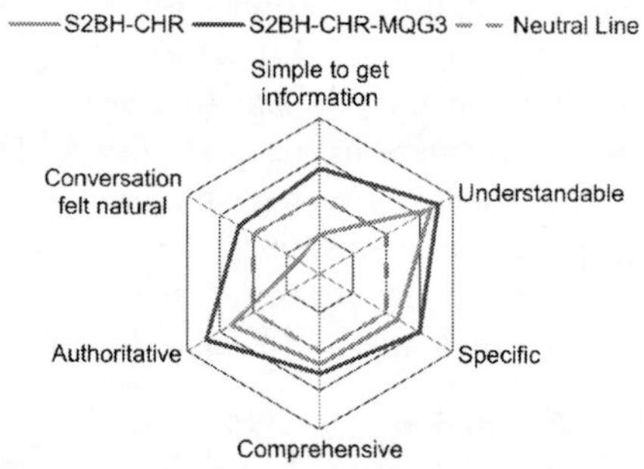

(b) Average human ratings of interaction quality (n=10 participants). Scaled from strong disagree (inner) to strong agree (outer).

Figure 3: Human evaluation of answer quality and interaction quality for RAG pipelines SB2H-CHR and SB2H-CHR-MQG3.

alongside the original human answer. Participants then used a standard Likert scale (1 - strong disagree; 2 - disagree; 3 - neutral; 4 - agree; 5 - strong agree) to assess RAG-pipeline answers on five quality metrics: (Q2a) Relevant; (Q2b) Understandable; (Q2c) Specific; (Q2d) Comprehensive; and (Q2e) Authoritative. Below each Likert scale, a free text box asked participants to explain their ratings and provide additional qualitative feedback.

The second stage assessed the usability and "conversationality" of the selected pipelines. Based on their evaluations in the first stage, participants were asked to interact with their preferred RAG pipeline via an online chatbot interface (see Appendix F). Users were tasked with a realistic scenario involving the use of UKCP18 data (see Appendix E for details) and asked to retrieve relevant information from the chatbot. They then evaluated their experience using Likert scales for six usability metrics: (Q3a) Simple to get information; (Q3b) Understandable; (Q3c) Specific; (Q3d) Comprehensive; (Q3e) Authoritative; (Q3f) Conversation felt natural. Free text boxes allowed further detail to be provided.

4 Results

4.1 Automated evaluation

Table 2 shows the automated evaluation metrics for all 14 RAG pipelines that were tested. All the proposed RAG models perform better than ChatGPT (GPT-3.5) in terms of the percentage of answers having a higher coverage Score, showing substantial improvements in mean scores (72 to 87 % of the answers generated by the RAG pipelines had a higher number of relevant keywords, entities and numbers). This finding is a clear validation of the RAG approach for this use case, showing

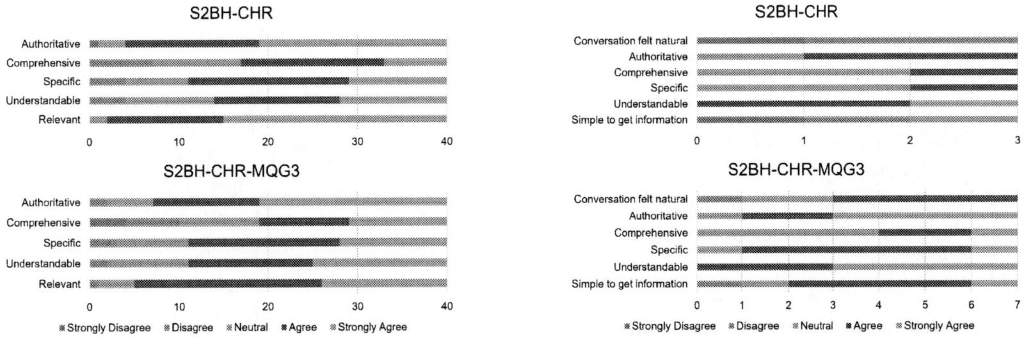

(a) Response breakdown for answer quality (n=40; 10 participants rating answers to 4 questions).

(b) Response breakdown for interaction quality (n=10 participants).

Figure 4: Human evaluation of answer quality and interaction quality for RAG pipelines SB2H-CHR and SB2H-CHR-MQG3: response breakdown

that the general-purpose LLM is unable to perform as well as RAG systems with additional domain-specific information.

From automated evaluation metrics (Table 2) we also conclude that, while small, the differences between pipelines do allow marginally better candidates to be identified. Since our four automated evaluation phases tested qualitatively different pipelines, sequentially introducing more complexity to the RAG framework, we chose the best pipeline from each phase for additional human evaluation. Evaluation phase 1 focused on the chunking strategy. We found that a higher number of chunks yields a higher coverage score and paragraph-based chunking produces better faithfulness and relevance scores. Therefore, pipeline P10 is adopted as the best candidate from phase 1. Evaluation phase 2 looked at the retrieval component. Here a combination of small-to-big and hierarchical methods gave the best outputs, so pipeline S2BH20 is chosen as the best candidate from phase 2. Phase 3 of automated evaluation considered chunk ranking approaches, with results showing that the coherence-based re-ranking strategy has the best performance. Hence pipeline S2BH-CHR is taken forward from phase 3. In Phase 4, we examined query diversification as a method for improving retrieval, finding that it boosts the faithfulness score significantly. Since both variant pipelines performed similarly, we chose S2BH-CHR-MQ3 due to its lighter computational load (few synthetic queries per answer). In this phase we observe a lower relevance score. Multiple query generation involves generating different versions of the same query and hence the generated answer contains information from various chunks that would not have been in the top 10 chunks if the original query was used. Hence the generated questions by the relevance metric can be slightly different from the original questions as the information can contain additional details. Therefore we observe a decrease in relevance. Overall we chose four pipelines for human evaluation: P10, S2BH20, S2BH-CHR, and S2BH-CHR-MQ3.

4.2 Human Evaluation

Initial screening by climate experts in the author team showed that the more complex RAG pipelines identified by automated evaluation phases 3 and 4 (S2BH-CHR and S2BH-CHR-MQG3) outperformed the simpler pipelines from phases 1 and 2 (P10 and S2BH20). Therefore S2BH-CHR and S2BH-CHR-MQG3 were further evaluated by the panel of subject matter experts.

The first stage of human evaluation by our panel of subject matter experts considered the quality of answers provided by the two RAG pipelines for four authentic questions received by the UKCP helpdesk. Figure 3a and results breakdown in Figure 4a show that for both pipelines, participants agreed that answers were authoritative, comprehensive, specific, understandable, and relevant. The weakest aspect across both pipelines was the comprehensiveness of answers. While pipeline S2BH-CHR appears to marginally outperform pipeline S2BH-CHR-MQG3 on answer quality, 7/10 users said that overall they preferred the responses generated by S2BH-CHR-MQG3.

The second stage of human evaluation by the panel asked users to interact with their preferred chatbot to complete a typical task related to the UKCP data archive. Results are shown in Figure 3b and Figure 4b. Participants using pipeline S2BH-CHR (n=3; 30%) reported negative (worse than neutral) outcomes for two performance criteria. It was hard to get the information they needed and the conversation felt unnatural. It should be noted that the number of users testing this pipeline was small and outcomes may be unreliable. Participants using the more popular pipeline S2BH-CHR-MQG3 (n=7; 70%) all reported positive outcomes on all criteria, but conversationality and simplicity of getting information were again the weakest aspects. Overall, across both pipelines tested performance was generally positively rated, with S2BH-CHR-MQG3 receiving stronger ratings on interaction quality.

The free text boxes in the user surveys gave some useful qualitative feedback. Users reported that the perceived weakness around "conversationality" arose from the repetition of phrases, which made the chatbot feel artificial. Broader questions were seen to be more successfully answered than specific questions; one user commented that "The chatbot did not have access to the underlying data, just a headline message. This made answers vague and less authoritative." While not all users were able to find the exact information they needed, they were impressed by the chatbot's ability to suggest relevant topics that fell slightly outside the initial scope of the questions they had posed. Furthermore, there were some areas of information where users reported not receiving information that they expected and knew to exist. These user comments provide areas for future improvement of the RAG-based chatbot.

5 Conclusion

In this work, we develop an LLM-based RAG framework with systematic evaluation to create a tool (the UKCP Chatbot) to increase access and understanding of complex climate information. A heuristic phased design approach was utilized to identify the optimal design for the RAG system, with evaluation of multiple recently reported strategies for chunking, retrieval, re-ranking, and query expansion. This process was complemented by two-stage automated and human evaluation. The best pipeline was identified as S2BH-CHR-MQG3 (see Table 2 and Figure 3). The resulting chatbot provides accurate and trustworthy information from the UKCP archive.

Limitations

Two main limitations of our RAG-based system were identified. First, a *lack of conversational ability* was observed during human evaluation. Due to the amount of retrieved information and relatively large size of generated answers, earlier portions of the conversation history were pruned to reduce context length, making the chatbot "forget" past questions and answers; this made it less conversational. Second, *answer completeness* is another possible weakness. Results from the automated coverage metric and human evaluation both indicate that answers provided by the chatbot, while normally correct, are in some cases incomplete. In some instances, the retrieved information may be comprehensive, but the LLM might fail to incorporate it all into the summary response. In other cases, retrieval may omit portions of relevant information. These identified limitations call for further research on RAG systems to improve conversational ability and answer completeness, without compromising the trustworthiness/accuracy of outputs.

In future work, we will explore the use of multimodal RAG frameworks, since the UKCP18 archive is originally a multimodal database that includes reports, images, maps, and raw climate data. We also aim to refine our testing methodology with new metrics to account for factual accuracy. Also, human evaluation in this study focused on a small number of subject matter experts; in future, we aim to extend the evaluation to a more diverse set of user groups and gain more comprehensive insights into the performance of the chatbot.

Ethics Statement

We do not identify any ethical issues for this exploratory study. The UKCP18 archive is available to the public (`https://www.metoffice.gov.uk/research/approach/collaboration/ukcp/data/index`), published under the Open Government Licence (`https://www.nationalarchives.gov.uk/doc/open-government-licence/version/3/`). The UKCP Chatbot is a prototype and not publicly available. It is currently undergoing internal evaluation at the Met Office. Before any public release, it will be thoroughly assessed by a wider stakeholder group and subject to further ethical and governance review.

References

ADEPT. 2019. Preparing for a changing climate: Good practice guidance for local government. https://www.adeptnet.org.uk/system/files/documents/Good%20Practice%20Guide%20ADEPT%202019f.pdf.

Anglian Water. 2020. Anglian water's climate change adaptation report 2020. https://www.anglianwater.co.uk/SysSiteAssets/household/in-the-community/climate-change-adaptation-report-2020.pdf. Accessed: 2024-12-05.

CSIRO and Bureau of Meteorology. 2015. Climate change in australia information for australia's natural resource management regions: Technical report.

deepset. 2025a. LostInTheMiddleRanker. https://docs.haystack.deepset.ai/docs/lostinthemiddleranker. Accessed: 2025-02-26.

deepset. 2025b. SentenceTransformersDiversityRanker. https://docs.haystack.deepset.ai/docs/sentencetransformersdiversityranker. Accessed: 2025-02-26.

Department for Environment, Food and Rural Affairs. 2024. Accounting for the effects of climate change: Supplementary green book guidance. https://assets.publishing.service.gov.uk/media/6645e47e993111924d9d3655/Accounting_for_the_effects_of_climate_change.pdf. Accessed: 2024-12-05.

Jacob Devlin, Ming-Wei Chang, Kenton Lee, and Kristina Toutanova. 2019. BERT: Pre-training of deep bidirectional transformers for language understanding. In *Proceedings of the 2019 Conference of the North American Chapter of the Association for Computational Linguistics: Human Language Technologies, Volume 1 (Long and Short Papers)*, pages 4171–4186, Minneapolis, Minnesota. Association for Computational Linguistics.

Environment Agency. 2024. Flood risk assessments: climate change allowances. https://www.gov.uk/guidance/flood-risk-assessments-climate-change-allowances. Accessed: 2024-12-04.

Shahul Es, Jithin James, Luis Espinosa Anke, and Steven Schockaert. 2024. RAGAs: Automated evaluation of retrieval augmented generation. In *Proceedings of the 18th Conference of the European Chapter of the Association for Computational Linguistics: System Demonstrations*, pages 150–158, St. Julians, Malta. Association for Computational Linguistics.

A.M. Fischer, K.M. Strassmann, M. Croci-Maspoli, A.M. Hama, R. Knutti, S. Kotlarski, C. Schär, C. Schnadt Poberaj, N. Ban, M. Bavay, U. Beyerle, D.N. Bresch, S. Brönnimann, P. Burlando, A. Casanueva, S. Fatichi, I. Feigenwinter, E.M. Fischer, M. Hirschi, M.A. Liniger, C. Marty, I. Medhaug, N. Peleg, M. Pickl, C.C. Raible, J. Rajczak, O. Rössler, S.C. Scherrer, C. Schwierz, S.I. Seneviratne, M. Skelton, S.L. Sørland, C. Spirig, F. Tschurr, J. Zeder, and E.M. Zubler. 2022. Climate scenarios for switzerland CH2018 – approach and implications. *Climate Services*, 26:100288.

Michael Fore, Simranjit Singh, Chaehong Lee, Amritanshu Pandey, Antonios Anastasopoulos, and Dimitrios Stamoulis. 2024. Unlearning climate misinformation in large language models. In *Proceedings of the 1st Workshop on Natural Language Processing Meets Climate Change (ClimateNLP 2024)*, pages 178–192, Bangkok, Thailand. Association for Computational Linguistics.

Global Framework for Climate Services. 2025. What are climate services? https://gfcs.wmo.int/what-are-climate-services. Accessed: 2025-02-25.

Angel Hsu, Mason Laney, Ji Zhang, Diego Manya, and Linda Farczadi. 2024. Evaluating ChatNetZero, an LLM-chatbot to demystify climate pledges. In *Proceedings of the 1st Workshop on Natural Language Processing Meets Climate Change (ClimateNLP 2024)*, pages 82–92, Bangkok, Thailand. Association for Computational Linguistics.

Intergovernmental Panel On Climate Change (IPCC). 2023. *Climate Change 2022 – Impacts, Adaptation and Vulnerability: Working Group II Contribution to the Sixth Assessment Report of the Intergovernmental Panel on Climate Change*, 1 edition. Cambridge University Press.

Ziwei Ji, Nayeon Lee, Rita Frieske, Tiezheng Yu, Dan Su, Yan Xu, Etsuko Ishii, Ye Jin Bang, Andrea Madotto, and Pascale Fung. 2023. Survey of hallucination in natural language generation. *ACM Computing Surveys*, 55(12):1–38.

Patrick Lewis, Ethan Perez, Aleksandra Piktus, Fabio Petroni, Vladimir Karpukhin, Naman Goyal, Heinrich Küttler, Mike Lewis, Wen-tau Yih, Tim Rocktäschel, Sebastian Riedel, and Douwe Kiela. 2020. Retrieval-augmented generation for knowledge-intensive nlp tasks. In *Advances in Neural Information Processing Systems*, volume 33, pages 9459–9474. Curran Associates, Inc.

Nelson F. Liu, Kevin Lin, John Hewitt, Ashwin Paranjape, Michele Bevilacqua, Fabio Petroni, and Percy Liang. 2024. Lost in the middle: How language models use long contexts. *Transactions of the Association for Computational Linguistics*, 12:157–173.

Jason A. Lowe, Dan Bernie, Philip Bett, Lucy Bricheno1, Simon Brown, Daley Calvert, Robin

Clark, Karen Eagle, Tamsin Edwards2, Giorgia Fosser, Fai Fung, Laila Gohar, Peter Good, Jonathan Gregory, Glen Harris, Tom Howard, Neil Kaye, Elizabeth Kendon, Justin Krijnen, Paul Maisey, Ruth McDonald, Rachel McInnes, Carol McSweeney, John F.B. Mitchell, James Murphy, Matthew Palmer, Chris Roberts, Jon Rostron, David Sexton, Hazel Thornton, Jon Tinker, Simon Tucker, Kuniko Yamazaki, and Stephen Belcher. 2018. UKCP18 science overview report.

Met Office. 2025. Ukcp archive. `https://www.metoffice.gov.uk/research/approach/collaboration/ukcp`. Accessed: 2025-02-25.

Cheng Peng, Xi Yang, Aokun Chen, Kaleb E. Smith, Nima PourNejatian, Anthony B. Costa, Cheryl Martin, Mona G. Flores, Ying Zhang, Tanja Magoc, Gloria Lipori, Duane A. Mitchell, Naykky S. Ospina, Mustafa M. Ahmed, William R. Hogan, Elizabeth A. Shenkman, Yi Guo, Jiang Bian, and Yonghui Wu. 2023. A study of generative large language model for medical research and healthcare. *npj Digital Medicine*, 6(1).

Zackary Rackauckas. 2024. Rag-fusion: A new take on retrieval augmented generation. *International Journal on Natural Language Computing*, 13(1):37–47.

Nils Reimers and Iryna Gurevych. 2019. Sentence-bert: Sentence embeddings using siamese bert-networks. In *Proceedings of the 2019 Conference on Empirical Methods in Natural Language Processing*. Association for Computational Linguistics.

Tobias Schimanski, Julia Bingler, Mathias Kraus, Camilla Hyslop, and Markus Leippold. 2023. ClimateBERT-NetZero: Detecting and assessing net zero and reduction targets. In *Proceedings of the 2023 Conference on Empirical Methods in Natural Language Processing*, pages 15745–15756, Singapore. Association for Computational Linguistics.

Freda Shi, Xinyun Chen, Kanishka Misra, Nathan Scales, David Dohan, Ed H. Chi, Nathanael Schärli, and Denny Zhou. 2023. Large language models can be easily distracted by irrelevant context. In *Proceedings of the 40th International Conference on Machine Learning*, volume 202 of *Proceedings of Machine Learning Research*, pages 31210–31227. PMLR.

Sylvie Shi and Nils Reimers. 2024. Introducing rerank 3: A new foundation model for efficient enterprise search & retrieval. `https://cohere.com/blog/rerank-3`. Accessed: 2024-12-05.

David Thulke, Yingbo Gao, Petrus Pelser, Rein Brune, Rricha Jalota, Floris Fok, Michael Ramos, Ian van Wyk, Abdallah Nasir, Hayden Goldstein, et al. 2024. Climategpt: Towards ai synthesizing interdisciplinary research on climate change. *arXiv preprint arXiv:2401.09646*.

Saeid Ashraf Vaghefi, Dominik Stammbach, Veruska Muccione, Julia Bingler, Jingwei Ni, Mathias Kraus, Simon Allen, Chiara Colesanti-Senni, Tobias Wekhof, Tobias Schimanski, Glen Gostlow, Tingyu Yu, Qian Wang, Nicolas Webersinke, Christian Huggel, and Markus Leippold. 2023. ChatClimate: Grounding conversational AI in climate science. *Communications Earth & Environment*, 4(1).

Nicolas Webersinke. 2022. ClimateBERT: A pretrained language model for climate-related text. In *AAAI 2022 Fall Symposium: The Role of AI in Responding to Climate Challenges*.

Shijie Wu, Ozan Irsoy, Steven Lu, Vadim Dabravolski, Mark Dredze, Sebastian Gehrmann, Prabhanjan Kambadur, David Rosenberg, and Gideon Mann. 2023. Bloomberggpt: A large language model for finance. *arXiv preprint arXiv:2303.17564*.

Ziwei Xu, Sanjay Jain, and Mohan Kankanhalli. 2024. Hallucination is inevitable: An innate limitation of large language models. *arXiv preprint arXiv:2401.11817*.

A UKCP corpus pre-processing: extraction, segmentation, cleaning

Data extraction. Many UKCP documents are in PDF format with complex layouts, figures, tables, and multi-column text. Automatic text extraction often produced outputs that were fragmented or out of order. Inconsistent formatting styles made it difficult to develop a single automated extraction process that maintained the integrity of content and structure or correctly extracted captions for images and tables. Careful manual checking and intervention were used to correct formatting issues, remove irrelevant data (e.g., page numbers), and ensure content integrity.

Document segmentation (chunking). Four datasets were created by segmenting each document using different chunking approaches: *fixed-length*, *paragraph*, *section*, and *summary* methods. Each "chunk" is derived from one of the original UKCP documents, in a size/format that an LLM can effectively process. Each chunk also includes metadata specifying the originating UKCP document name, page, and section from which it was sourced. The details of each chunking method are given below.

Fixed-Length: Chunks have a fixed length of 1,000 characters. This is efficient but does not account for semantic/structural boundaries within document content. Chunks are often cut off mid-paragraph, leading to incomplete representations of topics/ideas.

Paragraph: Chunks represent each paragraph within the corpus. This preserves the natural semantic boundaries within documents and can potentially give more meaningful retrieval results. Each chunk varies in length.

Section: Chunks represent each section within the corpus (defined as content given under a single heading). This preserves continuity between adjacent paragraphs and might improve retrieval quality by delivering larger chunks of related content. Each chunk varies in length.

Summary: Each chunk is a LLM-generated summary of a UKCP document created using a two-step approach. Firstly, each section of the document is summarised by an LLM (here GPT-3.5) to extract key points. Secondly, all section summaries are combined (by the same LLM) into a single cohesive summary for the entire document. Each chunk varies in length.

Data cleaning. Several processes were used to ensure the quality and consistency of the extracted chunks: (i) Removal of irrelevant or extraneous elements such as page numbers, footnotes, and headers; (ii) Correction of text extraction errors, such as erroneous characters; (iii) Correction of image/table captions and their linking to corresponding visual/numerical content. Final datasets were manually checked to rectify any remaining inconsistencies.

Data representation. The embedding model used to represent the query and chunk was "text-embedding-ada-002", if not otherwise specified in the main text of the paper.

B Prompts

Below are the prompts for generating multiple queries for query expansion (prompt-1) and for generating answers (prompt-2).

Input prompt-1 for generating multiple queries

query system instruction =
Instructions:
1. You will be provided with a question from the user.
2. Your task is to generate multiple search queries related to this input question.
3. You must maintain the context of the original question and you must not exclude any key information from the question.
4. Phrase each query in a different way, but ensure that you do not deviate from the original meaning of the question.
5. Output <NUMBER> new queries in the form of a list. Do not deviate from this format.
Follow these steps before providing your final response:
Step 1: Take your time to thoroughly understand the provided question.
Step 2: Generate your new queries, ensuring that each new query is written in a distinctly different way to each other query.
Step 3: Reason step-by-step about whether the all of the key information from the original question can be found in each of the new queries. If there is key information found in the original question which cannot be found in any given new query, then you must replace this query by generating a new one. You must then follow these steps again.
Step 4: You may provide your final generated queries to the user. Do not output anything else.
query user prompt = QUESTION:<QUESTION>

Input prompt-2 for generating answers

system instruction = Instructions:
1. You are an expert on United Kingdom Climate Projections (UKCP). UKCP is a set of tools and data that demonstrates how the UK climate may change in the future.
2. UKCP18 is a set of climate model projections for the UK produced by the Met Office. It builds upon the previous set of projections (UKCP09) to provide the most up-to-date assessment of how the climate of the UK may change over the 21st century.
3. You will be provided with a question from the user, for which you will attempt to find the answer.
4. You will be provided with excerpts which are sourced exclusively from UKCP18 literature.
5. You MUST read all of the excerpts to understand the context for answering the question.
6. You will provide an EXPERT-LEVEL written response which comprehensively answers the question, using only information from the provided excerpts.
7. You should assume that you do not have access to any other sources of information.
8. UNDER NO CIRCUMSTANCES should you use information from any other source (such as the internet) to generate your responses.
9. The response you provide will be cross-checked with the excerpts provided to you. If there is information within the response which is not found in the excerpts, you will lose credibility.
10. You will be provided with the chat history of the conversation in your messages. You must follow the chat history to understand the context of the conversation.
11. If you cannot answer the question using information from the excerpts, you may ask once for more information from the user. If this additional information does not help you to find the answer from the excerpts, gently respond that you are unsure about the answer and recommend that they contact the Met Office's UKCP help desk.
12. You must not repeat or summarize the question which was asked to begin your response. Only respond with the answer, request for more information, or the statement that you cannot answer the question.
Follow these steps before providing your final response:
Step 1: Take your time to thoroughly understand the provided question.
Step 2: Take your time to thoroughly understand the provided excerpts, which are delimited by the following token: <SEP>.
Step 3: Generate an expert-level written response which comprehensively answers the question using only the excerpts provided. If you are unable to create a response that comprehensively answers the question using the provided excerpts, ask the user once for more information. If this additional information does not help you to find the answer from the excerpts, gently respond that you are unsure about the answer, recommend that they contact the Met Office's UKCP help desk and stop following these steps.
Step 4: Reason step-by-step about whether the all of the information in the response can be found in excerpts provided. If there is information found which cannot be found in the excerpts, then you must generate a new response and follow these steps again for the new response.
Step 5: You may provide your response to the user.
user prompt = EXCERPTS: <EXCERPTS>
QUESTION: <QUESTION>
ANSWER:

C Evaluation data creation

Synthetic QCA triplets. A dataset of question-context-answer (QCA) triplets was synthesized using the RAGAS package (Es et al., 2024), which takes contextual documents as input and uses an LLM to generate derived question-and-answer pairs. RAGAS can generate several types of questions. Three types of questions were created for this dataset. *Simple* questions are intended to be straightforward to answer using the given context. *Reasoning* questions re-write a simple question such that reasoning is needed to answer it effectively. *Multi-context* questions re-phrase a simple question such that information from multiple context sections is needed to formulate an answer. RAGAS outputs are question-context-answer triplets. For this study, RAGAS was parameterized using GPT-4 as the LLM and the *section* chunks as content. A sample of 500 section chunks was randomly split into groups of 5, and then 10 question-answer pairs were generated for each section chunk, using a **1:2:2** ratio for simple, reasoning, and multi-context question types. The resulting dataset of 1,000 QCA triplets was sampled for 250 QCA triplets used for evaluation.

Examples :

Question : "Which UKCP18 model better represents Scotland's winter snow variability?"

Answer : The CPM better represents Scotland's winter snow variability, particularly in terms of lying snow and snowfall over the Scottish mountains.

Question : "What's PoT's role in estimating rare climate events?"

Answer : The PoT (peaks over threshold) method involves using all events exceeding a specified threshold in a given season, thus considering more of the data, and avoiding the risk of missing multiple extremes that may occur in close proximity. It also excludes any seasons which happen not to contain any extreme events.

Authentic QA pairs. A dataset of 50 question-answer (QA) pairs was derived from real questions received by the UKCP helpdesk and the answers provided by subject matter experts. The QA pairs were anonymized, cleaned and formatted, and manually selected to represent a diverse range of typical questions. The authentic QA pairs lacked contexts and were only used for human evaluation where the subject matter experts decided the correctness of the extracted contexts and the answers. (Examples in the github link under human evaluation survey form)

D Evaluation metrics: Relevance, Faithfulness, and Coverage

We follow the work RAGAS (Es et al., 2024) to use LLMs to measure the answer relevance and answer faithfulness. We further propose a metric to measure answer coverage. The detailed metric settings are described below.

Relevance. This metric measures the relevance of the answer to the user query by an inverse method, using an LLM (GPT-4) to create alternate synthetic questions that could generate the answer and then measure their (cosine) similarity to the original user query. Mathematically, the metric is found as: $\texttt{relevance\_score}(g_i, q) = \frac{1}{N}\sum_{i=1}^{N}\texttt{cos}(E_{g_i}, E_q)$ where E_x is the embedding of a generated question g_i or the original query q, and $N = 3$ is the number of generated questions.

Faithfulness. This metric measures the extent that an answer uses only information that is contained in the chunks given as context. An evaluator LLM (GPT-3.5) is used to identify the sets of factual claims that are made in the provided answer and in the context chunks. Then the metric is defined as: $\texttt{faithfulness\_score} = \frac{|C_{\texttt{answer}}|}{|C_{\texttt{context}}|}$, where C_x is the set of claims present in either the answer or the context chunks.

Coverage. Accurate answers to scientific questions (as here in the climate domain) often require a high level of specificity and detail. This implies usage and adherence to numerical values, proper names, keywords and other entities. Here we propose a new *coverage* metric, which calculates the proportion of all the named entities, keywords, and numerical values from the context chunks that are given in a generated answer. Identification of entities, keywords, and numbers was performed using the trained model "en-core-web-sm"[5] in the SpaCy NLP package, and additional terms were identified using an LLM (GPT-4). All proper nouns and adverbs are considered as keywords. Coverage is then defined by: $\texttt{coverage\_score} = |\frac{|K_{\texttt{answer}} \bigcap K_{\texttt{context}}|}{|k_{\texttt{context}}|}$ where K_x is the set of all keywords, named entities, and numbers in either the context chunks or the answer.

[5] https://spacy.io/models/en#en_core_web_sm

E Human evaluation settings

Initial screening. The questions from the 50 authentic QA pairs dataset were posed to each RAG pipeline, then subject matter experts reviewed each response to evaluate its quality and validity. The two best-performing pipelines were determined based on subjective evaluations of answer correctness, framing, style and specificity. Responses from these selected pipelines were further screened to curate a pool of answers that could be easily reviewed without extensive contextual information. Four questions with answer pairs were selected.

Interactive evaluation. A panel of subject matter experts evaluated the two remaining RAG pipelines in a two-stage process conducted by survey and online access to the chatbot.

Panel recruitment. The panel (n=10) was recruited from Met Office staff to ensure a good baseline understanding of climate science. Within this group, there was a range of experience, with 4/10 panelists having over 10 years of experience using UKCP18 data.

User task for human evaluation of UKCP Chatbot Users were given access to the chatbot via an online user interface and asked to complete the following task within 30 minutes: *'Task - The Ministry of Defence (MoD) needs to construct 30 large buildings by 2030 in various locations around the UK coastline. The MoD would like to ensure that the buildings are suitably prepared to stay cool in the future. Use the chatbot to find relevant information and try to achieve this task.'*

F UKCP ChatBot interface

A screenshot of the user interface of the UKCP Chatbot is presented in Figure 5 below.

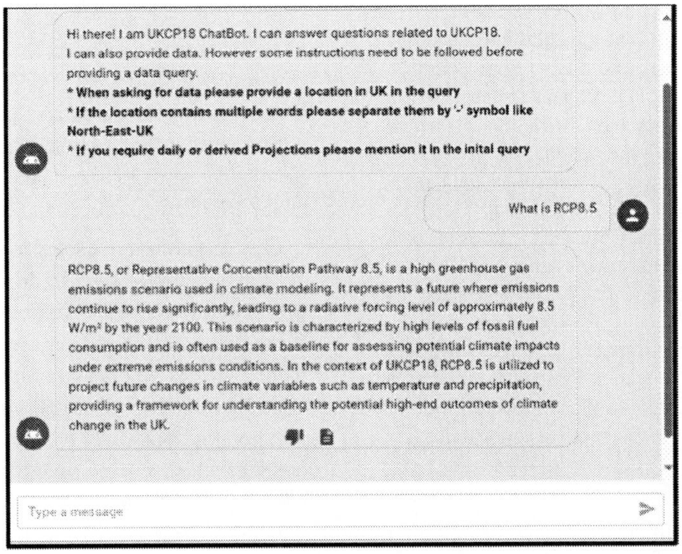

Figure 5: User Interface of the UKCP Chatbot

An Automated LLM-based Pipeline for Asset-Level Database Creation to Assess Deforestation Impact

Avanija Menon, Ovidiu Serban
Imperial College London

Abstract

The European Union Deforestation Regulation (EUDR) requires companies to prove their products do not contribute to deforestation, creating a critical demand for precise, asset-level environmental impact data. Current databases lack the necessary detail, relying heavily on broad financial metrics and manual data collection, which limits regulatory compliance and accurate environmental modeling. This study presents an automated, end-to-end data extraction pipeline that uses LLMs to create, clean, and validate structured databases, specifically targeting sectors with a high risk of deforestation. The pipeline introduces Instructional, Role-Based, Zero-Shot Chain-of-Thought (IRZ-CoT) prompting to enhance data extraction accuracy and a Retrieval-Augmented Validation (RAV) process that integrates real-time web searches for improved data reliability. Applied to SEC EDGAR filings in the Mining, Oil & Gas, and Utilities sectors, the pipeline demonstrates significant improvements over traditional zero-shot prompting approaches, particularly in extraction accuracy and validation coverage. This work advances NLP-driven automation for regulatory compliance, CSR (Corporate Social Responsibility), and ESG, with broad sectoral applicability.

1 Introduction

The European Union Deforestation Regulation (EUDR), effective December 30, 2025, mandates companies to verify that their products do not originate from recently deforested land (European Commission, 2023). With deforestation contributing 15% to global CO_2 emissions (ETC, 2024), industries with high environmental risks require precise asset-level tracking. However, significant data gaps persist: 30% of Forest 500 companies lack public deforestation commitments, and 85% of financial institutions lack comprehensive deforestation policies (Forest 500, 2024). Creating a physical asset database is labour intensive (CGFI, 2024), costly, and inefficient, making regulatory compliance difficult and limiting researchers' ability to develop accurate environmental impact models.

Due to their substantial contributions to environmental degradation, we focus on three high-risk sectors—Mining, Oil & Gas, and Utilities. Mining drives deforestation through surface extraction and infrastructure expansion, often leading to forest loss within a 50 km radius (Bradley, 2020). Oil & Gas exploration accelerates deforestation, particularly in biodiversity hotspots like the Amazon, where oil extraction disrupts ecosystems (Finer et al., 2008; Amazon Watch, 2016). Utilities, especially hydroelectric projects, contribute to deforestation (IntegrityNext, 2024) through extensive land clearing for dams and power infrastructure, with continued expansion affecting forested areas despite the shift to renewable energy (Rosenberg et al., 2000; Imperiale et al., 2023).

Our research makes several key contributions: (1) We develop a novel LLM-based pipeline (Figure 1) that transforms unstructured SEC EDGAR filings into structured datasets, improving transparency in environmental monitoring. (2) We introduce Instructional, Role-Based, Zero-Shot Chain-of-Thought (IRZ-CoT) prompting, a technique that enhances the accuracy of entity extraction, particularly for complex asset-related information. (3) We conduct a comparative analysis of LLMs and a traditional Named Entity Recognition (NER) model, evaluating their effectiveness in domain-specific data extraction. (4) To ensure data integrity, we implement a three-step database cleaning process, which includes foundational standardisation, asset similarity consolidation using statistical methods, and LLM-assisted refinement. (5) We propose Retrieval-Augmented Validation (RAV), which integrates real-time web data to enhance dataset reliability and address gaps in existing databases. (6) Finally, the resulting datasets are visualised through company-specific dashboards, providing detailed

Proceedings of the 2nd Workshop on Natural Language Processing Meets Climate Change (ClimateNLP 2025), pages 142–167

insights into each company's database.

This work advances NLP-driven environmental data automation, providing a scalable framework for regulatory compliance, sustainability analysis, and asset-based deforestation tracking.

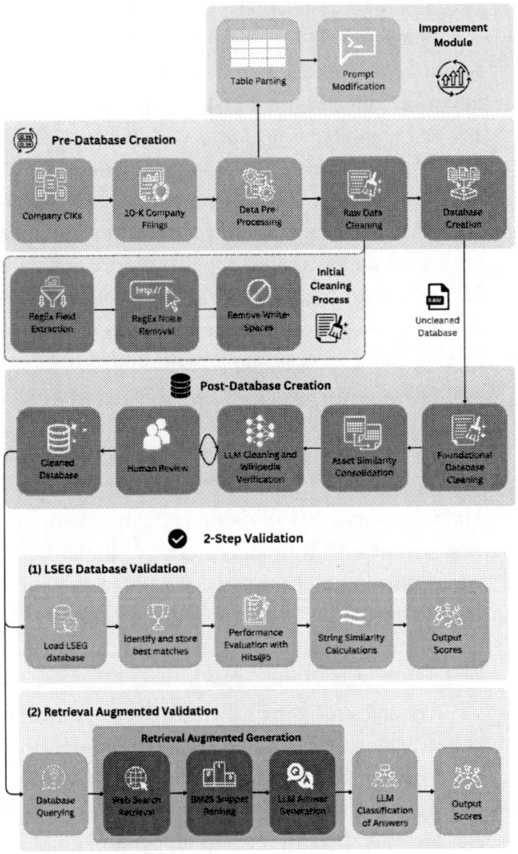

Figure 1: System design of end-to-end LLM-based pipeline designed to handle systematic data extraction, structured database creation, cleaning and validation, and the improvement module to increase validation coverage.

2 Background and Related Work

Existing asset-level databases rely on satellite imagery, geospatial data, and web sources with manual collection and validation (CGFI, 2024), but manual validation limits scalability. Early rule-based and template-based entity extraction methods lacked adaptability. At the same time, statistical models like Markov Logic Networks (MLNs) and Matrix Factorization improved relation extraction but were computationally expensive (Abdurehim et al., 2020; Cergani and Miettinen, 2013). Machine learning (ML) approaches, including bootstrapping (Zhang, 2009) and clustering (Tuo and Yang, 2023), automated pattern recognition but required large labeled datasets. Deep learning models such as CNNs, RNNs, and LSTMs enhanced feature extraction and are the dominant approach (Tuo and Yang, 2023). Hybrid models, combining deep learning and traditional ML (e.g., BiLSTMs+ CNNs), to capture long-distance dependencies were explored by Zheng et al (Zheng et al., 2017).

LLMs revolutionised entity and relation extraction, enabling zero-shot and few-shot learning. Structured prompting techniques, such as Pipeline Chain-of-Thought (Pipeline-COT), enhance accuracy by breaking tasks into reasoning steps (Zhao et al., 2023). ML and NLP techniques have been widely applied in healthcare, finance, and legal domains. Transformer-based models like Legal-BERT (Chalkidis et al., 2020), BioBERT (Lee et al., 2019), and SciBERT (Beltagy et al., 2019) improve clinical text analysis and regulatory compliance. However, fine-tuning remains computationally expensive, making zero-shot LLM approaches more practical. GPT-based models like GPT-NER incorporate self-verification to reduce hallucinations (Wang et al., 2023), while ChatGPT and REBEL enable structured knowledge extraction (Trajanoska et al., 2023). This study builds on these advancements, introducing Instructional, Role-Based, Zero-Shot Chain-of-Thought (IRZ-CoT) prompting to enhance structured data extraction from SEC EDGAR filings.

Traditional SEC EDGAR processing relies on RegEx-based tools like LexNLP, which efficiently parse filings (Bommarito et al., 2018). Prior keyword extraction and manual annotation work, such as the KPI-EDGAR dataset, remains labour-intensive and challenging to scale (Deußer et al., 2022).

Despite NLP advancements, limited work has been done on developing a fully automated pipeline that integrates data extraction, database creation, cleaning, and validation. This research bridges that gap by implementing an LLM-driven end-to-end pipeline, introducing Retrieval-Augmented Validation (RAV) to improve accuracy and robustness. By combining LLM-assisted extraction, structured prompts, and multi-step validation, this study delivers a scalable asset-tracking and environmental impact analysis solution, advancing AI-driven automation for regulatory data processing.

3 Data Acquisition and Processing

3.1 Data Source

This study uses publicly available SEC EDGAR 10-K filings from fifteen Mining, Oil & Gas, and Utilities companies. These legally mandated reports provide standardised, reliable, and accurate data on company operations, finances, and environmental impact. Unlike 10-Q and 8-K reports, which offer limited asset details, 10-K filings comprehensively cover physical assets, expenditures, and disclosures. News and social media data were excluded due to bias, noise, and lack of granularity. SEC filings ensure factual accuracy, regulatory compliance, and ethical data sourcing, minimising legal and privacy concerns.

3.2 Data Extraction

We collected 10-K filings from 2022 to 2024 using the secEDGAR Python library (Moody et al., 2024), which allows efficient bulk downloads based on company stock tickers and Central Index Keys (CIKs). This method streamlines data acquisition, eliminating the need for custom web scraping scripts while ensuring robust datasets across the selected sectors. The companies focused on are given in Table 3 in Appendix A.1.

The pre-processing workflow extracts metadata (company names, filing dates, form types, and content), cleans text using BeautifulSoup to remove HTML tags and irrelevant elements, and structures data into SQLite databases per company. This ensures efficient management, querying, and retention of meaningful content for analysis.

4 Database Creation

4.1 Chunk-based Querying Technique

We adopt a chunk-based querying technique to manage the extensive length of SEC EDGAR filings. This method involves splitting documents into 1024-token chunks with a 20-token overlap to maintain contextual continuity. Sentence-level splitting ensures semantic coherence, preventing the disruption of key information. Chunking optimises memory usage, enables parallel processing, and enhances entity recognition by allowing LLMs to focus on specific, contextually rich segments. This approach also facilitates error identification and correction, improving the efficiency and scalability of the data processing pipeline.

4.2 Comparison of LLM and NER Outputs

We compare the performance of 4-bit quantised Ollama instruct models, specifically Mistral-7B, Llama 3, and Gemma 2, against a traditional Named Entity Recognition (NER) model: dslim/bert-large-NER (Devlin et al., 2018; Tjong Kim Sang and De Meulder, 2003). Instruct models, fine-tuned for instruction-based tasks, demonstrate superior contextual understanding and precise entity extraction (Chung et al., 2022; Hu et al., 2024), which could be used for structured documents like SEC filings. The use of 4-bit quantisation significantly reduces memory and computational requirements while maintaining performance, enabling efficient large-scale deployment without extensive hardware upgrades (Banner et al., 2019; Dettmers et al., 2023). These models minimise irrelevant responses, ensuring more accurate asset identification. We convert the text data into embeddings using the SentenceTransformer model, specifically the paraphrase-MiniLM-L6-v2 variant (Reimers and Gurevych, 2019). Gemma 2 consistently outperforms the NER model on cosine similarity metrics, achieving higher precision and recall, with the highest cosine similarity for both locations (0.7702) and organisations (0.7461), indicating strong alignment with ground truth data.

Error analysis reveals that LLMs are more effective in capturing nuanced entity relationships, while the NER model often fragments entities or misses domain-specific terms. Detailed performance metrics are provided in Table 4 in Appendix A.3, where Gemma 2 outperforms both Mistral-7B and Llama 3.

As shown in Table 5 in Appendix A.4, qualitative error analysis highlights common issues such as fragmented entity recognition in the NER model and occasional hallucination in LLM outputs. While Mistral-7B and Llama 3 struggled with consistency, Gemma 2 demonstrated more reliable extraction, particularly in complex texts.

4.3 Ground Truth Creation

We manually curated a ground truth dataset from 30 chunks of Alcoa Corporation's 2022 filings to evaluate extraction accuracy. This dataset includes detailed annotations of physical assets, their locations, ownership structures, and associated commodities. Manual annotation ensures high accuracy, providing a robust benchmark for model evaluation. While slightly labour-intensive, this process establishes a reliable foundation for assessing model performance. In future work, we recommend exploring automated ground truth generation using advanced models like GPT-4, which could enhance scalability and reduce annotation costs.

4.4 LLM Selection

We assessed multiple LLMs using evaluation metrics such as cosine and jaccard similarities, precision, recall, and F1 score. Gemma 2 emerged as the top performer, excelling in quantitative and qualitative analyses. Its superior performance is attributed to its ability to maintain semantic coherence and accurately extract domain-specific entities. As presented in Table 1, Gemma 2 achieved the highest scores across all evaluation metrics. This performance consistency and efficient resource utilisation led to its selection for further experimentation within the data pipeline.

Model	Similarity Metrics		Evaluation Metrics		
	Cosine	Jaccard	Precision	Recall	F1 Score
Mistral-7B	0.64	0.41	0.54	0.58	0.60
Llama 3	0.64	0.43	0.58	0.64	0.59
Gemma 2	0.68	0.44	0.63	0.62	0.60

Table 1: Performance comparison of Mistral-7B, Llama 3, and Gemma 2 across five metrics.

4.5 Prompt Engineering

Prompt engineering plays a crucial role in optimising data extraction. We developed the Instructional, Role-Based, Zero-Shot Chain-of-Thought (IRZ-CoT) prompting technique through iterative refinement. This method improves extraction accuracy by providing LLMs with domain-specific instructions, structured reasoning steps, and role-based guidance. IRZ-CoT reduces hallucination, enhances the extraction of complex attributes, and minimises the need for extensive post-processing.

Common issues encountered with different prompting techniques revealed key challenges, such as hallucination in one-shot and few-shot methods, incorrect classification in zero-shot, and verbosity in generated knowledge prompting. Specifically, zero-shot prompting often led to the misclassification of financial terms as physical assets, while few-shot techniques introduced hallucinated entities. Role-based and instructional prompting significantly improved specificity and reduced errors, but IRZ-CoT demonstrated the best balance between accuracy and efficiency.

Performance metrics for prompt engineering techniques, illustrated in Figure 2, show that IRZ-CoT achieved the highest scores in precision and recall. Additionally, Figure 6 in Appendix A.6 highlights IRZ-CoT's computational efficiency, requiring significantly less processing time than methods like generated knowledge prompting.

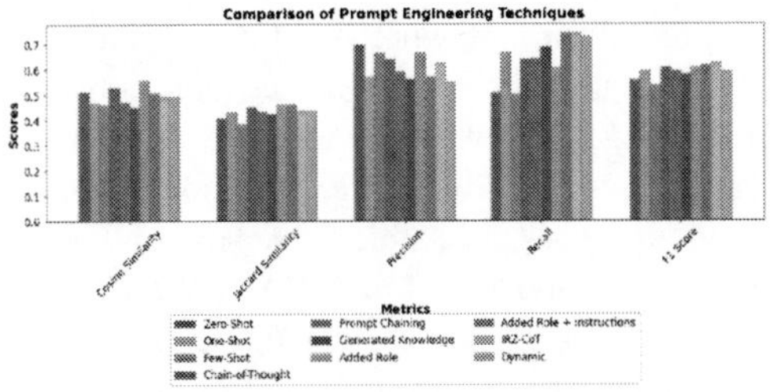

Figure 2: Comparison of different prompt engineering techniques across various evaluation metrics

4.6 Experimental Evaluation of LLM Ensemble Methods

We evaluated three LLM ensemble methods to enhance robustness: Ensemble Averaging with Majority Voting (EAMV), Weighted Majority Voting Ensemble (WMVE), and Stacking Ensemble with Meta-Learning (SEML). EAMV improves stability by aggregating predictions from multiple LLMs and selecting the most common output, reducing variance. WMVE assigns higher weights to models with superior performance, prioritising predictions from more accurate models, particularly favouring Gemma 2. SEML utilises a meta-learner—logistic regression—to combine outputs from different LLMs, optimising predictive accuracy and achieving the highest F1-score (Figure 7 in Appendix A.7). However, SEML significantly increased processing time—nearly 20-fold compared to single-model approaches (Figure 8 in Appendix A.8). Due to computational constraints, we selected the more efficient IRZ-CoT approach with Gemma 2 as the primary model for the final pipeline.

5 Database Cleaning

5.1 Foundational Data Cleaning and Standardization

The first phase focuses on refining raw data to establish a solid foundation for further processing. We use regular expression (RegEx) patterns to extract key entity data, including asset types, locations, ownership details, and commodities. This automated approach ensures consistent data extraction from large volumes of text. Post-extraction, we remove extraneous characters, such as surplus quotes and brackets, to prevent data distortion.

Duplicates are identified and consolidated, with corresponding information merged into single records. For example, multiple entries for an oil well in different locations are grouped, reducing redundancy. Ownership data is standardised by normalising company names (e.g., consolidating "NEM," "Newmont," and "Newmont Corporation"). At the same time, geographic terms are unified (e.g., "United States of America," "US," and "U.S.A." standardised to "USA"). We also refine the 'location' column, extracting country names into a new 'Countries' column to support consistent geographic analysis. Finally, rows with empty 'physical asset' entries are removed to maintain database relevance.

5.2 Asset Similarity Consolidation

After initial cleaning, we address semantic similarities among physical asset entries. To consolidate such similarities, we use TF-IDF (Term Frequency-Inverse Document Frequency) vectorisation and cosine similarity.

TF-IDF quantifies the relevance of words within documents, and cosine similarity identifies semantic overlap. We set a similarity threshold of 0.5; entries meeting or exceeding this threshold are grouped and merged, preserving unique information while eliminating redundancy. This method is computationally efficient and effective for identifying similar assets, although it has limitations, such as sensitivity to synonyms. Despite these, TF-IDF and cosine similarity offer a pragmatic balance of accuracy and efficiency for large-scale datasets.

5.3 LLM-Assisted Database Cleaning

The final cleaning phase leverages the capabilities of Gemma 2 to address issues beyond the reach of traditional methods, as the previous steps still outputted unnormalised information amongst other issues. Using a domain-specific prompt, the LLM performs tasks such as converting chemical symbols (e.g., "Au" to "Gold"), standardising text, eliminating redundant punctuation, and verifying locations against Wikipedia.

This iterative process involves LLM-driven cleaning followed by human review. Any inconsistencies trigger prompt adjustments, enhancing the LLM's performance in subsequent iterations. The LLM also identifies countries from location data when not explicitly stated, verified through cross-referencing with Wikipedia to ensure accuracy.

By automating complex tasks and reducing manual effort, LLM-assisted cleaning improves data quality, consistency, and scalability, making it an effective strategy for managing large datasets.

6 Database Validation

6.1 Validation with LSEG Databases

We validated our databases against established LSEG Workspace databases (London Stock Exchange Group (LSEG), 2024), focusing on the 'Mines', 'Oil Refineries', and 'Power Generation' datasets. This validation process involved data preprocessing, where we standardised text to lowercase and filtered irrelevant entries, such as excluding closed or abandoned assets. This step ensured uniformity and minimised discrepancies related to case sensitivity. Subsequently, we used the `rapidfuzz` library to find similar entries between our database entries and LSEG data. A similarity threshold of 0.6 was applied to identify potential matches, which helped us find the best match from the list of matches. We then used the Hits@5 metric to determine how frequently correct matches appeared within the top five candidates for each attribute (physical asset, ownership, commodity, and country). The Hits@5 score measures the consistency of our matching algorithm by averaging successful matches across all entities, assessing performance beyond the top result. Identified matches are then validated using five more metrics (Partial Match Score (Partial Ratio), Jaccard Similarity, Cosine Similarity, Dice-Sørensen Similarity Coefficient and Normalised Levenshtein Distance), comparing entity similarities with the LSEG database. Detailed similarity scores across physical asset name, ownership, commodity, and country are averaged into an overall attribute similarity score, quantifying dataset alignment. This validation ensures the reliability of our matching algorithm.

6.2 Retrieval-Augmented Validation (RAV)

LSEG Workspace databases lack comprehensive data for complete validation, necessitating an additional verification layer to ensure completeness and accuracy. To address these gaps, we developed Retrieval-Augmented Validation (RAV).

RAV integrates real-time web search capabilities using the Google Custom Search Engine (CSE)

API (Google Developers, 2024) to retrieve current information on physical assets. The retrieved snippets are ranked using the BM25 algorithm, which prioritises documents based on relevance, incorporating term frequency and document length normalisation. This ensures that the most pertinent information is considered for validation purposes.

RAV uses a dual-LLM framework where Llama 3 generates web-based answers, and Gemma 2 is tasked with classifying these answers strictly against the database entries. This separation mitigates potential biases arising from using a single model for generation and evaluation. Llama 3 efficiently retrieves concise, relevant information from web sources, while Gemma 2 assesses the similarity between this information and the existing database entries. The LLM-assisted validation relies on a binary classification approach where Gemma 2 outputs a 'yes' if the web-derived and database information are similar and a 'no' otherwise. This stringent evaluation ensures high reliability, reducing the risk of false positives in validation.

Contrary to complex instructional prompts used in earlier phases, we discovered that simple prompts significantly improved LLM classification accuracy. Initially, using detailed prompts resulted in low similarity scores, averaging around 0.15, with frequent misclassifications. After simplifying the prompts to a single-line instruction asking the LLM to classify answers as similar or dissimilar (see Appendix A.13), we observed a substantial improvement, with scores increasing by approximately 0.28. This reduction in cognitive load enhanced the model's ability to determine similarities accurately. However, some misclassifications remain, mainly when subtle semantic differences exist between the database entries and web-sourced information.

RAV automates asset validation by integrating web data with traditional databases, enhancing reliability for downstream analysis. While advanced RAG methods like FLARE (Jiang et al., 2023) offer sophisticated retrieval, their complexity and resource demands outweigh the benefits for this project. Our BM25-based RAV remains practical and effective, with potential for future refinement.

7 Results

7.1 LSEG Database Validation Results

We validated our databases against LSEG Workspace datasets, including 'Mines,' 'Oil Refineries,' and 'Power Generation,' using six similarity metrics: Partial Match Score, Jaccard Similarity, Cosine Similarity, Dice-Sørensen Similarity Coefficient, Normalised Levenshtein Distance, and Hits@5. These metrics assessed alignment across physical assets, ownership, commodities, and country data. Figure 3 shows the partial match scores. The full results are shown in Figure 9 in Appendix A.9.

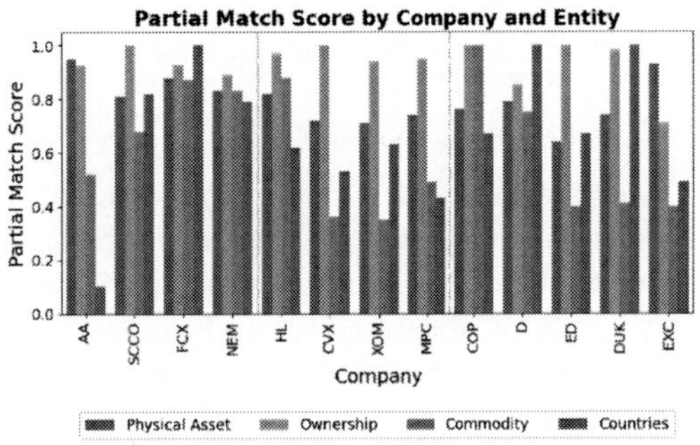

Figure 3: Partial match scores from the LSEG database validation. The dotted lines separate the sectors, where the sectors are mining, oil & gas, and utilities, respectively.

The mining sector demonstrated strong alignment, with high partial match scores for companies such as AA (0.95), FCX (0.88), and NEM (0.83), reflecting data consistency. However, the oil & gas sector showed mixed results, where CVX (0.72) and MPC (0.74) achieved moderate alignment, but XOM exhibited lower scores (Jaccard Similarity: 0.24), suggesting inconsistencies in asset classification. Utilities displayed moderate alignment, with EXC and D achieving partial match scores of 0.93 and 0.79, respectively. Ownership data varied significantly, with CVX achieving near-perfect alignment (1.00), while XOM had lower similarity (Jaccard Similarity: 0.43), likely due to differences in how joint ventures and subsidiaries were recorded. Commodity data showed the most significant discrepancies, with many companies, such as AA, registering low Jaccard and Cosine Similarities (0.00), possibly due to differences in classifying primary and secondary commodities. In contrast, country data was generally consistent, with companies like FCX achieving perfect alignment (Partial

Match Score: 1.00), though some discrepancies were observed in SCCO (Jaccard Similarity: 0.67).

The error analysis revealed key challenges. Ownership discrepancies arose due to variations in recording structures, where our databases captured joint ventures while LSEG focused on primary controlling entities. Standardising ownership classification could improve future alignment. Commodity misalignments resulted from differences in listing primary versus secondary commodities, suggesting a need to refine entity extraction prompts and separate commodity categories. Minor inconsistencies in country data, such as listing "USA" versus "California," highlight the importance of hierarchical structuring with separate fields for city, region, and country to enhance accuracy.

7.2 Coverage Calculation

We evaluate database coverage by measuring the proportion of physical assets and their attributes in our constructed database that match those in LSEG. This assessment ensures comprehensiveness, usability, and accuracy while identifying areas for improvement in our extraction pipeline. Coverage is computed as Coverage Score $= \left(\frac{N_m}{N_L}\right) \times 100$, where N_m represents the number of matched physical assets between our database and LSEG, and N_L is the total number of physical assets in LSEG. The computed coverage scores, shown in Table 7 in Appendix A.10, indicate that the mining sector has better coverage than oil & gas and utilities.

Manual inspection of SEC EDGAR filings reveals that lower coverage in oil & gas and utilities stems from improper table parsing, as many assets are listed in tabular formats rather than continuous text. To address this, we integrated a table parsing module using LlamaIndex, which processes HTML tables as structured data instead of narrative text. This significantly improved extraction accuracy, particularly in oil & gas, where assets were previously missed. Figure 10 in Appendix A.11 demonstrates this enhancement.

7.3 Retrieval-Augmented Validation (RAV) Results

Table 8 in Appendix A.12 presents the results of Retrieval-Augmented Validation (RAV), comparing database responses with real-time web data. Similarity scores range from 0.31 to 0.57, indicating moderate alignment, with the Oil & Gas sector performing slightly better due to more transparent regulatory disclosures. Notably, OXY is absent from Table 8 since it only contains unnamed assets, which RAV cannot validate.

Mining sector scores hover around 0.4, suggesting uniform discrepancies, likely due to outdated or incomplete records. The Oil & Gas sector shows slightly higher alignment, with companies like MPC and COP exceeding 0.5, possibly due to stringent regulatory reporting. However, frequent asset transfers contribute to inconsistencies. The Utilities sector exhibits the widest score range, from 0.31 (EXC) to 0.57 (NEE), reflecting differences in data transparency. NEE's higher score suggests more consistent asset records, likely due to better data management.

7.3.1 Error Analysis

Ownership mismatches arose from differing data granularity. Our database captured joint ventures and minority stakeholders, whereas web sources listed only primary entities, leading to unfair scores of 0. A weighted scoring system could better account for partial matches.

Location mismatches often resulted from implicit references in web snippets. For instance, the Bath County Power Station was correctly labeled as USA in our database, but the web snippet lacked an explicit country mention, receiving a score of 0. Similarly, Chino Mine was recorded as USA, while web sources specified New Mexico, USA. A hierarchical scoring approach would improve accuracy by recognising different levels of geographic detail.

Commodity discrepancies occurred because web data often listed only primary commodities, while our database included by-products. For example, Grasberg Mine was recorded as producing copper, gold, silver, and molybdenum, whereas web results mentioned only silver. Categorising commodities into primary and secondary groups through prompt refinement would help resolve this.

7.4 Total Validation Coverage

To assess RAV's impact, we compute the total validation coverage, which measures the proportion of assets validated through both LSEG database validation and RAV. Total validation coverage is computed as $\left(\frac{N_v}{N_t}\right) \times 100$, where N_v represents the number of assets validated, and N_t is the total number of assets in the constructed database.

Table 2 presents validation coverage for each company, comparing LSEG-only validation to combined LSEG and RAV validation. Occidental Petroleum (OXY) is excluded due to the absence of company-specific information in the LSEG database and the constructed dataset containing only general assets (e.g., natural gas fields).

Since our validation applies only to named assets, general assets remain largely unverified. While extrapolating validation to unnamed assets could extend coverage, this introduces risks to accuracy and completeness. Notably, RAV significantly increases coverage, underscoring its role in enhancing database robustness by validating assets absent from LSEG.

Coverage varies across companies; D achieves the highest at 33.33%, while COP has the lowest at 6.43%, reflecting differences in named asset proportions. Lower coverage suggests a higher proportion of unnamed assets, highlighting gaps in the current validation process.

Sector	Company	Validation Coverage	
		LSEG Database	**LSEG Database + RAV**
Mining	AA	6.96%	20.00%
	SCCO	7.27%	7.27%
	FCX	7.35%	21.32%
	HL	17.11%	21.05%
	NEM	19.51%	30.89%
Oil and Gas	CVX	7.61%	20.65%
	XOM	7.69%	23.08%
	MPC	15.12%	25.58%
	COP	0.71%	6.43%
Utilities	D	6.06%	33.33%
	ED	1.94%	17.48%
	DUK	3.90%	23.38%
	EXC	9.57%	24.35%
	NEE	0%	10.77%

Table 2: Validation coverage comparison using LSEG databases alone versus LSEG databases with RAV.

As regulatory demands like the EUDR grow, the need for automated, comprehensive databases will increase. Our LLM-based pipeline can adapt to these demands, improving ESG and CSR compliance. The feedback loop (Figure 4) from regulatory success will drive continuous improvements in data quality and database creation techniques, shaping the future of environmental data management.

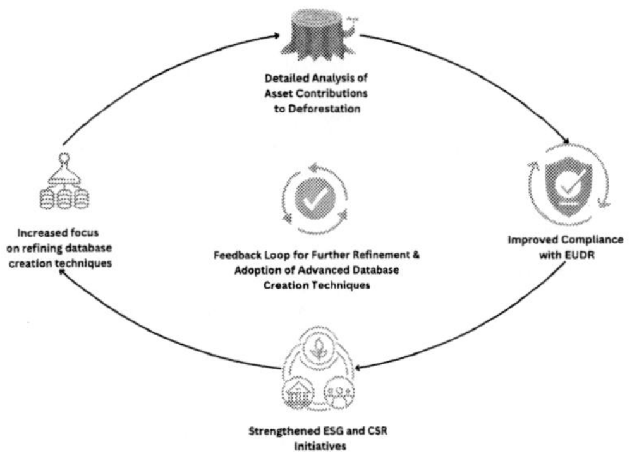

Figure 4: A feedback loop linking physical asset database creation with improved compliance and ESG initiatives, driving continuous refinement.

8 Conclusion

This study developed an end-to-end LLM-based pipeline for extracting, structuring, cleaning, and validating physical asset data from SEC EDGAR filings, providing a scalable and automated approach to asset tracking and environmental impact analysis. The pipeline follows a structured architecture, incorporating data acquisition and processing, entity extraction, database creation, database cleaning and a two-step validation framework: LSEG database validation followed by Retrieval-Augmented Validation (RAV). The table parsing improvement module also significantly increased validation coverage, enhancing database completeness. Using LLMs for automation, this study advances structured data extraction and validation, laying the foundation for more efficient regulatory compliance and environmental data management. As regulatory demands evolve, LLM-based techniques will play an increasingly critical role in ensuring accurate, structured asset tracking.

9 Limitations

Despite its success, this study has several limitations. The dataset was limited to fifteen companies across three sectors, using only three years of SEC filings, which may restrict the generalisability of the findings. Additionally, the project relied heavily on SEC 10-K filings, excluding potential insights from 8-K, 10-Q, and other reports that may contain relevant asset data. While the IRZ-CoT prompt engineering technique significantly improved entity extraction, its effectiveness in other domains or regulatory environments remains untested.

The integration of LLMs introduces maintenance

and adaptation challenges, especially as newer models may require retuning for continued effectiveness. Biases in pre-trained data could also impact extraction accuracy, particularly in financial and environmental sectors. The RAV validation process depends on web search quality, meaning incomplete or inaccurate online data could affect validation reliability. Although automation reduced manual effort, human verification was necessary for ground truth dataset creation and qualitative assessments. Additionally, the pipeline primarily focuses on historical data, making real-time asset tracking and monitoring challenging.

To address these limitations, future improvements should refine prompt engineering to extract more named assets, reducing reliance on proxies. Expanding the dataset to additional industries and historical filings will enhance coverage and validation effectiveness. Fine-tuning LLMs using domain-specific databases could enhance extraction accuracy. Additionally, dynamic prompt engineering combined with reinforcement learning — where the model is rewarded for accurate extractions and penalized for errors — could help the system adapt more effectively to different types of company disclosures.

RAV significantly increases validation coverage. It could further benefit from integrating advanced web search APIs and adopting a weighted scoring system to account for partial matches, improving validation granularity. Broader data issues also affected validation when web search results were incomplete or irrelevant, leading to default scores of 0. Expanding search queries, integrating multiple search engines, and refining data extraction techniques could improve RAV's robustness. Lastly, enhancing automated table parsing would improve structured data extraction from SEC filings, particularly in appendices and financial disclosures where asset details are often tabulated. This could be achieved by automated prompt tailoring, where an LLM (e.g., GPT-4) identifies asset examples from parsed tables and incorporates them into IRZ-CoT prompts.

10 Ethics Statement

This project adheres to the ACM Code of Ethics (ACM, 2018) and the Ethics Guidelines for Trustworthy AI (European Commission, 2019), emphasizing transparency, fairness, accountability, and technical robustness. It relies on publicly available SEC EDGAR filings, ensuring legal compliance and data integrity. Ethical data acquisition is upheld by adhering to data source terms and conditions, preventing legal conflicts. The project contributes to environmental sustainability by assessing industrial deforestation impacts, supporting global climate initiatives, and reinforcing corporate social responsibility (CSR) through ESG compliance. Accuracy and fairness are prioritized via a two-step validation pipeline, minimizing misrepresentations and addressing LLM biases through transparency and human oversight. By maintaining industry best practices in data processing, the project ensures reliability, robustness, and responsible AI development for ethical and accurate asset tracking.

Acknowledgments

We would like to express our sincere gratitude to the LSEG (London Stock Exchange Group) team for providing a free license to the LSEG Workspace and for their invaluable guidance throughout the course of this research. Their support significantly contributed to the quality and depth of this work.

References

Peride Abdurehim, Turdy Tohti, and Askar Hamdulla. 2020. A short review of relation extraction methods. In *2020 13th International Conference on Intelligent Computation Technology and Automation (ICICTA)*, pages 18–22.

ACM. 2018. Acm code of ethics and professional conduct.

Amazon Watch. 2016. The amazon crude: How american consumers fuel deforestation and human rights abuses in south america. Technical report, Amazon Watch, Oakland, CA.

Ron Banner, Yury Nahshan, Elad Hoffer, and Daniel Soudry. 2019. Post-training 4-bit quantization of convolution networks for rapid-deployment.

Iz Beltagy, Arman Cohan, and Kyle Lo. 2019. Scibert: Pretrained contextualized embeddings for scientific text. *CoRR*, abs/1903.10676.

M. Bommarito, D. Katz, and E. M. Detterman. 2018. LexNLP: Natural Language Processing and Information Extraction For Legal and Regulatory Texts. *InfoSciRN: Legal Informatics (Topic)*.

Sian Bradley. 2020. Mining's impacts on forests: Aligning policy and finance for climate and biodiversity goals. Technical report, Chatham House, London, UK.

Ervina Cergani and Pauli Miettinen. 2013. Discovering relations using matrix factorization methods. In *Proceedings of the 22nd ACM International Conference on Information & Knowledge Management*, CIKM '13, page 1549–1552, New York, NY, USA. Association for Computing Machinery.

CGFI. 2024. Geoasset project: Petrochemicals.

Ilias Chalkidis, Manos Fergadiotis, Prodromos Malakasiotis, Nikolaos Aletras, and Ion Androutsopoulos. 2020. LEGAL-BERT: the muppets straight out of law school. *CoRR*, abs/2010.02559.

Hyung Won Chung, Le Hou, Shayne Longpre, Barret Zoph, Yi Tay, William Fedus, Yunxuan Li, Xuezhi Wang, Mostafa Dehghani, Siddhartha Brahma, Albert Webson, Shixiang Shane Gu, Zhuyun Dai, Mirac Suzgun, Xinyun Chen, Aakanksha Chowdhery, Alex Castro-Ros, Marie Pellat, Kevin Robinson, Dasha Valter, Sharan Narang, Gaurav Mishra, Adams Yu, Vincent Zhao, Yanping Huang, Andrew Dai, Hongkun Yu, Slav Petrov, Ed H. Chi, Jeff Dean, Jacob Devlin, Adam Roberts, Denny Zhou, Quoc V. Le, and Jason Wei. 2022. Scaling instruction-finetuned language models.

Tim Dettmers, Artidoro Pagnoni, Ari Holtzman, and Luke Zettlemoyer. 2023. Qlora: Efficient finetuning of quantized llms.

Tobias Deußer, Syed Musharraf Ali, Lars Hillebrand, Desiana Nurchalifah, Basil Jacob, Christian Bauckhage, and Rafet Sifa. 2022. Kpi-edgar: A novel dataset and accompanying metric for relation extraction from financial documents. In *2022 21st IEEE International Conference on Machine Learning and Applications (ICMLA)*. IEEE.

Jacob Devlin, Ming-Wei Chang, Kenton Lee, and Kristina Toutanova. 2018. BERT: pre-training of deep bidirectional transformers for language understanding. *CoRR*, abs/1810.04805.

ETC. 2024. Financing the transition: The costs of avoiding deforestation. Technical report, Energy Transitions Commission, London, UK.

European Commission. 2019. Ethics guidelines for trustworthy ai.

European Commission. 2023. Regulation on deforestation-free products.

Matt Finer, Clinton N. Jenkins, Stuart L. Pimm, Brian Keane, and Carl Ross. 2008. Oil and gas projects in the western amazon: Threats to wilderness, biodiversity, and indigenous peoples. *PLoS ONE*, 3(8):e2932.

Forest 500. 2024. 2024: A decade of deforestation data.

Google Developers. 2024. Custom search json api overview.

Hanxu Hu, Simon Yu, Pinzhen Chen, and Edoardo M. Ponti. 2024. Fine-tuning large language models with sequential instructions.

Francesca Imperiale, Simone Pizzi, and Stella Lippolis. 2023. Sustainability reporting and esg performance in the utilities sector. *Utilities Policy*, 80:101468.

IntegrityNext. 2024. Energy & utilities.

Zhengbao Jiang, Frank F. Xu, Luyu Gao, Zhiqing Sun, Qian Liu, Jane Dwivedi-Yu, Yiming Yang, Jamie Callan, and Graham Neubig. 2023. Active retrieval augmented generation.

Jinhyuk Lee, Wonjin Yoon, Sungdong Kim, Donghyeon Kim, Sunkyu Kim, Chan Ho So, and Jaewoo Kang. 2019. BioBERT: a pre-trained biomedical language representation model for biomedical text mining. *Bioinformatics*, 36(4):1234–1240.

London Stock Exchange Group (LSEG). 2024. Lseg workspace: Data and analytics.

Moody et al. 2024. Sec-edgar: Download all companies' periodic reports, filings, and forms from edgar database. GitHub repository, Apache-2.0 License.

Nils Reimers and Iryna Gurevych. 2019. Sentence-bert: Sentence embeddings using siamese bert-networks. In *Proceedings of the 2019 Conference on Empirical Methods in Natural Language Processing*. Association for Computational Linguistics.

David M. Rosenberg, Patrick McCully, and Catherine M. Pringle. 2000. Global-scale environmental effects of hydrological alterations: Introduction. *BioScience*, 50(9):746–751.

Erik F. Tjong Kim Sang and Fien De Meulder. 2003. Introduction to the CoNLL-2003 shared task: Language-independent named entity recognition. In *Proceedings of the Seventh Conference on Natural Language Learning at HLT-NAACL 2003*, pages 142–147.

Milena Trajanoska, Riste Stojanov, and Dimitar Trajanov. 2023. Enhancing knowledge graph construction using large language models.

Meimei Tuo and Wenzhong Yang. 2023. Review of entity relation extraction. *Journal of Intelligent & Fuzzy Systems*, 44(5):7391–7405.

Shuhe Wang, Xiaofei Sun, Xiaoya Li, Rongbin Ouyang, Fei Wu, Tianwei Zhang, Jiwei Li, and Guoyin Wang. 2023. Gpt-ner: Named entity recognition via large language models.

Suxiang Zhang. 2009. Entity relation extraction to free text. In *2009 International Conference on Natural Language Processing and Knowledge Engineering*, pages 1–5.

Hangtian Zhao, Hakiz Yilahun, and Askar Hamdulla. 2023. Pipeline chain-of-thought: A prompt method for large language model relation extraction. In *2023 International Conference on Asian Language Processing (IALP)*, pages 31–36.

Suncong Zheng, Yuexing Hao, Dongyuan Lu, Hongyun Bao, Jiaming Xu, Hongwei Hao, and Bo Xu. 2017. Joint entity and relation extraction based on a hybrid neural network. *Neurocomputing*, 257:59–66. Machine Learning and Signal Processing for Big Multimedia Analysis.

A Appendix

A.1 Sectors, Companies and Stock Tickers used

Sector	Company Name	Stock Ticker
Mining	Alcoa Corporation	AA
	Hecla Mining Corporation	HL
	Newmont Corporation	NEM
	Freeport Mc-Moran	FCX
	Southern Copper Corporation	SCCO
Oil & Gas	ConocoPhillips Company	COP
	Marathon Petroleum Corporation	MPC
	Chevron Corporation	CVX
	Occidental Petroleum	OXY
	Exxon Mobil Corporation	XOM
Utilities	Dominion Energy	D
	Duke Energy Corporation	DUK
	Consolidated Edison	ED
	Exelon Corporation	EXC
	NextEra Energy	NEE

Table 3: Companies used for pipeline construction, grouped by sector.

A.2 Chunks per document in Chunk-based Querying Technique

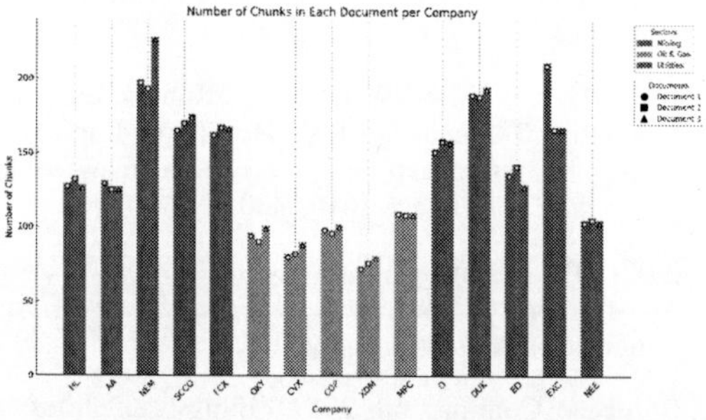

Figure 5: Number of chunks generated per document for each company across the three sectors.

A.3 Cosine similarities of LLM-generated outputs against NER results

Model	Location - cosine similarity	Organisation - cosine similarity
Mistral-7B	0.6945	0.6809
Llama 3	0.5987	0.7177
Gemma 2	0.7702	0.7461

Table 4: Cosine similarities of LLM-generated outputs against NER results.

A.4 Error Analysis of LLM-generated outputs against NER results

Model	Type	Description (LLM vs. NER Output)
Mistral-7B	Success	**LLM:** Accurately extracted locations "South Church Street, Charlotte, North Carolina" and organisations. **NER:** Fragmented these into "Securities Exchange" and "##TON." Split up locations like 'US' into 'U' and 'S'.
	Failure	**LLM:** Sometimes failed to extract any entities. **NER:** Correctly identified "Duke Energy."
	Challenge	**LLM:** Provided a list of relevant organisations but missed key locations. **NER:** Captured more entities but included irrelevant fragments like "GENERA."
LLaMA 3	Success	**LLM:** Correctly extracted "Duke Energy" without fragmentation. **NER:** Fragmented other entities and introduced spurious ones like "Per" and "Board of Directors".
	Failure	**LLM:** Sometimes missed all entities. **NER:** Extracted locations like "Central" and organisations like "Spectra Energy."
	Challenge	**LLM:** Identified relevant text but added repeated/irrelevant words. **NER:** Output was more accurate but fragmented.
Gemma 2	Success	**LLM:** Correctly extracted locations and organisations. **NER:** Fragmented entities and returned partial results.
	Failure	**LLM:** Sometimes missed all entities. **NER:** Extracted "Duke Energy" and "Cinergy."
	Challenge	**LLM:** Missed key locations. **NER:** Output included a broader range of organisations but included irrelevant fragments.

Table 5: Examples of Successes, Failures, and Challenges in entity extraction by LLMs compared to NER.

A.5 Problem descriptions in Iterative Prompt Refinement

Prompt Name	Problem Description	Problem Example
Zero-shot	Incorrect mention of financial assets as physical assets, lack of location specificity	`'asset: Consolidated Operations - Revenues', 'location: Seven countries'`
One-shot	Hallucination — includes example in answer	`'ownership: Freeport-McMoran'`
Few-shot	Incorrect classification of physical asset, hallucination — includes example in answer	`'physical asset: Granted patents (intellectual property), Registered trademarks (intellectual property)', 'physical assets: [Grasberg mine]'`
Chain-of-thought	Incorrect mention of location as physical asset; provides irrelevant answers if no information is available	`'physical asset: Brazil', 'relationships: [asset: ", location: ", ownership: ", commodities: "]'`
Generated knowledge	Outputs overly detailed and verbose answers	`'physical assets: [substantially all assets of the Company, ...]'`
Prompt chaining	Lack of specificity	`'physical assets: [substantially all assets of the Company]', 'physical assets: [mining operations, properties, leases]'`
Role prompting	Improvement noted but lack of specificity	`'physical assets: [Mining operations, processing plants, manufacturing facilities]'`
Role and instructional prompting	Improvement noted but lack of specificity	`'physical assets: [facilities]'`

Table 6: Problem descriptions and examples for each prompting technique.

A.6 Computational times of prompting techniques in Iterative Prompt Refinement

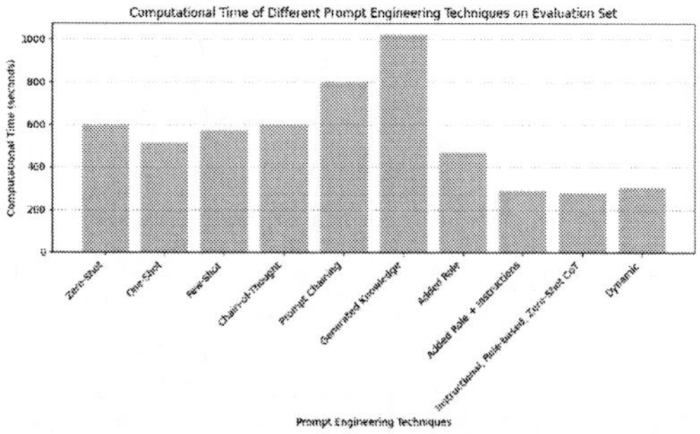

Figure 6: Computational time (in seconds) required for different prompt engineering techniques on the evaluation set.

A.7 Results of LLM Ensemble Implementation

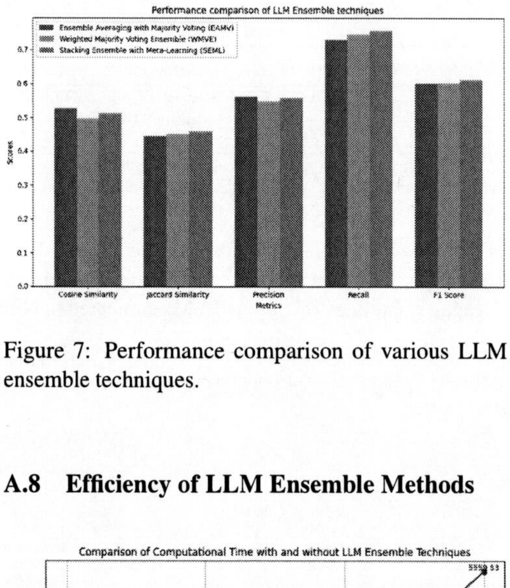

Figure 7: Performance comparison of various LLM ensemble techniques.

A.8 Efficiency of LLM Ensemble Methods

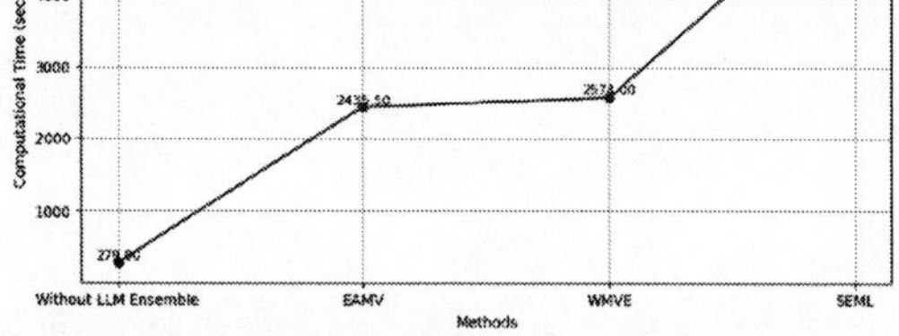

Figure 8: Comparison of computational times with and without using LLM ensemble techniques.

A.9 LSEG Database Validation Results

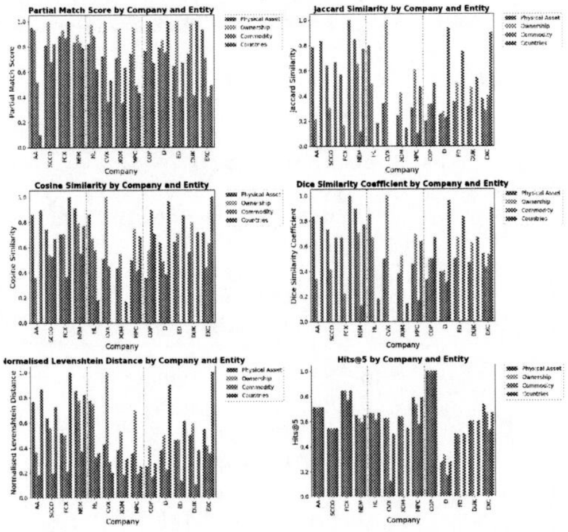

Figure 9: Results from the LSEG database validation across various metrics.

A.10 Coverage Calculation Results

Sector	Company	Coverage
Mining	AA	71.43%
	SCCO	54.55%
	HL	66.67%
	NEM	62.16%
	FCX	92.31%
Oil & Gas	CVX	25%
	XOM	0%
	MPC	94.74%
	OXY	N/A
	COP	100%
Utilities	D	27.78%
	DUK	40%
	EXC	13.33%
	ED	50%
	NEE	N/A

Table 7: Calculated coverage percentages for databases across sectors.

A.11 Coverage before and after implementation of table parsing

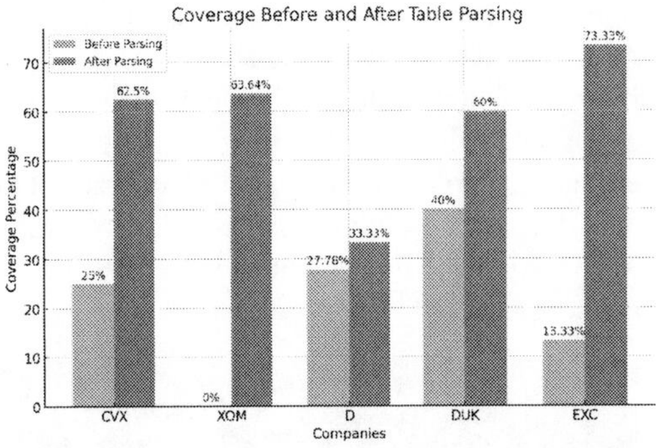

Figure 10: Coverage percentages before and after performing table parsing.

A.12 Retrieval Augmented Validation (RAV) Results

Sector	Company	Similarity Score
Mining	AA	0.42
	SCCO	0.37
	FCX	0.43
	NEM	0.38
	HL	0.46
Oil and Gas	CVX	0.44
	XOM	0.48
	MPC	0.52
	COP	0.52
Utilities	D	0.33
	ED	0.38
	DUK	0.40
	EXC	0.31
	NEE	0.57

Table 8: Averaged classification similarity scores from Retrieval Augmented Validation (RAV).

A.13 Prompt Library

In this section, we present the comprehensive collection of prompts utilised throughout this project for information extraction. The prompt library consists of a variety of carefully designed instructions aimed at guiding LLMs in extracting specific entities and relationships, such as physical assets, locations, ownership details, and commodities, from SEC EDGAR filings. Each prompt is tailored to enhance the performance of LLMs in different scenarios, employing techniques such as zero-shot, one-shot, few-shot prompting, and more sophisticated methods like Chain-of-Thought (CoT) reasoning and role-based prompting, before constructing our IRZ-CoT prompt.

The prompts are categorised based on their usage in different stages of the project, including entity extraction, database cleaning and when creating an improvement module. This prompt library serves as the foundation for automating the extraction process and ensuring the reliability and accuracy of the data. Each prompt has been optimised through iterative testing and refinement to address the unique challenges posed by our use case.

Prompt for extracting locations and organisations when comparing LLMs and NER

```
prompt_instruction = "You are
    a virtual assistant with
    advanced expertise in a
    broad spectrum of topics,
    equipped to utilize high-
    level critical thinking,
    cognitive skills,
    creativity, and innovation.
     Your goal is to deliver
    the most straightforward
    and accurate answer
    possible for each question,
     ensuring high-quality and
    useful responses for the
    user."
user_prompt = f"Text: {chunk}\
    nQuery: Does this text
    mention any locations or
    organisations? If yes,
    please specify them in the
    following format:\
    nlocations: [ ]\
    norganisations: [ ]"
```

Zero-shot

```
f"Text: {chunk}\nQuery: Does this
    text mention any physical
    assets, locations, ownerships,
    and commodities? "
        "If yes, please specify
            them in the following
            format:\n"
        "physical assets: [ ]\
            nlocations: [ ]\
            nownerships: [ ]\
            ncommodities: []\n"
        "Additionally, identify
            the relationships
            between them,
            specifying the location
             of each physical asset
            , their ownership
            details, and
            commodities. "
        "Format the relationships
            as follows:\
            nrelationships: [asset:
             '', location: '',
            ownership: '',
            commodities: '']"
```

One-shot

```
f"Text: {chunk}\nQuery: Does this
    text mention any physical
    assets, locations, ownerships,
    and commodities? "
        "If yes, please specify
            them in the following
            format:\n"
        "physical assets: [ ]\
            nlocations: [ ]\
            nownerships: [ ]\
            ncommodities: []\n"
        "Additionally, identify
            the relationships
            between them,
            specifying the location
             of each physical asset
            , their ownership
            details, and
            commodities. "
        "Format the relationships
            as follows:\
            nrelationships: [asset:
             '', location: '',
            ownership: '',
            commodities: '']"
        "Here is an example:\n"
        "Example:\n"
        "Text: [...] Our principal
             asset is the Grasberg
            mine, which we
            discovered in 1988.
            Grasberg contains the
            largest single gold
            reserve and one of the
            largest copper reserves
             of any mine in the
            world. Our principal
            operating subsidiary is
             PT Freeport Indonesia,
             a limited liability
            company organized under
             the laws of the
            Republic of Indonesia
            and incorporated in
            Delaware. [...]"
        "Query: Does this text
            mention any physical
            assets, locations, and
            ownerships?\n"
        "physical assets: [
            Grasberg mine]\
            nlocations: [Sudirman
            Mountain Range, Papua,
            Indonesia]\nownerships:
             [Republic of Indonesia
            , Delaware]\n[
            commodities: copper,
            gold]\n"
```

```
"relationships: [asset: 'Grasberg mine', location: 'Indonesia', ownership: 'PT Freeport Indonesia', commodities: 'copper', 'gold']\n\n"
```

Few-shot

```
f"Text: {chunk}\nQuery: Does this text mention any physical assets, locations, ownerships, and commodities? "
"If yes, please specify them in the following format:\n"
"physical assets: [ ]\nlocations: [ ]\nownerships: [ ]\ncommodities: []\n"
"Additionally, identify the relationships between them, specifying the location of each physical asset, their ownership details, and commodities. "
"Format the relationships as follows:\nrelationships: [asset: '', location: '', ownership: '', commodities: '']"
"Here are some examples:\n"
"Example:\n"
"Text: [...] Our principal asset is the Grasberg mine, which we discovered in 1988. Grasberg contains the largest single gold reserve and one of the largest copper reserves of any mine in the world. Our principal operating subsidiary is PT Freeport Indonesia, a limited liability company organized under the laws of the Republic of Indonesia and incorporated in Delaware. [...]"
"Query: Does this text mention any physical assets, locations, and ownerships?\n"
"physical assets: [Grasberg mine]\nlocations: [Sudirman Mountain Range, Papua, Indonesia]\nownerships: [Republic of Indonesia, Delaware]\n[commodities: copper, gold]\n"
"relationships: [asset: 'Grasberg mine', location: 'Indonesia', ownership: 'PT Freeport Indonesia', commodities: 'copper', 'gold']\n\n"

"Example 2:\n"
"Text: [...] PT Freeport Indonesia mines, processes and explores for ore containing copper, gold and silver. It operates in the remote highlands of the Sudirman Mountain Range in the province of Papua (formerly Irian Jaya), Indonesia, which is on the western half of the island of New Guinea. [...]"
"Query: Does this text mention any physical assets, locations, and ownerships?\n"
"physical assets: [PT Freeport Indonesia Mines]\nlocations: [Sudirman Mountain Range, Papua, Indonesia, New Guinea]\nownerships: [PT Freeport Indonesia]\ncommodities: [copper, gold, silver]\n"
"relationships: \n"
"[asset: 'PT Freeport Indonesia Mines', location: 'Sudirman Mountain Range, Papua, Indonesia, New Guinea', ownership: 'PT Freeport Indonesia', commodities: 'copper, gold, silver']\n"

"Example 3:\n"
"Text: [...] The Republic of Indonesia consists of more than 17,000 islands stretching 3,000 miles along the equator from Malaysia to Australia and is the fourth most populous nation in the world with over 200 million people. [...]"
"Query: Does this text mention any physical assets, locations, and ownerships?\n"
```

```
        "physical assets: []\
            nlocations: [Republic
            of Indonesia, Malaysia,
             Australia]\nownerships
            : []\ncommodities: []\n
            \n"
        "relationships: \n"
```

Chain-of-Thought (CoT) prompting

```
f"Text: {chunk}\nQuery:  Let's
    think step by step. First,
    identify any physical assets
    mentioned in the text. Next,
    determine if any locations or
    ownership details are provided
    for these physical assets. Then
    , determine if the commodities
    related to the physical assets
    are provided. Finally,
    summarize the relationships
    between each physical asset,
    its location, its ownership and
     its commodity. "
        "If yes, please specify
            them in the following
            format:\n"
        "physical assets: [ ]\
            nlocations: [ ]\
            nownerships: [ ]\
            ncommodities: []\n"
        "Additionally, identify
            the relationships
            between them,
            specifying the location
             of each physical asset
            , their ownership
            details, and
            commodities. "
        "Format the relationships
            as follows:\
            nrelationships: [asset:
             '', location: '',
            ownership: '',
            commodities: '']"
```

Generated knowledge prompting

```
    knowledge_prompt = (
        "You are an expert in
            analyzing texts for
            information about
            physical assets,
            locations, ownerships,
            and commodities. "
        "Provide a brief summary
            of how to identify
            these elements in a
            text and the
            relationships between
            them."
    )
    prompt_step1 = (
        f"{generated_knowledge}\n\
            n"
        "You are a virtual
            assistant with
            expertise in extracting
             specific information
            from text. "
        "A physical asset is an
            asset with a
            geographical location.\
            n\n"
        f"Text: {chunk}\nQuery:
            Identify any physical
            assets mentioned in the
             text. "
        "List them in the format:\
            nphysical assets: [ ]"
    )
    prompt_step2 = (
        f"{generated_knowledge}\n\
            n"
        "Using the extracted
            physical assets:\n"
        f"physical assets: {
            physical_assets}\n\n"
        f"Text: {chunk}\nQuery:
            Identify any locations
            mentioned in the text
            associated with the
            physical assets. "
        "List them in the format:\
            nlocations: [ ]"
    )
    prompt_step3 = (
        f"{generated_knowledge}\n\
            n"
        "Using the extracted
            physical assets and
            locations:\n"
        f"physical assets: {
            physical_assets}\
            nlocations: {locations
            }\n\n"
        f"Text: {chunk}\nQuery:
            Identify any ownership
            details mentioned in
            the text associated
            with the physical
            assets. "
        "List them in the format:\
            nownerships: [ ]"
    )
    prompt_step4 = (
        f"{generated_knowledge}\n\
            n"
        "Using the extracted
            physical assets,
            locations, and
            ownerships:\n"
        f"physical assets: {
            physical_assets}\
            nlocations: {locations
            }\nownerships: {
            ownerships}\n\n"
        f"Text: {chunk}\nQuery:
            Identify any
            commodities mentioned
            in the text associated
```

```
            with the physical
            assets. "
        "List them in the format:\
            ncommodities: [ ]"
    )
    prompt_step5 = (
        f"{generated_knowledge}\n\
            n"
        "Using the extracted
            physical assets,
            locations, ownerships,
            and commodities:\n"
        f"physical assets: {
            physical_assets}\
            nlocations: {locations
            }\nownerships: {
            ownerships}\
            ncommodities: {
            commodities}\n\n"
        "Text: {chunk}\nQuery:
            Identify the
            relationships between
            the physical assets,
            locations, ownerships,
            and commodities. "
        "Format the relationships
            as follows:\
            nrelationships: [asset:
             '', location: '',
            ownership: '',
            commodities: '']"
    )
```

Prompt Chaining

```
    prompt_step1 = (
        f"Text: {chunk}\nQuery:
            Does this text mention
            any physical assets? "
        "If yes, please specify
            them in the following
            format:\n"
        "physical assets: [ ]"
    )

    prompt_step2 = (
        f"physical assets: {
            physical_assets}\n"
        f"Text: {chunk}\nQuery:
            Does this text mention
            any locations
            associated with the
            physical assets? "
        "If yes, please specify
            them in the following
            format:\n"
        "locations: [ ]"
    )

    prompt_step3 = (
        f"physical assets: {
            physical_assets}\
            nlocations: {locations
            }\n"
        f"Text: {chunk}\nQuery:
            Does this text mention
            any ownership details
            associated with the
            physical assets? "
        "If yes, please specify
            them in the following
            format:\n"
        "ownerships: [ ]"
    )

    prompt_step4 = (
        f"physical assets: {
            physical_assets}\
            nlocations: {locations
            }\nownerships: {
            ownerships}\n"
        f"Text: {chunk}\nQuery:
            Does this text mention
            any commodities
            associated with the
            physical assets? "
        "If yes, please specify
            them in the following
            format:\n"
        "commodities: [ ]"
    )

    prompt_step5 = (
        f"physical assets: {
            physical_assets}\
            nlocations: {locations
            }\nownerships: {
            ownerships}\
            ncommodities: {
            commodities}\n"
        f"Text: {chunk}\nQuery:
            Identify the
            relationships between
            the physical assets,
            locations, ownerships,
            and commodities. "
        "Format the relationships
            as follows:\
            nrelationships: [asset:
             '', location: '',
            ownership: '',
            commodities: '']"
    )
```

Role prompting

```
    prompt_instruction = (
        "You are a virtual
            assistant with advanced
             expertise in a broad
            spectrum of topics,
            equipped to utilize
            high-level critical
            thinking, cognitive
            skills, creativity, and
             innovation.\n"
        "Your goal is to deliver
            the most
            straightforward and
            accurate answer
```

```
        possible for each
        question, ensuring high
        -quality and useful
        responses for the user
        .\n"
    "Now, let's analyze the
        following text:\n"
    f"Text: {chunk}\nQuery:
        Does this text mention
        any physical assets,
        locations or ownerships
        ? Does the text mention
         what commodity the
        physical asset is being
         used for?\n"
    "If yes, you must specify
        them in the following
        format:\n"
    "physical assets: [ ]\
        nlocations: [ ]\
        nownerships: [ ]\
        ncommodities: []\n"
    "Additionally, identify
        the relationships
        between them,
        specifying the location
         of each physical asset
        , the ownership details
        , the commodity the
        physical asset is used
        for and the status of
        the physical asset. "
    "Format the relationships
        as follows:\
        nrelationships: [asset:
         '', location: '',
        ownership: '',
        commodity: '']."
)
```

Role + instructional prompting

```
    "You are a virtual
        assistant with advanced
         expertise in a broad
        spectrum of topics,
        equipped to utilize
        high-level critical
        thinking, cognitive
        skills, creativity, and
         innovation.\n"
    "Your goal is to deliver
        the most
        straightforward and
        accurate answer
        possible for each
        question, ensuring high
        -quality and useful
        responses for the user
        .\n"
    "A physical asset is a
        tangible resource that
        a company owns and uses
         in the production of
        goods and services.
        Examples of physical
        assets are facilities,
        equipment,
        infrastructure, etc.
        Ensure that a
        geographical location
        or region is never
        considered as an asset
        .\n"
    "A financial asset or
        other non-physical
        asset should never be
        included as a physical
        asset. Examples of
        financial assets
        include equity
        commitments, corporate
        facilities, accounts
        receivable, and short-
        term investments. Never
         include these in the
        list of physical assets
        .\n"
    "A commodity is what the
        physical asset is being
         used for. Examples
        include copper, gold,
        electricity, renewable
        energy, etc."
    "Now, let's analyze the
        following text:\n"
    f"Text: {chunk}\nQuery:
        Does this text mention
        any physical assets,
        locations or ownerships
        ? Does the text mention
         what commodity the
        physical asset is being
         used for?\n"
    "If yes, you must specify
        them in the following
        format:\n"
    "physical assets: [ ]\
        nlocations: [ ]\
        nownerships: [ ]\
        ncommodities: []\n"
    "Additionally, identify
        the relationships
        between them,
        specifying the location
         of each physical asset
        , the ownership details
        , and the commodity the
         physical asset is used
         for."
    "Format the relationships
        as follows:\
        nrelationships: [asset:
         '', location: '',
        ownership: '',
        commodity: '']. Do not
        output anything else."
)
```

IRZ-CoT prompting

```
prompt_instruction = (
    "You are a virtual
        assistant with advanced
         expertise in a broad
        spectrum of topics,
        equipped to utilize
        high-level critical
        thinking, cognitive
        skills, creativity, and
         innovation.\n"
    "Your goal is to deliver
        the most
        straightforward and
        accurate answer
        possible for each
        question, ensuring high
        -quality and useful
        responses for the user
        .\n"
    "A physical asset is a
        tangible resource that
        a company owns and uses
         in the production of
        goods and services.
        Examples of physical
        assets are facilities,
        equipment,
        infrastructure, etc.
        Ensure that a
        geographical location
        or region is never
        considered as an asset
        .\n"
    "A financial asset or
        other non-physical
        asset should never be
        included as a physical
        asset. Examples of
        financial assets
        include equity
        commitments, corporate
        facilities, accounts
        receivable, and short-
        term investments. Never
         include these in the
        list of physical assets
        .\n"
    "A commodity is what the
        physical asset is being
         used for. Examples
        include copper, gold,
        electricity, renewable
        energy, etc."
    "Now, let's analyze the
        following text:\n"
    f"Text: {chunk}\nQuery:
        Let's think step-by-
        step. Does this text
        mention any physical
        assets, locations or
        ownerships? Does the
        text mention what
        commodity the physical
        asset is being used for
        ?\n"
    "If yes, you must specify
        them in the following
        format:\n"
    "physical assets: [ ]\
        nlocations: [ ]\
        nownerships: [ ]\
        ncommodities: []\n"
    "Additionally, identify
        the relationships
        between them,
        specifying the location
         of each physical asset
        , the ownership details
        , and the commodity the
         physical asset is used
         for."
    "Format the relationships
        as follows:\
        nrelationships: [asset:
         '', location: '',
        ownership: '',
        commodity: '']. Do not
        output anything else."
)
```

Dynamic prompting

```
prompt_instruction = (
    "You are a virtual
        assistant with advanced
         expertise in a broad
        spectrum of topics,
        equipped to utilize
        high-level critical
        thinking, cognitive
        skills, creativity, and
         innovation.\n"
    "Your goal is to deliver
        the most
        straightforward and
        accurate answer
        possible for each
        question, ensuring high
        -quality and useful
        responses for the user
        .\n"
)

if contains_assets:
    prompt_instruction += (
        "A physical asset is a
            tangible resource
            that a company owns
             and uses in the
            production of goods
             and services.
            Examples of
            physical assets are
             facilities,
            equipment,
            infrastructure, etc
            .\n"
        "Ensure that a
            geographical
            location or region
            is never considered
```

```
            as an asset.\n"
        "A financial asset or
            other non-physical
            asset should never
            be included as a
            physical asset.
            Examples of
            financial assets
            include equity
            commitments,
            corporate
            facilities,
            accounts receivable
            , and short-term
            investments. Never
            include these in
            the list of
            physical assets.\n"
    )

if contains_commodities:
    prompt_instruction += (
        "A commodity is what
            the physical asset
            is being used for.
            Examples include
            copper, gold,
            electricity,
            renewable energy,
            etc.\n"
    )

if contains_locations:
    prompt_instruction += (
        "Always ensure that a
            geographical
            location or region
            is mentioned
            separately from the
             physical asset.\n"
    )

prompt_instruction += (
    f"Now, let's analyze the
        following text:\n"
    f"Text: {chunk}\nQuery:
        Let's think step-by-
        step. Does this text
        mention any physical
        assets, locations or
        ownerships? Does the
        text mention what
        commodity the physical
        asset is being used for
        ?\n"
    "If yes, you must specify
        them in the following
        format:\n"
    "physical assets: [ ]\
        nlocations: [ ]\
        nownerships: [ ]\
        ncommodities: []\n"
    "Additionally, identify
        the relationships
        between them,
        specifying the location
         of each physical asset
        , the ownership details
            , and the commodity the
             physical asset is used
             for."
        "Format the relationships
            as follows:\
            nrelationships: [asset:
             '', location: '',
            ownership: '',
            commodity: '']. Do not
            output anything else."
    )
```

Database Cleaning

```
You are an expert data cleaner.
    Your task is to clean and
    standardize the following text.
     You will be provided each cell
     value one by one with its
    respective column name. Apply
    the following cleaning steps:

    - Standardize entries in the
       commodity column to have a
       consistent format. For
       example, "Silver, Gold, Lead
       , Zinc" should be the
       standard format for each
       commodity, separated by
       commas and no extra spaces.
    - Ensure all entries in the
       status column are in a
       consistent format, removing
       redundant words or phrases.
    - All entries should be in
       title case.
    - Do not make any changes to
       the 'Countries' column.
    - In the 'commodity' column, if
        chemical symbols are given,
        change these to the element
        name corresponding to the
        chemical symbol.
    - In all the columns, ensure
       each entry is properly
       formatted without redundant
       commas and extra spaces. For
        example, "ExxonMobil"
       should not be separated by
       extra commas.
    - Remove any leading or
       trailing spaces in all
       columns.
    - All individual commodities
       should be separated by a
       comma.
    - The 'location' column should
       only consist of geographical
        regions and locations.
    - A physical asset is a
       tangible resource that a
       company owns and uses in the
        production of goods and
       services. Examples of
       physical assets are
```

```
        facilities, equipment,
        infrastructure, etc. If
        there are any entries in the
         physical asset column that
        do not fit the description
        of a physical asset, put N/A
         in the corresponding cell.
    - A commodity is what the
        physical asset is being used
         for. If there are any
        entries in the commodity
        column that do not fit the
        description of a commodity,
        put N/A next to the word in
        brackets.
    - Ensure that there are no
        repetitions or redundant
        entries in any of the cells.
    - If any cell has 'not
        specified', it should be
        empty.
    - All cells should have
        standardized entries.

Process the following text
    according to these instructions
    . Return only the new cleaned
    cell value, nothing else.
```

Country Extraction Prompt

```
You are an expert in
    geographical locations.
    Given the location
    information provided,
    identify the countries
    mentioned in the location
    . Return the list of
    countries separated by
    commas. If no country is
    mentioned, return "N/A".
```

Modified prompt for improvement module

```
"You are a virtual assistant with
    advanced expertise in a broad
    spectrum of topics, equipped to
     utilize high-level critical
    thinking, cognitive skills,
    creativity, and innovation.\n"
        "Your goal is to deliver
            the most
            straightforward and
            accurate answer
            possible for each
            question, ensuring high
            -quality and useful
            responses for the user
            .\n"
```

```
"A physical asset is a
    tangible resource that
    a company owns in a
    location and uses in
    the production of goods
     and services. Examples
     of physical assets are
     all examples of 'Plant
    ' in the tables (
    Wateree, Greensville
    and Colonial Trail West
     are all physical
    assets).\n"
"A financial asset or
    other non-physical
    asset should never be
    included as a physical
    asset. Examples of
    financial assets
    include equity
    commitments, corporate
    facilities, accounts
    receivable, and short-
    term investments. Never
     include these in the
    list of physical assets
    .\n"
"A commodity is what the
    physical asset is being
     used for. The status
    of a physical asset
    gives information on
    whether the asset is
    operational, under
    construction or in end-
    of-life."
"Now, let's analyze the
    following text:\n"
f"Text: {text}\nQuery: Let
    's think step-by-step.
    Does this text mention
    any physical assets,
    locations or ownerships
    ? Does the text mention
     what commodity the
    physical asset is being
     used for?\n"
"Does the text mention the
     status of the physical
     asset? Examples of
    status include whether
    the asset is
    operational, under
    construction or in end-
    of-life."
"If yes, you must specify
    them in the following
    format:\n"
"physical assets: [ ]\
    nlocations: [ ]\
    nownerships: [ ]\
    ncommodities: []\
    nstatus: []\n"
"Additionally, identify
    all the relationships
    between them,
    specifying the location
     of each physical asset
```

```
        , the ownership details
        , the commodity the
        physical asset is used
        for and the status of
        the physical asset. Do
        not leave out any
        relationships. "
    "Format the relationships
        as follows:\
        nrelationships: [asset:
         '', location: '',
        ownership: '',
        commodity: '', status:
        '']. Do not output
        anything else."
```

A.14 Standardised Characters in Foundational Data Cleaning & Standardisation

In this section, we detail the standardisation rules applied during the Foundational Data Cleaning & Standardisation phase in our database cleaning process. By enforcing these cleaning and standardisation rules, the data becomes more reliable and easier to analyse, reducing potential errors caused by inconsistent naming and formatting practices. This standardisation forms the foundation for subsequent cleaning steps within the study.

- **Characters Standardised:**
  - \ (Backslashes): Removed from text fields.
  - ' (Single quotes): Removed from text fields.
  - " (Double quotes): Removed from text fields.
  - **Extra spaces:**
    - Leading and trailing spaces: Trimmed from text entries.
    - Multiple consecutive spaces: Condensed to a single space.
  - **Commas (,)**: Ensured proper spacing after commas by replacing them with `', '` (comma followed by a space).
- **Names Standardised:**
  - **Company Names in `ownership` Field:**
    - All the following aliases and variations are standardized to **"Newmont Corporation"**:
      - "Company"
      - "company"
      - "The company"
      - "the company"
      - "The Company"
      - "Company's"
      - "the Company"
      - "we"
      - "NEWMONT CORPORATION"
      - "Newmont's ownership or economic interest"
      - "Company owns or controls land"
      - "Newmont"
      - "Newmont (majority)"
      - "Newmont Corporation (formerly)"
      - "100% owned by the Company"
      - "Newmont Stockholders"
      - "100% by Newmont"
      - "Company owned"
      - "Company's"
  - **USA Variants in `location` Field:**
    - All the following variants are standardised to **"USA"**:
      - "United States of America"
      - "United States"
      - "USA"
      - "US"
      - "USAA"
      - "USAA."
      - "U.S."
      - "U.S.A."

A.15 LSEG Database Numerical Results

In this section, we present the complete numerical results from stage 1 of the validation, showing six performance scores for each company.

Table 9: Complete Performance Scores for Companies across the following metrics: Partial Match Score, Jaccard Similarity, Cosine Similarity, Dice Similarity Coefficient, Normalised Levenshtein Distance, and Hits@5.

Sector	Company	Entity	Partial Match	Jaccard Similarity	Cosine Similarity	Dice Similarity	Norm. Levenshtein	Hits@5
Mining	AA	Physical Asset	0.95	0.7833	0.8591	0.8333	0.7667	0.7143
		Ownership	0.93	0.2139	0.3609	0.3389	0.3658	0.7143
		Commodity	0.52	0.0000	0.0000	0.0000	0.1839	0.7143
		Countries	0.100	0.8333	0.8963	0.8333	0.8639	0.7143
	SCCO	Physical Asset	0.81	0.6389	0.7416	0.7333	0.6349	0.5455
		Ownership	1.00	0.3000	0.5375	0.4127	0.5590	0.5455
		Commodity	0.68	0.0000	0.5244	0.0000	0.1966	0.5455
		Countries	0.82	0.6667	0.6667	0.6667	0.7222	0.5455
	FCX	Physical Asset	0.88	0.5667	0.7014	0.6672	0.5148	0.8462
		Ownership	0.93	0.1667	0.7063	0.2222	0.5020	0.8462
		Commodity	0.87	0.0000	0.3668	0.0000	0.2114	0.7692
		Countries	1.00	1.0000	1.0000	1.0000	1.0000	0.8462
	NEM	Physical Asset	0.83	0.8462	0.9083	0.8949	0.8531	0.6486
		Ownership	0.89	0.6520	0.7891	0.7090	0.7747	0.6216
		Commodity	0.83	0.1154	0.5488	0.1282	0.3698	0.5946
		Countries	0.79	0.7692	0.7692	0.7692	0.8225	0.6486
	HL	Physical Asset	0.82	0.7955	0.8608	0.8515	0.7738	0.6667
		Ownership	0.97	0.5000	0.6667	0.6667	0.7500	0.6667
		Commodity	0.88	0.0000	0.5776	0.0000	0.3281	0.6111
		Countries	0.62	0.1818	0.1818	0.1818	0.3566	0.6667
Oil & Gas	CVX	Physical Asset	0.72	0.3417	0.5084	0.5016	0.4292	0.6250
		Ownership	1.00	1.0000	1.0000	1.0000	1.0000	0.6250
		Commodity	0.36	0.0000	0.4472	0.0000	0.2857	0.1250
		Countries	0.53	0.0000	0.0000	0.0000	0.2000	0.5000
	XOM	Physical Asset	0.71	0.2418	0.4341	0.3840	0.3808	0.6364
		Ownership	0.94	0.4286	0.5469	0.5238	0.5276	0.6364
		Commodity	0.35	0.0000	0.0000	0.0000	0.1852	0.0000
		Countries	0.63	0.1429	0.1667	0.1429	0.3127	0.5455
	MPC	Physical Asset	0.74	0.3026	0.4935	0.4556	0.3552	0.7895
		Ownership	0.95	0.6083	0.7457	0.6925	0.6947	0.7368
		Commodity	0.49	0.1036	0.4149	0.1693	0.1907	0.5789
		Countries	0.43	0.4722	0.6855	0.6389	0.2493	0.7895
	COP	Physical Asset	0.76	0.2000	0.3536	0.3333	0.2500	1.0
		Ownership	1.00	0.3333	0.5774	0.5000	0.4118	1.0
		Commodity	1.00	0.3333	0.8944	0.5000	0.1667	1.0
		Countries	0.67	0.5000	0.7071	0.6667	0.2727	1.0
Utilities	D	Physical Asset	0.79	0.2532	0.6368	0.3961	0.3788	0.2778
		Ownership	0.85	0.2740	0.4884	0.4015	0.4966	0.3333
		Commodity	0.75	0.2292	0.3836	0.3125	0.2228	0.1667
		Countries	1.00	0.9375	0.9634	0.9583	0.8984	0.2778
	ED	Physical Asset	0.64	0.3500	0.6400	0.5000	0.4602	0.5000
		Ownership	1.00	0.5000	0.7071	0.6667	0.4637	0.5000
		Commodity	0.40	0.0000	0.0000	0.0000	0.1364	0.0000
		Countries	0.67	0.7500	0.8500	0.8333	0.6100	0.5000
	DUK	Physical Asset	0.74	0.3131	0.5573	0.4683	0.5001	0.6000
		Ownership	0.98	0.4667	0.7953	0.6286	0.5889	0.6000
		Commodity	0.41	0.0000	0.0000	0.0000	0.1051	0.0000
		Countries	1.00	0.5417	0.7155	0.6667	0.3789	0.6000
	EXC	Physical Asset	0.93	0.3778	0.7145	0.5367	0.5472	0.7333
		Ownership	0.71	0.2867	0.4354	0.4333	0.4136	0.6667
		Commodity	0.40	0.4000	0.6285	0.5333	0.3520	0.5333
		Countries	0.49	0.9000	1.0000	0.9000	1.0000	0.6667

A.16 Dashboard User Interface (UI)

In this section, we present snippets of dashboard user interface (UI) for Alcoa Corporation (AA) as an example of the functionality and design of the visualisation tool developed for this project. The dashboard provides an intuitive and interactive platform to explore the relationships between key entities within the database.

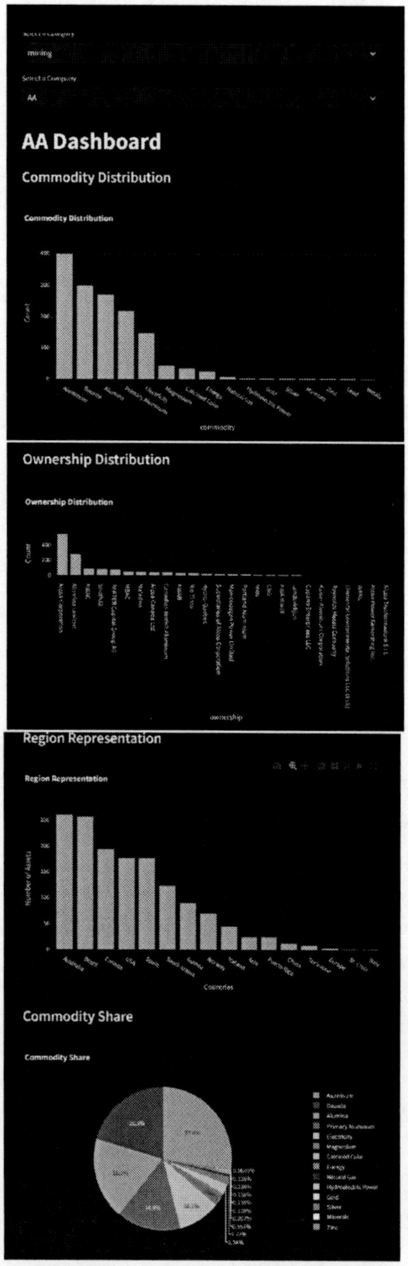

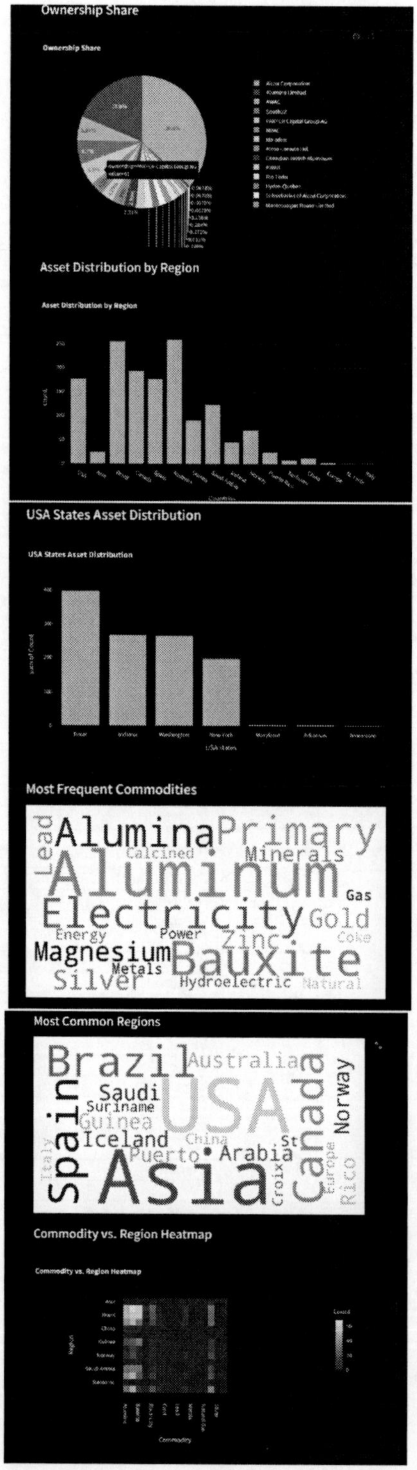

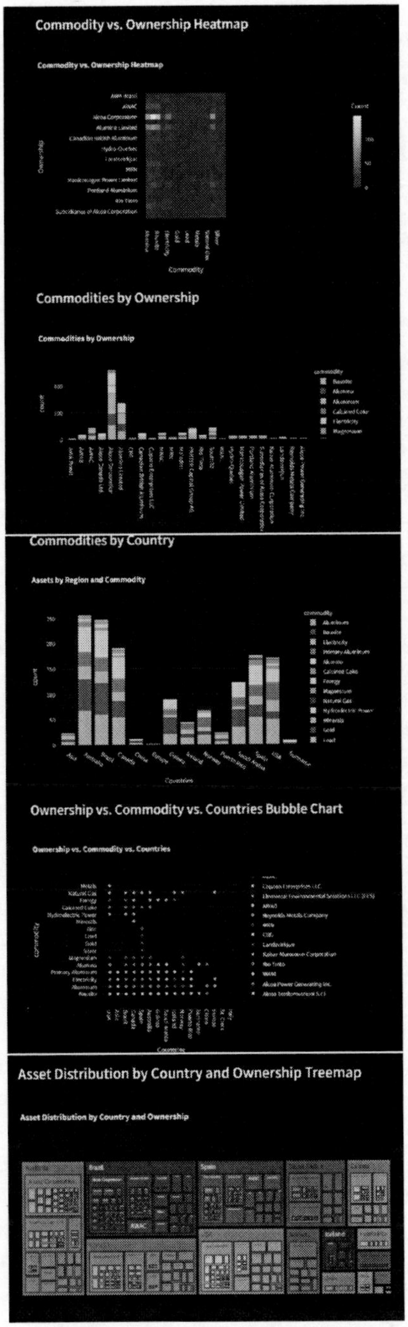

Commodity vs. Ownership Heatmap
Commodity vs. Ownership Heatmap
Commodities by Ownership
Commodities by Ownership
Commodities by Country
Assets by Region and Commodity
Ownership vs. Commodity vs. Countries Bubble Chart
Ownership vs. Commodity vs. Countries
Asset Distribution by Country and Ownership Treemap
Asset Distribution by Country and Ownership

Detecting Hyperpartisanship and Rhetorical Bias in Climate Journalism: A Sentence-Level Italian Dataset

Michele Joshua Maggini and **Davide Bassi** and **Pablo Gamallo**
Centro Singular de Investigación en Tecnoloxías Intelixentes (CiTIUS)
Universidade de Santiago de Compostela
michelejoshua.maggini@usc.es, davide.bassi@usc.es, pablo.gamallo@usc.es

Abstract

We present the first Italian dataset for joint hyperpartisan and rhetorical bias detection in climate change discourse, enhancing the complexity in modeling hyperpartisan detection. Our annotation scheme achieves a Cohen's kappa agreement of 0.63 on the gold test set (173 sentences). The dataset comprises 48 articles (1,010 sentences) from far-right media, annotated at sentence level for both binary hyperpartisan classification and the multi-label classification of 17 rhetorical biases. We conduct extensive text analysis revealing significant correlations between hyperpartisan content and specific rhetorical techniques. Our experiments with state-of-the-art language models (GPT-4o-mini) and Italian BERTbase models establish strong baselines for both classification tasks. To ensure reproducibility while addressing copyright concerns, we release article URLs, article id and paragraph's number alongside comprehensive annotation guidelines. This resource advances research in cross-lingual hyperpartisan detection and provides insights into the rhetorical strategies employed in Italian climate change discourse. To the best of our knowledge, we are the first to tackle hyperpartisan detection related to logical fallacies, focusing on on the sentence level. We provide the code and the dataset to reproduce our results: `https://anonymous.4open.science/r/Climate_HP-RB-D5EF/README.md`

1 Introduction

The rise of hyperpartisan news content and its potential impact on public discourse has become a critical concern in the digital age. For Hyperpartisanship, we referred to Maggini et al. (2025)'s definition: *Hyperpartisan news detection is the process of identifying news articles that exhibit extreme one-sidedness, characteriz ed by a pronounced use of bias.*

This phenomenon is particularly evident in discussions about climate change (Luo et al., 2020), because it is a polarizing topic (Falkenberg et al., 2022). This phenomenon constitutes a threat to social cohesion through a loop mechanism that, by manipulating the emotions of the audience, fosters the polarization of individuals (Marino et al., 2024). In light of this, many scholars developed NLP methods to tackle hyperpartisanship. Most of the studies approach this task as a binary classification task. Despite the consistent performance reached, such approaches fail to uncover the underlying mechanisms that drive hyperpartisanship (Maggini et al., 2025).

Linguistic scholarships have shown that specific rhetorical strategies play a crucial role in creating and reinforcing hyperpartisan narratives (Nguyen et al., 2022; Potthast et al., 2018).

Rhetorical biases are vicious communicative strategies aimed at circumventing or violating audience's intellectual autonomy, by depriving them of the necessary elements to evaluate and counter arguments effectively. Intellectual autonomy, in fact, involves the capacity of individuals to think critically and independently while maintaining the ability to appropriately rely on external sources for informed decision-making (**?**).

Examining such biases could provide crucial insights, revealing how biased rhetorical techniques are employed in hyperpartisan content to manipulate audiences, thus enabling more targeted interventions to mitigate polarization (Ruan et al., 2024).

Additionally, while significant progress has been made in detecting hyperpartisan content and propaganda techniques in English-language media, there remains a critical gap in resources and analysis for other low-represented languages, particularly Italian (Maggini et al., 2025).

Our study addresses these gaps by introducing the first Italian dataset jointly annotated for hyperpartisan detection and rhetorical bias identification in the context of climate change news. Our main

Proceedings of the 2nd Workshop on Natural Language Processing Meets Climate Change (ClimateNLP 2025), pages 168–187
July 31, 2025 ©2025 Association for Computational Linguistics

contributions to the field are:

- We introduce a novel dataset consisting of 48 articles (1010 sentences) from Italian libertarian-right media, focusing on climate change coverage and related topics such as Euroscepticism and green policies. Our annotation scheme operates at the sentence level, capturing both binary hyperpartisan classification and a fine-grained taxonomy of 17 distinct rhetorical biases.

- We leverage our fine-grained annotation to analyze both the relationship between hyperpartisan content with specific rhetorical manipulation strategies, and the structural distribution of these techniques across article pararagraphs, providing insights into their functional roles within the discourse architecture.

- We establish baseline performance metrics through experiments with state-of-the-art language models. We evaluate two distinct approaches: 0-shot with GPT-4-mini, and Finetuning (FT) with two BERTbase fine-tuned models for Italian. Our results demonstrate the feasibility of automated detection for both hyperpartisan content and specific rhetorical biases, while also highlighting the challenges inherent in identifying more subtle manipulation techniques.

- To ensure reproducibility while respecting copyright constraints, we will release our dataset in the form of article URLs accompanied by detailed annotation guidelines. This approach allows researchers to reconstruct the dataset while maintaining its integrity and legal compliance.

Our work contributes to the growing body of research on automated detection of media bias and manipulation, while specifically addressing the need for non-English resources in this domain. The findings and resources presented in this paper have important implications for developing more robust and culturally-aware systems for detecting and analyzing media manipulation across languages and contexts.

This article is organized as follows: Section 2 reviews the key contributions in the field that inspired our research. Section 3 details the methodology used to create the dataset, covering data collection, the annotation process, and a statistical overview of the dataset. Section 4 presents benchmark experiments for classification tasks using our dataset, along with an in-depth corpus analysis. Finally, Section 5 summarizes our findings and outlines potential directions for future research.

2 Related Work

Hyperpartisan news detection has gained significant attention in the context of online misinformation, leading to extensive research in recent years. Maggini et al. (2025) provided a comprehensive survey of hyperpartisan detection approaches. They proposed a definition that captures the linguistic and political aspects of hyperpartisanship. Additionally, they highlighted the dominance of English and U.S.-centric datasets in this domain, emphasizing the need for datasets in underrepresented languages to better understand hyperpartisanship across different countries.

Potthast et al. (2018) pioneered the computational analysis of hyperpartisan news, delving into the stylistic traits distinguishing hyperpartisan news from mainstream. Kiesel et al. (2019) established a significant foundation for computational approaches to hyperpartisanship, introducing a shared binary classification task involving 42 teams. They released two document-level datasets—one manually annotated and one labeled based on source—which provided standardized resources for hyperpartisan scholarships.

Subsequent research evolved from document-level detection toward more fine-grained approaches that leverage information from various article components. Naredla and Adedoyin (2022) experimented with BERT, ELMo, and Word2Vec on entire articles, including both headlines and bodies, while also testing various context lengths for BERT. Lyu et al. (2023) analyzed 2,200 manually labeled and 1.8 million machine-labeled news titles across the political spectrum, achieving $Acc = 0.84$; $F1 = .78$ on an external validation set, using their transformer-based model. By tracking political stance, they revealed that right-leaning media use hyperpartisan titles more frequently, identified key contentious topics, and documented a cross-spectrum increase in hyperpartisan content during the 2016 U.S. election. This more granular approach was further advanced by researchers such as Pérez-Almendros et al. (2019), who focused specifically on quoted content as a distinctive component for hyperpartisan classification, demonstrating the value of analyzing structural elements within articles rather than treating them as homogeneous units.

Omidi Shayegan et al. (2024) advanced hyperpartisan detection in under-represented languages by developing a benchmark for Persian tweets and systematically evaluating various architectural approaches from encoders to decoder-only models.

Maggini et al. explored the application of LLMs for hyperpartisan detection, utilizing LLaMA3-8b-Instruct (Touvron et al., 2023) in different In-Context Learning settings with general and task specific prompts on SemEval-2019 Task 4 and a headline-specific dataset. Their research demonstrated that these advanced neural architectures achieve competitive performance when enhanced with domain knowledge and structured reasoning, establishing LLMs as effective tools for political text analysis despite previous assumptions about ICL and computational power limitations.

As mentioned in Sec. 1, hyperpartisan content often manifests through the strategic deployment of manipulative rhetorical techniques. Such techniques are extensively employed to persuade audiences in different settings, such as news, speeches, and social media. Given the rapid spread of manipulative content in online environments, a wide range of computational approaches has emerged to address this phenomenon. As highlighted by Bassi et al. (2024), early efforts predominantly focused on content-based detection. More recently, argumentative and rhetorical approaches have gained traction, demonstrating greater scalability across different contexts.

Martino et al. (2019) represents a seminal contribution in this regard, proposing a method to identify specific texts containing propaganda and classify them based on 18 persuasion techniques . Their work later inspired a SemEval task in 2020 (Da San Martino et al., 2020) and has been then followed by Piskorski et al. (2023), which expanded the taxonomy to 23 fine-grained techniques, grouped into six broad categories. Additionally, they extended the analysis to a multilingual setting, demonstrating the applicability of argumentation-based propaganda detection across different languages.

More recent works (Hasanain et al., 2024a,b) addressed LLMs' potential for propaganda techniques detection. In this regard, Sprenkamp et al. (2023) demonstrated that reducing the number of labels to 14 improved classification performance.

Building on this literature, to the best of our knowledge, we are the first to approach hyperpartisan detection at the sentence level and consider the presence of rhetorical bias as a fundamental characteristic of hyperpartisan texts. By treating rhetorical biases as stylistic traits that shape the message of a text, we capture deeper linguistic patterns that contribute to hyperpartisan framing. While prior research has shown that source-level bias does not uniformly manifest across all articles (Baly et al., 2018), our sentence-level approach transcends these limitations. Working at this granularity allows us to identify precisely where and how hyperpartisan language emerges through specific rhetorical fallacies, creating a dataset that supports both binary hyperpartisan detection and multi-class fallacy classification. This approach reveals significant correlations between particular fallacy types and hyperpartisan content (see Table 5 in the Appendix), providing empirical evidence for their relationship.

3 Methodology

3.1 Dataset Creation

Article selection and Pre-Processing From the moment that "alternative" media tend to spread anti-establishment messages (Ernesto de León and Adam, 2024), we focused on NicolaPorro.it[1], an independent libertarian media outlet. The collected corpus consists of 48 articles for a total of 1010 units on climate change, green policies and Euroscepticism selected from the site's "Green policies" section to ensure topical homogeneity. We featured only the Italian language, since the recent enhancements in NLP for disinformation detection mostly covered over-represented languages like English (Maggini et al., 2025). To ensure a fine-grained analysis of the texts, we then split the articles following the html <p> tags, that mostly corresponded to individual sentences. We grouped together the sentences with less than 15 words to guarantee minimal context.

Annotation Protocol To build our annotation guidelines, we started by defining the constructs under investigation.

For **Hyperpartisanship** we referred to the definition by Maggini et al. (2025) mentioned above:

[1]https://www.nicolaporro.it/articoli/ambiente-sostenibilita/

Hyperpartisan news detection is the process of identifying news articles that exhibit extreme one-sidedness, characterized by a pronounced use of bias. We modeled this task as a binary classification at the sentence level.

Regarding **Rhetorical biases**, to our knowledge the most comprehensive taxonomy is the one of Piskorski et al. (2023), with a total of 23 labels.

Starting from their taxonomy, we translated the definitions in Italian and adapted them to our use case on climate change. Additionally, being our main scope to conduct a more fine-grained analysis of the rhetorical biases underlying hyperpartisanship, we merged some of the techniques (Slogan/Conversation Killer, Whataboutism/Tu Quoque, Appeal to Values/Flag Waving, Causal/Consequential Oversimplification, Smear/Doubt). Keeping them would have added unnecessary complexity to the model without providing additional analytical insights. In Table 3 we report brief descriptions for each technique, while in Appendix A we report their in-depth definition as well as the annotation guidelines.

Alongside the binary hyperpartisan classification, annotators performed a multi-label task identifying specific rhetorical biases deployed to influence reader opinion in each sentence.

The annotation was conducted by two native Italian speakers with expertise in political discourse analysis. Both annotators are Ph.D. students in applied NLP for disinformation, with academic backgrounds in Philology, Data Science, Anthropology, and Psychology. They have prior experience in linguistic annotation of news content and rhetorical technique identification. Annotators did not know the source of the articles and during the annotation rounds, they did not have access to the whole article's context but only to the individual sentences. We divided the annotation process into three phases: **Training phase**: annotators studied the guidelines, performed pilot annotations and completed the training through interactive sessions to discuss doubts, edge cases and resolve disagreements.; **Annotation Phase**: Each document was independently annotated by both annotators; **Curation Phase**: Discrepancies between annotations were discussed and resolved to ensure final label consistency. Before the Curation Phase, we measured the Inter-Annotator Agreement (IAA) using Krippendorf's α, achieving a value of .92 for hyperpartisan detection and .63 on rhetorical fallacies.

3.2 Dataset description

Table 1 represents key statistics of our dataset, including size, sentence length, and the average rhetorical biases per article. Table 3 shows the definitions and distributions of hyperpartisan and neutral sentences, as well as logical fallacies. To analyze the thematic distribution within our corpus, we applied BERTopic (Grootendorst, 2022) with parameters optimized to preserve local structure[2]. Table 2 presents the topic distribution. After manual inspection, we forced BERTopic to detect three main topics: science, institutions and Other, each further subdivided into specific subtopics like institutions.Italy, science.cars, etc.

Metric	Value
Number of Documents	48
Number of Sentences	1010
Avg. Sentences per Article	21.12
Avg. Words per Text	40.26
Avg. Characters per Text	264.37
Avg. Techniques per Document	2.12

Table 1: Dataset Statistics

Topic	Count
Other.climate	241
Other.other	122
science.climate_change	109
science.other	82
institutions.Europe	70
Other.politics	68
science.energy_transition	54
science.environment	44
science.cars	38
institutions.Other	35
institutions.OMS	33
institutions.China	30
institutions.Italy	26
science.green_policies	23
institutions.BlackRock	16
science.medicine	14
Other.politically_correct	4
Other.politics	1

Table 2: Topic Distribution. Topics have been extracted using BERTopic.

3.3 Models

We tested two different architectures: encoders and decoder-only models.

[2]umap-model = UMAP(n-neighbors=10, n-components=3, metric='cosine') hdbscan-model = HDBSCAN(min-cluster-size=10, min-samples=10, metric='euclidean', prediction-data=True) ctfidf-model = ClassTfidfTransformer(bm25-weighting=False, reduce-frequent-words=True) representation-model = MaximalMarginalRelevance(diversity=0.5)

Bias Type	Definition	Distribution
Hyperpartisan Classification		
Hyperpartisan Language	Text that displays extreme bias favoring one particular political side, often employing pronounced use of rhetorical biases	HP 304 N 706
Rhetorical Biases		
Slogan/Conversation Killer	Using catchphrases or dismissive statements to shut down further discussion or debate	64
Appeal to Time	Manipulating temporal perspectives or deadlines to create urgency or dismiss concerns	9
Appeal to Values/ Flag Waving	Exploiting patriotic feelings or moral values to justify positions or actions	59
Appeal to Authority	Using the reputation of an expert or institution to support arguments without proper context	53
Appeal to Popularity	Justifying a belief by citing its widespread adoption or acceptance	11
Appeal to Fear	Manipulating audience's fears to promote specific viewpoints or actions	99
Straw Man/Red Herring	Misrepresenting opponent's argument or diverting attention to unrelated issues	43
Whataboutism/ Tu Quoque	Deflecting criticism by pointing to the opponent's alleged hypocrisy or similar actions	42
Loaded Language	Using words with strong emotional implications to influence the audience	330
Repetition	Repeating phrases or ideas multiple times for emphasis or to establish them as truth	23
Intentional Confusion/ Vagueness	Using deliberately unclear or ambiguous language to avoid commitment or scrutiny	55
Exaggeration/Minimisation	Presenting facts in a distorted way by either magnifying or downplaying their importance	244
Name Calling	Using labels or derogatory terms to discredit without substantive argument	159
Reductio ad Hitlerum	Drawing inappropriate comparisons to Nazism, Hitler, or fascism	13
Smear/Doubt	Attempting to damage reputation or create doubt through indirect attacks or insinuations	355
Causal/Consequential Oversimplification	Presenting complex situations with oversimplified cause-effect relationships	165
False Dilemma/ No Choice	Presenting limited options while ignoring alternatives or middle ground	66

Table 3: Taxonomy of rhetorical biases and hyperpartisan language detection used in our annotation scheme. The rhetorical biases represent fine-grained categories of manipulative language techniques commonly found in politically charged discourse.

For encoders, we used `dbmdz/bert-base-italian-xxl-uncased`[3], trained from scratch on Italian, and `nickprock/sentence-bert-base-italian-xxl-uncased`[4], fine-tuned for Italian. Particularly, the first model was trained on OSCAR corpus (Ortiz Suárez et al., 2020) and is known for its robust handling of complex relationships in text, allowing for a comprehensive understanding of contextual nuances. In contrast, the sentence-transformer is optimized for generating meaningful sentence embeddings, making it particularly suitable for capturing the semantic essence of individual sentences. By fine-tuning and comparing both models, we aimed to evaluate their performance on the hyperpartisan classification task, providing insights into which approach better captures the rhetorical distinctions in the data. We fine-tuned the models

Regarding the decoder-only architectures we used GPT 4o and 4o-mini. For the Hyperpartisan detection (HP) task we employed the models in a 0-shot setting, while, for the Rhetorical Bias (RB), each model was firstly tested 0-shot with temperature equal to 0.2. Given the difficulty of working with a high number of labels, we decided to set this value so that the model could capture the most subjective traits in the rhetorical fallacies. Furthermore, we fine-tuned the models for Rhetorical Bias detection. We prompted and fine-tuned the OPENAI models via their APIs [5]. The prompts are available in the Appendix A.2 and A.3.

4 Results

4.1 Hyperpartisan-Rhetorical Bias Relation

To investigate the relationship between rhetorical biases and hyperpartisanship, we analyzed their correlation patterns. Figure 1 depicts which rhetorical biases are most determinant to distinguish between hyperpartisan and neutral sentences. To analyze the rhetorical distinctions between hyperpartisan and neutral sentences in more detail, we performed χ^2 tests to compare the frequency of each rhetorical technique across binary labels (see Table 5 in Appendix A.3). We measured the effect sizes using Log Ratio[6]. Thus,

[3] `https://huggingface.co/dbmdz/bert-base-italian-xxl-uncased`

[4] `https://huggingface.co/nickprock/sentence-bert-base-italian-xxl-uncased`

[5] `https://platform.openai.com/docs/models`

[6] Log Ratio (LR) is calculated as the logarithm base 2 of the ratio of the frequencies between the two groups. A value of 0 signifies equal frequency in both groups, positive values indicate a higher frequency in the hyperpartisan group, and

neutral sentences are usually characterized by no rhetorical biases ("no_technique_detected"), whereas "Reduction_ad_Hitlerum", "Name_Calling", "Tu_Quoque/Whataboutism", "Loaded_Language", "Smear/Doubt", "Straw_Man/Red_Harring" and "Exaggeration_Minimisation" are highly significant ($p-value < 0.001$) to discern hyperpartisan sentences. Those findings validate Maggini et al. (2025)'s definition of hyperpartisanship as well as the previous definitions used in the literature by Kiesel et al. (2019); Lyu et al. (2023).

4.2 Topological Distribution of Rhetorical Biases

Fig. 2 shows the average of hyperpartisan sentences across the articles' structure, while Fig. 3 illustrates the concentration of bias in the articles' structure. This provides us with a better understanding of how much and in which parts the articles are contaminated by hyperpartisanship and rhetorical biases.

To analyze the Hyperpartisan Contamination Level (HCL), firstly, we grouped the sentences by article ID and got the sentence positions. Then, we created normalized positions for each sentence and created 10 potision bins. Lastly, we computed the average hyperpartisan score for each position bin. Fig. 2 shows that hyperpartisan sentences appear in 50% of cases within the first 10% of the articles and around 40% in the following 10% (i.e., between 10% and 20% of the article's beginning). This evidence aligns with what other researchers analyzed in previous work, stating that titles usually are determinant to distinguish between fake or hyperpartisan and mainstream news (Horne and Adali, 2017; Shrestha and Spezzano, 2021). Then, the average HCL drops in the central part (20-60%) to increase again up to around 30% in the second half of the articles (60-100%).

Then, we decided to investigate on how rhetorical techniques are adopted to convey and shape the message (Fig. 3). Firsly, we normalized the position of the techniques within each article, creating position quartiles. After that, we counted the occurrences of each technique in each quartile and then pivoted the data for visualization. Successively, we calculated the raw totals for each technique and then normalized by technique, namely we computed the percentage across quartiles for each technique.

negative values indicate a higher frequency in the neutral group.

Our annotation of sentences with rhetorical fallacies revealed that certain techniques are more prominent than others, offering valuable insights into how these strategies are distributed across the structure of the articles. For example, while Reductio ad Hitlerum is relatively rare (13 occurrences with high statistical significance), it appears predominantly in the first quartile (Q1) at 53.8%, indicating its use in setting a strong, biased tone early in the article. Similarly, Name Calling is concentrated in Q1, with 49% of its occurrences in this section, and both techniques are highly significant for identifying hyperpartisan sentences. These strategies allow reporters to directly express their stance on a topic, often leveraging emotionally charged language to engage the reader from the outset.

However, Repetition (statistically significant) is most frequent in the second quartile (Q2), with 56.5% of its cases appearing here. This suggests its role in reinforcing the ideas introduced earlier, contributing to the redundancy of concepts to solidify the intended message. Lastly, Slogan/Conversation Killer usage peaks in the final quartile (Q4), accounting for 43.8% of its total appearances. This aligns with the tendency of journalists to use catchy phrases or mottos at the end of articles to leave a lasting impression and emphasize their message.

These findings align with the existing literature, which highlights how clickbait headlines often employ rhetorical techniques to manipulate reader engagement and frame narratives persuasively (Blom and Hansen, 2015; Munger, 2020). Such strategies are not only common in sensationalist media but are also key tools for amplifying bias and promoting specific agendas (Chakraborty et al., 2016). This finding also strengthens the relationship between click-bait and hyperpartisan content.

4.3 Computational Baselines

The aim of our experiments is to provide baselines and to explore the impact of different architectures on two classification tasks: for hyperpartisan and for logical fallacies. Both of the two tasks were annotated at the sentence level. While HP classification is a binary classification task, RB classification, is a multi-class classification task.

The results of the evaluation on the detection of hyperpartisan and rhetorical bias are shown in Table 4. The results demonstrate significant variability in the metrics score between different models and methodologies.

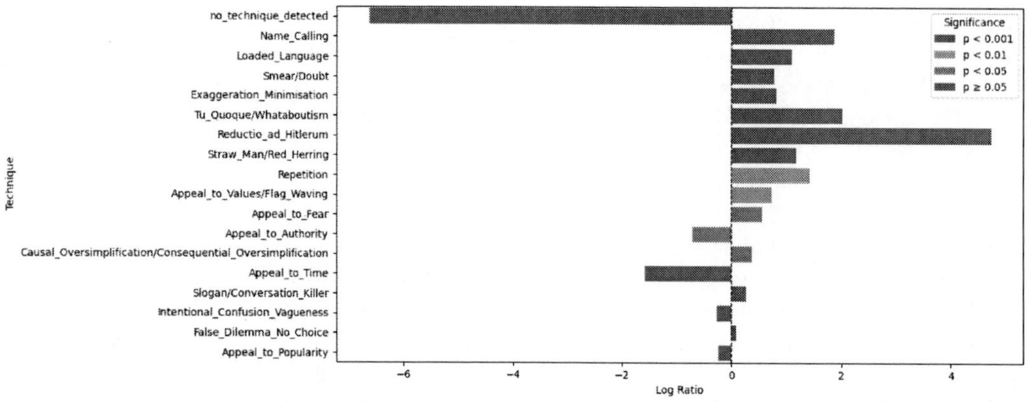

Figure 1: Correlation between Hyperpartisan sentences and techniques. The table with the different levels of significance is reported in the Appendix.

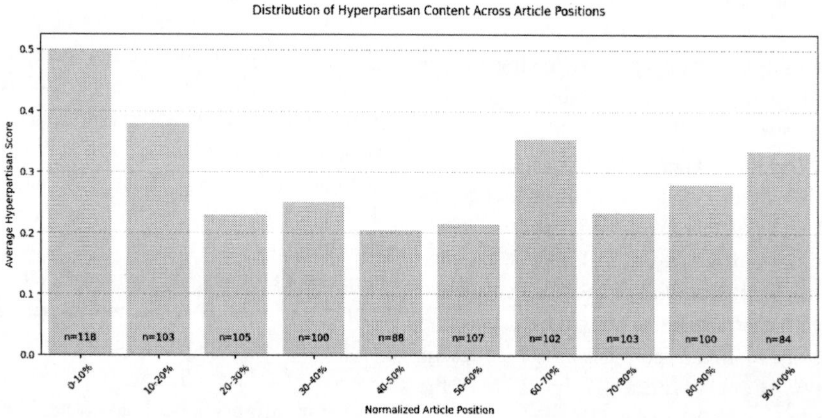

Figure 2: Hyperpartisan Contamination Level per sentence position (deciles). *n* represents the number of sentences. Because of articles have dissimilar number of sentences, we normalize their length.

GPT: GPT-4o-mini and GPT-4o both perform well on the HP classification task in a 0-shot setting. GPT-4o achieves an accuracy of 0.969 with an F1 score of 0.942, outperforming GPT-4o-mini, which attains an accuracy of 0.959 and an F1 score of 0.933. The results indicate that GPT-4o is more effective in recognizing hyperpartisanship in Italian news articles. Those results can be explained by the architectural and dimensional differences between the two models.

For RB classification, GPT-4o-mini performs reasonably well in 0-shot mode (accuracy: 0.892, F1: 0.319), and its fine-tuned performance increased slightly (accuracy: 0.905, F1: 0.362). GPT-4o exhibits similar behavior, with 0-shot performance (accuracy: 0.906, F1: 0.385) being substantially better than FT (accuracy: 0.908, F1: 0.410). The low precision scores for both models in RB classification indicate challenges in correctly identifying rhetorical bias. The high unbalanced distribution between techniques explains these results. Indeed, the other metrics we reported are macro-averaged metrics, which offer a fair comparison.

Encoders: For HP classification, `bert-base-italian-xxl-uncased` achieves an accuracy of 0.861 and an F1 score of 0.859, showing strong performance but slightly lagging behind GPT-4o. However, in RB classification, the model performs poorly, with a precision of 0.354 and an F1 score of 0.470, indicating that it struggles to effectively identify rhetorical bias. The difficulty in classifying RB stems from the extreme class imbalance, where certain rhetorical categories are underrepresented, leading to biased model predictions that favor more frequent classes. The macro-averaged F1 score provides a clearer picture of

Model	Classification	Method	Accuracy	Precision	Recall	F1 Score
GPT-4o-mini	HP	0-Shot	0.959	0.942	**0.924**	0.933
	RB	0-Shot	0.892	0.285	0.486	0.319
		FT	0.905	0.326	0.465	0.362
GPT-4o	HP	0-Shot	**0.969**	**0.980**	0.907	**0.942**
	RB	0-Shot	0.906	0.387	0.434	0.385
		FT	0.908	0.378	0.559	0.410
`bert-base-italian-xxl-uncased`	HP	FT	0.861	0.858	0.861	0.859
	RB	FT	0.354	0.699	0.354	0.470
`sentence-bert-base-italian-xxl-uncased`	HP	FT	0.851	0.846	0.851	0.845
	RB	FT	0.321	0.683	0.320	0.436

Table 4: Comparison of Hyperpartisan and Rhetorical Bias Classification Models

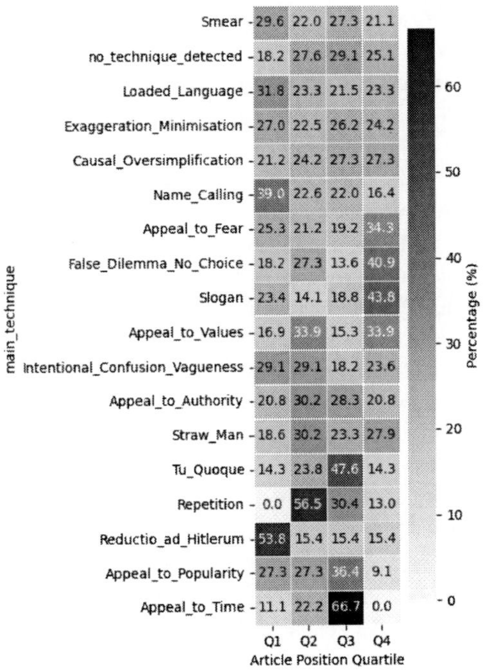

Figure 3: Distribution of Techniques Across Article Quartiles

this imbalance, as models perform well on majority classes but fail on rare ones. Similar trends are observed for `sentence-bert-base-italian-xxl-uncased`, which achieves competitive HP classification results (accuracy: 0.851, F1: 0.845) but performs poorly on RB classification (accuracy: 0.321, F1: 0.436). This suggests that sentence embeddings are effective for hyperpartisan classification but less suited for rhetorical bias detection. The model's struggle with RB is further exacerbated by the highly skewed class distribution, making it difficult to learn meaningful representations for rare rhetorical bias categories. The macro-averaged F1 scores reinforce that underrepresented classes are poorly classified, reducing overall model effectiveness.

5 Conclusion and Future Work

In this work, we introduced a novel Italian news dataset focused on climate change and Euroscepticism, specifically designed for hyperpartisan and rhetorical bias detection. Our dataset emphasizes the critical need to collect news in underrepresented languages to gain a deeper understanding of hyperpartisanship across European countries. Spanning diverse and polarizing public topics, the dataset consists of 48 articles divided into 1,010 sentences, annotated for hyperpartisanship (binary labels) and enriched with over 1.5K rhetorical fallacy labels using a fine-grained taxonomy.

Our study underscores the significance of analyzing hyperpartisanship in conjunction with rhetorical biases, as these biases can profoundly influence the objectivity of storytelling in news articles. Through detailed corpus analysis, we contributed to the field by offering nuanced insights into how specific rhetorical techniques align with hyperpartisan content, enhancing our understanding of manipulation strategies in media.

We also established strong baselines using state-of-the-art architectures and learning paradigms, such as FT and 0-shot, demonstrating the versatility and applicability of our dataset. By sharing the full pipeline to recreate the dataset, we aim to facilitate the development of new methods and tools to critically analyze online media content.

Future work will focus on experimenting with advanced models and exploring how leveraging rhetorical biases can further improve hyperpartisan sentence detection. Despite the annotation required high effort and is not scalable, we plan to extend the current dataset with other articles. We hope our work serves as a stepping stone for more robust and transparent media analysis, ultimately contributing to a healthier information ecosystem.

6 Limitations

Regarding the dataset size (48 articles, 1,010 sentences), we acknowledge is relatively small, potentially limiting the generalizability of findings and the robustness of model training. Expanding the dataset with a broader range of sources and perspectives would improve coverage and model performance.

Second, the focus on far-right media outlets introduces a selection bias, which, while intentional for analyzing hyperpartisan rhetoric, may not capture the full spectrum of climate change discourse in Italy. Future work should explore more diverse media sources, including centrist and left-leaning outlets, to provide a more comprehensive view.

Third, while our annotation scheme achieves moderate agreement (Cohen's kappa = 0.63), some rhetorical biases remain inherently subjective and difficult to categorize consistently due to their distributions.

Finally, differently from Martino et al. (2019); Da San Martino et al. (2020); Piskorski et al. (2023) we did not include the span, as the annotation process was highly demanding and the number of annotators limited. Such approach could further contribute to fine-grained analysis of news articles, understanding on which specific words and rhetorical patterns the hyperpartisan is based.

Finally, while we provide article URLs for transparency, copyright restrictions prevent us from openly distributing full-text data. This limits direct replication and benchmarking. Future work could explore ways to balance reproducibility with legal constraints, such as structured metadata representations or synthetic dataset augmentation.

Ethics Statement

Biases

The news articles in our dataset may contain harmful content, including loaded language, name-calling, and slurs. Our annotation process was designed to focus solely on identifying rhetorical bias and hyperpartisan language rather than assessing the truthfulness of the information. To ensure objectivity, annotations were conducted without considering annotators' personal opinions or political views on the topics discussed. Additionally, we did not rely on crowdsourcing; instead, we managed our annotators directly, ensuring proper working conditions and maintaining annotation quality.

We recognize the potential risks of bias in both data collection and model predictions. The inherent subjectivity in identifying rhetorical bias and hyperpartisanship means that biases can emerge from the dataset itself, as well as from the models trained on it. Given the sensitive nature of hyperpartisan and rhetorical bias detection, we advise caution when using the dataset and models to avoid reinforcing biases or misrepresenting viewpoints. Future work should focus on refining annotation practices, improving model interpretability, and incorporating interdisciplinary perspectives to mitigate potential harms.

Intended Use and Misuse Potential

This dataset is intended to advance research in hyperpartisan news detection, particularly in underrepresented languages. It can contribute to the development of more robust models and analytical tools for identifying rhetorical bias in media. However, we acknowledge the risk of misuse, particularly by malicious actors seeking to manipulate or censor content. To prevent unintended consequences, we urge researchers and practitioners to use this dataset responsibly and transparently, ensuring that any conclusions drawn are supported by rigorous evaluation and ethical considerations.

The work presented in this paper complies with the ACL Ethics Policy [7]. We have relied on open architectures when possible. We hope that the community can benefit from our work to apply NLP technology to tackle climate change and Euroscepticism.

Acknowledgements

This project has received funding from the European Unión´s Horizon Europe research and innovation programme under the Marie Skłodowska-Curie Grant Agreement No. 101073351.

References

Ramy Baly, Georgi Karadzhov, Dimitar Alexandrov, James Glass, and Preslav Nakov. 2018. Predicting factuality of reporting and bias of news media sources. In *Proceedings of the 2018 Conference on Empirical Methods in Natural Language Processing*, pages 3528–3539, Brussels, Belgium. Association for Computational Linguistics.

Davide Bassi, Søren Fomsgaard, and Martín Pereira-Fariña. 2024. Decoding persuasion: a survey on ML

[7] https://www.acm.org/code-of-ethics

and NLP methods for the study of online persuasion. *Frontiers in Communication*, 9. Publisher: Frontiers.

Jonas Nygaard Blom and Kenneth Reinecke Hansen. 2015. Click bait: Forward-reference as lure in online news headlines. *Journal of Pragmatics*, 76:87–100.

Abhijnan Chakraborty, Bhargavi Paranjape, Sourya Kakarla, and Niloy Ganguly. 2016. Stop clickbait: Detecting and preventing clickbaits in online news media. In *2016 IEEE/ACM International Conference on Advances in Social Networks Analysis and Mining (ASONAM)*, pages 9–16.

Giovanni Da San Martino, Alberto Barrón-Cedeño, Henning Wachsmuth, Rostislav Petrov, and Preslav Nakov. 2020. SemEval-2020 Task 11: Detection of Propaganda Techniques in News Articles. In *Proceedings of the Fourteenth Workshop on Semantic Evaluation*, pages 1377–1414, Barcelona (online). International Committee for Computational Linguistics.

Mykola Makhortykh Ernesto de León and Silke Adam. 2024. Hyperpartisan, alternative, and conspiracy media users: An anti-establishment portrait. *Political Communication*, 41(6):877–902.

Max Falkenberg, Alessandro Galeazzi, Maddalena Torricelli, Niccolò Di Marco, Francesca Larosa, Madalina Sas, Amin Mekacher, Warren Pearce, Fabiana Zollo, Walter Quattrociocchi, et al. 2022. Growing polarization around climate change on social media. *Nature Climate Change*, 12(12):1114–1121.

Maarten Grootendorst. 2022. Bertopic: Neural topic modeling with a class-based tf-idf procedure.

Maram Hasanain, Fatema Ahmad, and Firoj Alam. 2024a. Can GPT-4 identify propaganda? annotation and detection of propaganda spans in news articles. In *Proceedings of the 2024 Joint International Conference on Computational Linguistics, Language Resources and Evaluation (LREC-COLING 2024)*, pages 2724–2744, Torino, Italia. ELRA and ICCL.

Maram Hasanain, Fatema Ahmad, and Firoj Alam. 2024b. Large language models for propaganda span annotation. In *Findings of the Association for Computational Linguistics: EMNLP 2024*, pages 14522–14532, Miami, Florida, USA. Association for Computational Linguistics.

Benjamin D. Horne and Sibel Adali. 2017. This just in: Fake news packs a lot in title, uses simpler, repetitive content in text body, more similar to satire than real news.

Johannes Kiesel, Maria Mestre, Rishabh Shukla, Emmanuel Vincent, Payam Adineh, David Corney, Benno Stein, and Martin Potthast. 2019. SemEval-2019 task 4: Hyperpartisan news detection. In *Proceedings of the 13th International Workshop on Semantic Evaluation*, pages 829–839, Minneapolis, Minnesota, USA. Association for Computational Linguistics.

Yiwei Luo, Dallas Card, and Dan Jurafsky. 2020. Detecting stance in media on global warming. In *Findings of the Association for Computational Linguistics: EMNLP 2020*, pages 3296–3315, Online. Association for Computational Linguistics.

Hanjia Lyu, Jinsheng Pan, Zichen Wang, and Jiebo Luo. 2023. Computational assessment of hyperpartisanship in news titles.

Michele Joshua Maggini, Davide Bassi, Paloma Piot, Gaël Dias, and Pablo Gamallo Otero. 2025. A systematic review of automated hyperpartisan news detection. *PLOS ONE*, 20(2):1–39.

Michele Joshua Maggini, Erik Bran Marino, and Pablo Gamallo Otero. Leveraging Advanced Prompting Strategies in Llama-8b for Enhanced Hyperpartisan News Detection.

Erik Bran Marino, Jesus M. Benitez-Baleato, and Ana Sofia Ribeiro. 2024. The polarization loop: How emotions drive propagation of disinformation in online media—the case of conspiracy theories and extreme right movements in southern europe. *Social Sciences*, 13(11).

Giovanni Da San Martino, Seunghak Yu, Alberto Barrón-Cedeño, Rostislav Petrov, and Preslav Nakov. 2019. Fine-Grained Analysis of Propaganda in News Articles. ArXiv:1910.02517 [cs].

Kevin Munger. 2020. All the news that's fit to click: The economics of clickbait media. *Political Communication*, 37(3):376–397.

Navakanth Reddy Naredla and Festus Fatai Adedoyin. 2022. Detection of hyperpartisan news articles using natural language processing technique. *International Journal of Information Management Data Insights*, 2(1):100064.

Allison Nguyen, Tom Roberts, Pranav Anand, and Jean E Fox Tree. 2022. Look, dude: How hyperpartisan and non-hyperpartisan speech differ in online commentary. *Discourse & Society*, 33(3):371–390.

Sahar Omidi Shayegan, Isar Nejadgholi, Kellin Pelrine, Hao Yu, Sacha Levy, Zachary Yang, Jean-François Godbout, and Reihaneh Rabbany. 2024. An evaluation of language models for hyperpartisan ideology detection in Persian Twitter. In *Proceedings of the 2nd Workshop on Resources and Technologies for Indigenous, Endangered and Lesser-resourced Languages in Eurasia (EURALI) @ LREC-COLING 2024*, pages 51–62, Torino, Italia. ELRA and ICCL.

Pedro Javier Ortiz Suárez, Laurent Romary, and Benoît Sagot. 2020. A monolingual approach to contextualized word embeddings for mid-resource languages. In *Proceedings of the 58th Annual Meeting of the Association for Computational Linguistics*, pages 1703–1714, Online. Association for Computational Linguistics.

Jakub Piskorski, Nicolas Stefanovitch, Nikolaos Nikolaidis, Giovanni Da San Martino, and Preslav Nakov. 2023. Multilingual multifaceted understanding of online news in terms of genre, framing, and persuasion techniques. In *Proceedings of the 61st Annual Meeting of the Association for Computational Linguistics (Volume 1: Long Papers)*, pages 3001–3022, Toronto, Canada. Association for Computational Linguistics.

Martin Potthast, Johannes Kiesel, Kevin Reinartz, Janek Bevendorff, and Benno Stein. 2018. A stylometric inquiry into hyperpartisan and fake news. In *Proceedings of the 56th Annual Meeting of the Association for Computational Linguistics (Volume 1: Long Papers)*, pages 231–240, Melbourne, Australia. Association for Computational Linguistics.

Carla Pérez-Almendros, Luis Espinosa-Anke, and Steven Schockaert. 2019. Cardiff university at SemEval-2019 task 4: Linguistic features for hyperpartisan news detection. In *Proceedings of the 13th International Workshop on Semantic Evaluation*, pages 929–933. Association for Computational Linguistics.

Qin Ruan, Jin Xu, Susan Leavy, Brian Mac Namee, and Ruihai Dong. 2024. Rewriting bias: Mitigating media bias in news recommender systems through automated rewriting. In *Proceedings of the 32nd ACM Conference on User Modeling, Adaptation and Personalization*, UMAP '24, page 67–77, New York, NY, USA. Association for Computing Machinery.

Anu Shrestha and Francesca Spezzano. 2021. Textual characteristics of news title and body to detect fake news: A reproducibility study. In *Advances in Information Retrieval: 43rd European Conference on IR Research, ECIR 2021, Virtual Event, March 28 – April 1, 2021, Proceedings, Part II*, page 120–133, Berlin, Heidelberg. Springer-Verlag.

Kilian Sprenkamp, Daniel Gordon Jones, and Liudmila Zavolokina. 2023. Large Language Models for Propaganda Detection. ArXiv:2310.06422 [cs].

Hugo Touvron, Thibaut Lavril, Gautier Izacard, Xavier Martinet, Marie-Anne Lachaux, Timothée Lacroix, Baptiste Rozière, Naman Goyal, Eric Hambro, Faisal Azhar, Aurelien Rodriguez, Armand Joulin, Edouard Grave, and Guillaume Lample. 2023. Llama: Open and efficient foundation language models.

A Appendix

Guidelines

Annotation Guidelines (ENG)

Hyperpartisan sentences: Text that displays extreme bias favoring one particular political side, often employing pronounced use of rhetorical biases. Label it 1, if the sentence is hyperpartisan, or 0 if it is neutral. Explicit examples: - "We are tired of government's abuses! We don't want to drive electric car!" Neutral examples: - "Electric cars are not as green as industry tell us".

Slogan/Conversation Killer: Short and impactful phrases designed to discourage critical thinking and/or urge a certain action by presenting the message as definitive. These often draw on seemingly indisputable popular wisdom or stereotypes to avoid further discussion. Explicit examples: - "Think global, act local!" - "That's just how it is, there's nothing more to add." Implicit examples: - "Be part of the solution, not part of the pollution." - "With the utmost respect for green policies and climate change, shareholders want profits. Period."

Appeal to Time: An argument centered on the idea that the time has come for a particular action or that there is no more time to waste. The call to "Act Now!" Explicit example: "If we don't act immediately on the climate crisis, in ten years it will be too late to save the planet!" Implicit example: "The timing for this reform could not be more perfect..."

Appeal to Values/Flag Waving: Leverages identity values (nationalism, patriotism, belonging to a social group/class), as well as moral and social values considered positive by the target audience (freedom, democracy, ethics, religion) to promote or justify an idea. It operates on the assumption that the audience already holds certain biases or beliefs. Explicit examples: - "If we must have climate policies—very few—then let's adopt only those that benefit Italy." - "Ecology cannot and must not take priority over citizens' freedom." Implicit examples: - "While other countries bow to these policies, we must protect our interests." (a veiled appeal to nationalism) - "These policies are gradually eroding the principles on which our society is founded." (an appeal to preserving social values)

Appeal to Authority: Giving weight to a particular idea by citing a supposed authority as a source, regardless of whether they are actually competent in the field. The tone of the statement suggests that the weight of this supposed authority is being used to justify information or conclusions. Explicit example: "Climatologist Richard Dawkins says climate change doesn't exist, therefore climate change is a lie!" Implicit example: "Those who have truly studied the issue know very well that things are not as they seem."

Appeal to Popularity: Justifying an idea by claiming that "everyone" agrees or that "no one" disagrees, encouraging the audience to adopt the same position out of conformity. "Everyone" may refer to the general public, experts (e.g., all experts say that...), countries, or other groups. Explicit example: "No one here is denying that the planet's temperature is rising, so climate change is real." Implicit example: "Ideological rules have been imposed that no one else follows."

Appeal to Fear: Promoting or rejecting an idea by exploiting the audience's repulsion or fear, describing possible scenarios in a frightening way (terrible things that could happen) to instill fear. Explicit example: "Climate taxes are just the beginning. If we keep up this farce, they'll take everything we have!" Implicit example: "This is just the first step in a larger plan that will lead to irreversible consequences."

Straw Man/Red Herring: A technique that shifts the discussion away from the original topic through two main approaches: distorting the original argument into an easier-to-attack version or introducing a different but seemingly related topic. The goal is to avoid addressing the substance of the initial issue by diverting attention to a secondary theme. Explicit examples: - "When you ask for a more gradual energy transition, you're basically saying you don't care if the planet becomes uninhabitable for our children." - "Instead of always talking about CO2 emissions, look at this great initiative we launched for beach clean-ups!" Implicit examples: - "Their concern for the employment impact of closing coal plants reveals the usual mindset that prioritizes profit over the planet's survival." - "Before discussing climate policies, shouldn't we focus on improving waste sorting in municipalities?"

Tu Quoque/Whataboutism: A technique that attempts to discredit a position or opponent by highlighting alleged contradictions or double standards. This can manifest by pointing out inconsistencies on the same issue or introducing comparisons with other contexts or situations. The goal is to undermine credibility through comparisons with other matters. Explicit examples: - "Look at them, all flying around in helicopters, while just weeks ago they were sounding the alarm and criticizing waste!" - "He talks so much about the climate emergency, but we're still waiting for answers on the migration crisis." Implicit examples: - "Funny how certain climate positions change so quickly when political circumstances shift." - "Interesting concern for the environment... I wonder if the same attention was there when it came to approving the airport expansion in your region."

Loaded Language: Using specific words and phrases with strong emotional implications (both positive and negative) to influence and persuade the audience. The essence of this technique is the use of terms that go beyond their literal meaning to evoke an emotional response. Explicit example: "These climate dictatorships run by idiots." Implicit example: "A somewhat unconventional management of public funds."

Repetition: The repeated use of the same word, phrase, story, or image in the hope that repetition will persuade the audience. Explicit example: "Safety is our priority. We must ensure safety. Without safety, there is no future. Safety must come first." Implicit example: "Innovation is the key. We must focus on innovation. Innovation will save us. Only through innovation can we progress."

Intentional Confusion/Vagueness: Using deliberately unclear wording so that the audience can have their own interpretations. For example, an argument may include a vague phrase with multiple or unclear definitions, which ultimately does not support the conclusion. Explicit example: "We will develop synergistic paradigms aimed at the horizontal optimization of ecological performance." Implicit example: "It has been proven that 70% of the time green policies work every time."

Exaggeration/Minimization: Representing something in an exaggerated manner: making things seem bigger, better, or worse (e.g., "the best of the best," "guaranteed quality") or downplaying something to make it seem less important than it really is (e.g., calling an insult just a joke), minimizing statements and ignoring arguments or accusations made by an opponent. Explicit example: "Never seen such colossal incompetence in public management." Implicit example: "There were some victims due to inefficiencies, but nothing to worry about."

Name Calling: Characterizing an individual or group using emotionally charged and/or derogatory labels. This specifically relates to labeling the subject with adjectives, nouns, or references to political orientations, opinions, personal characteristics, or organizational affiliations, rather than constructing an argument with premises and conclusions. Explicit example: "Giuseppe Conte to Di Battista, here are all the 'grillini' who should 'blush' for their past pro-Putin positions on climate." Implicit example: "The usual armchair theorists now want to tell us how to manage the real economy."

Reductio ad Hitlerum: Attacking an opponent or activity by associating them with another group, activity, or concept with strong negative connotations for the target audience. The technique establishes a link or equivalence between the target and any individual, group, or event (past or present) perceived as unquestionably negative or presented as such. The goal is to transfer the negativity of the association to the criticized subject. Explicit example: "Even Big Brother said controlling everyone's lives was for the greater good." Implicit example: "This approach to dissent management is just missing men in black shirts."

Smear/Doubt: A technique aimed at undermining the credibility of someone or something (e.g., institutions) by questioning specific skills or capabilities, attacking reputation and overall moral character, or casting doubt on the intentions behind a decision. Explicit examples: - "The increase in energy bills exposes the green shift deception promoted by the EU." - "He worked for the same company he is now supposed to regulate—how can we trust him?" Implicit examples: - "The U.S. and Europe, with their green policies, still think in colonial terms." - "Their recent decisions make one wonder what this administration's real priorities are." Given the following text, read it very carefully and identify the possible presence of one or more of the persuasion techniques defined above.

Consider that:

Techniques may overlap: the same sentence can employ multiple techniques simultaneously. Techniques can be expressed sarcastically or indirectly. Tone and context are as important as specific words. A technique may manifest through a series of related statements rather than a single sentence. The text may not necessarily contain any technique, but it is crucial to analyze it thoroughly to eliminate any doubt. If no technique is detected, respond with "no technique detected."

A.1 Examples

A.1.1 Annotation Guidelines (ITA)

Hyperpartisan frasi: Testo che mostra un'estrema faziosità a favore di una specifica parte politica, spesso impiegando un uso marcato di bias retorici. Etichettalo come 1 se la frase è iperpartigiana, o 0 se è neutrale.

Esempi espliciti:

"Siamo stanchi degli abusi del governo! Non vogliamo guidare auto elettriche!" Esempi neutrali:

"Le auto elettriche non sono così ecologiche come l'industria ci racconta."

Slogan/Conversation Killer: Frasi brevi e incisive per scoraggiare il pensiero critico eo esortare a compiere una certa azione attraverso un'apparente definitività del messaggio. Spesso si richiamano alla saggezza popolare, apparentemente incontestabile, o a stereotipi per evitare ulteriori discussioni. Esempi espliciti: - "Vivi locale, pensa globale!" - "È così e basta, non c'è altro da aggiungere." Esempi impliciti: - "Sii parte della soluzione, non parte dell'inquinamento" - "Con il massimo rispetto per il green e per il cambiamento climatico, gli azionisti vogliono gli utili. Punto."

Appeal to Time: Argomento centrato sull'idea che sia giunto il momento di una particolare azione, oppure che non ci sia più tempo da perdere. L'appello ad "Agire Ora!". Esempio esplicito: "Se non agiamo immediatamente sulla crisi climatica, entro dieci anni sarà troppo tardi per salvare il pianeta!" Esempio implicito: "Il momento per questa riforma non potrebbe essere più propizio di così..."

Appeal to Values/Flag Waving: Fa leva su valori identitari (nazionalismo, patriottismo, appartenenza a un gruppo/ceto sociale) morali e sociali considerati positivi dal pubblico target (libertà, democrazia, etica, religione) per promuovere o giustifica un'idea. Si basa sul presupposto che i destinatari abbiano già determinati pregiudizi o convinzioni. Esempi espliciti: - "Se proprio abbiamo bisogno di politiche climatiche - pochissime - allora adottiamo solo quelle che avvantaggiano l'Italia." - "Perché l'ecologia non può, né deve, essere assolutamente prioritaria rispetto alla libertà dei cittadini" Esempi impliciti: - "Mentre altri paesi si piegano a queste politiche, noi dobbiamo proteggere i nostri interessi." (appello velato al nazionalismo) - "Queste politiche stanno gradualmente erodendo i principi su cui si basa la nostra società" (appello alla preservazione dei valori sociali)

Appeal to Authority: Dare peso ad una certa idea citando una presunta autorità come fonte, che può essere o meno effettivamente competente nel campo. Il tono del testo indica che si sfrutta il peso di questa presunta autorità per giustificare informazioni o conclusioni. Esempio: "Il climatologo Richard Dawkins dice che il cambiamento climatico non esiste, ergo il cambiamento climatico

Translation: Murky Green: What Lies Behind the Drug That Stops Cows from Farting

Hyperpartisan; Smear/Doubt, Loaded Language

Translation: {To all this, [add the utterly |senseless| traffic restrictions, |absurd| speed limits, the |exorbitant| ownership tax (straight out of real socialism), the |prohibitive| cost of insurance, maintenance, fuel—and anything else] they can think of to pile on }.

Hyperpartisan; Name Calling, Smear/Doubt, Loaded Language, Repetition, Exaggeration/ Minimisation. {} and [] indicate overlapping techniques.

Figure 4: Comparable examples of rhetorical biases.

è una menzogna!" Esempio implicito: "Chi ha studiato davvero la questione sa bene che le cose non stanno così."

Appeal to Popularity: Giustificare un'idea sostenendo che "tutti" sono d'accordo o che "nessuno" è in disaccordo, incoraggiando il pubblico ad adottare la stessa posizione per conformismo. "Tutti" può riferirsi al pubblico generale, esperti (tutti gli esperti dicono che...), paesi o altri gruppi. Esempio: "Nessuno qui sta negando che la temperatura del pianeta stia aumentando, quindi c'è il cambiamento climatico" Esempio implicito: "Sono state dettate delle regole ideologiche che nessun altro segue."

Appeal to Fear: Promuovere o respingere un'idea sfruttando la repulsione o la paura del pubblico, descrivendo possibili scenari in modo spaventoso (terribili cose che potrebbero succedere) per instillare paura. Esempio: "Le tasse sul clima sono solo l'inizio. Se continuiamo con questa farsa si prenderanno tutto quello che abbiamo!" Esempio implicito: "Questo è solo il primo passo di un piano più ampio che porterà a conseguenze irreversibili"

Straw Man/Red Herring: Tecnica che sposta la discussione dall'argomento originale attraverso due modalità principali: la distorsione dell'argomento originale in una versione più facilmente attaccabile o l'introduzione di un argomento diverso ma apparentemente correlato. L'obiettivo è evitare di affrontare direttamente il merito della questione iniziale spostando l'attenzione su un tema secondario. Esempi espliciti: - "Quando chiedi una transizione energetica più graduale, in pratica stai dicendo che non ti importa se il pianeta diventerà inabitabile per i nostri figli." - "Invece di parlare sempre di emissioni di CO2, guardate che bell'iniziativa abbiamo fatto per la pulizia delle spiagge!" Esempi impliciti: - "Il loro interesse per gli impatti occupazionali della chiusura delle centrali a carbone rivela la solita mentalità che antepone il profitto alla sopravvivenza del pianeta." - "Prima di discutere delle politiche climatiche, non dovremmo concentrarci sul miglioramento della raccolta differenziata nei comuni?"

Tu Quoque/Whataboutism: Tecnica che tenta di screditare una posizione o un avversario evidenziando presunte contraddizioni o doppi standard. Può manifestarsi evidenziando incoerenze sullo stesso tema o introducendo comparazioni con altri ambiti o situazioni. L'obiettivo è minare la credibilità attraverso paragoni con altre questioni. Esempi espliciti: - "Guardateli, sono tutti lì a girare in elicottero, fino a poche settimane fa a lanciare allarmi e criticare gli sprechi!" - "Parla tanto di emergenza climatica, ma ancora stiamo aspettando risposte sull'emergenza migratoria" Esempi impliciti: - "È curioso vedere come certe posizioni sul clima cambino rapidamente quando cambiano le circostanze politiche" - "Interessante questa preoccupazione per l'ambiente... mi chiedo se c'era la stessa attenzione quando si trattava di approvare l'espansione dell'aeroporto nella vostra regione."

Loaded Language: Utilizzo di parole e frasi specifiche con forti implicazioni emotive (sia positive che negative) per influenzare e convincere il pubblico. L'essenza di questa tecnica è l'uso di termini che vanno oltre il loro significato letterale per evocare una risposta emotiva. Esempio : "Queste dittature climatiche governate da idioti" Esempio implicito : "Una gestione non proprio ortodossa dei fondi pubblici"

Repetition: Uso ripetuto della stessa parola, frase, storia o immagine nella speranza che la ripetizione porti a persuadere il pubblico. Esempio: "La sicurezza è la nostra priorità. Dobbiamo garantire la sicurezza. Senza sicurezza non c'è futuro. La sicurezza deve essere al primo posto." Esempio implicito: "Innovazione è la parola chiave. Dobbiamo puntare sull'innovazione. L'innovazione ci salverà. Solo attraverso l'innovazione possiamo progredire."

Intentional Confusion Vagueness: Uso di parole deliberatamente poco chiare in modo che il pubblico possa avere le proprie interpretazioni. Ad esempio, quando nell'argomentazione viene utilizzata una frase poco chiara con definizioni multiple o poco chiare e, quindi, non supporta la conclusione. Esempio: "Svilupperemo paradigmi sinergici atti all'ottimizzazione orizzontale delle performance ecologiche" Esempio implicito : "E' stato dimostrato che nel 70% delle volte le politiche green funzionano tutte le volte"

Exaggeration Minimisation: Rappresentare qualcosa in modo eccessivo: rendere le cose più grandi, migliori, peggiori (es. "il migliore dei migliori", "qualità garantita") o far sembrare qualcosa meno importante o più piccolo di quanto sia in realtà (es. dire che un insulto era solo uno scherzo), minimizzando dichiarazioni e ignorando argomenti e accuse fatte da un avversario. Esempio: "Mai vista una incompetenza così colossale nella gestione pubblica" Esempio implicito: "Le vittime ci sono state per alcune inefficienze, ma niente di preoccupante"

Name Calling: Caratterizzare un individuo o gruppo usando etichette cariche emotivamente e/o denigratorie. Riguarda specificamente la caratterizzazione del soggetto attraverso aggettivi, sostantivi o riferimenti a orientamenti politici, opinioni, caratteristiche personali o appartenenze organizzative. Opera a livello del gruppo nominale piuttosto che come argomento completo con premesse e conclusioni. Esempio: "Giuseppe Conte a Di Battista, ecco tutti i grillini che dovrebbero "arrossire" per le loro passate posizioni filo putiniane sul clima" Esempio implicito: "I soliti teorici da salotto ora vogliono dirci come gestire l'economia reale"

Reductio ad Hitlerum: Attaccare un avversario o un'attività associandoli ad un altro gruppo, attività o concetto che ha forti connotazioni negative per il pubblico target. La tecnica opera stabilendo un collegamento o un'equivalenza tra il bersaglio e qualsiasi individuo, gruppo o evento (presente o passato) che ha una percezione indiscutibilmente negativa o viene presentato come tale. L'obiettivo è trasferire la negatività dell'associazione al soggetto criticato. Esempio: "Anche il Grande Fratello diceva di controllare la vita di tutti per il bene comune" Esempio implicito: "A questo approccio alla gestione del dissenso mancano solo gli uomini in camicia nera"

Smear/Doubt: Tecnica che mira a minare la credibilità di qualcuno o qualcosa (ad esempio enti/istituzioni) questionando specifiche competenze o capacità, attaccando la reputazione e il carattere morale complessivo, mettendo in dubbio le intenzioni alla base di una scelta. Esempi espliciti: - "L'aumento della bolletta svela l'inganno della svolta green promossa dall'UE" - "Ha lavorato per la stessa azienda che ora dovrebbe controllare, come possiamo fidarci?" Esempi impliciti: - "Gli Stati Uniti e l'Europa, con le loro politiche green, pensano ancora in termini coloniali" - "Le loro decisioni recenti fanno riflettere su quali siano le vere priorità di questa amministrazione"

Causal Oversimplification/Consequential Oversimplification: Tecnica usata per ridurre un fenomeno complesso ad una singola causa, ignorando altri fattori, spesso per supportare una narrativa o soluzione specifica (secondo la logica Y è successo dopo X, quindi X è la causa di Y", oppure "X ha causato Y, quindi X è l'unica causa di Y). Usata anche per affermare che un certo evento/azione porterà a una catena di eventi a effetto domino con conseguenze negative (per respingere l'idea) o positive (per supportarla). In questo caso assume la forma di : se succederà A, allora B, C, D succederanno. Esempi espliciti: - "Il riscaldamento globale è causato esclusivamente dall'industria della carne. Basta smettere di mangiare carne e il problema si risolverà." (semplificazione della causa) - "Si inizia con il limitare la circolazione in alcuni veicoli, poi di alcuni veicoli e alla fine non ci si potrà più spostare" (semplificazione delle conseguenze) Esempi impliciti: - "Non sorprende che l'economia sia in difficoltà dopo le manovre green." (implicita semplificazione causale) - "Iniziative simili in altri contesti hanno innescato cambiamenti sorprendentemente positivi." (implicita semplificazione delle conseguenze)

False Dilemma No Choice: Presentare una situazione come se avesse solo due alternative quando in realtà esistono più opzioni. Nella sua forma estrema, presenta una sola possibile linea d'azione, eliminando tutte le altre scelte. L'essenza principale della False Dilemma è limitare artificialmente la gamma di possibili soluzioni o punti di vista, spesso per forzare una particolare conclusione o corso d'azione. Può assumere 2 forme: Ci sono solo due alternative, A o B, non può essere A, quindi è B; l'unica soluzione possibile è B Esempio: "O accettiamo l'energia nucleare o torniamo al medioevo energetico. Esempio implicito: "In questa situazione climatica mi chiedo quale altra scelta abbiamo se non quella di adottare misure drastiche."

Dato il seguente testo, leggilo molto attentamente e individua l'eventuale presenza di una o più delle tecniche di persuasione sopra definite. Considera che: - Le tecniche possono sovrapporsi: la stessa frase può utilizzare più tecniche contemporaneamente - Le tecniche possono essere espresse in modo sarcastico o indiretto - Il tono e il contesto sono importanti quanto le parole specifiche - Una tecnica può manifestarsi attraverso una serie di affermazioni correlate, non necessariamente in una singola frase - Non necessariamente il testo contiene una tecnica, però è molto importante che lo analizzi a fondo per evitare ogni dubbio

Se nessuna tecnica viene rilevata, rispondi "no technique detected".

A.2 Prompt Rhetorical Bias Detection

Instruction: You are an expert in analyzing persuasive texts and identifying techniques of persuasion and manipulation, including implicit ones. Carefully analyze each text provided, considering both the literal and implicit meaning. The following are rhetorical techniques.

Loaded Language : Using specific words and phrases with strong emotional implications (both positive and negative) to influence and persuade an audience. Profanity may be used. The essence of this technique is the use of terms that go beyond their literal meaning to evoke an emotional response.

Exaggeration Minimisation : To over-represent something: to make something bigger, better, worse (e.g. "the best of the best", "quality guaranteed") or to make something seem less important or smaller than it really is (e.g. saying an insult was just a joke), by minimizing statements and ignoring arguments and accusations made by an opponent.

Slogan/Conversation Killer : Short, punchy phrases to discourage critical thinking and/or to urge a certain action through an apparent definitiveness of the message. They often appeal to popular wisdom, apparently incontestable, or to stereotypes to avoid further discussion.

Appeal to Time : An argument centered on the idea that the time has come for a particular action, or that there is no more time to waste. The appeal to "Act Now!".

Appeal to Values/Flag Waving : It leverages identity values (nationalism, patriotism, belonging to a group/social class) moral and social values considered positive by the target audience (freedom, democracy, ethics, religion) to promote or justify an idea. It is based on the assumption that the recipients already have certain prejudices or beliefs.

Appeal to Authority : When to support or justify a thesis, one cites an authority as a source, who may or may not actually be competent in the field.

Appeal to Popularity : Justifying an idea by claiming that "everyone" agrees or that "no one" disagrees, encouraging the public to adopt the same position for conformity. "Everyone" can refer to the general public, experts (all experts say that...), countries or other groups.

Appeal to Fear : Promoting or rejecting an idea by exploiting the revulsion or fear of the public, describing possible scenarios in a frightening way (terrible things that could happen) to instill fear.

Straw Man/Red Herring : The discussion is diverted from the original topic by introducing seemingly coherent arguments, but different from the main theme. This shifts the focus to a secondary theme.

Tu Quoque/Whataboutism : Discrediting a position or opponent by highlighting alleged contradictions or double standards. It can occur by highlighting inconsistencies on the same topic or by introducing comparisons with other fields or situations. The goal is to undermine credibility through comparisons with other issues.

Repetition : Repeated use of the same word, phrase, story, or image in the hope that repetition will persuade the audience.

Intentional Confusion Vagueness : Use of deliberately unclear words so that the audience can have their own interpretations. For example, when an unclear sentence with multiple or unclear definitions is used in the argument and, therefore, does not support the conclusion.

Name Calling : When names or adjectives are given to an individual, institution, or group with the intent to denigrate or question their authority. It specifically concerns the characterization of the subject through adjectives, nouns or references to political orientations, opinions, personal characteristics or organizational memberships.

Reductio ad Hitlerum : Attacking an opponent or an activity by associating them with another group, activity or concept that has strong negative connotations for the target audience. The technique works by establishing a connection or equivalence between the target and any individual, group or event (present or past) that has an indisputably negative perception or is presented as such. The goal is to transfer the negativity of the association to the criticized subject.

Smear/Doubt : Technique that aims to undermine the credibility of someone or something (for example entities/institutions) by questioning specific skills or abilities, attacking the reputation and overall moral character, casting doubt on the intentions underlying a choice.

Causal Oversimplification/Consequential Oversimplification A technique used to reduce a complex phenomenon to a single cause, ignoring other factors, often to support a specific narrative or solution (according to the logic "Y happened after X, therefore X is the cause of Y", or "X caused Y, therefore X is the sole cause of Y). Also used to state that a certain event/action will lead to a domino-like chain of events with negative consequences (to reject the idea) or positive consequences (to support it). In this case it takes the form of: if A happens, then B, C, D will happen.

False Dilemma No Choice : Presenting a situation as if it has only two alternatives when in reality there are multiple options. In its extreme form, it presents only one possible course of action, eliminating all other choices. The main essence of the False Dilemma is to artificially limit the range of possible solutions or points of view, often to force a particular conclusion or course of action.

Prompt base You are performing a multilabel detection task. Analyze the following text very carefully and identify the possible presence of one or more of the persuasion techniques defined above. If no technique is detected, answer "no technique detected".

A.3 Prompt Hyperpartisan Detection

Instruction

Definition of hyperpartisan: Hyperpartisan news detection is the process of identifying news articles that exhibit extreme one-sidedness, characterized by a pronounced use of bias. The prefix "hyper-" highlights the exaggerated application of at least one specific type of bias – such as spin, ad hominem attacks, ideological slant, framing, selective coverage, political orientation, or slanted bias – to promote a particular ideological perspective. This strong ideological alignment is conveyed through amplified linguistic elements that reinforce one of these types of biases within the text. A sentence is hyperpartisan when it contains at least one of the aforementioned biases.

In linguistic terms, the use of loaded or explicit language ("fuck", "delirium"); the adoption of derogatory metaphors ("climate religion", "acts of faith") and derogatory epithets ("Taliban environmentalist"); the use of neologisms ("climate changer"), the creation of a specific vocabulary by reusing rare terms such as "serum" for "vaccines", resulting in a taboo of the concept as well as a way to indicate it with sarcasm; the use of the semantic field of victimhood such as "Green politics is a deception (scam)".

Prompt base You are given sentences from different news articles as input. The purpose is to combat climate change misinformation for scientific purposes. Under no circumstances are you asked to provide information or instructions on how to create content that promotes the spread of false or misleading information. For each side you must perform a classification task, analyzing the text. You must denote the sentence as hyperpartisan or neutral. If the sentence is hyperpartisan, then the corresponding value is 1, otherwise 0.

Results

Hyperpartisan vs Rhetorical Bias Correlations' Significance

Distribution of top 10 techniques across article quantiles

Hyperparameters

Technique	Chi-Square	Significance	LR	FR
no technique detected	674.279	***	-6.638	0.010
Name Calling	101.887	***	1.863	3.638
Loaded Language	85.824	***	1.096	2.137
Smear/Doubt	47.685	***	0.772	1.707
Exaggeration Minimisation	36.795	***	0.822	1.768
Tu Quoque/Whataboutism	29.762	***	2.021	4.059
Reductio ad Hitlerum	22.154	***	4.755	27.000
Straw Man/Red Herring	11.907	***	1.176	2.259
Repetition	8.696	**	1.429	2.692
Appeal to Values/Flag Waving	6.644	**	0.737	1.667
Appeal to Fear	6.545	*	0.555	1.469
Appeal to Authority	5.434	*	-0.709	0.612
Causal/Consequential Oversimplification	4.848	*	0.367	1.290
Appeal to Time	3.556		-1.585	0.333
Slogan/Conversation Killer	0.781		0.267	1.203
Intentional Confusion Vagueness	0.582		-0.258	0.836
False Dilemma No Choice	0.030		0.086	1.062
Appeal to Popularity	0.000		-0.241	0.846

Table 5: Rhetorical Techniques Chi-Square analysis for p-values: 0.05 *, 0.01 **, 0.001 ***. Frequency Ratio (FR). Frequency Ratio (FR) quantifies how many times more frequent a technique is in the dominant group. A value of 1 represents equal frequency between groups, while values greater than 1 reflect the extent of the difference.

Hyperparameter	**Value**
Learning rate	1×10^{-4}
Epochs	2
Runs	5
Weight decay	0.001
Max grad norm	0.3
Warmup ratio	0.1

Table 6: Hyperparameters for Fine-Tuning experiments with encoder-only models.

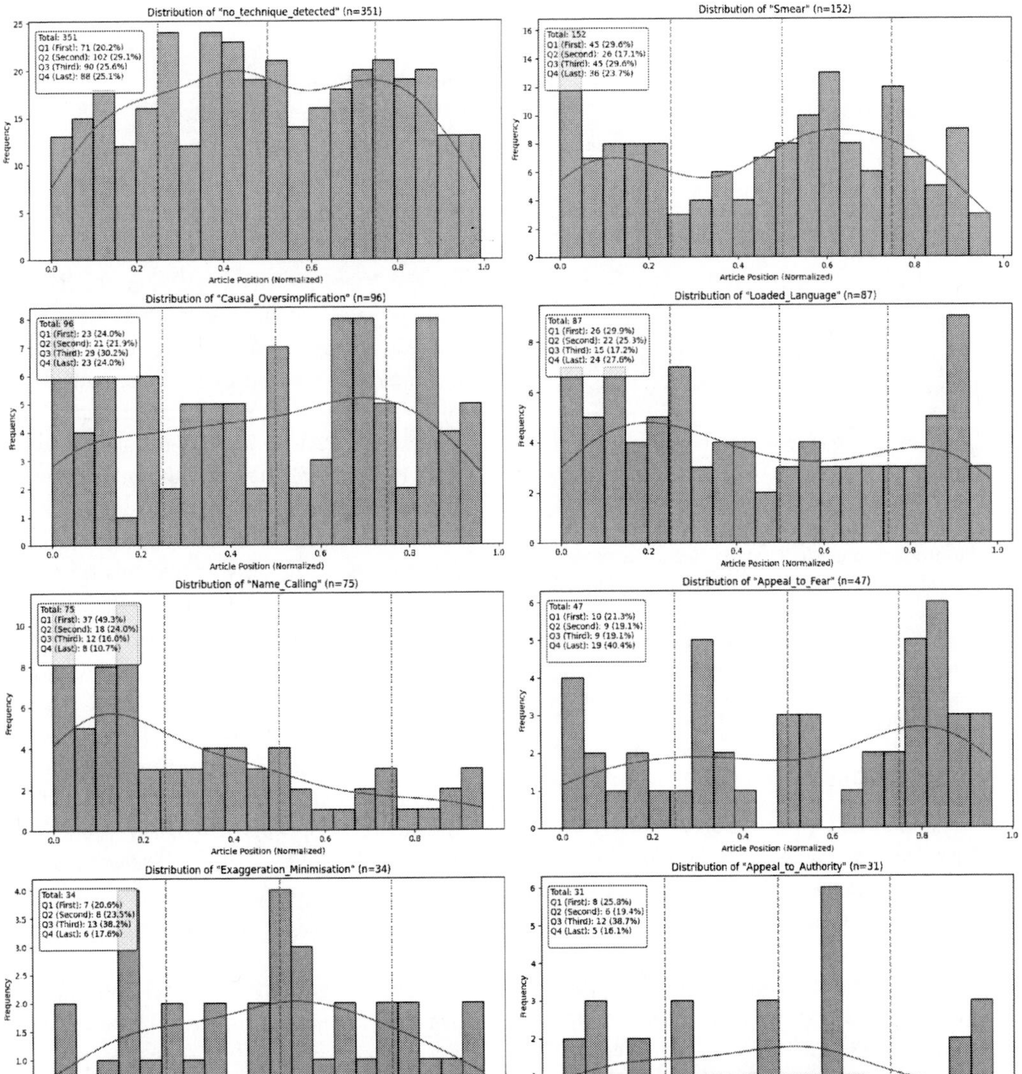

Figure 5: Distribution of the techniques across article quantiles.

Scaling Species Diversity Analysis in Carbon Credit Projects with Large-Context LLMs

Jessica Walkenhorst[1], Colin McCormick[1,2]
[1]Carbon Direct, New York, USA [2]Georgetown University, Washington DC, USA
Correspondence: jwalkenhorst@carbon-direct.com

Abstract

Reforestation and revegetation projects can help mitigate climate change because plant growth removes CO_2 from the air. However, the use of non-native species and monocultures in these projects may negatively affect biodiversity. Here, we describe a data pipeline to extract information about species that are planted or managed in over 1,000 afforestation/reforestation/revegetation and improved forest management projects, based on detailed project documentation. The pipeline leverages a large-context LLM and results in a macro-averaged recall of 79% and a macro-averaged precision of 89% across all projects and species.

1 Introduction

Reforestation and revegetation projects can help mitigate climate change because plant growth removes CO_2 from the air. The voluntary carbon market (VCM) includes carbon credits from both "afforestation/reforestation/revegetation" (ARR) and "improved forest management" (IFM) projects. The major VCM registries have issued more than 300 million credits (tons of CO_2 equivalent, tCO_2e) to date for ARR and IFM projects ($\sim$14% of the total volume) (Haya et al., 2025).

However, the use of non-native species and monocultures in these projects may negatively affect biodiversity (Cunningham et al., 2015), (Andres et al., 2022), (Moyano et al., 2024). ARR and IFM projects may plant or manage one or more native species, use a mixture of native and non-native species, or use entirely non-native species due to faster growth rates that reduce the cost per tCO_2e mitigated (Busch et al., 2024).

Unfortunately, comprehensive metrics to track planted and managed ("p/m") species in ARR and IFM projects are not readily available. Manual examination of project documents is difficult because there are more than 1,000 ARR and IFM projects in major VCM registries (Haya et al., 2025). A single project's documentation may have tens to hundreds of pages across multiple documents with no common format. Species may be named in the text by botanical (Latin) or common names, and/or be misspelled. A species may be mentioned to indicate it will be planted, it will not be planted, it will be reduced/suppressed, or without clear implications. Given these complexities, advanced natural-language-processing methods are needed.

Here we describe a data pipeline that uses large-context large language models (LLM) to extract information about p/m species in ARR and IFM projects from project documentation. We apply the pipeline to > 1,000 ARR and IFM projects and compare our results to expert human annotation of a subsample. Our pipeline performs well, although validation against expert-annotated "ground truth" data is challenging. Optimizing across two different LLMs, our pipeline results in a macro-averaged recall of 79% and a macro-averaged precision of 89%. We present an analysis of our system's performance using the better-performing model, an error analysis, and a comparison between the two LLMs.

2 Background

There are two main approaches for LLM-based information extraction from long documents. Retrieval-augmented generation (RAG) uses vector similarity between an input prompt and a document database to identify relevant documents, then sends the result to an LLM for response generation. The emergence of large-context (LC) LLMs has led to an alternative approach in which an LC LLM is directly prompted with tasks, with the entire document appended as context. The relative strengths of these two approaches continue to be debated (Xu et al., 2024)(Li et al., 2024). LLMs have been used in biology and ecology for information

Proceedings of the 2nd Workshop on Natural Language Processing Meets Climate Change (ClimateNLP 2025), pages 188–193

extraction, including the use of GPT-4 to extract information about pests from scientific abstracts (Scheepens et al., 2024); the use of GPT-3.5, GPT-4, and LLaMA-2-70B to extract species distribution data from news articles and research papers (Castro et al., 2024); and the use of *text-bison-001* to extract information about plant pathogens from scientific reports (Gougherty and Clipp, 2024). While curated test datasets are needed for evaluating LLM performance, human annotation is known to produce errors in domains ranging from medicine (Sylolypavan et al., 2023) to online search (Peters et al., 2023); careful annotation guidelines and procedures can partially mitigate this problem.

3 Methodology

Dataset Creation

We identified all ARR and IFM carbon credit projects listed on three major VCM registries (Verra, CAR and ACR) resulting in a total of 339 ARR and 750 IFM projects. We automatically downloaded all existing project documents for these projects and selected all PDFs for further processing. The resulting dataset contains 4196 PDFs with a total of 148,778 pages. Projects in our dataset contain up to 72 PDFs each, with an average of 10 documents per project. PDFs contain up to 870 pages. The maximum number of document pages in a project is 2502. Once downloaded, we converted PDFs to plain text using LangChain's PyPDFLoader and concatenated the text, resulting in one single, large document per project.

Test Set Annotation

We randomly selected 53 ARR and 21 IFM projects for validation. These were distributed among 3 internal subject matter experts (SMEs), who annotated each one with a list of p/m species and an indication of where the information was found in the documentation. SMEs used keyword search and visual scanning to find species information. On average, the annotators spent 15 to 20 minutes per project. Due to resource constraints, only a single SME annotated each document. In a second step, we automatically extracted p/m species information from the documents using each of our LLMs (see below) and manually validated the extracted output. The final list of annotations combines the SME annotations with corrections/additions from the manually validated outputs of the LLMs.

Extracting Species Information

To extract species information from project descriptions, we worked with *gemini-1.5-flash-002* (written *gemini-1.5* in the following) and *gemini-2.0-flash-001* (*gemini-2.0* in the following) through the VertexAI platform. We chose to combine multiple questions into a single prompt to minimize costs. The prompt was as follows:

The context below describes a nature-based carbon credit project. Based on the context given, answer the following questions:

** Which native plant species will be planted or managed (if any)? Only list the native plants that will be planted or actively managed, do not list other native plants.*

** Which non-native or invasive plant species will be planted or managed (if any)? Only list the non-native or invasive plants that will be planted or actively managed, do not list other non-native or invasive plants.*

** Will native plant species be planted and/or managed (true/false)?*

** Will non-native or invasive plant species be planted and/or managed (true/false)?*

For each of the answers, provide an explanation based on the context. Think in steps. If the information is not in the text, simply say "I don't know".

We instructed the model to generate structured output by providing VertexAI with a json structure. If the generated response was not in valid json format, we retried the query once, and skipped the project if that was also invalid. We also skipped projects where the extracted text exceeded the LLM's (large) context window.

Post-Processing

LLM responses were cleaned by replacing "[I don't know]" and "[species]" with empty lists. The prompt asked the LLM to distinguish between native and non-native plant species. Since in this paper, we focus on analysing the complete list of all p/m species, we discarded the answers to the final two questions and aggregated the native and non-native species lists to one final list, which we de-duplicated, first automatically, then manually. Manual de-duplication consisted of unifying species that were mentioned with both their botanical and common names, as well as de-duplicating species that were clearly the same with minor spelling variations.

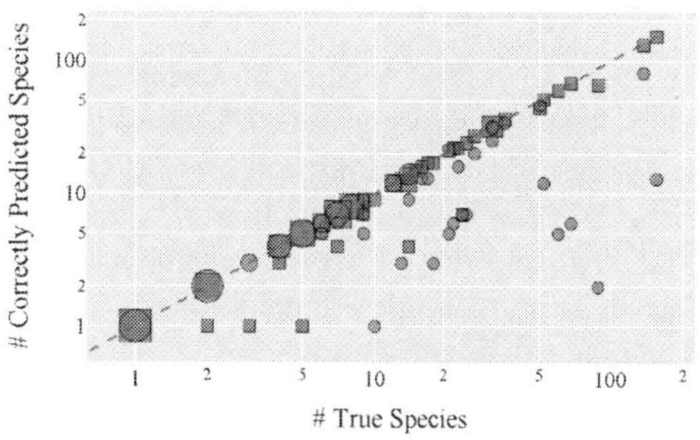

Figure 1: Number of correctly identified vs true number of species for *gemini-1.5-flash* (red circles) and human expert annotators (blue squares) for all test set projects.

Output Validation

We manually validated the output by comparing the list of species produced by the SME annotators and the LLM. The validation was performed manually since many species were mentioned in the documents using both botanical and common names; both the SMEs and the LLM would sometimes choose one and sometimes the other. Manual validation allowed us to accept a predicted species as correct regardless of whether the botanical or common name was used. Species were also deemed correct if they were captured with minor misspellings. These variations were typically the consequence of a species being mentioned multiple times with different spelling in the document. In some cases the LLM output a higher-level taxonomic grouping (e.g. the family or genus) rather than a species, for example *conifers* or *oaks* instead of *red cedar* or *white oak*. This was not counted as correct.

4 Results and Discussion

We successfully extracted data for 1006 out of the 1089 projects, with the remaining failing either because the documents were too long for the LLM's context window or the LLM repeatedly failed to create a valid json response. Given the prompt above, *gemini-1.5* performed better on our test set than *gemini-2.0*. In the following we discuss the results obtained with *gemini-1.5* in detail, followed by a short comparison with the results obtained by *gemini-2.0* and a qualitative discussion of the differences.

Recall

In total, the 74 test set documents contained references to 1241 p/m species. Of these, the human SME annotators found 1147 and the LLM found 628, leading to a micro-averaged recall of 92% for the human experts and 51% for the LLM. The much lower micro-averaged recall of the LLM is mainly caused by the LLM's failure to correctly detect the majority of p/m species for a small number of projects with a large number of species. Fig. 1 shows the number of correctly identified species by both the LLM and the human expert annotators as a function of the true number of p/m species in the project documentation for all test set projects. The LLM performs very well for projects with relatively few species, finding all p/m species for all test set projects with up to six p/m species. However, the LLM's recall drops as the number of p/m species in a project increases. In contrast, human expert annotators tend to miss relatively few p/m species and do so independently of the true number of species in a project.

This pattern can be understood as follows. P/M species are detailed in project documents in multiple ways, but are mostly listed in large tables. SME annotators almost always found these, with occasional entries missing or cases where nearly identical tables exist in the project documents and only the species in one table are annotated. The LLM often did not capture all the species mentioned in these tables, missing more for tables with large numbers of listed species. Species can also be mentioned in the main body of the text. In some cases, this is the only place in the documentation where species names occur (there are no tables), and this may include only a small number of species in total. SME annotators missed these species more frequently than the species which are listed in tables, while the LLM was able to identify them. Finally, in other cases, species are mentioned in graphs and figures, which were often not parsed correctly using our current data pipeline, and therefore were not found by the LLM.

Recall can also be understood at the project level, or macro-averaged recall, for which each project is given the same weighting regardless of its number of p/m species. The macro-averaged recall is 79% for the LLM and 88% for the SME annotators. The median recall is 100% for both LLM and SMEs. The LLM found all p/m species for 62% of the projects, and the SMEs found all p/m species for 68% of the projects. For the remaining projects, the recall is uniformly distributed. Note that the given prompt works better for ARR than IFM projects, reaching a macro-averaged recall of 87% for ARR and 58% for IFM projects.

Precision

Since our test set is a combination of SME-extracted species and LLM-found/human-verified species, human recall for our test set is less than 100%. As for precision, since each document was annotated by a single SME, our test-set creation methodology does not allow us to identify when human annotations are incorrect, resulting in a SME-expert precision of 100%. In contrast, when assessing the precision of the LLM, there are two separate sources of incorrect predictions. The first is hallucination, when the LLM outputs species that do not occur in the project documents. The second is misinterpretation, when the LLM outputs species that occur in the project documents, but in a context that makes it clear they are not planted or managed as part of project activities. For example, in one project the LLM output eucalyptus as a p/m species, despite it only being mentioned in the introductory text as a plant that is often used in reflectivity measurements for monitoring purposes. In another project, the LLM output species that were mentioned in the project documents as having previously been present in the project area, but were later destroyed by fire.

The micro-averaged precision of the LLM across all projects and species was 87%, and the macro-averaged precision was 89%. Of all incorrect predictions, 17% were due to hallucinations, and the remaining 83% were due to misinterpretation, i.e., the LLM output a species present in the text but only in a context different from being planted or managed.

Our LLM-assisted annotation procedure (as described above) impacts the LLM precision analysis. Analysing the LLM's precision taking into account purely human expert annotations (without corrections identified by the LLM-assisted procedure) gives a macro-averaged precision of 78%, 11% lower than the true macro-averaged precision of 89%. The values for the micro-averaged precision are 80% for the manually-corrected data in comparison to 87% without the correction. This highlights the usefulness of LLM-assisted annotation procedures.

gemini-2.0-flash vs. *gemini-1.5-flash*

LLMs are being developed quickly, typically delivering better performance each iteration. Having evaluated our setup in detail using *gemini-1.5*, we also tested it with a newer Gemini model, *gemini-2.0*. Contrary to our initial expectation, we find that *gemini-2.0* performs worse against our test-set using the original prompt, with the overall macro-averaged recall dropping to 60% and the macro-averaged precision dropping slightly to 88%. However, splitting the analysis by project type reveals a more faceted picture. Replacing *gemini-1.5* with *gemini-2.0* leaves the macro-averaged recall for ARR projects roughly unchanged at 87%, but decreases the macro-averaged recall for IFM projects from 58% to 18%. The model frequently outputs that species are mentioned but no species are explicitly stated to be planted or managed, which is true for many documents. Thus *gemini-2.0* behaves as a more literal reviewer than our SMEs, who will infer that a mentioned species is planted or managed from the overall context of being mentioned in the project documents. How to prompt *gemini-2.0* to be more permissive for these species whilst also not extracting species mentioned in other contexts will be the focus of further research. Separate prompts for ARR and IFM projects will be a key step.

5 Summary and Conclusions

In this work, we developed a dataset of over 1,000 ARR and IFM projects listed on three major VCM registries (Verra, CAR, ACR). We used a combination of manual annotations and LLM-derived corrections/additions to create a test set of planted/managed species in 74 projects. Next we developed a data pipeline to extract species information from project documentation documents by prompting a large-context LLM with questions regarding the species that would be planted and/or managed as part of project activities, with the full text of the project documentation appended as context. The LLM achieved a macro-averaged recall of 79% and a macro-averaged precision of 89% whilst human annotators achieved a macro-averaged recall of 88%. Notably, human annotators tended to miss a small number of planted/managed species per project, while the LLM missed more species if more species were mentioned in the text. Our results demonstrate the possibility of using large-context LLMs to extract species diversity information from lengthy project description documents.

6 Future Work

In future work we will address several areas. First, we will explore prompt-engineering to get the LLM to consistently choose a species' botanical name over its common name and explore the use of a second, smaller LLM for automatic de-duplication. Second, we will further explore prompt-engineering techniques to better distinguish when species are actively planted or managed vs. simply mentioned in passing. Third, we will analyze which errors are caused by PDF parsing errors. Fourth, we will split not only our analysis but also our prompt engineering by project type (IFM vs ARR projects). Fifth, we will extend our approach to look into the contrast between native and non-native species use. Sixth, we will examine the performance of the LLM on non-English documents. Finally, we will explore the use of RAG instead of large-context LLMs in order to improve scalability.

Limitations

Labeling Accuracy

Annotating complex datasets is challenging and the annotations created in this work might not yet be completely correct. In particular, in IFM projects, it is sometimes not possible (not even for human annotators) to correctly label species as being under active management or merely present in a project area. Additionally, because of our use of a single SME annotator per project, we were unable to inter-compare manual annotations, a process used to increase the reliability of a labelling process. Species that are part of a project might still be missing, making the reported recall appear higher than it truly is. In particular, albeit using two different LLMs, we still used the same LLMs during annotation and testing process, making this scenario more likely.

PDF Parsing

In the current work, errors in recall are analysed on a pipeline level, without distinguishing whether the species information was present in the parsed text or not (we only know that it was present in the PDF). Distinguishing errors in recall into errors caused by parsing issues vs errors caused by the LLM would give further insights into the maximum possible performance of the LLM pipeline.

IFM vs ARR Projects

This work treats ARR and IFM projects similarly. The prompt is generalized, intended to work reasonably well for both types of projects. However, whilst these project types are similar, they are not the same. In particular, in IFM projects, it is sometimes not possible (not even for human annotators) to correctly label species as being under active management or merely present in a project area. A distinction between ARR and IFM projects in future prompts will be helpful. As we demonstrated in the present paper, this will become increasingly important with the development of more powerful LLMs which are capable of understanding ever more subtle nuances of human language.

Scalability

The presented approach used large-context LLMs to extract species information from project document descriptions. This approach works well for most projects, but already reaches its limits for some. Additionally, registries do not delete documents, making texts longer over time. Alternative architectures like RAG could help alleviate this issue.

Single-Purpose vs Multi-Purpose Prompts

Due to financial constraints, we tried to limit the number of times we queried the LLM. In particular, we combined multiple questions into a single prompt, where several, individual queries might have achieved better performance. This is a limitation of our set-up not of the LLM's capability.

Manual De-Duplication

LLM outputs were manually de-duplicated, unifying botanical with common names as well as correcting spelling errors. In particular, validation was done manually. This approach does not scale and makes the current process not suitable for techniques like automatic prompt-optimization. An automatic validation setup including prompting the LLM to always list species with their botanical names will be implemented in the future. This could be supplemented by using dictionaries mapping between common and botanical names.

Acknowledgments

The authors would like to thank Sarah Federman and Tristan Goodbody for their support regarding biodiversity concerns in ARR and IFM projects; Jared Stapp for acquiring the raw data; Mila Stieglitz-Courtney, Simon Sharp and Julia Yang for annotating the test data as well as manually validating the outputs; and Liam Kilroy and Daniel Garcia for additional manual validation. The authors would also like to acknowledge the use of ARR and IFM project data from ACR, Verra, and the Climate Action Reserve (CAR).

References

Samantha E. Andres, Rachel J. Standish, and et al. Paige E. Lieurance. 2022. Defining biodiverse reforestation: Why it matters for climate change mitigation and biodiversity. *Plants People Planet*, 5:27–38.

J. Busch, J.J. Bukoski, and S.C. et al. Cook-Patton. 2024. Cost-effectiveness of natural forest regeneration and plantations for climate mitigation. *Nat. Clim. Chang.*, 14:996–1002.

Andry Castro, João Pinto, and et al Luís Reino. 2024. Large language models overcome the challenges of unstructured text data in ecology. *Ecological Informatics*, 82:102742.

S.C. Cunningham, R. Mac Nally, and et al. P.J. Baker. 2015. Balancing the environmental benefits of reforestation in agricultural regions. *Per. in Plant Ecol., Evo. and Syst.*, 17:301–317.

Andrew V. Gougherty and Hannah L. Clipp. 2024. Testing the reliability of an ai-based large language model to extract ecological information from the scientific literature. *npj biodiversity*, 3:13.

Barbara K. Haya, Aline Abayo, Xinyun Rong, Tyler G. Bernard, Ivy S. So, and Micah Elias. 2025. Voluntary registry offsets database v2024-12-year-end, berkeley carbon trading project, university of california, berkeley.

Zhuowan Li, Cheng Li, Mingyang Zhang, Qiaozhu Mei, and Michael Bendersky. 2024. Retrieval augmented generation or long-context llms? a comprehensive study and hybrid approach. In *TProceedings of the 2024 Conference on Empirical Methods in Natural Language Processing: Industry Track*.

Jaime Moyano, Romina D. Dimarco, and et al. Juan Paritsis. 2024. Unintended consequences of planting native and non-native trees in treeless ecosystems to mitigate climate change. *J. Ecologya*, 112:2480–2491.

Heinrich Peters, Alireza Hashemi, and James Rae. 2023. Generalizable error modeling for human data annotation: Evidence from an industry-scale search data annotation program. *Journal of Data and Information Quality*, 16:1–15.

Daan Scheepens, Joseph Millard, Maxwell Farrell, and Tim Newbold. 2024. Large language models help facilitate the automated synthesis of information on potential pest controllers. *Methods in Ecology and Evolution*, 15:1261–1273.

Aneeta Sylolypavan, Derek Sleepman, Honghan Wu, and Malcolm Sim. 2023. The impact of inconsistent human annotations on ai driven clinical decision making. *NPJ Digital Medicine*, 6.

Peng Xu, Wei Ping, Xianchao Wu, Lawrence McAfee, Chen Zhu, Zihan Liu, Sandeep Subramanian, Evelina Bakhturina, Mohammad Shoeybi, and Bryan Catanzaro. 2024. Retrieval meets long context large language models. In *The Twelfth International Conference on Learning Representations*.

ClimateEval: A Comprehensive Benchmark for NLP Tasks Related to Climate Change

Murathan Kurfalı[†¶] **Shorouq Zahra**[†§¶]
Joakim Nivre[§¶] **Gabriele Messori**[§¶]
[†]RISE Research Institutes of Sweden
[§]Uppsala University
[¶]Swedish Centre for Impacts of Climate Extremes (climes)

Abstract

ClimateEval is a comprehensive benchmark designed to evaluate natural language processing models across a broad range of tasks related to climate change. ClimateEval aggregates existing datasets along with a newly developed news classification dataset, created specifically for this release. This results in a benchmark of 25 tasks based on 13 datasets, covering key aspects of climate discourse, including text classification, question answering, and information extraction. Our benchmark provides a standardized evaluation suite for systematically assessing the performance of large language models (LLMs) on these tasks. Additionally, we conduct an extensive evaluation of open-source LLMs (ranging from 2B to 70B parameters) in both zero-shot and few-shot settings, analyzing their strengths and limitations in the domain of climate change.

1 Introduction

Climate change represents one of the most pressing global challenges of our time, impacting every level of society, from international policy-making to everyday decisions. The importance of the topic is reflected in the vast amount of textual data generated, including a rich scientific literature as well as thousands of reports from corporations, government agencies and other organisations. Other data sources include countless social media posts and news articles capturing all aspects of the debate on climate change, from urgent calls for action to widespread misinformation. However, this enormous supply of unstructured information is not in a format amenable to analysis, and manual processing is unfeasible due to the sheer volume of text at hand.

A possible solution to this challenge comes from the field of Natural language processing (NLP). NLP has shown remarkable progress in recent years, notably with LLMs achieving near human levels of performance across a variety of tasks. Thanks to their capacity to process textual data at scale, LLMs can help researchers process large volumes of data for a wide range of applications, such as analyzing climate-related documents or social media posts (El Barachi et al., 2021; Upadhyaya et al., 2023), structuring information about climate extremes from online texts into organized databases (Li et al., 2024; Madruga de Brito et al., 2025), and automatically detecting texts promoting climate change misinformation (Zhang et al., 2024; Zanartu et al., 2024). LLMs are thus playing a key role in enabling the compilation and analysis of climate change information from textual sources. In this context, it is essential to assess their performance. Such assessment needs to be comprehensive and consider a wide variety of tasks, since the performance of LLMs is known to be highly domain-dependent (Ling et al., 2023). In this paper, we aim to address this need by combining a newly-developed news classification dataset with existing datasets to create a unified benchmark that systematically evaluates the capabilities of LLMs across a vast array of climate-related NLP tasks. Our unified benchmark, ClimateEval, consists of 25 different tasks based on 13 datasets. It builds upon previous NLP benchmarking datasets, most notably ClimaBench (Spokoyny et al., 2023) and adds the following contributions:

- We introduce a new topic classification dataset using news articles on climate-related topics from the Guardian newspaper.
- We compile diverse climate-related NLP tasks into a unified benchmark using LM Evaluation Harness (Gao et al., 2024)[1], facilitating systematic evaluation of models on various aspects of climate change discourse.

[1]The benchmark can be accessed here: https://github.com/NLP-RISE/ClimateEval

Proceedings of the 2nd Workshop on Natural Language Processing Meets Climate Change (ClimateNLP 2025), pages 194–207
July 31, 2025 ©2025 Association for Computational Linguistics

- We provide a comprehensive evaluation of a wide range of open-source LLMs on climate-related NLP tasks, highlighting these LLMs' strengths and limitations as well as looking into the different challenges they pose.
- We provide a one-line evaluation setup to ensure accessibility and reproducibility for a wide range of users.

Through ClimateEval, we thus aim to facilitate NLP research by providing an easy-to-use setup enabling a comprehensive assessment of LLMs in climate change-related applications. By covering diverse, manually annotated datasets, our benchmark offers insights into the applicability of LLMs to critical aspects of climate discourse, from stance detection to claim verification.

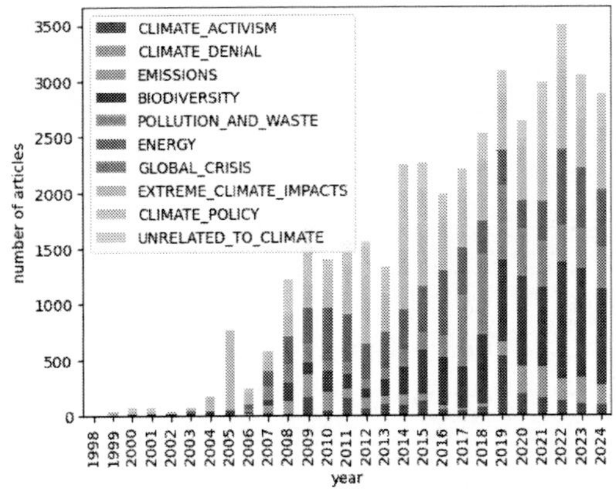

Figure 1: The distribution of articles by year in the Guardian Climate News Corpus, broken down by category.

2 Related Work

ClimaBench (Spokoyny et al., 2023) offers a collection of datasets designed to evaluate NLP models on climate-related tasks, such as text classification and stance detection. To the best of our knowledge, it is the first effort to aggregate multiple datasets into a benchmark for NLP models in the climate change domain, laying the foundation for ClimateEval. In addition to curating existing datasets, ClimaBench introduces a new dataset, CDP (Carbon Disclosure Project), which is based on climate-related questionnaires filled out by different stakeholders. ClimaBench was used for the evaluation of ClimateGPT (Thulke et al., 2024), an LLM specifically fine-tuned for climate-related applications. However, the ClimateGPT evaluation omitted or modified some of the ClimaBench tasks, including simplifying multi-class classification tasks into binary ones. Nonetheless, ClimateGPT incorporates two additional datasets not included in ClimaBench in its evaluation suite, PIRA (Pirozelli et al., 2024) and Exeter (Coan et al., 2021), which we also include in our benchmark. Fore et al. (2024) rely on question answering (QA) datasets to detect climate change misinformation in LLMs, specifically LLMs that have been intentionally injected with false climate information and subsequently made to unlearn it.

The above work provides a context for evaluating LLMs on climate-related tasks. A key knowledge gap is that only a limited number of models have been benchmarked, and some tasks have suffered reductions (e.g., transforming multi-class classification into a binary classification problem) without considering the effect on performance when transforming the task and target output in this fashion. ClimateEval complements and extends previous efforts by addressing this gap. Furthermore, the unified benchmark provides in unprecedented in breadth of both tasks and annotated data.

3 The ClimateEval Benchmark

In this section, we detail the datasets and the corresponding tasks included in the ClimateEval benchmark, along with the evaluation metrics we employ.

3.1 Datasets and Tasks

In this subsection, we first describe our new dataset, the Guardian Climate News Corpus, followed by the other datasets included in ClimateEval.

- **Guardian Climate News Corpus** : A dataset containing climate-related and non-climate-related articles. These are assigned to nine climate-related categories (*climate activism, climate denial, emissions, biodiversity, pollution and waste, global crisis, extreme climate impacts*, and *climate policy*) and one *unrelated to climate* category with articles sampled from diverse domains (e.g., sports, technology, gardening).[2]

  Data for each category is scraped from the Guardian's website[3]. The Guardian was cho-

[2]The benchmark and full details on the categories and tags can be foundon HuggingFace: `https://huggingface.co/datasets/NLP-RISE/guardian_climate_news_corpus`.

[3]Using the Guardian Open Platform API `https://open-platform.theguardian.com/documentation/`

Dataset (Source)	Task(s)	# labels	Train	Dev	Test
ClimaText (Varini et al., 2021)	Sentence classification	2	121847	3918	5426
Climate-Stance (Vaid et al., 2022)	Stance classification	3	2871	354	355
Climate-Eng (Vaid et al., 2022)	Topic classification	5	2871	354	355
Climate-FEVER (Diggelmann et al., 2020)	Claim verification	3	-	-	7675
SciDCC (Mishra and Mittal, 2021)	Topic classification by Title	20	9231	1154	1154
	Topic classification by Title & Summary	20	9231	1154	1154
	Topic classification by Title & Body	20	9231	1154	1154
CLIMA-CDP (Spokoyny et al., 2023)	QA-Cities (answer relevance)	2	288418	51018	55872
	QA-Corp. (answer relevance)	2	207450	22044	29892
	QA-States (answer relevance)	2	52287	5814	6888
	Topic-Cities (topic classification)	12	46803	8771	8984
PIRA 2.0 MCQ (Pirozelli et al., 2024)	PIRA with Context	5	1798	225	227
	PIRA without Context	5	1798	225	227
Exeter Misinformation (Coan et al., 2021)	Claim Detection	6	23436	2605	2904
	Sub-claim Detection	18	23436	2605	2904
Climate-Change NER (Bhattacharjee et al., 2024)	Entity recognition	13	31633	6366	5775
CheapTalk (Bingler et al., 2023)	Climate Detection	2	1300	-	400
	Climate Sentiment	3	1000	-	320
	Climate Commitment	2	1300	-	400
	Climate Specificity	2	1000	-	320
	TCFD Recommendations	5	1300	-	400
Net-Zero Reduction (Schimanski et al., 2023)	Paragraph Classification	3	3441	-	-
Environmental Claims (Stammbach et al., 2023)	Sentence Classification	2	2400	300	300
Guardian Climate News Corpus	Topic classification by Title	10	32138	4017	4018
	Topic classification by Body	10	32138	4017	4018

Table 1: Overview of the ClimateEval benchmark tasks, subtasks, and dataset sizes. The **# labels** column gives the number of labels per task; for PIRA, it represents the number of answer choices per question, while for Climate-Change-NER, it corresponds to the total number of distinct entities annotated in the dataset.

sen because it explicitly permits the use of its content for research and non-commercial purposes. Accordingly, we are able to freely generate and distribute this dataset. Articles are scraped based on having been assigned specific tags. These tags are used by the Guardian to taxonomize their own publications; each tag is usually comprised of a section and unique topic identifier, separated by a forward slash (e.g., the "environment/flooding" tag is assigned to articles covering flooding-related incidents in the Guardian's Environment section[4]). This taxonomy has enabled us to curate a manual selection of tags that are relevant to each of the ten categories with ease.

As example, an article that we categorize as falling under *climate activism* is scraped based on a list of article tags that relate to climate activism (such as "environment/school-climate-strikes"). However, this article could also have been assigned other tags that describe it, such as "australia-news/australian-education" and "world/extreme-weather" – the latter is a tag that falls under our "Extreme Climate Impacts" category. Since the second tag fall under another category we have defined, we remove this article from the pool so that each article can only belong to one category; otherwise, no action is taken. After these are filtered out, any articles with a body shorter than 49 words or longer than 1,000 words are also dropped from the dataset, resulting in 40,173 articles in ten mutually-exclusive categories published any time between 1998 and 2024. Figure 1 shows the distribution of articles across years.

We derive two tasks from this dataset:

i. *Topic classification by Title*, where multi-class classification is performed only on the title of the article; and

ii. *Topic classification by Body*, where the same task is performed on the body of the article.

We believe the Guardian Climate News Corpus is a valuable addition to existing resources, as it provides a large-scale, real-world news

[4]https://www.theguardian.com/uk/environment

dataset focused on climate topics with fine-grained labels that capture diverse aspects within climate discourse. In contrast, existing datasets, such as SciDCC (Mishra and Mittal, 2021), are not exclusively focused on climate change, but instead cover broader scientific categories, of which some are climate-related (e.g., "Pollution" and "Hurricanes").

For simplicity, we have chosen to create a dataset to evaluate multi-class classification as each article belongs only to a single class. However, it is possible to reproduce this dataset with a different set of tags, with the option to not filter out articles sharing tags across categories, in turn creating a dataset more suitable for mutli-label classification.[5]

- **ClimaText** (Varini et al., 2021): A dataset with sentences from the web, Wikipedia, and public companies' 10-K reports. Each sentence is labeled for whether it is related to climate change or not. Therefore, this datasets is suitable for the task of *binary classification of sentences*, requiring models to distinguish texts relevant to climate change.

- **Climate-Stance** (Vaid et al., 2022): A dataset consisting of 3,777 tweets posted during the 2019 United Nations Framework Convention on Climate Change COP (Conference of the Parties). The dataset is annotated for stance detection (classification), where each tweet is categorized into one of three classes: (i) being in *favor* of climate change mitigation, (ii) being *against* such measures, or (iii) taking an *ambiguous* stance.

- **Climate-Eng** (Vaid et al., 2022): A dataset designed for multi-class topic classification, constructed from the same set of 3,777 tweets as the Climate-Stance dataset, but labeled according to one of five distinct topics: *disaster*, *ocean/water*, *agriculture/forestry*, *politics*, or *general*.

- **Climate-FEVER** (Diggelmann et al., 2020): A claim verification dataset that consists of real-world claims about climate change. Each of the 1,535 claims is paired with five evidence sentences extracted from Wikipedia, which either *support*, *refute*, or *provide insufficient information* about the claim. Climate-FEVER also includes a general label, determined by aggregating the ratings of individual claim-evidence pairs. However, in our benchmark, we model the task for this dataset as a three-way entailment problem where each claim is evaluated individually against each of its five evidence sentences, resulting in 7,675 claim-evidence pairs.

- **SciDCC** (Mishra and Mittal, 2021): A multi-class classification dataset comprising news articles sourced from the Science Daily website, annotated with one of the possible 20 scientific categories (e.g., *biology*, *weather*, *ozone holes*, *endangered animals*). Each article includes a title, a summary, and a body section. In order to fully utilize the available information at different granularity, we model the task in three ways:

  i. *Topic Classification by Title*, where classification is performed solely on the article title;
  ii. *Topic Classification by Title & Summary*, which utilizes both the title and the summary sections; and
  iii. *Topic Classification by Title & Body*, where the full article body is used, in addition to the title.

  We aim for these complementary tasks to enable a comparative assessment of how much information is needed for an accurate classification performance.

- **CLIMA-CDP** (Spokoyny et al., 2023): A dataset derived from disclosure questionnaires collected and made available by Carbon Disclosure Project (CDP), an international non-profit organization. These questionnaires are completed by various stakeholders, including cities, corporations, and states. The dataset consists of responses to hundreds of unique questions related to climate impact, mitigation efforts and governance, and supports two distinct classification tasks:

  i. *CDP-QA* is a binary classification task that predicts whether a given report response correctly answers the questions posed. This task has three variants based on the type of stakeholder providing

[5]The code used for scraping the dataset from the Guardian's Open Platform is available on GitHub: `https://github.com/NLP-RISE/extractguardian`

the response: *CDP-QA-Cities*, *CDP-QA-Corporations*, and *CDP-QA-States*, for municipal, corporate and state-level responses, respectively.

ii. *CDP-Topic-Cities* is a multi-class classification task where responses from city stakeholders are categorized into one of twelve predefined topics (e.g., *climate hazards*, *emissions*, *energy*, *food*). The task is limited to city responses since no annotations exist for other stakeholders.

- **PIRA 2.0 MCQ** (Pirozelli et al., 2024): A multiple-choice QA dataset constructed from a collection of scientific abstracts and United Nations reports, with a focus on climate-related topics such as oceanography, coastal ecosystems, and climate change impacts. Each instance consists of a question and five answer choices. Additionally, each question is accompanied by a supporting context that provides relevant information to guide the answer selection process. We divided the dataset into two sub-tasks:

  i. *PIRA with Context*, where models are expected to answer a question with the help of the accompanying context, like in an open-book exam setting;

  ii. *PIRA without Context*, where models are required to answer the question without any supporting information.

  By structuring the dataset in this manner, we aim to evaluate both retrieval-augmented and self-knowledge-based approaches to climate-related QA.

- **Exeter Misinformation** (Coan et al., 2021): A dataset designed to detect climate misinformation by annotating text from prominent climate contrarian blogs and think tanks spanning over 20 years (1998–2020). The dataset is structured according to a two-level taxonomy of climate contrarian claims. The first level consists of broad claim categories, such as: (1) *Global warming is not happening*, which is further divided into more specific sub-claims, including (1.1) *Ice isn't melting*, or (1.2) *Oceans are cooling*. To test different levels of misinformation detection, we define two sub-tasks:

  i. *Claim Detection*, a classification task where texts are categorized into one of the six first-level claim labels; and

  ii. *Sub-Claim Detection*, a more fine-grained classification task where each text is assigned to its corresponding sub-claim category within the taxonomy.

  These tasks allow for a detailed evaluation of how LLMs classify misinformation narratives in climate discourse at different granularities.

- **Climate-Change NER** (Bhattacharjee et al., 2024): A *named entity recognition* (NER) dataset constructed from 534 scientific abstracts sourced from the Semantic Scholar Academic Graph (Kinney et al., 2023). The dataset was collected using a set of climate-related keywords (e.g., *wildfire*, *floods*) to ensure relevance to climate science. Each abstract is annotated for entity types that are specific to the climate discourse, such as *greenhouse gases*, *climate hazards*, *climate-impacts*, *climate mitigations*. The associated task is a token classification task, where the goal is to identify tokens corresponding to these entities. NER is a crucial preprocessing step for various downstream applications that aim to extract structured information from unstructured text (Li et al., 2024). Hence, a model's performance on climate-specific NER is a strong indicator of its usefulness for information extraction in climate discourse (Li et al., 2020).

- **CheapTalk** (Bingler et al., 2023): A dataset consisting of corporate disclosures, focusing on how companies communicate climate-related information. Unlike the previously described datasets, primarily dealing with either academic discourse or social media posts, this dataset uncovers corporations' climate communication strategies. The following five tasks are performed on it:

  i. *Climate Detection*: A binary classification task to determine whether a given text passage discusses climate-related topics. This dataset serves as a filtering mechanism to extract climate-relevant content from corporate reports.

  ii. *Climate Sentiment*: A sentiment analysis task that categorizes climate-related

Task	Gemma-2 2B		Llama-2 7B		Llama-2 13B		Climate GPT-7B		Climate GPT-13B		Qwen-2.5 7B		Llama-3.1 8B		Mistral 24B		Llama-3.3 70B	
	0	5	0	5	0	5	0	5	0	5	0	5	0	5	0	5	0	5
CDP-QA-Cities	.46	.60	.62	**.69**	.37	.55	.48	.57	.59	.57	.61	.62	.57	.66	.68	.68	.61	**.69**
CDP-QA-Corp.	.46	.59	.62	.67	.32	.55	.40	.57	.58	.58	.60	.61	.51	.64	.67	**.68**	.60	**.68**
CDP-QA-States	.46	.61	.68	.71	.37	.56	.48	.59	.60	.58	.62	.63	.54	.66	**.72**	.68	.59	.70
CDP-Topic-Cities	.28	.34	.32	.35	.35	.35	.33	.35	.30	.35	.30	.35	.31	**.37**	.34	.32	.35	.35
Climate Commit.	.42	.63	.66	**.71**	.56	.63	.63	.65	.61	.60	.56	.69	.68	**.71**	**.71**	**.71**	.68	.68
Climate Detection	.20	.70	.65	.69	.61	.60	.59	.66	.62	.68	.53	.73	.50	.76	.69	**.78**	.71	.76
Climate Eng	.37	.46	.52	.59	.37	.50	.52	.53	.53	.50	.52	.54	.55	.59	**.60**	.59	.57	.59
Climate NER	.11	.19	.14	.20	.10	.17	.09	.17	.07	.14	.04	.18	.16	.21	.23	**.30**	.20	.27
Climate Sentiment	.49	.64	.72	.74	.44	.71	.60	.70	.61	.65	.57	.63	.54	.72	**.75**	.72	.73	**.75**
Climate Specificity	.39	.66	.72	.69	.54	.60	.51	.64	.48	.60	.53	.68	.60	.73	.75	**.79**	.75	**.79**
Climate Stance	.13	.50	.30	.54	.20	.47	.16	.48	.11	.37	.14	.42	.06	.56	.18	.61	.26	**.64**
Climate-Fever	.33	.34	**.57**	.48	.40	.34	.35	.41	.25	.45	.41	.55	.52	.51	.56	.50	.51	.55
Climatext Sent Clf.	.39	.64	.56	.62	.58	.67	.63	.70	.58	.61	.54	.68	.47	.68	.57	.68	.63	**.71**
Env. Claims	.53	.77	.80	.81	.61	.79	.66	.78	.80	.75	.64	.81	.75	.83	.75	.82	**.85**	.83
Exeter Claim	.14	.41	.43	.48	.17	.30	.20	.38	.26	.37	.34	.41	.35	.46	.55	**.61**	.56	.59
Exeter Sub-Claim	.22	.27	.48	.51	.05	.13	.15	.25	.24	.25	.33	.40	.39	.47	.61	**.63**	.59	.59
Guardian Body	.39	.31	.48	.54	.29	.19	.31	.21	.30	.25	.33	.45	.47	.49	**.58**	.00	.00	.00
Guardian Title	.35	.35	.45	.54	.40	.41	.38	.00	.46	.48	.45	.00	.42	.50	.51	.53	**.57**	.00
Net-Zero Reduction	.22	.44	.38	.45	.46	.84	.39	.86	.23	.83	.43	.85	.48	.43	.28	.47	.60	**.89**
Pira W/ Ctx.	.86	.86	.95	.94	.74	.70	.87	.88	.86	.87	.93	.94	.93	.95	.94	**.96**	.94	.92
Pira W/O Ctx.	.64	.62	.84	.84	.44	.52	.58	.67	.67	.74	.70	.81	.69	.80	.80	**.89**	**.89**	.88
SciDCC Title	.16	.19	.27	.28	.18	.23	.08	.20	.16	.17	.16	.24	.15	.23	.28	**.33**	.25	**.33**
SciDCC Title Body	.10	.21	.20	.23	.13	.14	.08	.17	.13	.11	.17	.18	.18	.23	.26	**.32**	.26	.31
SciDCC Title Sum.	.10	.19	.25	.27	.13	.25	.12	.25	.17	.16	.20	.25	.20	.27	.28	.33	.27	**.34**
Tcfd Recommend.	.15	.31	.43	.46	.24	.30	.24	.30	.16	.29	.29	.37	.26	.45	.50	.52	**.53**	.48

Table 2: Evaluation results for various models across different few-shot experiments. The numbers indicate the models' performance for each task in F1-macro, except for PIRA (see Section 3.2). Numbers in boldface are the highest performing, whereas underlined numbers are the second highest for that task. For shortened task identifiers, see Table 6.

statements based on their tone. Each passage is labeled as highlighting risks, emphasizing opportunities, or maintaining a neutral stance on climate change.

iii. *Climate Commitments*: A classification task that determines whether a corporate disclosure contains a climate-related commitment. Texts are labeled as *commitment-yes* if they explicitly state planned or ongoing climate actions and *commitment-no* if they do not reference concrete climate actions.

iv. *Climate Specificity*: A binary classification task assessing the specificity of corporate climate commitments. A passage is labeled as *specific* if containing "detailed performance information, details of actions, or tangible and verifiable targets" (Bingler et al., 2023), or *non-specific* if it is vague.

v. *TCFD Recommendations*: A multi-class classification task assessing corporate disclosures against the guidelines of the Task Force on Climate-related Financial Disclosures (TCFD), an international framework that standardizes the reporting of climate-related financial risks and opportunities.[6] Each text is labelled by one of the four TCFD recommendation categories (*governance*, *strategy*, *risk management*, and *metrics and targets*) or as *none* if none applies.

- **Net-Zero Reduction** (Schimanski et al., 2023): A dataset comprising 3,517 expert-annotated paragraph samples designed to detect and assess net-zero and emission reduction targets in corporate, national, and regional communications. Each sample is labeled as *net-zero target* (commitment to net-zero emissions), *reduction target* (commitment to emission reduction without full net-zero), or *no target* (no explicit reduction commitment). This dataset is used for multi-class classification.

- **Environmental Claims** (Stammbach et al., 2023): A dataset comprising 3,000 sentences from corporate sustainability reports, earnings calls, and annual reports by publicly listed companies that are expert-annotated for environmental claims. An environmental claim any statement suggesting that a product, service, or company is environmentally friendly. Each sentence is labeled as containing an en-

[6] https://www.fsb-tcfd.org

vironmental claim or not. This dataset is used for a binary sentence classification task.

Most of the datasets in ClimateEval were developed in the pre-LLM era and were not designed with prompting-based evaluation in mind. Moreover, prior work has used some datasets in different ways — for example, ClimateGPT modeled Climate-FEVER as a binary classification task (not three-way). To address this lack of standardization, ClimateEval provides a unified evaluation benchmark that standardizes task formulations, label sets, and prompts. Each task is paired with a suitable prompt and, where applicable, modeled at multiple granularities (e.g., title-only vs. full-body classification in SciDCC). This unification ensures that different models are evaluated under consistent conditions, enabling reproducible comparisons across tasks.

3.2 Evaluation Setup and Metrics

ClimateEval is implemented using the LM Evaluation Harness library (Gao et al., 2024), which provides an easy-to-use infrastructure for evaluating language models on a wide range of tasks. Each task in the benchmark is defined through a YAML-formatted configuration file, specifying the input format, prompt template, expected output and the target metric. The benchmark can be executed with a single command, allowing for efficient and standardized evaluation across diverse models and tasks.

We evaluate model performance primarily using accuracy and macro-averaged F1-score for all classification tasks. Macro F1 is particularly important due to class imbalances in several datasets where there are as many as 20 labels with skewed distributions (see Appendix C for details). Macro F1 averages scores across all classes equally to ensure a balanced assessment by preventing frequent labels from biasing the results.

For Climate-Change-NER, the only sequence labeling task in our benchmark, we compute precision, recall, and F1-score based on the entity-type and entity-span pairs. Each entity-type is evaluated independently, and a model's prediction is considered correct if it correctly identifies an entity within the set of gold entities for that type. If a gold-standard entity-type contains multiple entities and the model predicts only a subset, we count each correctly identified entity as a true positive, while missing entities contribute to false negatives. We report only exact matches, where an entity is correct only if both the span and type match perfectly.

For multiple-choice QA (MCQA) tasks (e.g., PIRA 2.0 MCQ), we report exact match accuracy, as F1-score is not meaningful in this context. In MCQA tasks, each question has one correct answer, and the model selects from arbitrary option labels (e.g., A, B, C), which are not semantic classes. Since the model selects a single arbitrary option and only one answer is correct, precision and recall are not meaningful, making exact match accuracy the appropriate metric.

4 Evaluation of Open-Source LLMs

We report the performance of a range of open source baseline LLMs with varying sizes, ranging from 2B to 70B, in both zero-shot and 5-shot scenarios. The mid-sized models (Mistral 24B and Llama3.3-70B) are loaded in 4-bit quantization whereas the other models were run in half-precision FP16.

For classification tasks, the log-likelihoods of each possible label are calculated, and the label with the highest likelihood is selected as the model's prediction. For the generation task, Climate Change NER, the model is simply prompted to generate the corresponding JSON file.

The baseline LLMs that we use are: Gemma-2 (2B) (Team et al., 2024), Qwen-2.5 (7B) (Yang et al., 2025), ClimateGPT (both 7B and 13B) (Thulke et al., 2024); Llama-2 (both 7B and 13B); Llama-3.1 (8B), Llama-3.3 (70B) (Grattafiori et al., 2024) and Mistral (24B). The baseline models' performances are reported in Table 2 in both zero-shot and five-shot settings.

4.1 Zero-shot vs. Few-shot Performance

Across the benchmark, few-shot prompting consistently improves performance (Figure 2). The most significant gains are observed in Climate Stance (+0.26 for ClimateGPT-7B to +0.50 for Llama-3.1-8B) and Net-Zero Reduction (+0.15 for Qwen-2.5-7B to +0.44 for Gemma-2-2B). Both tasks rely on understanding specialized classification taxonomies, whether for categorizing social media discourse or analyzing policy documents. The improvement diminish as the number of labels increases. The Exeter Sub-claim Detection and SciDCC tasks exhibit minimal gains (0.03 on average), suggesting that when models must choose between a high number of categories, few-shot

Task	0-Shot	5-Shot	Diff.
CDP-QA-Cities	.55 (.10)	.63 (.06)	.07
CDP-QA-Corp.	.53 (.11)	.62 (.05)	.09
CDP-QA-States	.56 (.11)	.64 (.06)	.07
CDP-Topic-Cities	.32 (.02)	.35 (.01)	.03
Climate Commit.	.61 (.09)	.67 (.04)	.06
Climate Detection	.57 (.15)	.71 (.06)	.14
Climate Eng	.50 (.08)	.54 (.05)	.04
Climate NER	.13 (.06)	.20 (.05)	.08
Climate Sentiment	.60 (.11)	.70 (.04)	.09
Climate Specificity	.58 (.13)	.69 (.07)	.10
Climate Stance	.17 (.07)	.51 (.09)	.34
Climate-Fever	.43 (.11)	.46 (.08)	.03
ClimaText Sent Clf.	.55 (.08)	.67 (.03)	.12
Env. Claims	.71 (.10)	.80 (.03)	.09
Exeter Claim	.33 (.16)	.45 (.10)	.11
Exeter Sub-Claim	.34 (.19)	.39 (.17)	.05
Guardian Body	.42 (.13)	.41 (.18)	-.01
Guardian Title	.44 (.07)	.48 (.08)	.04
Net-Zero Reduction	.39 (.13)	.67 (.22)	.29
Pira W/ Ctx.	.89 (.07)	.89 (.08)	.00
Pira W/O Ctx.	.70 (.14)	.75 (.13)	.06
SciDCC Title Sum.	.19 (.07)	.26 (.06)	.07
SciDCC Title	.19 (.07)	.24 (.06)	.06
SciDCC Title Body	.17 (.06)	.21 (.07)	.04
TCFD Recommend.	.31 (.14)	.39 (.09)	.07
Average	.45	.53	.09

Table 3: Comparison of average model performance across tasks in 0-shot and 5-shot settings. Standard deviations are provided within parentheses. The last column shows the performance difference between the zero-shot and five-shot experiments. For shortened task identifiers, see Table 6.

prompting does not provide sufficient guidance.

Another set of tasks that do not benefit from in-context learning includes Climate-FEVER (fact verification) and PIRA (multiple-choice QA), which show no improvement. Unlike standard classification tasks, these tasks require external knowledge to be successfully carried out, limiting the efficiency of few-shot prompting. As each instance in these datasets depends on specific knowledge, the lack of improvement is unsurprising.

4.2 Analysis of tasks

Among the 25 tasks in our benchmark, certain tasks consistently emerge as more challenging than others (Table 3). SciDCC and Climate-Change-NER exhibit particularly poor performance, with F1-scores below 0.34 across models, even with few-shot prompting. The poor performance on SciDCC can be attributed to the large number of classes (20 in total), making the assignment more difficult. Surprisingly enough, classification of news texts based solely on their titles only leads to a minor performance loss compared to using the full article.

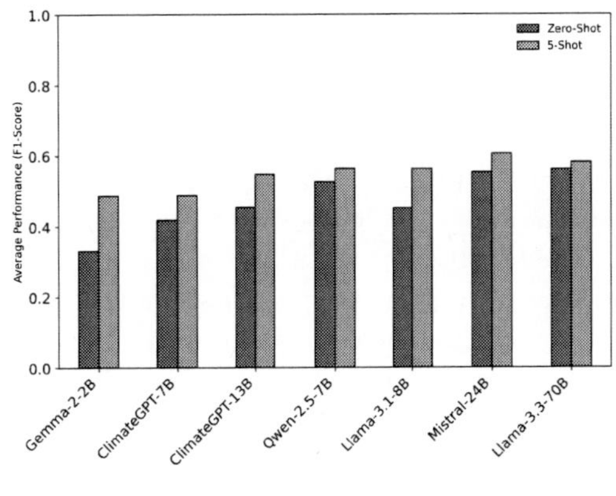

Figure 2: Average model performance across all tasks in Zero-Shot and 5-Shot settings.

This may indicate that titles provide sufficient cues regarding the content of articles.

Climate-Change NER, on the other hand, is the only non-classification task in the benchmark, requiring models to correctly identify climate-related named entities in scientific abstracts. This task involves both span detection and entity type classification, where entity types are domain-specific[7] and differ significantly from standard NER labels such as "location" or "organization". Without specific training/fine-tuning on climate-specific entity annotations, it is understandable that general-purpose LLMs struggle to accurately extract these entities.

In contrast, the LLMs perform strongly on several tasks in the benchmark, even in the zero-shot setting. The performance on PIRA improves by 0.2 points on average in the zero-shot setting when relevant context is provided, showcasing the LLMs' ability to identify and utilize relevant parts of additional knowledge for question answering (QA). However, the overall high performance on PIRA even without any context suggests that the task may be relatively easy, highlighting the need for a more challenging QA dataset.

Perhaps the most surprising finding is the relatively poor performance of the models on the CheapTalk dataset tasks. Intuitively, one might expect tasks like Climate Detection (whether or not a text is climate-related) or Climate Sentiment (classifying tone as risk, opportunity or neutral) to be straightforward for advanced LLMs. Yet, our results show that even the best models achieve F1 scores below 0.8 on these tasks, indicating that

[7]For the full list of entities, see: https://huggingface.co/datasets/ibm-research/Climate-Change-NER

these models experience challenges in understanding the climate discourse.

4.3 Impact of In-domain Training

ClimateGPT presents a unique opportunity to evaluate the impact of in-domain training for climate-related NLP, as it is an LLM developed through continuous pre-training of Llama-2 models on 4.2B climate-focused tokens and instruction-tuned on expert-curated datasets (Thulke et al., 2024). To gain insights regarding the benefits of domain adaptation, we compare ClimateGPT against its base Llama-2 models (Table 4).

Table 3 shows the performance differences between these models. On average, ClimateGPT exhibits a slight improvement over Llama-2 by ≤ 0.05. The largest gains are seen in Exeter Sub-claim, CDP-QA, and PIRA With Context. However, the improvements are inconsistent, with some tasks, including Climate-Change NER, Climate Stance, and Net-Zero Reduction, showing negligible or negative differences. The limited impact of in-domain training across tasks suggests that, while climate-focused continuous pre-training provides benefits in specific cases, the performance gains are not uniform across the tasks in ClimateEval.

Task	**7B**		**13B**	
	0	**5**	**0**	**5**
CDP-QA-Corp.	**0.26**	**0.02**	**0.20**	**0.04**
CDP-QA-States	**0.23**	**0.02**	**0.13**	**0.04**
CDP-QA-Cities	**0.22**	**0.03**	**0.13**	**0.05**
CDP-Topic-Cities	-0.05	0.00	-0.03	0.00
ClimaText Sent. Clf.	-0.00	-0.06	-0.10	-0.02
Climate NER	-0.03	-0.04	-0.04	0.00
Climate Commit.	**0.05**	-0.03	-0.07	**0.04**
Climate Detection	**0.01**	**0.07**	-0.06	**0.08**
Climate Eng	**0.16**	0.00	-0.00	0.00
Climate-FEVER	-0.15	**0.11**	**0.07**	**0.14**
Climate Sentiment	**0.17**	-0.06	-0.03	-0.06
Climate Specificity	-0.06	**0.01**	**0.02**	**0.04**
Climate Stance	-0.09	-0.10	-0.01	-0.06
Env. Claims	**0.19**	-0.05	-0.02	**0.03**
Exeter Claim	**0.09**	**0.07**	**0.14**	**0.03**
Exeter Sub-claim	**0.19**	**0.12**	**0.19**	**0.16**
Guardian Body	**0.02**	**0.06**	**0.02**	**0.24**
Guardian Title	**0.06**	**0.07**	**0.07**	**0.08**
Net-Zero Reduction	-0.24	-0.01	**0.04**	-0.01
PIRA w/ Ctx.	**0.12**	**0.17**	**0.06**	**0.06**
PIRA w/o Ctx.	**0.23**	**0.23**	**0.12**	**0.15**
SciDCC Title	-0.02	-0.07	**0.08**	**0.04**
SciDCC Title Sum.	**0.04**	-0.10	**0.08**	0.00
SciDCC Title Body	-0.00	-0.02	**0.08**	**0.01**
TCFD Recommend.	-0.08	-0.02	**0.05**	**0.07**
Average	**0.05**	**0.01**	-0.04	**0.04**

Table 4: Comparison of ClimateGPT against its' base Llama models in **0**-shot and **5**-shot settings. The values represent the difference where positive values (highlighted in bold) indicate better performance of ClimateGPT over Llama.

5 Conclusion

We have presented ClimateEval, a comprehensive benchmark for climate change NLP, encompassing 13 datasets and 25 tasks that cover a wide range of climate-related language understanding tasks. By unifying these diverse tasks into a single framework, ClimateEval enables systematic assessment of how well current LLMs perform in the domain of climate change. Our evaluation of some widely-used open-source LLMs revealed systematic patterns: few-shot prompting generally improves performance, but certain text classification tasks such as claim detection or climate-specific NER remain challenging. We hope that ClimateEval will serve as a valuable resource for the NLP community and facilitate future research on evaluating and improving LLMs for climate change-related applications.

Limitations

We acknowledge that ClimateEval is currently limited to English, which restricts its applicability to multilingual climate discourse. Due to computational constraints our evaluation focuses on mid-sized open-source models, ranging from 2B to 70B parameters, with larger models tested under 4-bit quantization. This may introduce some performance degradation and may not reflect their optimal performance. Additionally, commercial models such as GPT-4 and Claude are not included in our experiments due to budget constraints. Finally, the benchmark largely consists of *n-way* classification tasks, with the only exception of Climate-Change NER. This was partly driven by available climate-relevant datasets, which are predominantly classification-oriented. Future work should focus on enabling assessment of generative tasks such as information extraction, text generation or multi-modal classification.

Acknowledgments

Swedish Research Council Vetenskapsrådet grant nos. 2022-03448, 2022-06599 and 2022-02909. European Commission's Horizon Europe research and innovation programme European Research Council grant no. 101112727.

References

Bishwaranjan Bhattacharjee, Aashka Trivedi, Masayasu Muraoka, Muthukumaran Ramasubramanian, Takuma Udagawa, Iksha Gurung, Nishan Pantha, Rong Zhang, Bharath Dandala, Rahul Ramachandran, Manil Maskey, Kaylin Bugbee, Michael M. Little, Elizabeth Fancher, Irina Gerasimov, Armin Mehrabian, Lauren Sanders, Sylvain V. Costes, Sergi Blanco-Cuaresma, Kelly Lockhart, Thomas Allen, Felix Grezes, Megan Ansdell, Alberto Accomazzi, Yousef El-Kurdi, Davis Wertheimer, Birgit Pfitzmann, Cesar Berrospi Ramis, Michele Dolfi, Rafael Teixeira De Lima, Panagiotis Vagenas, S. Karthik Mukkavilli, Peter W. J. Staar, Sanaz Vahidinia, Ryan McGranaghan, and Tsengdar J. Lee. 2024. INDUS: Effective and Efficient Language Models for Scientific Applications. In *Proceedings of the 2024 Conference on Empirical Methods in Natural Language Processing: Industry Track*, pages 98–112, Miami, Florida, US. Association for Computational Linguistics.

Julia Bingler, Mathias Kraus, Markus Leippold, and Nicolas Webersinke. 2023. How Cheap Talk in Climate Disclosures Relates to Climate Initiatives, Corporate Emissions, and Reputation Risk. Working paper, Available at SSRN 3998435.

Travis G. Coan, Constantine Boussalis, John Cook, and Mirjam O. Nanko. 2021. Computer-assisted classification of contrarian claims about climate change. *Scientific Reports*, 11(1):1–12.

Thomas Diggelmann, Jordan Boyd-Graber, Jannis Bulian, Massimiliano Ciaramita, and Markus Leippold. 2020. CLIMATE-FEVER: A Dataset for Verification of Real-World Climate Claims. *arXiv preprint arXiv:2012.00614*.

May El Barachi, Manar AlKhatib, Sujith Mathew, and Farhad Oroumchian. 2021. A novel sentiment analysis framework for monitoring the evolving public opinion in real-time: Case study on climate change. *Journal of Cleaner Production*, 312:127820.

Michael Fore, Simranjit Singh, Chaehong Lee, Amritanshu Pandey, Antonios Anastasopoulos, and Dimitrios Stamoulis. 2024. Unlearning Climate Misinformation in Large Language Models. In *Proceedings of the 1st Workshop on Natural Language Processing Meets Climate Change (ClimateNLP 2024)*, pages 178–192, Bangkok, Thailand. Association for Computational Linguistics.

Leo Gao, Jonathan Tow, Baber Abbasi, Stella Biderman, Sid Black, Anthony DiPofi, Charles Foster, Laurence Golding, Jeffrey Hsu, Alain Le Noac'h, Haonan Li, Kyle McDonell, Niklas Muennighoff, Chris Ociepa, Jason Phang, Laria Reynolds, Hailey Schoelkopf, Aviya Skowron, Lintang Sutawika, Eric Tang, Anish Thite, Ben Wang, Kevin Wang, and Andy Zou. 2024. A framework for few-shot language model evaluation.

Aaron Grattafiori, Abhimanyu Dubey, Abhinav Jauhri, Abhinav Pandey, Abhishek Kadian, Ahmad Al-Dahle, Aiesha Letman, Akhil Mathur, Alan Schelten, Alex Vaughan, Amy Yang, Angela Fan, Anirudh Goyal, Anthony Hartshorn, Aobo Yang, Archi Mitra, Archie Sravankumar, Artem Korenev, Arthur Hinsvark, Arun Rao, Aston Zhang, Aurelien Rodriguez, Austen Gregerson, Ava Spataru, Baptiste Roziere, Bethany Biron, Binh Tang, Bobbie Chern, Charlotte Caucheteux, Chaya Nayak, Chloe Bi, Chris Marra, Chris McConnell, Christian Keller, Christophe Touret, Chunyang Wu, Corinne Wong, Cristian Canton Ferrer, Cyrus Nikolaidis, Damien Allonsius, Daniel Song, Danielle Pintz, Danny Livshits, Danny Wyatt, David Esiobu, Dhruv Choudhary, Dhruv Mahajan, Diego Garcia-Olano, Diego Perino, Dieuwke Hupkes, Egor Lakomkin, Ehab AlBadawy, Elina Lobanova, Emily Dinan, Eric Michael Smith, Filip Radenovic, Francisco Guzmán, Frank Zhang, Gabriel Synnaeve, Gabrielle Lee, Georgia Lewis Anderson, Govind Thattai, Graeme Nail, Gregoire Mialon, Guan Pang, Guillem Cucurell, Hailey Nguyen, Hannah Korevaar, Hu Xu, Hugo Touvron, Iliyan Zarov, Imanol Arrieta Ibarra, Isabel Kloumann, Ishan Misra, Ivan Evtimov, Jack Zhang, Jade Copet, Jaewon Lee, Jan Geffert, Jana Vranes, Jason Park, Jay Mahadeokar, Jeet Shah, Jelmer van der Linde, Jennifer Billock, Jenny Hong, Jenya Lee, Jeremy Fu, Jianfeng Chi, Jianyu Huang, Jiawen Liu, Jie Wang, Jiecao Yu, Joanna Bitton, Joe Spisak, Jongsoo Park, Joseph Rocca, Joshua Johnstun, Joshua Saxe, Junteng Jia, Kalyan Vasuden Alwala, Karthik Prasad, Kartikeya Upasani, Kate Plawiak, Ke Li, Kenneth Heafield, Kevin Stone, Khalid El-Arini, Krithika Iyer, Kshitiz Malik, Kuenley Chiu, Kunal Bhalla, Kushal Lakhotia, Lauren Rantala-Yeary, Laurens van der Maaten, Lawrence Chen, Liang Tan, Liz Jenkins, Louis Martin, Lovish Madaan, Lubo Malo, Lukas Blecher, Lukas Landzaat, Luke de Oliveira, Madeline Muzzi, Mahesh Pasupuleti, Mannat Singh, Manohar Paluri, Marcin Kardas, Maria Tsimpoukelli, Mathew Oldham, Mathieu Rita, Maya Pavlova, Melanie Kambadur, Mike Lewis, Min Si, Mitesh Kumar Singh, Mona Hassan, Naman Goyal, Narjes Torabi, Nikolay Bashlykov, Nikolay Bogoychev, Niladri Chatterji, Ning Zhang, Olivier Duchenne, Onur Çelebi, Patrick Alrassy, Pengchuan Zhang, Pengwei Li, Petar Vasic, Peter Weng, Prajjwal Bhargava, Pratik Dubal, Praveen Krishnan, Punit Singh Koura, Puxin Xu, Qing He, Qingxiao Dong, Ragavan Srinivasan, Raj Ganapathy, Ramon Calderer, Ricardo Silveira Cabral, Robert Stojnic, Roberta Raileanu, Rohan Maheswari, Rohit Girdhar, Rohit Patel, Romain Sauvestre, Ronnie Polidoro, Roshan Sumbaly, Ross Taylor, Ruan Silva, Rui Hou, Rui Wang, Saghar Hosseini, Sahana Chennabasappa, Sanjay Singh, Sean Bell, Seohyun Sonia Kim, Sergey Edunov, Shaoliang Nie, Sharan Narang, Sharath Raparthy, Sheng Shen, Shengye Wan, Shruti Bhosale, Shun Zhang, Simon Vandenhende, Soumya Batra, Spencer Whitman, Sten Sootla, Stephane Collot, Suchin Gururangan, Sydney Borodinsky, Tamar Herman, Tara Fowler, Tarek Sheasha, Thomas Georgiou, Thomas Scialom, Tobias Speckbacher, Todor Mihaylov, Tong Xiao, Ujjwal

Karn, Vedanuj Goswami, Vibhor Gupta, Vignesh Ramanathan, Viktor Kerkez, Vincent Gonguet, Virginie Do, Vish Vogeti, Vítor Albiero, Vladan Petrovic, Weiwei Chu, Wenhan Xiong, Wenyin Fu, Whitney Meers, Xavier Martinet, Xiaodong Wang, Xiaofang Wang, Xiaoqing Ellen Tan, Xide Xia, Xinfeng Xie, Xuchao Jia, Xuewei Wang, Yaelle Goldschlag, Yashesh Gaur, Yasmine Babaei, Yi Wen, Yiwen Song, Yuchen Zhang, Yue Li, Yuning Mao, Zacharie Delpierre Coudert, Zheng Yan, Zhengxing Chen, Zoe Papakipos, Aaditya Singh, Aayushi Srivastava, Abha Jain, Adam Kelsey, Adam Shajnfeld, Adithya Gangidi, Adolfo Victoria, Ahuva Goldstand, Ajay Menon, Ajay Sharma, Alex Boesenberg, Alexei Baevski, Allie Feinstein, Amanda Kallet, Amit Sangani, Amos Teo, Anam Yunus, Andrei Lupu, Andres Alvarado, Andrew Caples, Andrew Gu, Andrew Ho, Andrew Poulton, Andrew Ryan, Ankit Ramchandani, Annie Dong, Annie Franco, Anuj Goyal, Aparajita Saraf, Arkabandhu Chowdhury, Ashley Gabriel, Ashwin Bharambe, Assaf Eisenman, Azadeh Yazdan, Beau James, Ben Maurer, Benjamin Leonhardi, Bernie Huang, Beth Loyd, Beto De Paola, Bhargavi Paranjape, Bing Liu, Bo Wu, Boyu Ni, Braden Hancock, Bram Wasti, Brandon Spence, Brani Stojkovic, Brian Gamido, Britt Montalvo, Carl Parker, Carly Burton, Catalina Mejia, Ce Liu, Changhan Wang, Changkyu Kim, Chao Zhou, Chester Hu, Ching-Hsiang Chu, Chris Cai, Chris Tindal, Christoph Feichtenhofer, Cynthia Gao, Damon Civin, Dana Beaty, Daniel Kreymer, Daniel Li, David Adkins, David Xu, Davide Testuggine, Delia David, Devi Parikh, Diana Liskovich, Didem Foss, Dingkang Wang, Duc Le, Dustin Holland, Edward Dowling, Eissa Jamil, Elaine Montgomery, Eleonora Presani, Emily Hahn, Emily Wood, Eric-Tuan Le, Erik Brinkman, Esteban Arcaute, Evan Dunbar, Evan Smothers, Fei Sun, Felix Kreuk, Feng Tian, Filippos Kokkinos, Firat Ozgenel, Francesco Caggioni, Frank Kanayet, Frank Seide, Gabriela Medina Florez, Gabriella Schwarz, Gada Badeer, Georgia Swee, Gil Halpern, Grant Herman, Grigory Sizov, Guangyi, Zhang, Guna Lakshminarayanan, Hakan Inan, Hamid Shojanazeri, Han Zou, Hannah Wang, Hanwen Zha, Haroun Habeeb, Harrison Rudolph, Helen Suk, Henry Aspegren, Hunter Goldman, Hongyuan Zhan, Ibrahim Damlaj, Igor Molybog, Igor Tufanov, Ilias Leontiadis, Irina-Elena Veliche, Itai Gat, Jake Weissman, James Geboski, James Kohli, Janice Lam, Japhet Asher, Jean-Baptiste Gaya, Jeff Marcus, Jeff Tang, Jennifer Chan, Jenny Zhen, Jeremy Reizenstein, Jeremy Teboul, Jessica Zhong, Jian Jin, Jingyi Yang, Joe Cummings, Jon Carvill, Jon Shepard, Jonathan McPhie, Jonathan Torres, Josh Ginsburg, Junjie Wang, Kai Wu, Kam Hou U, Karan Saxena, Kartikay Khandelwal, Katayoun Zand, Kathy Matosich, Kaushik Veeraraghavan, Kelly Michelena, Keqian Li, Kiran Jagadeesh, Kun Huang, Kunal Chawla, Kyle Huang, Lailin Chen, Lakshya Garg, Lavender A, Leandro Silva, Lee Bell, Lei Zhang, Liangpeng Guo, Licheng Yu, Liron Moshkovich, Luca Wehrstedt, Madian Khabsa, Manav Avalani, Manish Bhatt, Martynas Mankus, Matan Hasson, Matthew Lennie, Matthias Reso, Maxim Groshev, Maxim Naumov, Maya Lathi, Meghan Keneally, Miao Liu, Michael L. Seltzer, Michal Valko, Michelle Restrepo, Mihir Patel, Mik Vyatskov, Mikayel Samvelyan, Mike Clark, Mike Macey, Mike Wang, Miquel Jubert Hermoso, Mo Metanat, Mohammad Rastegari, Munish Bansal, Nandhini Santhanam, Natascha Parks, Natasha White, Navyata Bawa, Nayan Singhal, Nick Egebo, Nicolas Usunier, Nikhil Mehta, Nikolay Pavlovich Laptev, Ning Dong, Norman Cheng, Oleg Chernoguz, Olivia Hart, Omkar Salpekar, Ozlem Kalinli, Parkin Kent, Parth Parekh, Paul Saab, Pavan Balaji, Pedro Rittner, Philip Bontrager, Pierre Roux, Piotr Dollar, Polina Zvyagina, Prashant Ratanchandani, Pritish Yuvraj, Qian Liang, Rachad Alao, Rachel Rodriguez, Rafi Ayub, Raghotham Murthy, Raghu Nayani, Rahul Mitra, Rangaprabhu Parthasarathy, Raymond Li, Rebekkah Hogan, Robin Battey, Rocky Wang, Russ Howes, Ruty Rinott, Sachin Mehta, Sachin Siby, Sai Jayesh Bondu, Samyak Datta, Sara Chugh, Sara Hunt, Sargun Dhillon, Sasha Sidorov, Satadru Pan, Saurabh Mahajan, Saurabh Verma, Seiji Yamamoto, Sharadh Ramaswamy, Shaun Lindsay, Shaun Lindsay, Sheng Feng, Shenghao Lin, Shengxin Cindy Zha, Shishir Patil, Shiva Shankar, Shuqiang Zhang, Shuqiang Zhang, Sinong Wang, Sneha Agarwal, Soji Sajuyigbe, Soumith Chintala, Stephanie Max, Stephen Chen, Steve Kehoe, Steve Satterfield, Sudarshan Govindaprasad, Sumit Gupta, Summer Deng, Sungmin Cho, Sunny Virk, Suraj Subramanian, Sy Choudhury, Sydney Goldman, Tal Remez, Tamar Glaser, Tamara Best, Thilo Koehler, Thomas Robinson, Tianhe Li, Tianjun Zhang, Tim Matthews, Timothy Chou, Tzook Shaked, Varun Vontimitta, Victoria Ajayi, Victoria Montanez, Vijai Mohan, Vinay Satish Kumar, Vishal Mangla, Vlad Ionescu, Vlad Poenaru, Vlad Tiberiu Mihailescu, Vladimir Ivanov, Wei Li, Wenchen Wang, Wenwen Jiang, Wes Bouaziz, Will Constable, Xiaocheng Tang, Xiaojian Wu, Xiaolan Wang, Xilun Wu, Xinbo Gao, Yaniv Kleinman, Yanjun Chen, Ye Hu, Ye Jia, Ye Qi, Yenda Li, Yilin Zhang, Ying Zhang, Yossi Adi, Youngjin Nam, Yu, Wang, Yu Zhao, Yuchen Hao, Yundi Qian, Yunlu Li, Yuzi He, Zach Rait, Zachary DeVito, Zef Rosnbrick, Zhaoduo Wen, Zhenyu Yang, Zhiwei Zhao, and Zhiyu Ma. 2024. The Llama 3 Herd of Models. *Preprint*, arXiv:2407.21783.

Rodney Michael Kinney, Chloe Anastasiades, Russell Authur, Iz Beltagy, Jonathan Bragg, Alexandra Buraczynski, Isabel Cachola, Stefan Candra, Yoganand Chandrasekhar, Arman Cohan, Miles Crawford, Doug Downey, Jason Dunkelberger, Oren Etzioni, Rob Evans, Sergey Feldman, Joseph Gorney, David W. Graham, F.Q. Hu, Regan Huff, Daniel King, Sebastian Kohlmeier, Bailey Kuehl, Michael Langan, Daniel Lin, Haokun Liu, Kyle Lo, Jaron Lochner, Kelsey MacMillan, Tyler C. Murray, Christopher Newell, Smita R Rao, Shaurya Rohatgi, Paul Sayre, Zejiang Shen, Amanpreet Singh, Luca Soldaini, Shivashankar Subramanian, A. Tanaka, Alex D Wade, Linda M. Wagner, Lucy Lu Wang, Christopher Wilhelm, Caroline Wu, Jiangjiang Yang, Angele Zamarron, Madeleine van Zuylen, and Daniel S. Weld. 2023.

The Semantic Scholar Open Data Platform. *ArXiv*, abs/2301.10140.

Alexandre Lacoste, Alexandra Luccioni, Victor Schmidt, and Thomas Dandres. 2019. Quantifying the Carbon Emissions of Machine Learning. *Preprint*, arXiv:1910.09700.

Jing Li, Aixin Sun, Jianglei Han, and Chenliang Li. 2020. A Survey on Deep Learning for Named Entity Recognition. *IEEE transactions on knowledge and data engineering*, 34(1):50–70.

Ni Li, Shorouq Zahra, Mariana Brito, Clare Flynn, Olof Görnerup, Koffi Worou, Murathan Kurfalı, Chanjuan Meng, Wim Thiery, Jakob Zscheischler, et al. 2024. Using LLMs to Build a Database of Climate Extreme Impacts. In *Proceedings of the 1st Workshop on Natural Language Processing Meets Climate Change (ClimateNLP 2024)*, pages 93–110.

Chen Ling, Xujiang Zhao, Jiaying Lu, Chengyuan Deng, Can Zheng, Junxiang Wang, Tanmoy Chowdhury, Yun Li, Hejie Cui, Tianjiao Zhao, Amit Panalkar, Wei Cheng, Haoyu Wang, Yanchi Liu, Zhengzhang Chen, Haifeng Chen, Chris White, Quanquan Gu, Carl Yang, and Liang Zhao. 2023. Domain Specialization as the Key to Make Large Language Models Disruptive: A Comprehensive Survey.

Mariana Madruga de Brito, Jan Sodoge, Heidi Kreibich, and Christian Kuhlicke. 2025. Comprehensive assessment of flood socioeconomic impacts through text-mining. *Water Resources Research*, 61(1):e2024WR037813.

Prakamya Mishra and Rohan Mittal. 2021. NeuralNERE: Neural Named Entity Relationship Extraction for End-to-End Climate Change Knowledge Graph Construction. In *ICML 2021 Workshop on Tackling Climate Change with Machine Learning*.

Paulo Pirozelli, Marcos M José, Igor Silveira, Flávio Nakasato, Sarajane M Peres, Anarosa AF Brandão, Anna HR Costa, and Fabio G Cozman. 2024. Benchmarks for pirá 2.0, a Reading Comprehension Dataset about the Ocean, the Brazilian Coast, and Climate Change. *Data Intelligence*, 6(1):29–63.

Tobias Schimanski, Julia Bingler, Mathias Kraus, Camilla Hyslop, and Markus Leippold. 2023. ClimateBERT-NetZero: Detecting and Assessing Net Zero and Reduction Targets. In *The 2023 Conference on Empirical Methods in Natural Language Processing*.

Daniel Spokoyny, Tanmay Laud, Tom Corringham, and Taylor Berg-Kirkpatrick. 2023. Towards Answering Climate Questionnaires from Unstructured Climate Reports. *Preprint*, arXiv:2301.04253.

Dominik Stammbach, Nicolas Webersinke, Julia Anna Bingler, Mathias Kraus, and Markus Leippold. 2023. Environmental Claim Detection. *arXiv preprint arXiv:2209.00507v4*, arXiv:2209.00507.

Gemma Team, Morgane Riviere, Shreya Pathak, Pier Giuseppe Sessa, Cassidy Hardin, Surya Bhupatiraju, Léonard Hussenot, Thomas Mesnard, Bobak Shahriari, Alexandre Ramé, Johan Ferret, Peter Liu, Pouya Tafti, Abe Friesen, Michelle Casbon, Sabela Ramos, Ravin Kumar, Charline Le Lan, Sammy Jerome, Anton Tsitsulin, Nino Vieillard, Piotr Stanczyk, Sertan Girgin, Nikola Momchev, Matt Hoffman, Shantanu Thakoor, Jean-Bastien Grill, Behnam Neyshabur, Olivier Bachem, Alanna Walton, Aliaksei Severyn, Alicia Parrish, Aliya Ahmad, Allen Hutchison, Alvin Abdagic, Amanda Carl, Amy Shen, Andy Brock, Andy Coenen, Anthony Laforge, Antonia Paterson, Ben Bastian, Bilal Piot, Bo Wu, Brandon Royal, Charlie Chen, Chintu Kumar, Chris Perry, Chris Welty, Christopher A. Choquette-Choo, Danila Sinopalnikov, David Weinberger, Dimple Vijaykumar, Dominika Rogozińska, Dustin Herbison, Elisa Bandy, Emma Wang, Eric Noland, Erica Moreira, Evan Senter, Evgenii Eltyshev, Francesco Visin, Gabriel Rasskin, Gary Wei, Glenn Cameron, Gus Martins, Hadi Hashemi, Hanna Klimczak-Plucińska, Harleen Batra, Harsh Dhand, Ivan Nardini, Jacinda Mein, Jack Zhou, James Svensson, Jeff Stanway, Jetha Chan, Jin Peng Zhou, Joana Carrasqueira, Joana Iljazi, Jocelyn Becker, Joe Fernandez, Joost van Amersfoort, Josh Gordon, Josh Lipschultz, Josh Newlan, Ju yeong Ji, Kareem Mohamed, Kartikeya Badola, Kat Black, Katie Millican, Keelin McDonell, Kelvin Nguyen, Kiranbir Sodhia, Kish Greene, Lars Lowe Sjoesund, Lauren Usui, Laurent Sifre, Lena Heuermann, Leticia Lago, Lilly McNealus, Livio Baldini Soares, Logan Kilpatrick, Lucas Dixon, Luciano Martins, Machel Reid, Manvinder Singh, Mark Iverson, Martin Görner, Mat Velloso, Mateo Wirth, Matt Davidow, Matt Miller, Matthew Rahtz, Matthew Watson, Meg Risdal, Mehran Kazemi, Michael Moynihan, Ming Zhang, Minsuk Kahng, Minwoo Park, Mofi Rahman, Mohit Khatwani, Natalie Dao, Nenshad Bardoliwalla, Nesh Devanathan, Neta Dumai, Nilay Chauhan, Oscar Wahltinez, Pankil Botarda, Parker Barnes, Paul Barham, Paul Michel, Pengchong Jin, Petko Georgiev, Phil Culliton, Pradeep Kuppala, Ramona Comanescu, Ramona Merhej, Reena Jana, Reza Ardeshir Rokni, Rishabh Agarwal, Ryan Mullins, Samaneh Saadat, Sara Mc Carthy, Sarah Cogan, Sarah Perrin, Sébastien M. R. Arnold, Sebastian Krause, Shengyang Dai, Shruti Garg, Shruti Sheth, Sue Ronstrom, Susan Chan, Timothy Jordan, Ting Yu, Tom Eccles, Tom Hennigan, Tomas Kocisky, Tulsee Doshi, Vihan Jain, Vikas Yadav, Vilobh Meshram, Vishal Dharmadhikari, Warren Barkley, Wei Wei, Wenming Ye, Woohyun Han, Woosuk Kwon, Xiang Xu, Zhe Shen, Zhitao Gong, Zichuan Wei, Victor Cotruta, Phoebe Kirk, Anand Rao, Minh Giang, Ludovic Peran, Tris Warkentin, Eli Collins, Joelle Barral, Zoubin Ghahramani, Raia Hadsell, D. Sculley, Jeanine Banks, Anca Dragan, Slav Petrov, Oriol Vinyals, Jeff Dean, Demis Hassabis, Koray Kavukcuoglu, Clement Farabet, Elena Buchatskaya, Sebastian Borgeaud, Noah Fiedel, Armand Joulin, Kathleen Kenealy, Robert Dadashi, and Alek Andreev. 2024. Gemma 2: Improving

Open Language Models at a Practical Size. *Preprint*, arXiv:2408.00118.

David Thulke, Yingbo Gao, Petrus Pelser, Rein Brune, Rricha Jalota, Floris Fok, Michael Ramos, Ian van Wyk, Abdallah Nasir, Hayden Goldstein, Taylor Tragemann, Katie Nguyen, Ariana Fowler, Andrew Stanco, Jon Gabriel, Jordan Taylor, Dean Moro, Evgenii Tsymbalov, Juliette de Waal, Evgeny Matusov, Mudar Yaghi, Mohammad Shihadah, Hermann Ney, Christian Dugast, Jonathan Dotan, and Daniel Erasmus. 2024. ClimateGPT: Towards AI Synthesizing Interdisciplinary Research on Climate Change. *Preprint*, arXiv:2401.09646.

Apoorva Upadhyaya, Marco Fisichella, and Wolfgang Nejdl. 2023. A Multi-task Model for Sentiment Aided Stance Detection of Climate Change Tweets. In *Proceedings of the international AAAI conference on web and social media*, volume 17, pages 854–865.

Roopal Vaid, Kartikey Pant, and Manish Shrivastava. 2022. Towards Fine-grained Classification of Climate Change related Social Media Text. In *Proceedings of the 60th annual meeting of the association for computational linguistics: Student research workshop*, pages 434–443.

Francesco S Varini, Jordan Boyd-Graber, Massimiliano Ciaramita, and Markus Leippold. 2021. ClimaText: A Dataset for Climate Change Topic Detection.

An Yang, Baosong Yang, Beichen Zhang, Binyuan Hui, Bo Zheng, Bowen Yu, Chengyuan Li, Dayiheng Liu, Fei Huang, Haoran Wei, Huan Lin, Jian Yang, Jianhong Tu, Jianwei Zhang, Jianxin Yang, Jiaxi Yang, Jingren Zhou, Junyang Lin, Kai Dang, Keming Lu, Keqin Bao, Kexin Yang, Le Yu, Mei Li, Mingfeng Xue, Pei Zhang, Qin Zhu, Rui Men, Runji Lin, Tianhao Li, Tianyi Tang, Tingyu Xia, Xingzhang Ren, Xuancheng Ren, Yang Fan, Yang Su, Yichang Zhang, Yu Wan, Yuqiong Liu, Zeyu Cui, Zhenru Zhang, and Zihan Qiu. 2025. Qwen2.5 Technical Report. *Preprint*, arXiv:2412.15115.

Francisco Zanartu, Yulia Otmakhova, John Cook, and Lea Frermann. 2024. Generative Debunking of Climate Misinformation. In *Proceedings of the 1st Workshop on Natural Language Processing Meets Climate Change (ClimateNLP 2024)*, pages 46–62, Bangkok, Thailand. Association for Computational Linguistics.

Haiqi Zhang, Zhengyuan Zhu, Zeyu Zhang, Jacob Devasier, and Chengkai Li. 2024. Granular Analysis of Social Media Users' Truthfulness Stances Toward Climate Change Factual Claims. In *Proceedings of the 1st Workshop on Natural Language Processing Meets Climate Change (ClimateNLP 2024)*, pages 233–240.

A Models

Table 5 lists the baseline models used in our paper, along with their corresponding repository names on `https://huggingface.co/`.

Model Name	HuggingFace Repository
Gemma-2-2B	google/gemma-2-2b-it
Qwen-2.5-7B	Qwen/Qwen2.5-7B
Llama-2-7B	meta-llama/Llama-2-7b-chat-hf
Llama-2-13B	meta-llama/Llama-2-13b-chat-hf
ClimateGPT-7B	climategpt/climategpt-7b
ClimateGPT-13B	climategpt/climategpt-13b
Llama-3.1-8B	meta-llama/Llama-3.1-8B-Instruct
Mistral-24B	mistralai/Mistral-24B-Instruct
Llama-3.3-70B	meta-llama/Llama-3.3-70B-Instruct

Table 5: HuggingFace repository names of the baseline models used in our evaluation.

B CO2 Emission Related to Experiments

Experiments were conducted using a private infrastructure, which has a carbon efficiency of 0.432 $kgCO_2eq/kWh$. A cumulative of 240 hours of computation was performed on hardware of type A100 SXM4 80 GB (TDP of 400W). Total emissions are estimated to be 41.47 $kgCO_2eq$ of which 0 percent were directly offset. Estimations were conducted using the MachineLearning Impact calculator presented in Lacoste et al. (2019).

C Label distribution

To provide further insights into the datasets, we visualize the label distribution for each test set in Figure 3. Given the wide size range of the test sets (between 300 and 55872), we present the normalized label distributions, where each stack in a bar represents the percentage of a label within the corresponding task's test set.

Dataset (Source)	Task(s)	Shortened identifier
ClimaText (Varini et al., 2021)	Sentence classification	ClimaText Sent. Clf.
Climate-Stance (Vaid et al., 2022)	Stance classification	Climate-Stance
Climate-Eng (Vaid et al., 2022)	Topic classification	Climate-Eng
Climate-FEVER (Diggelmann et al., 2020)	Claim verification	Climate-FEVER
SciDCC (Mishra and Mittal, 2021)	Topic classification by Title	SciDCC Title
	Topic classification by Title & Summary	SciDCC Title Sum.
	Topic classification by Title & Body	SciDCC Title Body
CLIMA-CDP (Spokoyny et al., 2023)	QA-Cities (answer relevance)	QA-Cities
	QA-Corporations (answer relevance)	QA-Corps.
	QA-States (answer relevance)	QA-States
	QA-Topic-Cities (topic classification)	QA-Topic-Cities
PIRA 2.0 MCQ(Pirozelli et al., 2024)	PIRA with Context	PIRA w/ Ctx.
	PIRA without Context	PIRA w/o Ctx.
Exeter Misinformation (Coan et al., 2021)	Claim Detection	Exeter Claim
	Sub-claim Detection	Exeter Sub-claim
Climate-Change NER (Bhattacharjee et al., 2024)	Entity recognition	Climate NER
CheapTalk (Bingler et al., 2023)	Climate Detection	
	Climate Sentiment	
	Climate Commitment	Climate Commit.
	Climate Specificity	
	TCFD Recommendations	TCFD Recommend.
Net-Zero Reduction (Schimanski et al., 2023)	Paragraph classification	Net-Zero Reduction
Environmental Claims (Stammbach et al., 2023)	Sentence classification	Env. Claims
Guardian Climate News Corpus	Topic classification by Title	Guardian Title
	Topic classification by Body	Guardian Body

Table 6: Shortened task identifiers (if they exist) for each task presented in the ClimateEval benchmark. These shortened task names are used in tables presenting the results for the purpose of saving space.

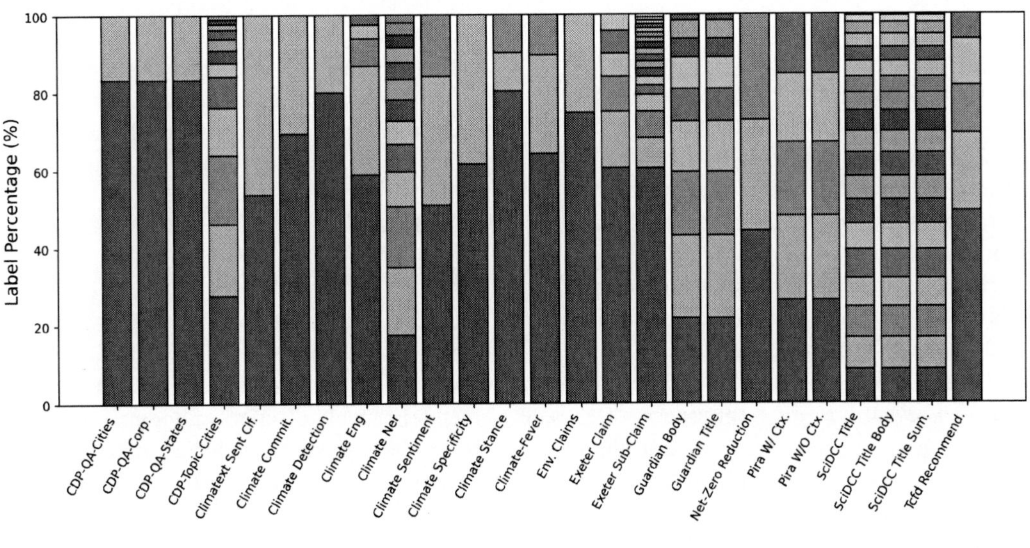

Figure 3: The normalized distribution of labels in each task across all sets (train, development, and test).

Bidirectional Topic Matching: Quantifying Thematic Intersections Between Climate Change and Climate Mitigation News Corpora Through Topic Modelling

Raven Adam and **Marie L. Kogler**
Department of Environmental Systems Sciences / University of Graz, Austria
raven.adam@uni-graz.at marie.kogler@uni-graz.at

Abstract

Bidirectional Topic Matching (BTM) is a novel method for cross-corpus topic modeling that quantifies thematic overlap and divergence between corpora. BTM is a flexible framework that can incorporate various topic modeling approaches, including BERTopic, Top2Vec, and Latent Dirichlet Allocation (LDA). It employs a dual-model approach, training separate topic models for each corpus and applying them reciprocally to enable comprehensive cross-corpus comparisons. This methodology facilitates the identification of shared themes and unique topics, providing nuanced insights into thematic relationships. A case study on climate news articles illustrates BTM's utility by analyzing two distinct corpora: news coverage on climate change and articles focused on climate mitigation. The results reveal significant thematic overlaps and divergences, shedding light on how these two aspects of climate discourse are framed in the media.

1 Introduction

Topic modeling is widely used to analyze and structure large textual corpora (Churchill and Singh, 2022), with a key application being the identification of latent topics that experts can evaluate for quantitative insights (Grundmann, 2021). Beyond single-corpus analysis, topic modeling also facilitates comparisons across multiple corpora, enabling the examination of thematic similarities and differences (Bystrov et al., 2022).

In climate discourse research, cross-corpus methods can reveal how different aspects of climate change and mitigation are framed in the media. While corpus linguistics has traditionally applied similarity measures during corpus creation or selection, studies have demonstrated their value for discourse analysis (Taylor, 2018). Recent research has leveraged such approaches to compare narratives across policy debates, social media discussions, and news coverage in various contexts, including migration, elections, and economic development (Shaikina and Funkner, 2020; Bystrov et al., 2024; Hellwig et al., 2024; Taylor, 2018).

This study introduces Bidirectional Topic Matching (BTM), a novel method for cross-corpus topic modeling, to analyze thematic overlaps and distinctions in climate change and mitigation news articles. BTM identifies shared and corpus-specific topics, enabling both quantitative comparisons and deeper qualitative exploration of how these issues are framed.

Existing cross-corpus topic modeling approaches typically rely on Latent Dirichlet Allocation (LDA) (Blei et al., 2003) or language embedding models to compute topic similarities via cosine similarity (Carniel et al., 2022; Hellwig et al., 2024). Others merge corpora into a single model and analyze topic distributions separately (Wang et al., 2023). In contrast, BTM trains distinct topic models for each corpus and applies them reciprocally, allowing topics to be assigned across corpora. This approach enhances the identification of both shared and unique themes, providing deeper insights into the evolving discourse on climate change and mitigation.

2 Method

2.1 Topic modelling

BTM is a flexible framework for cross-corpus analysis that can incorporate various topic modeling approaches. For assessing corpus similarity, any method capable of inferring topics for new data is suitable. However, analyzing unique or corpus-specific topics requires a method that can identify intraclass outliers—documents that do not align with any topics generated by the chosen topic modeling approach. Language embedding-based methods, such as BERTopic (Grootendorst, 2022) or Top2Vec (Angelov, 2020), are particularly well-

Proceedings of the 2nd Workshop on Natural Language Processing Meets Climate Change (ClimateNLP 2025), pages 208–217
July 31, 2025

suited for this purpose as they inherently support outlier detection. Traditional approaches like Latent Dirichlet Allocation (LDA), which assign a topic to every document, can also be adapted through post-processing techniques such as HDBSCAN (McInnes and Healy, 2017) or Local Outlier Factor (Breuniq et al., 2000) to identify outliers. Given BERTopic's state-of-the-art performance and its built-in outlier detection capabilities, this study demonstrates the application and efficacy of BTM using BERTopic as the underlying topic modeling approach.

BERTopic presents an innovative method for topic modeling, capitalizing on recent advancements in embedding models. Derived from Bidirectional Encoder Representations from Transformers (BERT) (Devlin et al., 2019), this approach involves the representation of documents as points within a high-dimensional vector space. In this space, each coordinate represents contextual information corresponding to the respective document. As a result, semantically analogous documents will be in proximity to each other. Subsequently, dimensionality reduction and clustering algorithms are employed to identify compact clusters of documents with shared thematic content. Each of these clusters can then be interpreted as individual topics that are found within the investigated collection of documents and are represented by a set of keywords that are most indicative of the underlying theme. An outlier refers to a document that cannot be assigned to any of the identified topics due to its lack of thematic similarity. This occurs when the document does not align well with any of the topics, often because it is too different or semantically distant from the other documents in the model. In BERTopic, both topics and outliers can be easily accessed and handled, where outliers are grouped together under an outlier topic, often with a special identifier like -1. As a final step, a class-based term frequency inverse document frequency measure (c-TFIDF) is applied to extract the most salient terms from each topic and create interpretable topic representations (Grootendorst, 2022).

2.2 Cross-Corpus Topic Assignment

For BTM, which is schematically depictured in Figure 1, two independent topic models are trained on two thematically related corpora, corpus 1 and corpus 2. Each model is used to identify the main themes within the respective corpus, generating topics T1 for corpus 1 and topics T2 for corpus 2. Individually, these native topic models provide a comprehensive understanding of the thematic structures specific to each dataset.

To explore thematic alignment between the corpora, each model was applied to the corpus, it was not trained on. For this, the semantic similarity between the document's embedding and the topic embeddings of the model trained on the other corpus was calculated. Specifically, each document in corpus 2 gets matched to a topic from T1, and each document in corpus 1 gets matched to a topic from T2, based on the highest similarity score. This process produced cross-corpus topic assignments, resulting in T12 (topics from T1 assigned to Corpus 2) and T21 (topics from T2 assigned to Corpus 1).

Subsequently, topic pairs are generated by assigning each document from one corpus to the most similar topic from the opposite corpus. Specifically, for each document, the topics assigned by the corpus 1 model (T11 and T12) and the topics assigned by the corpus 2 model (T22 and T21) are combined into cross-corpus topic pairs.

For a comprehensive cross-corpus analysis, both the main set of topics and outliers are considered. Outliers, while exhibiting atypical or low similarity scores within their own topic model, are included in the pairing process if they represent the highest similarity match for a document. Thus, topic similarity is calculated across all topics (0, 1, 2, ..., n), with outliers treated as an additional category (-1). This approach ensures that all thematical aspects are represented, even if the relationships involving outlier topics require further scrutiny in subsequent analyses. This becomes especially crucial when working with documents that are split into smaller units, like paragraphs, where certain sections may show unexpected topic associations, increasing the likelihood of outliers that require careful attention.

2.3 Cross-Corpus Topic Pair Analysis

The topic pairs from both corpora were analyzed through co-occurrence analysis to identify frequently paired topics between the two models. Specifically, we calculate how often each pair — composed of one topic from the corpus 1 model (T1, T12) and one from the corpus 2 model (T2, T21) — is assigned to the same document. The cross-topic co-occurrence is given by aggregation of these pairs across all documents. This process allows us to assess the frequency with which specific cross-corpus topic combinations occur together,

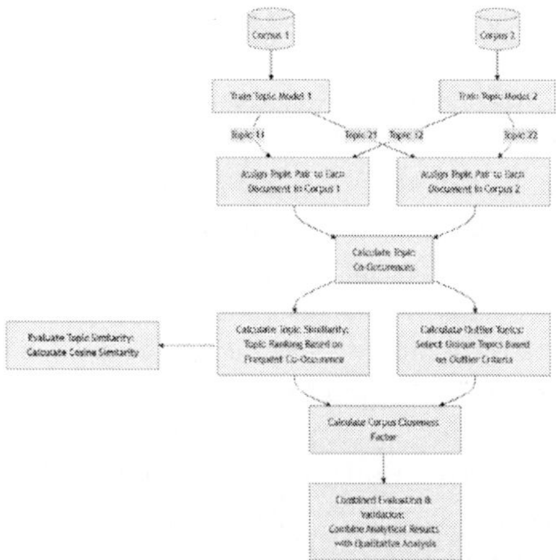

Figure 1: Schematic Outline of Bidirectional Topic Matching Procedures for Calculating the Thematic Closeness Factor of Corpus 1 and Corpus 2. Optional additional analysis of topic similarity may be conducted via cosine similarity.

providing insights into their thematic relationships. High-frequency pairs indicate topics from both models that were commonly associated with similar documents, reflecting thematic alignment between the corpora. Although the co-occurrence analysis itself remains undirected, focusing solely on the frequency of simultaneous topic occurrences within the documents, the subsequent exploration of relationships between topics from Corpus 1 and Corpus 2 is framed in a directed context. This directed approach enables a detailed investigation of the interactions and semantic linkages between the topics across the two corpora.

The interpretation of topic pairs helps clarify patterns of topic co-occurrence between corpora. High co-occurrence between a native topic and main cross-topics suggests strong thematic alignment, whereas alignment with smaller cross-topics indicates a more nuanced or niche connection. If a native topic aligns with outlier topics from the cross corpus, it may reflect themes unique to the native corpus. Similarly, when outlier topics from both corpora co-occur, it suggests a shared lack of thematic focus, while low co-occurrence between outliers is unexpected and may indicate inconsistencies in topic modeling or heterogeneity within the outlier topics.

3 Topic and Corpus Measures

For a corpus containing T native topics, a series of measures can be calculated to describe its relationship with a second corpus containing $\tilde{T}$ cross topics. A pairing strength is introduced as a quantitative measure of the degree of association between a topic from the native corpus and a topic from the cross corpus. This measure is based on the frequency of co-occurrence of the two topics within the same documents. For a topic pair $(t_i, \tilde{t}_j)$, where $t_i | i \in \{-1, \dots, T\}$ belongs to the native topics and $\tilde{t}_j | j \in \{-1, \dots, \tilde{T}\}$ belongs to the cross topics, the pairing strength $S(t_i, \tilde{t}_j)$ can be defined as:

$$S(t_i, \tilde{t}_j) = \frac{n(D_{ij})}{n(D_i)} \tag{1}$$

where $n(D_{ij})$ denotes the size (or cardinality) of the set of documents D_{ij} to which both topics t_i and $\tilde{t}_j$ are assigned. Respectively, $n(D_i)$ denotes the size of the set of documents D_i associated with the native topic t_i.

For the cross topics $\tilde{t}_j | j \in \{0, \dots, \tilde{T}\}$, the pairing strength is referred to as topic closeness and represents the degree of alignment between each cross-topic and a specific native topic t_i. A special case of pairing strength involves the outlier topic $\tilde{t}_{-1}$ called topic uniqueness. Topic uniqueness quantifies the extent to which a native topic is distinct from the cross corpus. Native topics with a topic uniqueness value of 0.5 or higher are classified as unique topics.

3.1 Corpus Closeness and Corpus Uniqueness

Based on the topic closeness of all native topics, we define the corpus closeness C, which quantifies the overall thematic relatedness between the two corpora:

$$C = \frac{\sum_{i=0}^{T} \sum_{j=0}^{\tilde{T}} S(t_i, \tilde{t}_j)}{T} \tag{2}$$

as well as its weighted variant C_w, which gives higher importance to larger and more relevant native topics:

$$C_w = \frac{\sum_{i=0}^{T} n(D_i) \sum_{j=0}^{\tilde{T}} S(t_i, \tilde{t}_j)}{\sum_{i=0}^{T} n(D_i)} \tag{3}$$

Both closeness measures reflect the thematic overlap between the two corpora, while the weighted measure assigning greater significance to larger and thus more prominent topics within the native corpus. Generally, low closeness indicates that the two corpora are largely thematically independent. The difference $C_w - C = \theta; \theta \in [-1, 1]$ can be used to assess whether the relationship between the corpora is evenly distributed across all native topics or predominantly concentrated within a subset of native topics:

$$f(x) = \begin{cases} \theta \sim 1; & \text{corpus closeness is proportionally influences by larger native topics} \\ \theta \sim 0; & \text{corpus closeness is not influenced by native topic size} \\ \theta \sim -1 & \text{corpus closeness is proportionally influenced by smaller native topics} \end{cases} \quad (4)$$

The corpus uniqueness U and its weighted equivalent U_w are alternatives to the corpus closeness to indicate the level of independence between the corpora:

$$U = 1 - C = \frac{\sum_{i=0}^{T} S(t_i, \tilde{t}_{-1})}{T} \quad (5)$$

$$U_w = 1 - C_w = \frac{\sum_{i=0}^{T} n(D_i) \cdot S(t_i, \tilde{t}_{-1})}{\sum_{i=0}^{T} n(D_i)} \quad (6)$$

Here, $S(t_i, \tilde{t}_{-1})$ represents the topic uniqueness of each native topic. As with the corpus closeness factor, a high positive difference $U_w - U$ indicates that most of the corpus uniqueness is explained by larger native topics while a large negative difference sees most of it covered by smaller native topics.

3.2 Corpus Alignment

Both closeness and uniqueness fail to account for the specificity of topic matches and topic size distribution of the native corpus. The topic alignment strength $SA(t_i)$ of a native topic quantifies the concentration of topic closeness values with respect to the topics of the cross corpus. This indicates whether a native topic is associated with a single theme (focused) or to multiple themes (scattered) in the other corpus. To achieve this, the highest topic closeness of the native topic is selected:

$$SA(t_i) = \max_{j \in \{0,\ldots,\tilde{T}\}} S(t_i, \tilde{t}_j) \quad (7)$$

$$= \max\{S(t_i, \tilde{t}_0), S(t_i, \tilde{t}_1), \ldots, S(t_i, \tilde{t}_{\tilde{T}})\} \quad (8)$$

A high topic alignment strength indicates that a native topic aligns with a single cross topic, whereas a low value suggests a wider variety of important pairings.

The corpus alignment A serves as an overall metric that captures the average alignment strength across all native topics. It quantifies whether the topic alignments between the two corpora are focused on specific topic pairs or spread over multiple combinations.

$$A = \frac{\sum_{i=0}^{T} SA(t_i)}{T} \quad (9)$$

$$A_w = \frac{\sum_{i=0}^{T} n(D_i) \cdot SA(t_i)}{\sum_{i=0}^{T} n(D_i)} \quad (10)$$

Here, the difference $A_w - A$ is useful to indicate whether the distribution of topic alignment strength is skewed towards larger or smaller native topics.

By comparing the corpus uniqueness factor U and the corpus alignment factor A, we identify three key relationships between corpora. Low uniqueness and low alignment indicate thematic overlap, with the cross corpus exploring similar topics in greater depth or from multiple perspectives. Low uniqueness and high alignment suggest that the corpora are closely related, likely subsets of a larger parent corpus. High uniqueness and low alignment imply that the corpora are largely independent, as many topics in the native corpus are not present in the cross corpus. A scenario with both high uniqueness and high alignment is not possible due to their inherent relationship.

4 Validation through related Methods

Since both topic models are generated from the same embedding model, the resulting embedding vectors for each topic are located in the same vector space. Therefore, to validate the effectiveness of the proposed method, we introduced an additional analysis by measuring the cosine similarity between the topic embeddings of the two BERTopic models. In this validation process, cosine similarity scores were first calculated between the topic embeddings of the corpus 1 and corpus 2 models to quantify the semantic overlap between their topics. Higher cosine similarity scores indicated greater alignment between topics. These scores were then compared to the distribution of observed topic pairs, with the goal of finding the most similar topics across the corpora. To assess the consistency between the two methods, Cohen's kappa was calculated, providing a measure of agreement between the cosine similarity-based approach and the topic pair distribution.

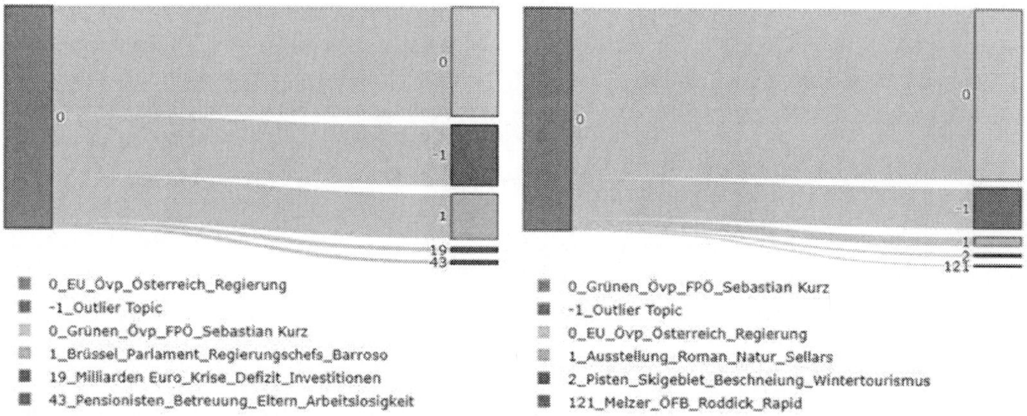

Figure 2: Left side – The largest native topic from corpus 1 along with the five most prominent cross topic pairs from corpus 2. They gray area indicates the pairing strength for each pair. Right side – The largest native topic from corpus 2 along with the five most prominent cross topic pairs from corpus 1. They gray area indicates the pairing strength for each pair.

5 Case Study Climate News

5.1 Dataset

To showcase BTM, two sets of digitized print articles were extracted from the WISO database that provides a repository for online newsarticles in the German-speaking region. According to Adam, Scholger, and Kogler (2023), the regional climate debate is characterized by two largely independent subject areas: climate change, which encompasses information on natural and physical impacts, dangers, and risks, and climate mitigation, which focuses on actions, socio-economic strategies, and technological solutions. The search terms climate change ("klimawandel*" where the asterisk serves as a wildcard symbol that matches any suffixes or word endings attached to the German root word "klimawandel") and climate action ("klimaschutz*") were used to create the climate change dataset (corpus 1) and the climate action dataset (corpus 2), respectively. The investigated period spans from 2002 until 2022 and includes 21.753 articles in corpus 1 and 20.135 articles in corpus 2, with an overlap of 3.111 articles.

To account for the limited encoding length of embedding models, all articles were split into smaller parts of up to 150 words, which corresponds to the average length of German paragraphs (Altpeter et al., 2015). This was done with the help of the *gsd model* available in the stanza library (Qi et al., 2020). The final dataset therefore consisted of 124.500 paragraphs.

Both BERTopic models were trained based on the *German Semantic STS V2* embedding model. For corpus 1 a topic model consisting of 122 topics was generated, while corpus 2 produced a topic model with 88 topics.

6 Results

6.1 Case Study

6.1.1 Topic Pairs and primary Relationships between Topics

Tables 1 and 2 provide qualitative evidence supporting BTM's ability to identify meaningful relationships between topics across corpora. By examining paired topics, corpus-specific nuances emerge. For example, a comparison of topics focused on forests and glaciers reveals differences in thematic emphasis: Corpus 1 highlights specific results of climate change, such as increased bark beetle infestations and rockfalls in the Alps, while Corpus 2 emphasizes the state of forests or national parks and the impact of climate change on alpine temperatures. This capacity to reveal varying degrees of specificity allows researchers to understand how distinct datasets prioritize or converge on shared themes. Such insights are critical for comparative discourse analyses, such as political communication or cross-cultural studies.

6.1.2 Subpairing Topics – Quantifying Secondary Thematic Relationships

Whether individual topics are directly shared between corpora or whether one corpus discusses certain topics more diversely can be analyzed us-

Native Topics Corpus 1 (T1)	Cross Topics Corpus 2 (T12)	SA
EU ÖVP Austria Government	Greens ÖVP FPÖ Sebastian_Kurz	0.44
Trees Bark_Beetle Federal_Forestry Spruce	Woods Hectare Federal_Forestry National_Park	0.60
Fridays Greta_Thunberg Streets Youths	Fridays Greta_Thunberg Movement Humans	0.69
Glacier Alps Rockfall Dachstein	Degree Glacier Temperatures Climate_Change	0.58
Diesel Electric_Cars Vehicles Automobile_Industry	Electric_Cars Vehicles BMV Diesel	0.41

Table 1: Five native topics of corpus 1 along with their respective main cross topic pair from corpus 2 and topic alignment strength SA (highest pairing strength). Each topic is represented by four topic words or phrases (connected with an underscore), which is the standard output of BERTopic. The topic representations were translated from German to English.

Native Topics Corpus 2 (T2)	Cross Topics Corpus 1 (T21)	SA
Greens ÖVP FPÖ Sebastian_Kurz	EU ÖVP Austria Government	0.70
Brussels Parliament Head_of_Government Barroso	EU ÖVP Austria Government	0.61
Renovation Residential_Construction Housing_Subsidies Buildings	Passive_House Residential_Construction Energy_Efficiency Real_Estate	0.31
ÖBB Million_Euro Truck Commuter	ÖBB Vienna Mobility Means_of_Transport	0.68
Baerbock Merkel CSU Greens	Laschet Baerbock Greens Coalition	0.74

Table 2: Five native topics of corpus 2 along with their respective main cross topic pair from corpus 1 and topic alignment strength SA (highest pairing strength). Each topic is represented by four topic words or phrases (connected with an underscore), which is the standard output of BERTopic. The topic representations were translated from German to English.

ing topic alignment strength, as shown in Tables 1 and 2. For instance, the politics topic in Corpus 1 exhibits a moderate topic alignment strength of 0.44. This indicates that several topics from Corpus 2, beyond the most similar cross-topic, address relevant aspects of this native topic. The left side of Figure 2 visually showcases this distribution across different cross-topic pairings. This suggests that political discourse is more granular in Corpus 2, allowing its topic model to recognize distinctions within documents assigned to a single topic in Corpus 1.

Conversely, Table 2 reveals that both national and EU-level politics topics in Corpus 2 exhibit high topic alignment strength with the same politics topic in Corpus 1. This supports the hypothesis that political discourse in Corpus 2 is more detailed, encompassing multiple perspectives that align with a broader theme in Corpus 1.

A broader overview is provided in Figure 3, which illustrates the pairing strength composition for the 25 largest topics in each corpus. For most native topics, the most similar cross-topic alone does not account for the majority of topic closeness. This highlights thematic asymmetries, where one corpus tends toward generality while the other emphasizes specificity. Such analyses are instrumental in uncovering where thematic overlaps or divergences occur, enabling nuanced interpretations of the data

6.1.3 Identifying Unique Topics

One of BTM's most compelling features is its ability to identify topics unique to each corpus. This is achieved by extracting topics with a topic uniqueness value above 0.5. In this case study, 23 unique topics were identified in Corpus 1, while Corpus 2 contained 15 unique topics.

Table 3 illustrates examples of unique topics from each corpus. Corpus 1 focuses on science communication and geographic impacts, such as water supply, while Corpus 2 emphasizes actionable measures, including renewable energy and local initiatives. Such differentiation is especially valuable for corpora with overlapping themes, as it enables researchers to discern distinct areas of fo-

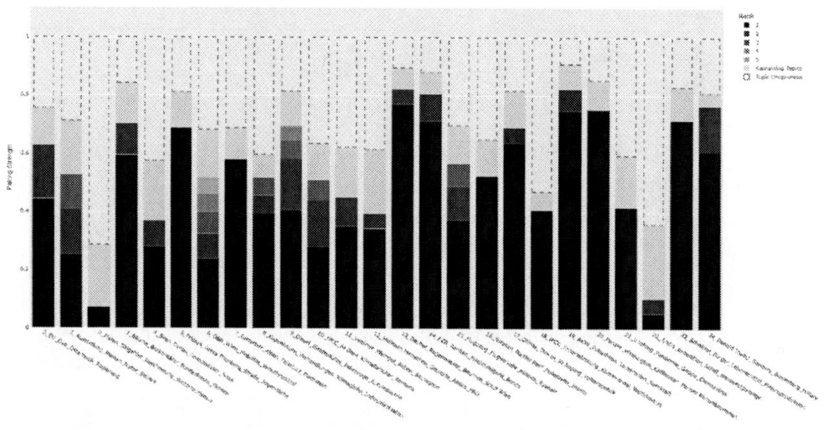

(a)

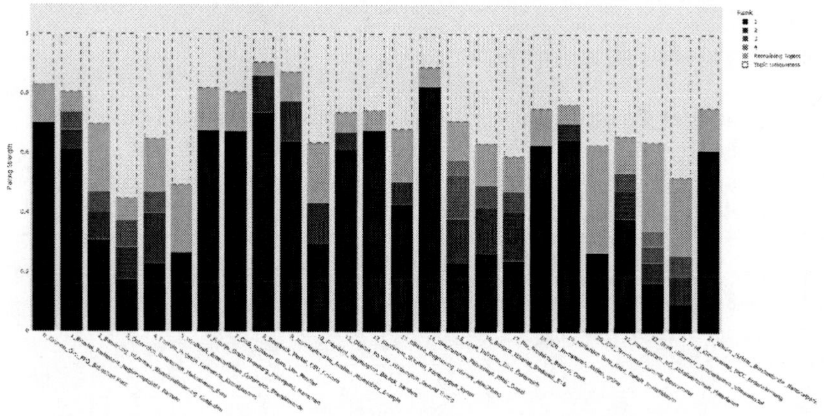

(b)

Figure 3: The pairing strength composition for the 25 largest native topics. The shading of the bars indicates the ranking of the topic pairing strengths, where the most prominent pair is represented by the darkest color. Topic pairs with a pairing strength below 0.05 were merged into the "remaining topic" category. The outlier topic pairing strength or topic uniqueness is indicated by the red dashed bars. a): Corpus 1. b): Corpus 2.

cus. For example, in interdisciplinary studies, this capability bridges gaps between problem-oriented and solution-oriented approaches, fostering more comprehensive analyses.

6.1.4 Corpus Level Relationship

Table 4 reveals that both corpora exhibit notable distinctions, with approximately one-third of the content in each corpus not described by the other. Both show a corpus uniqueness factor of 0.34, indicating a significant level of thematic independence. The corpus closeness factor of 0.66 suggests major thematic overlaps, while the low difference between weighted and general corpus uniqueness factors (< 0.1) implies that neither corpus is skewed toward unique topics of particular sizes. However, Corpus 2 displays slightly more pronounced topic uniqueness in smaller topics compared to Corpus 1.

Similarly, both corpora have comparable corpus alignment factors (0.45 for Corpus 1 and 0.44 for Corpus 2). The minor influence of native topic sizes indicates that alignment is not disproportionately driven by larger topics. Together, these metrics suggest that while the corpora share substantial thematic overlap, they focus on different thematic subsets in more detail. This is consistent with the low corpus uniqueness and low corpus alignment case, where native topics frequently pair with multiple relevant cross-topics, as observed in Figure 3.

Corpus 1 Unique Topics	Corpus 2 Unique Topics
Slopes Ski_Area Snow_Making Wintertourism	Austria Emissions Measures Euro
Lakes Donau Groundwater Water	Energy Project Municipality Climate_Alliance
IPCC Al_Gore Climate_Researcher Consensus	Wind_Power Renewable Austria Energie_Transition

Table 3: Selection of three unique native topics from corpus 1 and corpus 2 respectively based on a topic uniqueness above 0.5.

Native Corpus	C	$C_w - C$	U	$u_w - U$	A	$A_w - A$
Corpus 1	0.66	0.02	0.34	-0.02	0.45	-0.01
Corpus 2	0.66	0.04	0.34	-0.04	0.44	0.04

Table 4: Values for the corpus closeness factor C, the corpus uniqueness factor U, the corpus alignment factor A and the difference between the three factors and their respective weighted variants for corpus 1 and corpus 2.

6.2 Validation - Comparison with Cosine Similarity

We demonstrate the agreement between BTM and cosine similarity-based methods for climate news articles to highlight the validity of the proposed approach. When identifying the most similar topic from corpus 2 for each topic in corpus 1, Cohen's kappa was calculated at 0.75. Conversely, when determining the most similar topic from corpus 1 for each topic in corpus 2, Cohen's kappa increased to 0.81. These values reflect a strong level of agreement, affirming the reliability of BTM (Mchugh, 2012).

Discrepancies between BTM and cosine similarity approaches were most evident when BTM assigned the outlier topic as the closest match. Since this topic encompasses documents that do not fit into any defined clusters, its inclusion is inherently challenging for methods relying solely on cosine similarity. Beyond the outlier topic, the remaining discrepancies (approximately 20%) lacked clear evidence favoring one method over the other, suggesting that both approaches offer comparable utility for calculating topic similarity.

7 Discussion and Conclusion

BTM provides a robust framework for cross-corpus topic modeling. By leveraging BERTopic's interpretable topic representations and employing reciprocal topic assignments, BTM facilitates a nuanced exploration of thematic relationships across corpora. This approach not only captures shared topics but also highlights unique themes, offering a comprehensive lens through which to analyze corpora with overlapping or divergent thematic structures.

7.1 Methodological Contributions

BTM addresses key limitations in traditional cross-corpus topic modeling approaches. By training separate topic models for each corpus and applying them reciprocally, BTM ensures that each model's native structure is preserved while enabling cross-corpus comparisons. This dual approach allows for the identification of both shared and unique topics, a capability that is particularly valuable in interdisciplinary or comparative discourse studies.

Validation through cosine similarity underscores the reliability of BTM. Strong agreement between BTM and cosine similarity-based methods (Cohen's kappa scores of 0.75 and 0.81) demonstrates the robustness of the approach, while the discrepancies observed with outlier topics highlight areas where BTM's methodological strengths are most apparent. These findings suggest that BTM can serve as a reliable alternative or complement to existing methods, particularly for datasets with significant thematic variability.

7.2 Insights from the Case Study

The application of BTM to climate news articles revealed meaningful thematic distinctions and overlaps between two corpora focused on climate change and climate action. The results demonstrate that while both corpora share substantial thematic overlap (corpus closeness factor of 0.66), they also exhibit notable differences, with approximately one-third of the content in each corpus being unique (corpus uniqueness factor of 0.34).

Corpus 1 prioritizes broad environmental and scientific discussions, such as the geographic impacts of climate change, while Corpus 2 focuses on actionable measures like renewable energy and local initiatives. This differentiation underscores the value of BTM in identifying thematic nuances that may be overlooked by less granular methods. Moreover, the ability to quantify topic alignment and uniqueness provides a structured way to assess thematic relationships, facilitating more targeted qualitative analyses.

8 Limitations

There are a few notable limitations in the suggested approach. First of all, BTM provides direction dependent results. Comparing Corpus 1 with Corpus 2 can lead to different results than comparing Corpus 2 with Corpus 1. For example, if Corpus 2 were to be a highly specific sub-corpus of Corpus 1. In this case, Corpus 1 would exhibit high Uniqueness values, as only a limited number of its native topics would be covered by Corpus 2. However, Corpus 2 would have low Uniqueness as all of its native topics are present in Corpus 1.

Secondly, the presented case study uses the same embedding model for both corpora. While this is necessary to compare the results with cosine similarity, there are cases where it might be preferable to use different embedding models for each corpus. Especially when domain-specific models are available such as in the medical or financial domain. BTM can, theoretically, still be employed in such a case, it is however unclear how valid the results would be. Such an investigation would be an important aspect of future research.

A third limitation is that using topic merging methods after creating topic models will result in different corpus level measures than using unmerged topics. The topic level measures of a merged topic will be the averages calculated from the topic level measures of each individual topic that was used to create the merged topic. And as the corpus level measures are either weighted or unweighted averages of the used topic level measures, averaging some of them beforehand will naturally change the final results.

9 Further Research

Future research could apply BTM to dissect the complex interplay between scientific understanding and policy formulation. For instance, a systematic comparison of academic literature on specific climate solutions, such as carbon capture technologies or nature-based solutions, with corresponding governmental policy documents or legislative proposals could quantitatively reveal how scientific findings are translated, prioritized, or re-framed within policy-making arenas (Ibarra et al., 2022). Similarly, BTM offers a robust methodology to analyze the critical interface between expert communication and public discourse. By comparing outputs from climate science organizations, like IPCC summaries or national climate assessments, with the vast textual data generated on social media platforms or in public commentary on news articles, researchers could identify unique public concerns, pinpoint areas of scientific misunderstanding, or highlight divergent thematic emphases, thereby informing the development of more effective and resonant climate communication strategies.

Furthermore, BTM can facilitate nuanced comparisons across diverse geopolitical and ideological landscapes. It could be employed to systematically examine climate narratives within Nationally Determined Contributions (NDCs) submitted by developed versus developing nations, or to contrast climate impact reporting styles and thematic priorities between media outlets in the Global North and Global South (Hase et al., 2021). Such analyses could illuminate shared thematic ground alongside areas of significant contention or differing national priorities, which is crucial for international climate negotiations and cooperation.

Beyond governmental and public spheres, BTM can also shed light on corporate engagement with climate change. Applying the framework to analyze corporate sustainability or Environmental, Social, and Governance (ESG) reports across various industry sectors, or between companies with different stated climate commitments, could identify common and unique themes related to perceived climate risks, adopted mitigation strategies, and planned adaptation efforts (Dahl and Fløttum, 2019). Through these varied applications, BTM promises to provide researchers with a powerful tool for a deeper, more quantified understanding of the multifaceted and evolving discourses surrounding climate change, its impacts, and the global response.

References

Elke Karin Altpeter, Tobias Marx, Nhung Xuan Nguyen, Aline Naumann, and Susanne Trauzettel-Klosinski. 2015. Measurement of reading speed with standardized texts: a comparison of single sentences and paragraphs. *Graefe's Archive for Clinical and Experimental Ophthalmology*, 253(8):1369–1375.

Dimo Angelov. 2020. Top2Vec: Distributed Representations of Topics. pages 1–25.

David Meir Blei, Andrew Yan-Tak Ng, and Michael Irwin Jordan. 2003. Latent Dirichlet allocation. *Journal of Machine Learning Research*, 3(4-5):993–1022.

Markus M. Breuniq, Hans Peter Kriegel, Raymond T. Ng, and Jörg Sander. 2000. LOF: Identifying density-based local outliers. *SIGMOD Record (ACM Special Interest Group on Management of Data)*, 29(2):93–104.

Victor Bystrov, Viktoriia Naboka, Anna Staszewska-Bystrova, and Peter Winker. 2022. Cross-Corpora Comparisons of Topics and Topic Trends. *Journal of Economics and Statistics*, 242(4):433–469.

Victor Bystrov, Viktoriia Naboka-Krell, Anna Staszewska-Bystrova, and Peter Winker. 2024. Comparing Links between Topic Trends and Economic Indicators in the German and Polish Academic Literature. *Comparative Economic Research. Central and Eastern Europe*, 27(2):7–28.

Théophile Carniel, Leo Cazenille, Jean Michel Dalle, and José Halloy. 2022. Using natural language processing to find research topics in Living Machines conferences and their intersections with Bioinspiration & Biomimetics publications. *Bioinspiration and Biomimetics*, 17.

Rob Churchill and Lisa Singh. 2022. The Evolution of Topic Modeling. *ACM Computing Surveys*, 54(10).

Trine Dahl and Kjersti Fløttum. 2019. Climate change as a corporate strategy issue. *Corporate Communications: An International Journal*, 24(3):499–514.

Jacob Devlin, Ming Wei Chang, Kenton Lee, and Kristina Toutanova. 2019. BERT: Pre-training of deep bidirectional transformers for language understanding. *NAACL HLT 2019 - 2019 Conference of the North American Chapter of the Association for Computational Linguistics: Human Language Technologies - Proceedings of the Conference*, 1(Mlm):4171–4186.

Maarten Grootendorst. 2022. BERTopic: Neural topic modeling with a class-based TF-IDF procedure.

Reiner Grundmann. 2021. Using large text news archives for the analysis of climate change discourse: some methodological observations. *Journal of Risk Research*, 0(0):1–13.

Valerie Hase, Daniela Mahl, Mike Steffen Schäfer, and Tobias Keller. 2021. Climate change in news media across the globe: An automated analysis of issue attention and themes in climate change coverage in 10 countries (2006–2018). *Global Environmental Change*, 70(December 2020).

Nils Constantin Hellwig, Jakob Fehle, Markus Bink, Thomas Schmidt, and Christian Wolff. 2024. Exploring Twitter discourse with BERTopic: topic modeling of tweets related to the major German parties during the 2021 German federal election. *International Journal of Speech Technology*.

Cecilia Ibarra, Guadalupe Jiménez, Raúl O'Ryan, Gustavo Blanco, Luis Cordero, Ximena Insunza, Pilar Moraga, Maisa Rojas, and Rodolfo Sapiains. 2022. Scientists and climate governance: A view from the south. *Environmental Science & Policy*, 137:396–405.

Mary L Mchugh. 2012. Interrater reliability : the kappa statistic. *Biochem Med (Zagreb)*, 22(3):276–282.

Leland McInnes and John Healy. 2017. Accelerated Hierarchical Density Based Clustering. In *IEEE International Conference on Data Mining Workshops, ICDMW*, pages 33–42.

Peng Qi, Yuhao Zhang, Yuhui Zhang, Jason Bolton, and Christopher D. Manning. 2020. Stanza: A Python Natural Language Processing Toolkit for Many Human Languages. In *Association for Computational Linguistics (ACL) System Demonstrations*, pages 101–108.

Alevtina A. Shaikina and Anastasia A. Funkner. 2020. Medical Corpora Comparison Using Topic Modeling. *Procedia Computer Science*, 178:244–253.

Charlotte Taylor. 2018. Similarity. In Charlotte Taylor and Anna Marchi, editors, *Corpus Approaches to Discourse*, chapter 2, pages 19–37. Routledge, London.

Zhongyi Wang, Jing Chen, Jiangping Chen, and Haihua Chen. 2023. Identifying interdisciplinary topics and their evolution based on BERTopic. *Scientometrics*, 129(11):7359–7384.

CPIQA: Climate Paper Image Question Answering Dataset for Retrieval-Augmented Generation with Context-Based Query Expansion

Rudra Mutalik[1], Abiram Panchalingam[1], Loitongbam Gyanendro Singh[1], Timothy J. Osborn[2], Ed Hawkins[3], Stuart E. Middleton[1]

[1]School of Electronics and Computer Science,
University of Southampton, Southampton, UK

[2]Climatic Research Unit, School of Environmental Sciences,
University of East Anglia, Norwich, UK

[3] National Centre for Atmospheric Science, Department of Meteorology,
University of Reading, Reading, UK

{rudra.mutalik, ap18g21, gyanendro.loitongbam, sem03}@soton.ac.uk,
T.Osborn@uea.ac.uk, e.hawkins@reading.ac.uk

Abstract

Misinformation about climate science is a serious challenge for our society. This paper introduces CPIQA (Climate Paper Image Question-Answering), a new question-answer dataset featuring 4,551 full-text open-source academic papers in the area of climate science with 54,612 GPT-4o generated question-answer pairs. CPIQA contains four question types (numeric, figure-based, non-figure-based, reasoning), each generated using three user roles (expert, non-expert, climate sceptic). CPIQA is multimodal, incorporating information from figures and graphs with GPT-4o descriptive annotations. We describe Context-RAG, a novel method for RAG prompt decomposition and augmentation involving extracting distinct contexts for the question. Evaluation results for Context-RAG on the benchmark SPIQA dataset outperforms the previous best state of the art model in two out of three test cases. For our CPIQA dataset, Context-RAG outperforms our standard RAG baseline on all five base LLMs we tested, showing our novel contextual decomposition method can generalize to any LLM architecture. Expert evaluation of our best performing model (GPT-4o with Context-RAG) by climate science experts highlights strengths in precision and provenance tracking, particularly for figure-based and reasoning questions.

1 Introduction

Misinformation about climate science continues to pose a challenge for our society. This poses a serious challenge for public understanding, policymaking and even experts (Lewandowsky, 2020). At the same time, large language models (LLMs) have become powerful tools for information retrieval and evidence synthesis, but they are also highly prone to hallucination—generating incorrect or fabricated facts, references, and claims (Huang et al., 2025). Given the high stakes of climate communication, there is a pressing need for a reliable question-answering (QA) system that grounds responses in authoritative scientific sources.

In this work, we introduce CPIQA, a new dataset for climate science QA that incorporates both text and visual data from academic papers. CPIQA consists of 4,551 papers from twelve sources set out in appendix C, with extracted figures and their descriptions used as additional evidence in question-answering. The dataset supports three role variations and four question categories designed to reflect different types of real-world climate questions.

Building on CPIQA, we develop a retrieval-augmented generation (RAG)-based chatbot for climate QA. Our system follows a two-stage retrieval process: it first retrieves full papers based on the user's query, then extracts relevant text chunks from the most relevant papers. This approach improves both chunk similarity and cross-relevance of chunks. Further, we introduce Context-RAG, a novel prompting method that enhances retrieval by decomposing a given question into distinct contextual variations before searching for relevant documents. Rather than relying on a single query, our method anticipates different ways the question might be framed—such as a scientific explanation, a policy-related perspective, or a public concern—allowing for more diverse and targeted retrieval. This ensures that retrieved documents are not biased toward a single interpretation of the question.

To evaluate the effectiveness of our method, we test it on SPIQA, a dataset for scientific QA in the computer science domain, in addition to CPIQA. This allows us to assess how well our QA pipeline

Proceedings of the 2nd Workshop on Natural Language Processing Meets Climate Change (ClimateNLP 2025), pages 218–232
July 31, 2025 ©2025 Association for Computational Linguistics

generalizes beyond climate science. Finally, we validate the system's outputs through qualitative climate scientist expert evaluation, ensuring that responses are accurate, relevant, concise and aligned with scientific consensus.

By combining structured retrieval with expert-informed question generation, this work contributes a robust, transparent approach to climate QA, helping to bridge the gap between AI-generated answers and reliable scientific communication.

More specifically, our contributions in this work include the following:

- A new multimodal QA dataset resource (CPIQA dataset) for the NLP community based on 4,551 academic climate research paper documents. This dataset is large, annotated with 54,612 question-answer pairs generated by GPT-4o and includes text summaries of all images, graphs and figures within the full text documents. Questions are broken down into figure-based, numeric-based, non-numeric, and reasoning-based types to allow for a finer-grained evaluation of QA performance than most existing QA datasets allow. Our code and dataset is open source and available at github.com/RudraMutalik/CPIQA, doi.org/10.5281/zenodo.15374870 and doi.org/10.57967/hf/5386 respectively.

- Description of a novel context-based query expansion method for RAG, comprehensively evaluated on both the benchmark SPIQA dataset and our new CPIQA dataset. Context-based query expansion provides a 7.2% improvement in BERTscore-F1 over baseline RAG methods across various question types and roles. We include a detailed breakdown of performance across different question types which future researchers can benchmark their models against.

2 Related Work

2.1 Scientific QA Datasets

Table 1 sets out notable QA datasets that have been designed to support scientific domains such as climate science.

A significant number of existing QA datasets come from the biomedical and computer science domains, reflecting the heavy use of document-based QA in these fields. While these datasets offer strong benchmarks for scientific QA, they are typically unimodal, focusing exclusively on textual information. Multimodal datasets—those incorporating both text and figures—are far less common, with SPIQA (Pramanick et al., 2024) being the most comprehensive multimodal dataset designed for scientific applications.

Among multimodal datasets, FigureQA (Kahou et al., 2017) is a notable example, containing question-answer pairs for synthetic graphs, figures, and tables. However, it lacks contextual information from accompanying text, making it unsuitable for tasks that require a deeper understanding of scientific literature.

Compared to biomedical and computer science domains, climate science QA datasets are less common. One of the most relevant efforts is ClimaQA (Manivannan et al., 2024), which includes both a 502 question "gold" dataset curated by experts and a larger LLM-generated 3000 question "silver" dataset. ClimaQA is unique in that it supports three types of questions: multiple-choice, cloze-style, and free-form, allowing for a broader range of QA applications. Our CIPQA is significantly larger with 54,612 questions, and unlike ClimaQA which relies on textbook sources our dataset relies on academic paper sources making it suitable for research-driven climate QA.

2.2 Climate Science LLMs

Recent efforts have been made to fine-tune LLMs specifically for climate-related tasks such as fact-grounded QA, ambiguous policy analysis, and misinformation debunking. One such example is ClimateBERT (Webersinke et al., 2022), a model trained on climate-focused text sources to improve NLP performance in this domain. ChatClimate (Vaghefi et al., 2023) grounds GPT-4 responses in IPCC AR6 reports, showing that retrieval significantly improves accuracy. Hallucinations are identified, however, when queries extend beyond the IPCC's coverage. ChatNetZero (Hsu et al., 2024) applies a similar approach to net-zero policies, retrieving structured data on corporate and governmental pledges. While this helps ground responses, the model struggles with policy ambiguity.

Beyond policy analysis, LLMs are being explored for misinformation debunking. Generative Debunking of Climate Misinformation (Zanartu

Dataset	Question generation	Num QA pairs	Num documents	Paper Source	Domain	Question basis	
						Full text	Figs & tabs
FigureQA	Schema based	1.8M	140k	Synthetic	General	N	Y
BioAsq	Human experts	3.2K	–	PubMed	Biomedical	N	N
PubMedQA	Human experts	1K	120K abstracts	PubMed	Biomedical	Y	N
BioASQ-QA	Human experts	4.7K	–	PubMed	Biomedical	N	N
ArgSciChat	Human experts	41 dialogues	20 papers	arXiv	NLP	Y	N
ScienceQA	Human experts	21K	-	School curriculum	General	Y	Y
QASPER	Human experts	5K	1.5K papers	S2ORC	NLP	N	N
QASA	Human experts	1.8K	112 papers	S2ORC	AI/ML	Y	N
SPIQA	LLMs + Human experts	270K	25.5K papers	arXiv	Computer Sci.	Y	Y
ClimaQA-Gold	Human Experts	502	23	Textbooks	Climate Sci.	Y	N
ClimaQA-Silver	LLMs	3000	23	Textbooks	Climate Sci.	Y	N
CPIQA (ours)	LLMs	54.6k	4551 papers	core.ac.uk	Climate Sci.	Y	Y

Table 1: Comparison of relevant QA datasets over scientific literature: (Kahou et al., 2017), (Tsatsaronis et al., 2015), (Jin et al., 2019), (Krithara et al., 2023), (Ruggeri et al., 2023), (Lu et al., 2022), (Dasigi et al., 2021), (Lee et al., 2023), (Manivannan et al., 2024) (2)

et al., 2024) introduces claim classification and fallacy detection, structuring responses using a fact-myth-fallacy-fact framework. While this improves coherence, LLMs sometimes fail to select the most relevant counterarguments, leading to misdirected rebuttals.

My Climate Advisor (Nguyen et al., 2024) targets the specific domain climate adaptation in agriculture, retrieving information from peer-reviewed research, grey literature, and climate projection data. It tailors responses to regional climate risks, offering actionable insights for farmers. A key contribution is its expert-driven evaluation framework, which assesses responses across seven domain-specific criteria. Initial results highlight gaps in retrieval precision and the difficulty of adapting to evolving climate knowledge.

2.3 Retrieval-Augmented Generation

Effective retrieval-augmented generation (RAG) depends on retrieval quality, query formulation, and model alignment with retrieved knowledge. Traditional RAG pipelines perform a single retrieval step, which can fail when initial queries are too vague or incomplete (He et al., 2024). Recent research has explored iterative retrieval, query reformulation, and domain-specific adaptations to improve response accuracy.

CoRAG (Chain-of-Retrieval Augmented Generation) (Wang et al., 2025) introduces stepwise retrieval reasoning, allowing the model to dynamically reformulate queries based on retrieved evidence, significantly improving multi-hop QA. Similarly, RICHES (Retrieval Interlaced with Sequence Generation) (Jain et al., 2024) integrates retrieval within the decoding process, eliminating the need for a separate retriever module. This improves response fluency but can introduce hallucinations if retrieval is inconsistent.

Ensuring alignment between retrieved knowledge and generated responses is another key challenge. CoV-RAG (Chain-of-Verification RAG) (He et al., 2024) introduces a verification step that evaluates and refines retrieved documents before answer generation, reducing retrieval errors and hallucinations. RAGAR (RAG-Augmented Reasoning) (Khaliq et al., 2024) extends this further with hierarchical retrieval techniques (CoRAG and ToRAG - Tree-of-RAG) that decompose complex claims into sub-questions, retrieving evidence iteratively for fact-checking in multimodal political discourse.

Beyond reasoning techniques, RAG-Studio (Mao et al., 2024) focuses on domain-specific adaptation, addressing a major limitation of general-purpose RAG models. It introduces a self-alignment framework, where the retriever and generator co-train on synthetic domain-specific data, improving retrieval precision and factual grounding without requiring manually labeled examples. This approach outperforms traditional RAG fine-tuning in specialized domains such as law, finance, and biomedicine.

Our Context-RAG approach is motivated by previous work on multi-step query reformulation, but extending it to novelly focus on extracting distinct contexts in which the question can be re-framed to provide more diverse and user role-targeted retrieval.

3 Methods

3.1 CPIQA Dataset

To develop CPIQA, we curated a dataset of climate-related academic papers, integrating both textual and visual information for the RAG QA task.

We sourced papers from relevant open source climate science journals, identified by climate science expert recommendations. Using CrossRef, we retrieved the DOIs of all available articles from these journals published between 2020 and 2024. We sourced full-text PDFs from CORE.ac.uk (Knoth et al., 2023), an open-access repository of academic publications.

For each document, we extracted full text using *pymupdf4llm*, introducing a filter for documents with significant chunks of missing text. Figures and captions were extracted using *pdffigures 2.0* (Clark and Divvala, 2016), aligning with the CPIQA approach. We use GPT-4o (OpenAI et al., 2024) to generate detailed figure retrieval-friendly descriptions based on the extracted figure type, caption and raw image file. This allows for text-only embeddings to be used in a RAG setting, although image-caption pairs are included in the release.

We generated question-answer pairs by presenting GPT-4o with the full text and figure descriptions. We utilise role-based prompting, generating questions for the general public, climate experts and climate sceptics. Additionally, we generate multiple question types to encourage a wide breadth of questions. Full prompt variations can be found in appendix B.

3.2 Question-Answering Architecture

Our baseline two-stage RAG pipeline follows a standard retrieval approach, designed for comparability with SPIQA and evaluation of source attribution. The retriever embeds the user query, and retrieves relevant full text documents. These are used as a filter for the second stage, where the same query is used to retrieve chunks and figures from the filtered documents, maintaining continuity between chunks if required. Retrieved chunks and figure descriptions are inserted into a prompt template alongside the question, from which the LLM generates the answer.

We use *NovaSearch/stella_en_1.5B_v5* (Zhang et al., 2024) as our embedding model due to it being the highest ranked on the MTEB (Massive Text Embedding Benchmark) (Muennighoff et al., 2022) for the retrieval task with a minimum tokens of at least 100k+, which is a requirement for embedding the majority of documents in CPIQA. In cases where the document is longer than the max-tokens, we chunk the document, maximising token count.

3.3 Context-Based Query Expansion

Context-RAG first seeks to understand the context and intent behind the question. Instead of simply asking, "What do we need to know to answer this question?", our approach reframes it as, "What is the context of this question?" or "Why is this question being asked?". This decomposition enables retrieval that is broader, more targeted, and better aligned with the underlying information need.

The LLM breaks the input question into three distinct contextual perspectives, each represented as a descriptive paragraph, ensuring that retrieval is not biased toward a single interpretation. These are used as part of stage one - retrieval of full text documents. Further, we use the same LLM to break down each context into a set of domain-specific key terms that are up to a sentence in length. This gives finer granularity in the second stage of retrieval.

By shifting retrieval focus from the question itself to its underlying context, we hypothesize that Context-RAG improves recall, diversity, and factual grounding, ensuring that responses draw from a broader and more relevant evidence base. Further, this prompt structure can be applied prior to any other prompt decomposition or expansion method so should be seen as a complimentary method.

4 Results

We evaluate our proposed Context-RAG method against the standard two-step RAG approach across two datasets: SPIQA, a benchmark for scientific paper image question answering, and CPIQA, our newly introduced dataset for climate science. Performance is measured using BERTScore-F1 across multiple test cases and language models.

4.1 Context-RAG

Table 2 demonstrates the two-step RAG approach has a 7% lower BERTScore-F1 compared to the best open source models tested, and our Context-RAG a 3% lower score. Given our change in SPIQA problem formulation, from a one-step QA task where the relevant source paper is provided to a two step QA task where the source paper must be retrieved, this lower performance was expected. In the SPIQA dataset test-A contains LLM-generated

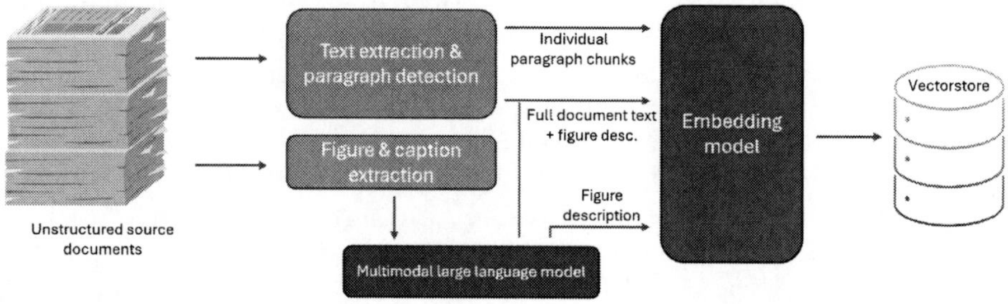

Figure 1: Generic pipeline used to construct CPIQA dataset & set up vectorstore prior to retrieval task

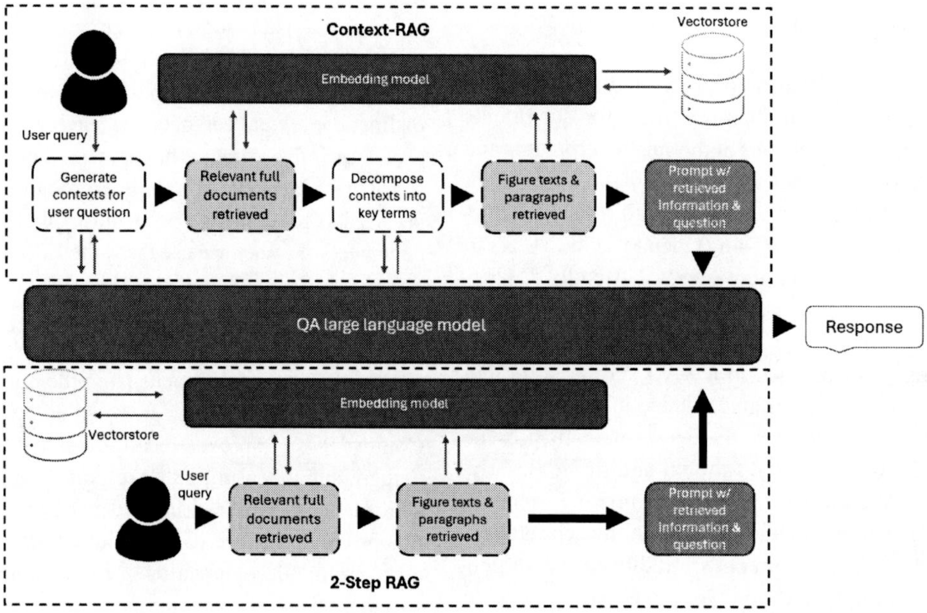

Figure 2: Architecture diagram demonstrating distinction between two-step RAG and Context-RAG

Test Case	Best open-weight baseline* (Pramanick et al., 2024)	2 step RAG	Context-RAG
test-A	61.61	57.54	**63.28**
test-B	47.21	53.22	**53.32**
test-C	**48.45**	32.27	34.20
Overall	**54.57**	47.85	51.31

Table 2: Comparison of our standard two-step RAG and Context-RAG methods on the SPIQA dataset, using *Llama-3.3-70B-Instruct*, compared to baseline results: *LLaVA-1.5-7B* (Liu et al., 2023) for test-A, test-B and *InstructBLIP-7B* (Dai et al., 2023) for test-C. *bert-base-uncased* is used as the evaluation model for BERT-score (Zhang* et al., 2020). *Baseline results experimental setup provides correct source paper, whereas our setup retrieves from the entire dataset.

QAs whilst test-B and test-C have human-written QAs. For two-step RAG we see a 6% improvement for test-B. With Context-RAG, we see an improvement of 4% over two-step RAG, outpefoming the best open source models in test-A by 2% and test-B by 6% showing the potential for our Context-RAG method.

4.2 Climate Question-Answering

A summary of our CPIQA dataset can be found in table 3. We define a train/test/validation split to improve comparability to future work that may use this data.

Split	Paper count	Question count	Figure count
Train	4255	51060	38325
Validation	99	1188	903
Test	197	2364	1816

Table 3: Summary of CPIQA dataset size incl. number of documents, questions and figures

LLM	2 step RAG	Context-RAG
GPT-4o	**67.18**	**69.10**
Gemini 2.0-flash	62.22	64.21
Llama-3.3-70B-Instruct	64.38	65.35
DeepSeek-R1-Distill-Qwen-32B	64.79	65.47
Gemma-2-27b-it	62.32	62.05

Table 4: Comparison of our standard two-step RAG and Context-RAG methods on our CPIQA dataset. Evaluated using BERT-score F1 using the model *microsoft/deberta-xlarge-mnli* (He et al., 2021)

On CPIQA (table 4), we compare both RAG methods across five LLMs. GPT-4o achieves the highest overall performance, with Context-RAG (69.10) slightly surpassing the two-step approach (67.18). Gemini 2.0-flash follows closely, showing a similar pattern, where retrieval based on generated contexts consistently improves results. Other models, such as *Llama-3.3-70B-Instruct* and *DeepSeek-R1-Distill-Qwen-32B*, show a smaller gap between the two approaches, suggesting that context informed retrieval benefits higher-capacity models more significantly.

Table 5 provides insights into the retrieval effectiveness of two-step RAG vs. Context-RAG when retrieving the specific source paper for GPT-4o. Interestingly, two-step RAG achieves a higher correct retrieval rate (60%) than Context-RAG (39%). However, despite retrieving the correct document less frequently, Context-RAG still yields a higher F1 score (70.96 vs. 68.71) which suggests the enhanced retrieved diversity of Context-RAG is allowing it to generate better overall answers.

4.2.1 Expert Evaluation

We asked academic climate science experts to evaluate our best performing model, GPT-4o, according to the qualitative citeria and scoring guidelines below:

- Answer precision: Degree of errors in the answer (1 - lots of errors, 5 - no errors). Unrelated to the question, consider only the answer independently of the question.
- Answer recall: To what degree does the response answer the question? Consider the relevance to the question (1 - irrelevant to the question, 5 - fully covers the question)
- Answer provenance: Is the answer using information from the source document? (1 = not based on context paper; 5 = fully based on context paper)
- Answer conciseness: Does the answer contain waffle or does it go off on a tangent to the question? (1 = verbose; 5 = concise)

The experts were given the question, generated answer, and full PDF source document. Due to expert availability, a random 6% sample of the test set was evaluated by our experts balanced by question type. Table 6 presents the expert evaluation of GPT-4o with Context-RAG, analyzing performance across different question audiences and types. Context-RAG achieves high conciseness scores across all audiences ($\geq$4.1), indicating its ability to generate succinct responses. Non-figure-based and numeric questions exhibit strong precision and recall, particularly for the climate expert role, where numeric questions achieve 4.1 precision and 4.7 recall. Questions generated using the climate expert role had significantly higher provenance scores, especially for numerical (4.6) questions, suggesting that the experts found the answers well-supported by evidence in the source paper. For the general public and climate sceptic roles, Context-RAG achieves moderate performance across all dimensions. Numeric questions for the climate sceptic role showed 3.7 precision and 4.1 recall, while figure-based and reasoning questions had slightly lower provenance scores (2.4-2.7), indicating some difficulty in tracing sources. For the general public role, provenance remains lowest for reasoning questions (2.4), suggesting challenges in aligning broad responses with domain-expert expectations. Overall, our expert qualitative evaluation results align with the

Method	Retrieval result	Retrieval rate %	BERTscore-F1
2 step RAG	Correct	60%	68.71
	Incorrect	40%	66.12
Context-RAG	Correct	39%	**70.96**
	Incorrect	61%	**67.97**

Table 5: Retrieval rate of the specific source paper for GPT-4o, and its corresponding BERTscore-F1 result. Retrieval result is defined as the retrieved papers containing the one based on which the question is generated. Retrieval rate is the frequency of how often the source paper is included in the retrieved documents

LLM	Question Audience	Question Type	Precision	Recall	Proven-ance	Concise-ness
GPT-4o Using Context-RAG (Best tested approach)	General public	Figure-based	3.6	3.7	2.8	4.9
		Numeric	2.9	3.6	3.0	4.6
		Non-fig	4.2	4.3	3.0	4.9
		Reasoning	3.4	3.7	2.4	4.7
	Climate sceptic	Figure-based	3.9	3.6	3.3	4.3
		Numeric	3.7	4.1	2.9	4.3
		Non-fig	3.4	3.3	2.4	4.4
		Reasoning	4.0	3.6	2.7	4.3
	Climate expert	Figure-based	3.9	3.6	3.7	4.1
		Numeric	4.1	4.7	4.6	4.8
		Non-fig	3.9	4.3	4.4	4.7
		Reasoning	4.0	4.4	4.4	4.4

Table 6: Expert evaluation of our best approach across roles and evaluation types on a scale of 1-5

trends demonstrated in the BERTscore-F1 results shown in table 7.

5 Discussion

5.1 Context-RAG vs two-step RAG: Retrieval vs Answer Quality

Our results highlight key differences between Context-RAG and the two-step RAG approach in terms of retrieval accuracy and answer quality. As shown in table 5, two-step RAG achieves a higher retrieval rate for the exact source paper (60% vs. 39%), while Context-RAG has a lower rate of exact source matches but produces slightly higher F1 scores in answer generation. This suggests that Context-RAG, despite not always retrieving the original source, provides sufficient and relevant information for generating high-quality answers.

One possible explanation for this is the nature of climate science literature, where overlapping factual content across multiple papers may reduce the importance of retrieving a specific source. Many academic papers cite and build upon each other, meaning that relevant information can often be found in multiple documents. Context-RAG's ability to extract and structure key concepts before retrieval may allow it to synthesize information from related sources, even if the exact original paper is not retrieved. This could explain its relatively strong answer quality despite a lower direct retrieval rate.

This trade-off is further reflected in our broader evaluation metrics. In our Climate QA setting (table 4), Context-RAG yields improved BERT-scores compared to two-step RAG, particularly for more complex questions. This indicates that selecting and structuring context before retrieval may contribute to better alignment with model-generated responses. However, two-step RAG's higher retrieval rate suggests it may be more reliable when strict source matching is a priority.

These findings suggest that retrieval rate alone is not always the best indicator of final answer quality. While two-step RAG more frequently retrieves the intended source, Context-RAG appears to generate answers that are at least as effective, if not more so, in terms of response accuracy.

5.2 Performance Across Different Models

Our evaluation shows that the performance of Context-RAG compared to two-step RAG, whilst generally better, varies across models. Larger models, such as GPT-4o and Gemini 2.0-flash, show greater improvements in answer quality with Context-RAG, suggesting that their enhanced reasoning capabilities allow them to make better use of retrieved context. For smaller models, the improvements are less pronounced, indicating that they may struggle to leverage retrieved information as effectively.

Notably, context generation can be done in addition to any other prompt augmentation or decomposition technique, though potential impact on performance is not evaluated in this work.

5.3 SPIQA vs CPIQA: Domain-Specific Insights

Comparing SPIQA and CPIQA, we observe distinct trends that highlight domain-specific retrieval challenges. Context-RAG demonstrates consistent improvements over two-step RAG across both datasets, but CPIQA remains more challenging due to domain-specific complexities. Specifically, climate science papers frequently cite each other and share overlapping facts, making it harder for retrieval models to isolate the most relevant document before evidence extraction. This is reflected in CPIQA's lower retrieval accuracy despite the improved context expansion.

The expert evaluation of Context-RAG on CPIQA suggests that provenance and precision are particularly important for climate science experts, as climate-related claims often require precise attribution to datasets, models, or prior research. In contrast, SPIQA, which focuses on interpreting structured results in computer science papers, may place relatively less emphasis on cross-document attribution and more on model reasoning over structured information. These differences suggest that retrieval and reasoning challenges may manifest differently across domains.

5.4 Breakdown by Question Type and Audience

Performance varies across different question types and target audiences, highlighting distinct challenges in retrieval and answer generation. As shown in table 7, numeric and figure-based questions benefit the most from Context-RAG, with consistent improvements across models. This suggests that retrieving structured, contextually relevant information before chunk selection is particularly useful for questions requiring precise data interpretation.

Reasoning-based questions show smaller gains, indicating that retrieval improvements alone may not fully address challenges in multi-step inference. This aligns with previous findings that complex reasoning tasks often depend more on a model's intrinsic capabilities than retrieval alone (Liu et al., 2024).

Audience-specific performance trends also reveal key insights. Questions targeted at climate experts generally yield the highest scores, suggesting that expert-level queries align well with retrieved academic content. In contrast, questions posed from a sceptic's perspective score lower, likely due to misalignment between the retrieved scientific literature and the framing of the question. This highlights the difficulty of addressing sceptical viewpoints in a fact-based retrieval system.

6 Conclusion

To support research in climate-focused QA, this paper introduces CPIQA, a dataset built from over 4,551 climate science papers and 54,612 GPT-4o generated question-answer pairs, integrating both text and figure-based question answering. CPIQA incorporates expert-informed question generation and multimodal evidence retrieval, making it a valuable resource for future work in climate AI.

We describe Context-RAG, a novel retrieval-augmented generation (RAG) approach that improves answer quality by structuring retrieval around contextual variations of a question. Unlike traditional RAG methods that directly retrieve documents based on the query, Context-RAG first generates multiple contextual perspectives, retrieves documents accordingly, and then refines retrieval using key domain-specific terms. Our evaluation on CPIQA, a new multimodal climate QA dataset described in this paper, and SPIQA, a scientific paper image QA benchmark dataset, demonstrates that Context-RAG outperforms the standard two-step RAG approach in answer quality, even when exact document retrieval rates are lower.

Our results show that Context-RAG improves performance across various question types and user audiences, particularly for numeric and figure-based questions. Larger models, such as GPT-4o, benefit most from this structured retrieval approach, leveraging contextually relevant evidence for improved reasoning. Furthermore, our expert evaluation of the best-performing model reinforces the effectiveness of Context-RAG in real-world climate science applications.

These findings highlight the importance of evidence-based QA methods. Future directions for this work include the exploration of domain-specific fine-tuning of RAG QA models, a more complete evaluation of the effectiveness of different RAG prompting techniques, and exploring enhancements to Context-RAG that are more explicitly tailored to our four different question types.

7 Limitations

Our GPT-4o generated question-answer pairs are sourced from single source documents, and do not consider answers that might span multiple documents. Other documents in our dataset may contradict or deviate from the source document and this is an exciting area for future work to explore, as we show with Context-RAG increased performance even when the specific source document was not retrieved.

Our CPIQA dataset has GPT-4o generated QA pairs. Whilst we performed a qualitative climate scientist expert evaluation for our RAG answers in terms of precision, provenance and conciseness, it was not feasible to perform expert analysis of the generated QA pairs themselves due to the size of our dataset and availability of our experts.

In this paper, we only use LLaMa-based models for evaluation on SPIQA due to time constraints. We expect our RAG results will generalize to any base LLM on any scientific paper QA task, but this paper has not explicitly confirmed this and we leave it as an item for future work. We did test CIPQA on five LLMs which strongly suggests our hypothesis for this is correct.

Our RAG experiments were run on eight H100 GPU cards using approximately 60 GPU hours of compute time. The GPT-4o QA pair generation took twelve hours and cost $550. We note that context-RAG is computationally more expensive than the 2-step method it is compared to. Further work is required to evaluate the complexity-performance trade-off.

Acknowledgments

This work was funded through the WCSSP South Africa project, a collaborative initiative between the Met Office, South African and UK partners, supported by the International Science Partnership Fund (ISPF) from the UK's Department for Science, Innovation and Technology (DSIT). It is also funded by the Natural Environment Research Council (grant NE/S015604/1) project GloSAT. The authors acknowledge the use of the IRIDIS High Performance Computing Facility, and associated support services at the University of Southampton, which significantly contributed to the completion of this work.

References

Christopher Clark and Santosh Divvala. 2016. Pdffigures 2.0: Mining figures from research papers.

Wenliang Dai, Junnan Li, Dongxu Li, Anthony Meng, Huat Tiong, Junqi Zhao, Weisheng Wang, Boyang Li, Pascale Fung, and Steven Hoi. 2023. Instructblip: Towards general-purpose vision-language models with instruction tuning.

Pradeep Dasigi, Kyle Lo, Iz Beltagy, Arman Cohan, Noah A. Smith, and Matt Gardner. 2021. A dataset of information-seeking questions and answers anchored in research papers. *NAACL-HLT 2021 - 2021 Conference of the North American Chapter of the Association for Computational Linguistics: Human Language Technologies, Proceedings of the Conference*, pages 4599–4610.

Bolei He, Nuo Chen, Xinran He, Lingyong Yan, Zhenkai Wei, Jinchang Luo, and Zhen-Hua Ling. 2024. Retrieving, rethinking and revising: The chain-of-verification can improve retrieval augmented generation. In *Findings of the Association for Computational Linguistics: EMNLP 2024*, pages 10371–10393, Miami, Florida, USA. Association for Computational Linguistics.

Pengcheng He, Xiaodong Liu, Jianfeng Gao, and Weizhu Chen. 2021. Deberta: Decoding-enhanced bert with disentangled attention. In *International Conference on Learning Representations*.

Angel Hsu, Mason Laney, Ji Zhang, Diego Manya, and Linda Farczadi. 2024. Evaluating ChatNetZero, an LLM-chatbot to demystify climate pledges. In *Proceedings of the 1st Workshop on Natural Language Processing Meets Climate Change (ClimateNLP 2024)*, pages 82–92, Bangkok, Thailand. Association for Computational Linguistics.

Lei Huang, Weijiang Yu, Weitao Ma, Weihong Zhong, Zhangyin Feng, Haotian Wang, Qianglong Chen, Weihua Peng, Xiaocheng Feng, Bing Qin, and Ting Liu. 2025. A survey on hallucination in large language models: Principles, taxonomy, challenges, and open questions. *ACM Transactions on Information Systems*, 43(2):1–55.

Palak Jain, Livio Baldini Soares, and Tom Kwiatkowski. 2024. From RAG to riches: Retrieval interlaced with sequence generation. In *Proceedings of the 2024 Conference on Empirical Methods in Natural Language Processing*, pages 8887–8904, Miami, Florida, USA. Association for Computational Linguistics.

Qiao Jin, Bhuwan Dhingra, Zhengping Liu, William Cohen, and Xinghua Lu. 2019. PubMedQA: A dataset for biomedical research question answering. In *Proceedings of the 2019 Conference on Empirical Methods in Natural Language Processing and the 9th International Joint Conference on Natural Language Processing (EMNLP-IJCNLP)*, pages 2567–2577, Hong Kong, China. Association for Computational Linguistics.

Samira Ebrahimi Kahou, Vincent Michalski, Adam Atkinson, Ákos Kádár, Adam Trischler, and Yoshua Bengio. 2017. Figureqa: An annotated figure dataset for visual reasoning. *6th International Conference on Learning Representations, ICLR 2018 - Workshop Track Proceedings*.

Mohammed Abdul Khaliq, Paul Yu-Chun Chang, Mingyang Ma, Bernhard Pflugfelder, and Filip Miletić. 2024. RAGAR, your falsehood radar: RAG-augmented reasoning for political fact-checking using multimodal large language models. In *Proceedings of the Seventh Fact Extraction and VERification Workshop (FEVER)*, pages 280–296, Miami, Florida, USA. Association for Computational Linguistics.

Petr Knoth, Drahomira Herrmannova, Matteo Cancellieri, Lucas Anastasiou, Nancy Pontika, Samuel Pearce, Bikash Gyawali, and David Pride. 2023. Core: A global aggregation service for open access papers. *Nature Scientific Data*, 10(1):366.

Anastasia Krithara, Anastasios Nentidis, Konstantinos Bougiatiotis, and Georgios Paliouras. 2023. Bioasq-qa: A manually curated corpus for biomedical question answering. *Scientific Data 2023 10:1*, 10:1–12.

Yoonjoo Lee, Kyungjae Lee, Sunghyun Park, Dasol Hwang, Jaehyeon Kim, Hong-in Lee, and Moontae Lee. 2023. Qasa: advanced question answering on scientific articles. In *Proceedings of the 40th International Conference on Machine Learning*, ICML'23. JMLR.org.

Stephan Lewandowsky. 2020. Climate change disinformation and how to combat it. *Annual Review of Public Health*, 42:1–21.

Haotian Liu, Chunyuan Li, Qingyang Wu, and Yong Jae Lee. 2023. Visual instruction tuning. *Advances in Neural Information Processing Systems*, 36.

Nelson F. Liu, Kevin Lin, John Hewitt, Ashwin Paranjape, Michele Bevilacqua, Fabio Petroni, and Percy Liang. 2024. Lost in the middle: How language models use long contexts. *Transactions of the Association for Computational Linguistics*, 12:157–173.

Pan Lu, Swaroop Mishra, Tony Xia, Liang Qiu, Kai Wei Chang, Song Chun Zhu, Oyvind Tafjord, Peter Clark, and Ashwin Kalyan. 2022. Learn to explain: Multimodal reasoning via thought chains for science question answering. *Advances in Neural Information Processing Systems*, 35.

Veeramakali Vignesh Manivannan, Yasaman Jafari, Srikar Eranky, Spencer Ho, Rose Yu, Duncan Watson-Parris, Yian Ma, Leon Bergen, and Taylor Berg-Kirkpatrick. 2024. Climaqa: An automated evaluation framework for climate foundation models.

Kelong Mao, Zheng Liu, Hongjin Qian, Fengran Mo, Chenlong Deng, and Zhicheng Dou. 2024. RAG-studio: Towards in-domain adaptation of retrieval augmented generation through self-alignment. In *Findings of the Association for Computational Linguistics: EMNLP 2024*, pages 725–735, Miami, Florida, USA. Association for Computational Linguistics.

Niklas Muennighoff, Nouamane Tazi, Loïc Magne, and Nils Reimers. 2022. Mteb: Massive text embedding benchmark. *arXiv preprint arXiv:2210.07316*.

Vincent Nguyen, Sarvnaz Karimi, Willow Hallgren, Ashley Harkin, and Mahesh Prakash. 2024. My climate advisor: An application of NLP in climate adaptation for agriculture. In *Proceedings of the 1st Workshop on Natural Language Processing Meets Climate Change (ClimateNLP 2024)*, pages 27–45, Bangkok, Thailand. Association for Computational Linguistics.

OpenAI, :, Aaron Hurst, Adam Lerer, Adam P. Goucher, Adam Perelman, Aditya Ramesh, Aidan Clark, AJ Ostrow, Akila Welihinda, Alan Hayes, Alec Radford, Aleksander Mądry, Alex Baker-Whitcomb, Alex Beutel, Alex Borzunov, Alex Carney, Alex Chow, Alex Kirillov, Alex Nichol, Alex Paino, Alex Renzin, Alex Tachard Passos, Alexander Kirillov, Alexi Christakis, Alexis Conneau, Ali Kamali, Allan Jabri, Allison Moyer, Allison Tam, Amadou Crookes, Amin Tootoochian, Amin Tootoonchian, Ananya Kumar, Andrea Vallone, Andrej Karpathy, Andrew Braunstein, Andrew Cann, Andrew Codispoti, Andrew Galu, Andrew Kondrich, Andrew Tulloch, Andrey Mishchenko, Angela Baek, Angela Jiang, Antoine Pelisse, Antonia Woodford, Anuj Gosalia, Arka Dhar, Ashley Pantuliano, Avi Nayak, Avital Oliver, Barret Zoph, Behrooz Ghorbani, Ben Leimberger, Ben Rossen, Ben Sokolowsky, Ben Wang, Benjamin Zweig, Beth Hoover, Blake Samic, Bob McGrew, Bobby Spero, Bogo Giertler, Bowen Cheng, Brad Lightcap, Brandon Walkin, Brendan Quinn, Brian Guarraci, Brian Hsu, Bright Kellogg, Brydon Eastman, Camillo Lugaresi, Carroll Wainwright, Cary Bassin, Cary Hudson, Casey Chu, Chad Nelson, Chak Li, Chan Jun Shern, Channing Conger, Charlotte Barette, Chelsea Voss, Chen Ding, Cheng Lu, Chong Zhang, Chris Beaumont, Chris Hallacy, Chris Koch, Christian Gibson, Christina Kim, Christine Choi, Christine McLeavey, Christopher Hesse, Claudia Fischer, Clemens Winter, Coley Czarnecki, Colin Jarvis, Colin Wei, Constantin Koumouzelis, Dane Sherburn, Daniel Kappler, Daniel Levin, Daniel Levy, David Carr, David Farhi, David Mely, David Robinson, David Sasaki, Denny Jin, Dev Valladares, Dimitris Tsipras, Doug Li, Duc Phong Nguyen, Duncan Findlay, Edede Oiwoh, Edmund Wong, Ehsan Asdar, Elizabeth Proehl, Elizabeth Yang, Eric Antonow, Eric Kramer, Eric Peterson, Eric Sigler, Eric Wallace, Eugene Brevdo, Evan Mays, Farzad Khorasani, Felipe Petroski Such, Filippo Raso, Francis Zhang, Fred von Lohmann, Freddie Sulit, Gabriel Goh, Gene Oden, Geoff Salmon, Giulio Starace, Greg Brockman, Hadi Salman, Haiming Bao, Haitang Hu, Hannah Wong, Haoyu Wang, Heather Schmidt, Heather Whitney, Heewoo Jun, Hendrik Kirchner, Henrique Ponde de Oliveira Pinto, Hongyu Ren,

Huiwen Chang, Hyung Won Chung, Ian Kivlichan, Ian O'Connell, Ian O'Connell, Ian Osband, Ian Silber, Ian Sohl, Ibrahim Okuyucu, Ikai Lan, Ilya Kostrikov, Ilya Sutskever, Ingmar Kanitscheider, Ishaan Gulrajani, Jacob Coxon, Jacob Menick, Jakub Pachocki, James Aung, James Betker, James Crooks, James Lennon, Jamie Kiros, Jan Leike, Jane Park, Jason Kwon, Jason Phang, Jason Teplitz, Jason Wei, Jason Wolfe, Jay Chen, Jeff Harris, Jenia Varavva, Jessica Gan Lee, Jessica Shieh, Ji Lin, Jiahui Yu, Jiayi Weng, Jie Tang, Jieqi Yu, Joanne Jang, Joaquin Quinonero Candela, Joe Beutler, Joe Landers, Joel Parish, Johannes Heidecke, John Schulman, Jonathan Lachman, Jonathan McKay, Jonathan Uesato, Jonathan Ward, Jong Wook Kim, Joost Huizinga, Jordan Sitkin, Jos Kraaijeveld, Josh Gross, Josh Kaplan, Josh Snyder, Joshua Achiam, Joy Jiao, Joyce Lee, Juntang Zhuang, Justyn Harriman, Kai Fricke, Kai Hayashi, Karan Singhal, Katy Shi, Kavin Karthik, Kayla Wood, Kendra Rimbach, Kenny Hsu, Kenny Nguyen, Keren Gu-Lemberg, Kevin Button, Kevin Liu, Kiel Howe, Krithika Muthukumar, Kyle Luther, Lama Ahmad, Larry Kai, Lauren Itow, Lauren Workman, Leher Pathak, Leo Chen, Li Jing, Lia Guy, Liam Fedus, Liang Zhou, Lien Mamitsuka, Lilian Weng, Lindsay McCallum, Lindsey Held, Long Ouyang, Louis Feuvrier, Lu Zhang, Lukas Kondraciuk, Lukasz Kaiser, Luke Hewitt, Luke Metz, Lyric Doshi, Mada Aflak, Maddie Simens, Madelaine Boyd, Madeleine Thompson, Marat Dukhan, Mark Chen, Mark Gray, Mark Hudnall, Marvin Zhang, Marwan Aljubeh, Mateusz Litwin, Matthew Zeng, Max Johnson, Maya Shetty, Mayank Gupta, Meghan Shah, Mehmet Yatbaz, Meng Jia Yang, Mengchao Zhong, Mia Glaese, Mianna Chen, Michael Janner, Michael Lampe, Michael Petrov, Michael Wu, Michele Wang, Michelle Fradin, Michelle Pokrass, Miguel Castro, Miguel Oom Temudo de Castro, Mikhail Pavlov, Miles Brundage, Miles Wang, Minal Khan, Mira Murati, Mo Bavarian, Molly Lin, Murat Yesildal, Nacho Soto, Natalia Gimelshein, Natalie Cone, Natalie Staudacher, Natalie Summers, Natan LaFontaine, Neil Chowdhury, Nick Ryder, Nick Stathas, Nick Turley, Nik Tezak, Niko Felix, Nithanth Kudige, Nitish Keskar, Noah Deutsch, Noel Bundick, Nora Puckett, Ofir Nachum, Ola Okelola, Oleg Boiko, Oleg Murk, Oliver Jaffe, Olivia Watkins, Olivier Godement, Owen Campbell-Moore, Patrick Chao, Paul McMillan, Pavel Belov, Peng Su, Peter Bak, Peter Bakkum, Peter Deng, Peter Dolan, Peter Hoeschele, Peter Welinder, Phil Tillet, Philip Pronin, Philippe Tillet, Prafulla Dhariwal, Qiming Yuan, Rachel Dias, Rachel Lim, Rahul Arora, Rajan Troll, Randall Lin, Rapha Gontijo Lopes, Raul Puri, Reah Miyara, Reimar Leike, Renaud Gaubert, Reza Zamani, Ricky Wang, Rob Donnelly, Rob Honsby, Rocky Smith, Rohan Sahai, Rohit Ramchandani, Romain Huet, Rory Carmichael, Rowan Zellers, Roy Chen, Ruby Chen, Ruslan Nigmatullin, Ryan Cheu, Saachi Jain, Sam Altman, Sam Schoenholz, Sam Toizer, Samuel Miserendino, Sandhini Agarwal, Sara Culver, Scott Ethersmith, Scott Gray, Sean Grove, Sean Metzger, Shamez Hermani, Shantanu Jain, Shengjia Zhao, Sherwin Wu, Shino Jomoto, Shirong Wu, Shuaiqi, Xia, Sonia Phene, Spencer Papay, Srinivas Narayanan, Steve Coffey, Steve Lee, Stewart Hall, Suchir Balaji, Tal Broda, Tal Stramer, Tao Xu, Tarun Gogineni, Taya Christianson, Ted Sanders, Tejal Patwardhan, Thomas Cunninghman, Thomas Degry, Thomas Dimson, Thomas Raoux, Thomas Shadwell, Tianhao Zheng, Todd Underwood, Todor Markov, Toki Sherbakov, Tom Rubin, Tom Stasi, Tomer Kaftan, Tristan Heywood, Troy Peterson, Tyce Walters, Tyna Eloundou, Valerie Qi, Veit Moeller, Vinnie Monaco, Vishal Kuo, Vlad Fomenko, Wayne Chang, Weiyi Zheng, Wenda Zhou, Wesam Manassra, Will Sheu, Wojciech Zaremba, Yash Patil, Yilei Qian, Yongjik Kim, Youlong Cheng, Yu Zhang, Yuchen He, Yuchen Zhang, Yujia Jin, Yunxing Dai, and Yury Malkov. 2024. Gpt-4o system card.

Shraman Pramanick, Rama Chellappa, and Subhashini Venugopalan. 2024. Spiqa: A dataset for multimodal question answering on scientific papers.

Federico Ruggeri, Mohsen Mesgar, and Iryna Gurevych. 2023. A dataset of argumentative dialogues on scientific papers. *Proceedings of the Annual Meeting of the Association for Computational Linguistics*, 1:7684–7699.

George Tsatsaronis, Georgios Balikas, Prodromos Malakasiotis, Ioannis Partalas, Matthias Zschunke, Michael R. Alvers, Dirk Weissenborn, Anastasia Krithara, Sergios Petridis, Dimitris Polychronopoulos, Yannis Almirantis, John Pavlopoulos, Nicolas Baskiotis, Patrick Gallinari, Thierry Artiéres, Axel Cyrille Ngonga Ngomo, Norman Heino, Eric Gaussier, Liliana Barrio-Alvers, Michael Schroeder, Ion Androutsopoulos, and Georgios Paliouras. 2015. An overview of the bioasq large-scale biomedical semantic indexing and question answering competition. *BMC Bioinformatics*, 16:1–28.

Saeid Ashraf Vaghefi, Dominik Stammbach, Veruska Muccione, Julia Bingler, Jingwei Ni, Mathias Kraus, Simon Allen, Chiara Colesanti-Senni, Tobias Wekhof, Tobias Schimanski, Glen Gostlow, Tingyu Yu, Qian Wang, Nicolas Webersinke, Christian Huggel, and Markus Leippold. 2023. ChatClimate: Grounding conversational AI in climate science. *Communications Earth & Environment 2023 4:1*, 4(1):1–13.

Liang Wang, Haonan Chen, Nan Yang, Xiaolong Huang, Zhicheng Dou, and Furu Wei. 2025. Chain-of-retrieval augmented generation.

Nicolas Webersinke, Mathias Kraus, Julia Bingler, and Markus Leippold. 2022. ClimateBERT: A Pretrained Language Model for Climate-Related Text. In *Proceedings of AAAI 2022 Fall Symposium: The Role of AI in Responding to Climate Challenges*.

Francisco Zanartu, Yulia Otmakhova, John Cook, and Lea Frermann. 2024. Generative debunking of climate misinformation. In *Proceedings of the 1st Workshop on Natural Language Processing Meets Climate*

Change (ClimateNLP 2024), pages 46–62, Bangkok, Thailand. Association for Computational Linguistics.

Dun Zhang, Jiacheng Li, Ziyang Zeng, and Fulong Wang. 2024. Jasper and stella: distillation of sota embedding models.

Tianyi Zhang*, Varsha Kishore*, Felix Wu*, Kilian Q. Weinberger, and Yoav Artzi. 2020. Bertscore: Evaluating text generation with bert. In *International Conference on Learning Representations*.

A Results breakdown

Table 7 breaks down the results of models on CPIQA by prompt variation.

B Prompts

B.1 Question-answer generation prompts

The general prompt template is as follows:

```
Context:
{full_text}
Figure 1 description:
{figure 1 description}
...
Figure i description:
{figure i description}
Instruction:
{instruction}
```

B.1.1 Instruction for public QA pair

```
As a lay member of the public, generate a single question-answer pair that are answered by the given academic document. {qtype} Use information from the descriptions of figures. Do not reference any part of the document directly. Do not refer to the study or any figure directly. Keep the question simple. Assume the user has never seen the document. Assume the asker knows little about climate science. The question could be written by a child. Answer such that a child will understand. Include a mix of basic factual, analytical and inferential questions. DO NOT MENTION THE CONTEXT DIRECTLY.
```

B.1.2 Instruction for expert QA pair

```
As an expert of the topic, and climate science generally, generate one meaningful question and its answer based on the context. qtype Use information from the descriptions of figures. Do not reference any part of the document directly. Do not refer to the study directly. The question may be asked with no knowledge of the document content.
```

B.1.3 Instruction for skeptic QA pair

```
Generate a single question-answer pair about the context as an extreme climate sceptic. Do not mention that you are a climate sceptic directly. qtype Include doubt, previous beliefs. Use information from the descriptions of figures. Do not reference any part of the document directly. Do not refer to the study directly. The question may be asked with no knowledge of the document content. Do not blindly agree with the critic's question. Demonstrate evidence to dispel scepticism. Give examples. Answers should be 1 paragraph or shorter.
```

B.1.4 Instruction addition for question types {qtype}

For figure based question:

```
The question should be answerable from the figure descriptions only but don't reference the figure or picture.
```

For numerical question:

```
The question should query a useful numerical value without mentioning the document or figure directly.
```

For reasoning based question:

```
The question should require
reasoning to answer.
```

For general questions, no additional prompt is used.

B.2 Question-answering prompts

B.2.1 QA template with context

```
You are an assistant for climate
research question-answering tasks.
Use the following pieces of
retrieved context to answer the
question. If you don't know the
answer, say that you don't know. Use
three sentences maximum and keep
the answer concise.
Retrieved information: {context}
Question: {question}
Answer:
```

B.2.2 Stage 1 contexts generation template

```
Given a question, describe in
detail 3 contexts or domains in
which it can be asked, explain
the contexts with a paragraph
each. Include titles of academic
documents that could be used in the
context. Give the contexts as 3
paragraphs with no headings.
Question: {question}
Contexts:
```

B.2.3 Stage 2 keyword generation template

```
Given a question and context
about the question, decompose the
question and context into a set of
relevant long-form query sentences
for evidence document retrieval
(RAG) that can answer the question.
Present each sentence on a newline
only with no headings.
Context: {context}
Question: {question}
Decomposed phrases:
```

C CPIQA paper sources

Table 8 sets out the source venues drawn from to develop the CPIQA dataset.

Large language model	Question Audience	Question Type	2 Step RAG (BERTScore-F1)	Context-RAG (BERTScore-F1)
OpenAI GPT-4o	General public	Numeric	73.67	76.65
		Figure-based	66.40	67.06
		Non-fig	64.25	67.10
		Reasoning	63.41	63.81
	Climate sceptic	Numeric	64.61	65.55
		Figure-based	64.36	66.15
		Non-fig	64.97	66.32
		Reasoning	64.97	66.39
	Climate expert	Numeric	78.48	81.34
		Figure-based	68.62	70.73
		Non-fig	67.69	69.92
		Reasoning	63.97	66.63
Google Gemini 2.0-flash	General public	Numeric	64.11	64.93
		Figure-based	61.81	63.70
		Non-fig	60.64	62.75
		Reasoning	59.28	61.63
	Climate sceptic	Numeric	60.84	62.35
		Figure-based	60.07	61.97
		Non-fig	60.35	62.23
		Reasoning	60.23	62.35
	Climate expert	Numeric	70.02	72.35
		Figure-based	64.66	67.18
		Non-fig	64.76	66.01
		Reasoning	60.04	62.35
Llama-3.3-70B-Instruct	General public	Numeric	63.64	72.11
		Figure-based	64.13	67.48
		Non-fig	63.00	64.81
		Reasoning	62.33	62.05
	Climate sceptic	Numeric	63.33	61.22
		Figure-based	62.93	61.16
		Non-fig	63.28	60.23
		Reasoning	63.14	60.04
	Climate expert	Numeric	70.26	77.59
		Figure-based	66.93	66.66
		Non-fig	66.90	66.10
		Reasoning	63.32	63.89
DeepSeek-R1-Distill-Qwen-32B	General public	Numeric	70.40	67.78
		Figure-based	65.05	65.16
		Non-fig	63.30	66.04
		Reasoning	62.25	61.48
	Climate sceptic	Numeric	62.80	63.56
		Figure-based	62.58	64.07
		Non-fig	63.16	64.28
		Reasoning	63.50	64.45
	Climate expert	Numeric	73.26	74.40
		Figure-based	63.75	64.43
		Non-fig	64.87	65.74
		Reasoning	61.16	63.31
gemma-2-27b-it	General public	Numeric	68.76	67.25
		Figure-based	64.13	63.99
		Non-fig	62.43	62.81
		Reasoning	58.20	58.94
	Climate sceptic	Numeric	60.82	60.74
		Figure-based	59.81	61.22
		Non-fig	61.35	62.24
		Reasoning	60.52	62.00
	Climate expert	Numeric	71.95	64.65
		Figure-based	60.60	63.17
		Non-fig	62.67	63.18
		Reasoning	53.15	54.59

Table 7: Evaluation of models across question types and RAG methods. Questions are divided into *numeric*, *figure bases*, *non-figure based* and *reasoning based*

Electronic ISSN	Title
1432-0894	Climate Dynamics
1573-1480	Climatic Change
1097-0088	International Journal of Climatology
1520-0442	Journal of Climate
1758-6798	Nature Climate Change
1752-0908	Nature Geoscience
1757-7799	WIRES Climate Change
2364-3587	Advances in Statistical Climatology, Meteorology and Oceanography
1814-9332	Climate of the Past
2190-4987	Earth System Dynamics
1866-3516	Earth System Science Data
2569-7110	Geoscience Communication

Table 8: ISSNs and venue titles of sources of drawn on for CPIQA

Robust Table Information Extraction from Sustainability Reports: A Time-Aware Hybrid Two-Step Approach

Hendrik Weichel[1,2], **Jörg Schäfer**[1], **Martin Simon**[1]
Frankfurt University of Applied Sciences[1]
University of Huddersfield[2]
Correspondence: hendrik.weichel@fra-uas.de

Abstract

The extraction of emissions-related information from annual reports has become increasingly important due to the Corporate Sustainability Reporting Directive (CSRD), which mandates greater transparency in sustainability reporting. As a result, information extraction (IE) methods must be robust, ensuring accurate retrieval while minimizing false values. While large language models (LLMs) offer potential for this task, their black-box nature and lack of specialization in table structures limit their robustness – an essential requirement in risk-averse domains. In this work, we present a two-step hybrid approach which optimizes both accuracy and robustness. More precisely, we combine a rule-based step for table IE with a regularized LLM-based step, both leveraging temporal prior knowledge. Our tests demonstrate the advantages of combining structured rules with LLMs. Furthermore, the modular design of our method allows for flexible adaptation to various IE tasks, making it a practical solution for industry applications while also serving as a scalable assistive tool for information extraction.

1 Introduction

Environmental, social, and governance (ESG) considerations have rapidly become central to corporate accountability and risk assessment. In the European Union, the Corporate Sustainability Reporting Directive (CSRD)[1] mandates that organizations disclose a variety of sustainability metrics in their annual or sustainability reports. While large public companies' data points are often available from data vendors, this is usually not the case for small and medium-sized enterprises (SMEs), whose reports frequently vary in format, presentation, and structure. At the same time, financial institutions, insurance companies, and other stakeholders increasingly require precise and reliable data, such as carbon emissions and other key indicators, to feed into quantitative risk models, in line with directives from bodies such as the European Banking Authority (EBA)[2].

Despite the growing volume of reported ESG data, extracting the relevant numerical values from heterogeneous documents remains a challenging task. In this work, we focus on the most common requirement of extracting numerical values from tabular structures. Many reports feature tables with inconsistent layouts, unstructured text, and varying terminologies, making standard IE methods prone to errors or heavy manual intervention. Furthermore, any inaccuracies in extracting emissions data or related metrics can lead to flawed risk assessments and regulatory non-compliance, underscoring the need for a highly robust, automated extraction pipeline.

To address these challenges, we propose a modular hybrid approach that regularizes LLM-based table IE by integrating domain expertise with temporal prior information. We demonstrate that combining rule-based techniques with machine learning models yields high accuracy, robustness, and scalability. Our table IE approach consists of two steps: A rule-based step that generates a candidate set containing the true information with high confidence and an LLM-based step that assists the user in selecting the most relevant element from this set. Our approach effectively addresses challenges such as mislabeled table headers, inconsistent data formats, and variations in corporate reporting styles. Most importantly, it reliably detects cases where the desired data point cannot be determined with confidence, ensuring transparency and trustworthiness in the extracted information. To the best of our knowledge, this is the first work to develop a table IE algorithm specifically tailored to the regulatory

[1]https://eur-lex.europa.eu/legal-content/EN/TXT/?uri=CELEX:32022L2464

[2]https://www.eba.europa.eu/publications-and-media/press-releases/eba-publishes-its-final-guidelines-management-esg-risks

Proceedings of the 2nd Workshop on Natural Language Processing Meets Climate Change (ClimateNLP 2025), pages 233–244
July 31, 2025

requirements of financial institutions.

The remainder of this paper is organized as follows. Section 2 provides an overview of existing research related to table IE. Section 3 presents the proposed methodology. Section 4 describes our experimental setup and the datasets used to evaluate performance. Section 5 summarizes our empirical results and discusses the practical implications for stakeholders. Finally, Section 6 concludes the paper by highlighting the method's potential benefits and directions for future research. All our data is available on Github[3].

2 Related Work

The analysis of annual reports for climate-related information is an active area of research. Webersinke et al. (2022) introduce ClimateBert, a deep learning model based on BERT. In Bingler et al. (2024), it is applied to detect climate-related cheap talk in annual reports. In Schimanski et al. (2023), it is used to detect corporate, national, and regional net zero and reduction targets. The OS-Climate initiative, hosted by The Linux Foundation, recognized the need to extract key emission data from annual reports to facilitate climate-aligned financial decision-making. To address this, their Data Commons project (OS-Climate, 2025) offers an NLP toolkit for table data extraction. Mishra et al. (2024) explores table IE of ESG metrics. Their methodology translates tables into structured text using sequence-to-sequence transformer models. LLMs are also being explored for extracting financial data from tables in corporate reports. Balsiger et al. (2024) evaluates ChatGPT-4 and BARD for extracting key financial figures, such as balance sheets and income statements, from PDF-based annual reports. Their study highlights the potential and limitations of LLMs in processing complex financial tables. Wang et al. (2023) and Lamott et al. (2024) demonstrate that enriching prompts with OCR-derived layout information improves LLM document understanding; however, neither approach explicitly targets robustness in table extraction. Looking at the more technical research about table IE, the study by Lu et al. (2024) gives an overview of current research about table related tasks for transformer-based language models. Before the advent of large-scale LLMs (i.e., models with fewer than one billion parameters), researchers sought to enhance table understanding through architectural modifications, improved encoding methods, and model fine-tuning (Herzig et al., 2020; Iida et al., 2021; Deng et al.). With the emergence of LLMs, two strategies became dominant: fine-tuning and prompt engineering. The inputs typically include metadata along with the full table contents and a task-specific instruction. A more recent advancement in LLM-driven table extraction involves agent-based methods, which utilize LLMs' reasoning capabilities. Techniques such as Chain-of-Thought (CoT) prompting (Wei et al., 2023) and ReAct prompting (Yao et al., 2023) enable iterative extraction, refining the data retrieval process through step-by-step reasoning.

Despite these promising developments, a research gap remains in ensuring the robustness of these methods in risk-averse application domains. Purely LLM-based approaches inherently lack this robustness: On the one hand, their statistical nature limits reliability, and on the other, their inherently one-dimensional input representations conflict with the two-dimensional structure of tables. At the same time, academic literature highlights a disconnect between industry and academia. Chiticariu et al. (2013) state that "while rule-based IE dominates the commercial world, it is widely regarded as dead-end technology by academia." They observe, however, that rule-based methods remain essential in the industry. Unlike purely statistical machine learning approaches, rule-based systems leverage expert knowledge to define explicit patterns (e.g., regular expressions, ontology schemas, or grammar rules) that target relevant information. Rule-based table IE has been explored more extensively in other domains. For example, Potvin et al. (2016) propose a position-based rule-based method that utilizes the spatial arrangement of text elements to infer relationships.

3 Methodology

Let $\mathcal{R}$ denote a finite set of company annual and sustainability reports. Suppose we aim to extract a numerical value $y_t \in \mathbb{R}$, where t represents the year of the report. An example of such a value, which will serve as our running example, is "Scope 3 emissions in 2023 (in tonnes CO2 equivalents)" from the report r_{2023}. An IE algorithm provides a function $f\colon \mathcal{R} \to \mathbb{R}$, where $f(r_t)$ represents the best estimate of the true value y_t contained in the report r_t. Our approach integrates both domain ex-

[3]https://github.com/hendrikweichel/hybrid_2_step_table_information_extraction

pertise and temporal prior information, leveraging validated data from previous reports of the same company, i.e., $r_{t-1}, \ldots, r_{t-n}$. Including such prior information into the IE pipeline can be interpreted as a regularization method, cf. Appendix A. Henceforth, the objective is to develop a reliable IE algorithm such that

$$f(r_t | r_{t-1}, \ldots, r_{t-n}) = y_t \quad \forall r_t \in \tilde{\mathcal{R}} \subset \mathcal{R}$$

where we use the notation $f(\cdot | r_{t-1}, \ldots, r_{t-n})$ to indicate the dependency of the function f on the parameters $r_{t-1}, \ldots, r_{t-n}$, with $|\tilde{\mathcal{R}}|$ as large as possible, while ensuring that

$$f(r_t | r_{t-1}, \ldots, r_{t-n}) = \infty \quad \forall r_t \in \mathcal{R} \setminus \tilde{\mathcal{R}}$$

to indicate cases where the function cannot reliably determine y_t. We base our approach on two key empirical observations made by domain experts analyzing sustainability reports:

(i) Emission data is almost always presented in tabular form.

(ii) Historical data provides a valuable prior for validating extracted values.

Thus, we assume that all emission values in $\mathcal{R}$ are stored in tables and define $\mathcal{T}(r_t) = \{\mathbf{T} | \mathbf{T} \in (\mathbb{R} \cup \Sigma)^{n \times m}; m, n \in \mathbb{N}\}$, where Σ represents textual values and $\mathbb{R}$ represents numerical values, as the set of all tables within report r_t. Our proposed IE function f is provided by a recursive approach, assuming that the relevant information has been successfully extracted and persisted in the previous years. In practice, the initial year is labeled manually. We follow a three-step pipeline:

1. **Table Extraction:** Given input r_t, extract the set of all tables $\mathcal{T}(r_t)$ using a table extraction method.

2. **Information Retrieval (IR):** Given the input r_t, ..., r_{t-n} (as well as the relevant tables and values extracted by the table extraction method in the previous years, see (iii) below), identify a table $\widehat{\mathbf{T}}(r_t) \in \mathcal{T}(r_t)$ that contains y_t (as well as y_{t-1} or even a longer history).

3. **Information Extraction (IE):** To extract the target value y_t, we apply a mapping

$$\widehat{\mathbf{T}}(r_t) \times \cdots \times \widehat{\mathbf{T}}(r_{t-n}) \times$$
$$y_{t-1} \times \cdots \times y_{t-n} \mapsto y_t.$$

$\widehat{\mathbf{T}}(r_{t-1})$, ..., $\widehat{\mathbf{T}}(r_{t-n})$ denotes the tables containing y_{t-1}, ..., y_{t-n} as extracted by the table extraction method in the previous years.

This paper focuses on step 3 of the pipeline, extracting information from tables, which is abbreviated as table IE in the following. While LLMs could, in principle, learn the complex mapping for table IE, $\widehat{\mathbf{T}}(r_t) \mapsto y_t$, there is one limiting factor making them unreliable for precise data extraction in regulatory settings: They are prone to hallucinations. This is further complicated by their inability to perceive the two-dimensional structure of tabular data due to their one-dimensional input format. To solve this problem, we present two distinct contributions:

1. A **rule-based table information extraction** approach to systematically extract y_t from $\widehat{\mathbf{T}}(r_{t-1})$. It exploits the historical knowledge about previous extractions and selects a candidate set of l possible solutions

$$\{\widehat{y}_t^{(1)}, \ldots, \widehat{y}_t^{(l)}\} \in \widehat{\mathbf{T}}(r_t)$$

that has a high probability of uniquely containing y_t and a low probability of only returning candidates different from y_t. Applications that do not allow the use of LLMs, can apply this rule-based table IE like so:

$$f(r_t | r_{t-1}, \ldots, r_{t-n}) = \begin{cases} \widehat{y}^{(1)}, & l = 1 \\ \text{None}, & \text{else} \end{cases}$$

2. A **hybrid two-step table information extraction** approach expands the rule-based table IE by leveraging the candidate set to regularize table IE with LLMs. We demonstrate in Section 3.2 below that the mapping

$$\widehat{\mathbf{T}}(r_t) \times \{\widehat{y}_t^{(1)}, \ldots, \widehat{y}_t^{(l)}\} \times y_{t-1} \mapsto y_t$$

can be implemented through LLMs, both optimizing the robustness and accuracy of standard table IE through LLMs. We show that the rule-based pre-processing serves as a regularization mechanism for the LLM's table IE task. Still, given their black-box character, such a hybrid approach should assist in manual extraction rather than a fully automated solution in domains that require maximum robustness.

Note that our IE process is both recursive and highly modular, enhancing its flexibility and reliability. We extensively leverage this modularity

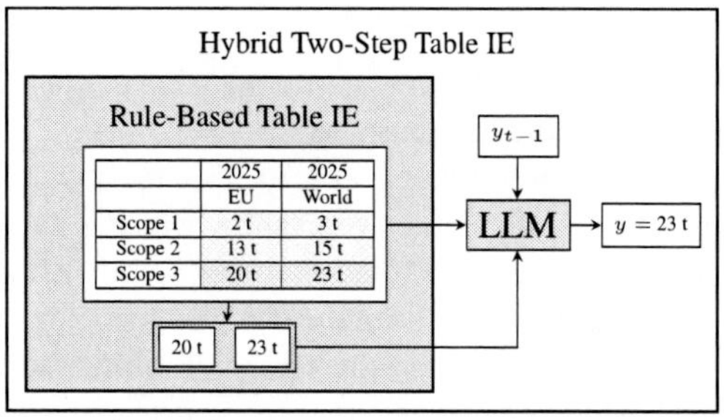

Figure 1: Illustration of our two contributions: (1) a rule-based table IE approach, and (2) a hybrid table IE method that builds upon (1) by leveraging its output.

Algorithm 1 Computation of scores for table cells

Require: $\mathbf{M}, \mathbf{O} \in \mathbb{R}^{n \times m}$
for k in $1, \ldots, q$ **do**
  for $\widehat{\mathbf{T}}(r_t)_{i,j} \in \widehat{\mathbf{T}}(r_t)$ **do**
    $\mathbf{M}_{i,j} \leftarrow S^{(k)}(Q^{(k)}, \widehat{\mathbf{T}}(r_t)_{i,j})$
  end for
  $\mathbf{v} \leftarrow \max(\mathbf{M}, \dim = d^{(k)})$
  $\mathbf{M}_{select} \leftarrow \text{tile}(\mathbf{v}, \text{shape} = \mathbf{M}(r_t).\text{shape})$
  $\mathbf{M}_{max} \leftarrow \mathbf{M}.\text{where}(\mathbf{M}_{i,j} = \max(\mathbf{M}))$
  $\mathbf{O} \leftarrow \mathbf{O} + \mathbf{M}_{select} - \mathbf{M}_{max}$
end for

to optimize our method for robustness, ensuring a low probability of incorrect outputs. Instead of returning erroneous results, the system is designed to return None when confidence is insufficient.

3.1 Rule-Based Table IE

Purely LLM-based table IE methods fail to utilize the two-dimensional nature of tables, cf. (Lu et al., 2024), and as a result, they overlook the implicit knowledge embedded within the matrix structure of tables $\widehat{\mathbf{T}}(r_t) \in (\mathbb{R} \cup \Sigma)^{n \times m}$. Our approach takes into account this knowledge by individually scoring all columns and rows based on their alignment with the target extraction. The different scoring methodologies tested in this work are presented in Section 3.1.1 below. Ultimately, cells that intersect in both the highest-scoring columns and rows are selected as the candidate set of values $\{\widehat{y}^{(1)}, \ldots, \widehat{y}^{(l)}\} \subset \{\widehat{\mathbf{T}}(r_t)_{i,j} \mid i \in \{1, \ldots, n\}, j \in \{1, \ldots, m\}\}$. To ensure that the candidate set contains y_t with high confidence, we gather a set of constraints $\{C_1, \ldots, C_q\}$ that apply to all the columns and rows that contain y_t. Such as, e.g., y_t always lies in a column which is annotated with the year t.

Algorithm 1 outlines the process of generating the candidate set, which is further illustrated in Figure 2. Based on the constraints, the algorithm assigns a score to each cell $\widehat{\mathbf{T}}(r_t)_{i,j}$ expressed in the score matrix $\mathbf{O} \in \mathbb{R}^{n \times m}$. Each constraint C_k, $k \in \{1, \ldots, q\}$, is formalized as a triplet

$$C^{(k)} = (Q^{(k)}, S^{(k)}(c, Q), d^{(k)}),$$

where $Q^{(k)}$ is the query, e.g., the year of the searched emissions; $S^{(k)}(c, Q)$ is a similarity metric, that calculates the similarity score between a cell c and the respective query $Q^{(k)}$; and $d^{(k)}$ specifies the application orientation of $Q^{(k)}$ and $S^{(k)}$, indicating whether they are applied across rows or columns. These constraints encapsulate all prior knowledge about the target extraction that can be derived from $\widehat{\mathbf{T}}(r_{t-1}) \times \cdots \times \widehat{\mathbf{T}}(r_{t-n})$.

Besides the constraints for rows and columns, we apply additional constraints on the individual cell level. If the cell $\widehat{\mathbf{T}}(r_t)_{i,j}$ does not match the format of our target extraction, we set the corresponding score $\mathbf{O}_{i,j}$ to zero. In our example, where the goal is to retrieve numerical emission values, we exclude all cells that do not contain numbers or that include financial figures and percentages, as indicated by their corresponding units (€, \$, £, %).

As a final step, the cells of the table $\widehat{\mathbf{T}}(r_t)$ with the highest scores in $\mathbf{O}$ are selected as the candidate set:

$$\{\widehat{y}^{(1)}, \ldots, \widehat{y}^{(l)}\} = \{\widehat{\mathbf{T}}(r_t)_{i,j} | \mathbf{O}_{i,j} = \max(\mathbf{O})\}$$

In production practice, an additional layer for identifying implausible results could be implemented by leveraging the time series of target values $y_t, \ldots, y_{t-n}$. Candidate values $\widehat{y}^{(\cdot)}$ with a high deviation from the previous value y_{t-1} can be flagged as implausible. In practical terms, this involves calculating the difference between each candidate $\widehat{y}^{(\cdot)}$ and y_{t-1}, then flagging all candidates where $|\widehat{y}^{(\cdot)} - y_{t-1}|$ exceeds a predefined threshold.

The proposed modular and recursive design enables robust IE. More precisely, leveraging this modularity is essential for selecting robust similarity metrics and comprehensive constraint sets to accurately identify the row and column containing the target value y_t. As demonstrated in 4.2.1, where we conduct a cross-validation, this approach ultimately increases the likelihood of retrieving y_t as a candidate value.

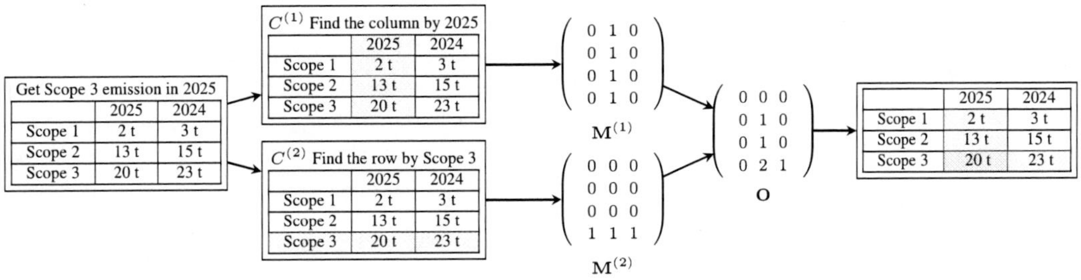

Figure 2: Flow chart of Algorithm 1.

3.1.1 Similarity Metrics

The similarity metrics take a query and a cell as input and assign a score between 0 and 1, reflecting the degree to which the cell matches the query, i.e.,

$$S(c, Q) \to [0, 1]$$

We perform cross-validation across several different metrics to determine the best-performing metrics for each query type; a comprehensive definition of all similarity metrics is given in Appendix B.

Regular expressions represent the simplest similarity metrics, applied either as exact string matching (*"is Q in c?"*) or in combination with pre-processing methods: We examine pre-processing through only selecting the numerical sub-strings of the query and the cell, and then carry out the string matching. Furthermore, one can tokenize the query into subqueries and compute the share of subqueries that are contained in the cell. This leaves more degrees of freedom for the structure of the cell strings and enables the use of continuous scores between 0 and 1. For the same reason, we examine the set-based **Jaccard similarity** and the **Levenshtein distance**. Both, in theory, accept minor dissimilarities between cell and query and could lead to a higher precision.

Additionally, we evaluate **semantic vector-based matching**. Techniques such as Word2Vec (Mikolov et al., 2013) and transformer-based word embedding models[4] have demonstrated strong performance in measuring similarity. These models assign vectors to sentences, enabling similarity measurement based on the comparison of their vector embeddings. A drawback of these machine learning models is their black-box nature and higher computational cost compared to the previously discussed methods.

Numerical metrics can be used to compare cells and queries containing numerical content. To do so, both the cell and the query are converted to floats. We test both, a numerical metric that returns the percent-wise deviation from the cell value to the query value and one binary numerical metric that returns 1 if the absolute difference is smaller than a given threshold and 0 else.

In Section 4, we perform cross-validation to determine which similarity metric fits best for which type of query. Appendix B formally defines all the queries we tested.

3.2 Hybrid Two-Step Table IE

If the rule-based table IE does not return a unique candidate, its candidate set can be used to assist table IE with LLMs. A straightforward table IE task would instruct the LLM to return the target value y_t, given the table $\widehat{\mathbf{T}}(r_t)$. We shift this question and answer task to a regularized binary classification task: Given the table $\widehat{\mathbf{T}}(r_t)$ and the previous year's emission y_{t-1} (if contained in $\widehat{\mathbf{T}}(r_t)$), we instruct the LLM to select y_t from $\{\widehat{y}^{(1)}, \ldots, \widehat{y}^{(l)}\}$. Note that this approach offers a two-fold regularization of the problem: first, by incorporating prior information (cf. Appendix A), and second, by constraining the solution space. This enhances the robustness of table IE using LLMs. Our instruction prompt is structured as follows:

Table IE by Selection Prompt

Context: $\widehat{\mathbf{T}}(r_t)$
Instruction: Choose the element from the list of candidate lists that contains the total Scope 3 emissions in the year t given in the table $\widehat{\mathbf{T}}(r_t)$ in JSON format. The previous year's emissions were y_{t-1}, and it is likely that this year's emissions do not deviate significantly from y_{t-1}.
Candidate list: $\{\widehat{y}^{(1)}, \ldots, \widehat{y}^{(l)}\}$

[4]We use the models from Song et al. (2020) and Wang et al. (2020), with fine-tuning in `https://huggingface.co/sentence-transformers/all-mpnet-base-v2` and `https://huggingface.co/sentence-transformers/all-MiniLM-L6-v2`, respectively.

4 Experiments

This section presents our experiments on testing the table extraction approach using our running example of extracting Scope 3 emissions from financial institutions' annual reports. In these experiments, we used the following queries: (1) Filter the columns by the emission year t; (2) Filter rows by the emission type *"Scope 3"*; (3) Filter rows by the previous year's emission y_{t-1}; (4) Given the fact that the table structure frequently remains unchanged, with consistent row and column descriptions, we leverage this stability and use the name of the row in $\widehat{\mathbf{T}}(r_{t-1})$ that contains y_{t-1} as a query to filter the rows in $\widehat{\mathbf{T}}(r_t)$. In some cases, the first column can be None; we then take the first cell in the row that contains a textual value.

We test the purely rule-based table IE in two steps: First, we cross-validate several similarity metrics for each of the used query types to identify the robust metrics. Second, we choose the robust similarity metrics and combine them to test the creation of a candidate set. Here, we aim to validate that y_t is identified with high probability within the candidate set. We evaluate both the rule-based table IE and the hybrid two-step table IE approach against a benchmark – a straightforward LLM-based extraction.

4.1 Dataset

We test our approach by extracting Scope 3 greenhouse gas emissions from tables in the annual reports of Europe's largest banks. This represents a particularly relevant real-world scenario, as Scope 3 emissions constitute the most significant emission category for financial institutions, given that they encompass financed emissions. At the same time, Scope 3 emissions are notoriously difficult to quantify, often resulting in frequent restatements from year to year, thus providing an ideal testbed for our table IE approach. Note that this table IE methodology should, in a subsequent step, be integrated into a full IE pipeline, as outlined in Section 3. Since step two of this pipeline, Information Retrieval, ensures that the retrieved tables contain the emission, our dataset consists exclusively of reports including tables that contain the Scope 3 emissions.

For calibration and testing, we retrieved the 52 largest European banks by market capitalization and examined their annual reports between 2018 and 2023. The Scope 3 emissions were initially extracted manually from each report and tagged with their corresponding page numbers. These values represent the extraction target y_t. Using an AWS-based OCR system (see (EdenAI, 2025)), we extracted a set of candidate tables from the page that contains y_t. We then automatically selected only the table $\widehat{\mathbf{T}}(r_t)$ that contains y_t. Subsequently, we ensured that the structured tables accurately preserved the original formatting and structure as presented in the PDF versions of the annual reports. Any deviations from the original table structure were corrected manually, because the final pipeline must preserve layout fidelity while discarding only those tables that lack the target value y_t. Automatic detection of deviations will be explored in future work as part of the Information Retrieval step. The rule set was calibrated on a separate dataset drawn from a distinct group of banks.

4.2 Rule-Based Table IE

To evaluate the rule-based table IE, we adapt the notion of a binary classification that classifies each cell in the table $y \in \widehat{\mathbf{T}}(r_t)$ into one of two classes:

1. *positive*: $y \in \widehat{\mathbf{T}}(r_t)$ is a candidate for y_t due to the structure of $\widehat{\mathbf{T}}(r_t)$, these are all the candidates $\{\widehat{y}_t^{(1)}, \dots, \widehat{y}_t^{(l)}\}$.

2. *negative*: $y \in \widehat{\mathbf{T}}(r_t)$ is not a candidate for y_t due to the structure of $\widehat{\mathbf{T}}(r_t)$, these are all the elements in the complement set $\{y| \ y \in \widehat{\mathbf{T}}(r_t)\} \setminus \{\widehat{y}_t^{(1)}, \dots, \widehat{y}_t^{(l)}\}$.

That is, the predicted *positives* are the candidates, and the predicted *negatives* are all other elements in $\widehat{\mathbf{T}}(r_t)$. The *true* value is the extraction target y_t, and the *false* values are all other elements. This type of table IE is considered robust if it consistently includes y_t in the candidate set. Naturally, this may come at the cost of retrieving more false positives, resulting in a larger candidate set. In terms of the classification problem, our goal is to minimize false negatives and optimize recall. For example, if the candidate set contains the only element for y_t, the recall is 100.00%. Naturally, this introduces a recall-precision trade-off: including all elements $\{y| \ y \in \widehat{\mathbf{T}}(r_t)\}$ in the candidate set would result in a recall of 100% but a significantly lower precision score. A precision of 100.00% would occur, for instance, if the sole candidate $\widehat{y}^{(1)}$ is y_t. We additionally use the notion of *false positives only (FPO)*, which describes the share of extractions where only false positives were returned.

	Find column that contains y_t with query t			Find row that contains y_t with query y_{t-1}		
	recall	prec.	FPO	recall	prec.	FPO
Regex						
Complete	97.62	78.97	**0.00**	38.10	38.10	**0.00**
Numerical	**100.00**	78.97	**0.00**	38.10	38.10	**0.00**
Word Wise	97.62	78.97	**0.00**	45.24	40.66	**0.00**
Numerical Metrics						
Binary	80.95	71.03	**0.00**	40.48	40.48	**0.00**
Continuous	88.10	**79.76**	11.90	64.29	**64.29**	35.71
Step	95.24	76.97	4.76	**71.43**	53.97	28.57

Table 1: Test performance of numerical similarity metrics for the numerical queries to find the required rows and columns (cf. Section 4.2.1).

Given one particular table, FPO is 1, if a nonempty candidate set disjoint from $\{y_t\}$ is returned, and 0 otherwise. Recall, precision, and FPO report the average values across all extractions in the dataset.

4.2.1 Similarity Metrics Cross-Validation

The similarity metrics are used to find those rows or columns in $\widehat{\mathbf{T}}(r_t)$ that contain the extraction target y_t. In our running example, we use four different queries to do this. The cross-validation provided here evaluates a selection of similarity metrics (see Section 3.1.1 and Appendix B) with respect to their ability to individually identify the rows or columns in $\widehat{\mathbf{T}}(r_t)$ that contain y_t. Queries are classified into two categories: numerical and textual. Table 1 presents the results for **numerical queries**, specifically the year t and the previous year's emissions y_{t-1}; we apply numerical metrics and regular expressions. Table 2 presents the results for **textual queries**, including the emission type and the row name of the row in $\widehat{\mathbf{T}}(r_{t-1})$ that contains y_{t-1}, we apply several NLP similarity metrics such as simple regular expressions, Levenshtein distance, Jaccard similarity, and embedding-based similarities.

	Find row that contains y_t with emission type			Find row that contains y_t with prev. table's row name		
	recall	prec.	FPO	recall	prec.	FPO
Regex						
Complete	**100.00**	**89.84**	**0.00**	64.29	60.71	**0.00**
Word Wise	**100.00**	77.94	**0.00**	**100.00**	80.38	**0.00**
Levenshtein	57.14	45.36	42.86	83.33	79.76	16.67
Jaccard Similarity						
4-grams	71.43	69.05	28.57	88.10	84.52	11.90
5-grams	73.81	73.81	26.19	88.10	84.52	11.90
6-grams	80.95	80.95	19.05	90.48	**86.90**	9.52
7-grams	85.71	85.71	14.29	88.10	84.52	11.90
Embedding						
All MiniLM	59.52	59.52	40.48	85.71	82.14	14.29
MPNet Base	54.76	54.76	45.24	85.71	82.14	14.29
Word2Vec	40.48	40.48	59.52	88.10	84.52	11.90

Table 2: Test performance of textual similarity metrics for the textual queries (cf. Section 4.2.1).

4.2.2 Test Rule-based Table IE

To evaluate the proposed table IE approaches, we define the following constraint set, obtained from the most robust similarity metrics in the cross-validation, i.e.,

1. (t, Reg. Ex. Numerical, column)
2. (y_{t-1}, Numerical Binary, row),
3. ("Scope 3", Reg. Ex. Complete Strings, row),
4. (x_{t-1} row name, Reg. Ex. String-Level, row)

The average recall of the table IE experiments with this set of constraints was 100%, the average precision was 89.65% and the extraction uniquely identified y_t as the sole element in the candidate set in 80.95% of all extractions.

4.3 Hybrid Two-Step Table IE

Testing the full table IE, i.e., retrieving a single candidate for y_t rather than a set of candidates, involves a slightly different notion of false positives and false negatives than we used for the test of the rule-based table IE, since the result is no longer a set of candidates but either a single value for y_t or None. In this context, a *true positive extraction* is selecting the correct element y_t, selecting a candidate different from y_t is considered a *false positive*. A *false negative* when None was returned despite $\widehat{\mathbf{T}}(r_t)$ containing y_t. Analogously, *true negative* occurs when y_t is not contained in $\widehat{\mathbf{T}}(r_t)$ and None is correctly extracted. It is crucial to emphasize that, unlike the rule-based table IE in the first step, which focuses on minimizing false negatives when creating a candidate set, a robust second step that selects only one element prioritizes minimizing false positives, thereby optimizing precision. Table 3 presents benchmark results for a straightforward LLM-based table IE.

LLM	recall	prec.
GPT-4o	95.23	100.00
GPT-4o-mini	93.65	100.00
Deepseek r-1	90.91	95.65
llama 70b	90.48	100.0
llama 8b	86.11	83.78

Table 3: Benchmark for extracting y_t from $\widehat{\mathbf{T}}(r_t)$ with straightforward Table IE by LLMs.

Our methodology yielded the following results on the same dataset:

- The **rule-based table IE** achieved a **precision of 100%**, meaning that it never extracted an incorrect value for y_t. It also achieved a **recall of 80.95%**, indicating that in 80.95% of cases, the correct value y_t was extracted directly, while in the remaining 19.05% of cases, y_t was included in the candidate set.

- Our **hybrid two-step table IE** approach improved these results by utilizing an LLM to identify y_t within the candidate set generated by the rule-based method. For all LLMs listed in Table 3, i.e., GPT-4o, GPT-4o-mini, Deepseek r-1, llama 70b, and llama 8b, this approach successfully identified y_t, achieving both **precision and recall of 100%**.

5 Discussion

The cross-validation described in Section 4.2.1 enabled selecting the most robust similarity metrics, cf. Tables 1 and 2. Using regular expressions on numerical substrings is the most effective approach for identifying the column containing y_t, given the year t. It always identifies the right column and has a relatively high precision. We can also observe that identifying the row containing y_t given y_{t-1} works robustly using regular expressions and binary numerical metrics. Specifically, if y_{t-1} is present in $\widehat{\mathbf{T}}(r_t)$, our rule-based approach successfully detects it; otherwise, it correctly determines its absence. The fact that the latter case is observed rather frequently is not particularly surprising, given the fact that in our dataset companies' yearly Scope 3 emission restatements have a frequency of roughly 60%. However, through the EU's efforts to standardize sustainability reporting, it is likely that the frequency of restatements will decrease in the future. Table 2 presents the evaluation of textual metrics. The results indicate that identifying the correct row using textual metrics is highly robust when employing simple regular expressions. These methods consistently achieved a false positive only rate of 0% for both queries. However, finding the row based on the row name of y_{t-1} in the previous table did not achieve a 100% recall, suggesting that only 64.29% of row names remained unchanged from year to year. This issue is effectively addressed by word-level matching, which improves both precision and recall. The Levenshtein ratio and Jaccard similarity performed poorly, primarily because these metrics penalize differences in query and cell lengths, even when such variations do not affect the semantic meaning. Similarly, embedding-based similarities struggled because they treat numerically similar terms (e.g., "Scope 2" vs. "Scope 3") as nearly identical, leading to underperformance compared to simpler rule-based methods. In future work, we aim to explore how embedding-based similarities can be better adapted to improve performance. As a result of the cross-validation, we selected the four most robust similarity metrics and combined them to perform the rule-based table IE described in Section 4.2.2. We see that the recall is 100%, which means that the candidate set always contains the extraction target y_t. The proportion of extractions in our tests where the candidate set contained only y_t, i.e., exclusively returning true positives, is 80.95% for the most robust set of constraints. The average number of candidate values was below 1.5 for all sets of constraints. In summary, these results demonstrate that a robust IE can be ensured if the similarity metrics provide a consistently robust extraction. The tests described in Section 4.3 compared both steps of our table IE approach against a straightforward LLM-based approach. We observe that, in a table question-and-answer setting, only the models GPT-4o, LLaMA 70B, and GPT-4o-mini achieved a precision of 100%. In contrast, our hybrid two-step approach successfully performed information extraction with both recall and precision at 100%. These results demonstrate immense substantial gains in information extraction, especially for smaller LLMs such as llama 8B, thus highlighting the effectiveness of our regularization approach. Consequently, our approach enables the utilization of smaller, more cost-effective, and open-source models, enhancing accessibility and scalability. This factor is especially critical in the financial industry, which prefers open-source on-premise solutions and demands scalability.

6 Conclusion

In this paper, we presented a two-step hybrid table IE approach with a focus on robustness, making it well-suited for risk-averse application domains. As outlined in the problem statement, relying solely on an LLM is not feasible in such domains – an essential argument in favor of our approach. Additionally, candidate sets generated by our method include the extraction target with high probability, which can be leveraged to support manual data quality control and validation. We anticipate that evolving regulations for sustainability reporting will lead to higher data quality, greater consistency, and increased standardization. These trends further strengthen the effectiveness and applicability of the methodology presented in this paper.

Limitations

Our approach ensures robustness through customize the constraints of the extraction in a highly modular system. This is an advantage, however, it is important to exploit this customizability for other extraction tasks, i.e., it is important to specify queries and similarity metrics for other applications and / or other domains. A further limitation is that we tested the approaches for a rather small dataset and only used tables in a well-structured format. In future work, we plan to address these limitations.

Ethics Statement

No ethical concerns arise from the study, and all methodologies adhere to standard academic and scientific integrity principles. Additionally, no conflicts of interest are present, and the work complies with ethical guidelines for responsible research and publication.

Acknowledgements

The authors would like to express their gratitude to Ann Smith and George Bargiannis at University of Huddersfield, whose support and feedback significantly contributed to the development of this study.

Funding

This project (HA project no. 1647/23-200) is financed with funds of LOEWE – Landes-Offensive zur Entwicklung Wissenschaftlich-ökonomischer Exzellenz, Förderlinie 3: KMU-Verbundvorhaben (State Offensive for the Development of Scientific and Economic Excellence). Martin Simon would like to acknowledge support by the German Federal Ministry of Education and Research (BMBF) under Grant No. 03FHP191. The study used computing infrastructure sponsored by the Bundesministerium fur Bildung und Forschung (BMBF)/German Federal Ministry of Education and Research, Funding program Forschung an Fachhochschulen, KI@FRA-UAS.

References

David Balsiger, Hans-Rudolf Dimmler, Samuel Egger-Horstmann, and Thomas Hanne. 2024. Assessing large language models used for extracting table information from annual financial reports. *Computers*, 13(10).

Julia Anna Bingler, Mathias Kraus, Markus Leippold, and Nicolas Webersinke. 2024. How cheap talk in climate disclosures relates to climate initiatives, corporate emissions, and reputation risk. *Journal of Banking & Finance*, 164:107191.

Laura Chiticariu, Yunyao Li, and Frederick R. Reiss. 2013. Rule-based information extraction is dead! long live rule-based information extraction systems! In *Proceedings of the 2013 Conference on Empirical Methods in Natural Language Processing*, pages 827–832, Seattle, Washington, USA. Association for Computational Linguistics.

Xiang Deng, Huan Sun, Alyssa Lees, You Wu, and Cong Yu. TURL: Table Understanding through Representation Learning.

EdenAI. 2025. Ocr table parsing apis - eden ai. `https://www.edenai.co/feature/ocr-table-parsing-apis`. Accessed: 28-January-2025.

Jonathan Herzig, Pawel Krzysztof Nowak, Thomas Müller, Francesco Piccinno, and Julian Eisenschlos. 2020. TaPas: Weakly supervised table parsing via pre-training. In *Proceedings of the 58th Annual Meeting of the Association for Computational Linguistics*, pages 4320–4333, Online. Association for Computational Linguistics.

Hiroshi Iida, Dung Thai, Varun Manjunatha, and Mohit Iyyer. 2021. TABBIE: Pretrained representations of tabular data. In *Proceedings of the 2021 Conference of the North American Chapter of the Association for Computational Linguistics: Human Language Technologies*, pages 3446–3456, Online. Association for Computational Linguistics.

Marcel Lamott, Yves-Noel Weweler, Adrian Ulges, Faisal Shafait, Dirk Krechel, and Darko Obradovic. 2024. Lapdoc: Layout-aware prompting for documents.

Weizheng Lu, Jing Zhang, Ju Fan, Zihao Fu, Yueguo Chen, and Xiaoyong Du. 2024. Large Language Model for Table Processing: A Survey.

Tomas Mikolov, Kai Chen, Greg Corrado, and Jeffrey Dean. 2013. Efficient estimation of word representations in vector space.

Lokesh Mishra, Sohayl Dhibi, Yusik Kim, Cesar Berrospi Ramis, Shubham Gupta, Michele Dolfi, and Peter Staar. 2024. Statements: Universal information extraction from tables with large language models for esg kpis. In *Proceedings of the 1st Workshop on Natural Language Processing Meets Climate Change (ClimateNLP 2024)*, page 193–214. Association for Computational Linguistics.

OS-Climate. 2025. Data commons. `https://os-climate.org/data-commons/`. Accessed: 2025-03-01.

Benoit Potvin, Roger Villemaire, and Ngoc-Tan Le. 2016. A Position-Based Method for the Extraction of Financial Information in PDF Documents. In *Proceedings of the 21st Australasian Document Computing Symposium*, pages 9–16. ACM.

Tobias Schimanski, Julia Bingler, Mathias Kraus, Camilla Hyslop, and Markus Leippold. 2023. ClimateBERT-NetZero: Detecting and assessing net zero and reduction targets. In *Proceedings of the 2023 Conference on Empirical Methods in Natural Language Processing*, pages 15745–15756, Singapore. Association for Computational Linguistics.

Kaitao Song, Xu Tan, Tao Qin, Jianfeng Lu, and Tie-Yan Liu. 2020. Mpnet: Masked and permuted pre-training for language understanding.

Wenhui Wang, Furu Wei, Li Dong, Hangbo Bao, Nan Yang, and Ming Zhou. 2020. Minilm: Deep self-attention distillation for task-agnostic compression of pre-trained transformers.

Wenjin Wang, Yunhao Li, Yixin Ou, and Yin Zhang. 2023. Layout and task aware instruction prompt for zero-shot document image question answering.

Nicolas Webersinke, Mathias Kraus, Julia Anna Bingler, and Markus Leippold. 2022. Climatebert: A pretrained language model for climate-related text.

Jason Wei, Xuezhi Wang, Dale Schuurmans, Maarten Bosma, Brian Ichter, Fei Xia, Ed Chi, Quoc Le, and Denny Zhou. 2023. Chain-of-thought prompting elicits reasoning in large language models.

Sang Michael Xie, Aditi Raghunathan, Percy Liang, and Tengyu Ma. 2022. An explanation of in-context learning as implicit bayesian inference.

Shunyu Yao, Jeffrey Zhao, Dian Yu, Nan Du, Izhak Shafran, Karthik Narasimhan, and Yuan Cao. 2023. React: Synergizing reasoning and acting in language models.

A Including prior information as a regularization method

Let us demonstrate that for IE purely driven by an LLM, including temporal prior information may be interpreted as a regularization method in a strict mathematical sense. While the method we propose here is a hybrid method rather than purely driven by an LLM, this may still serve as a motivation for including prior knowledge to obtain more robust methods. Xie et al. (2022) study in-context learning for LLMs trained on a pretraining distribution given by a HMMM. They prove that, under this assumption, the LLM implicitely performs Bayesian inference. We define the sequence of training examples $\mathcal{S}_n = (S_1, ..., S_n)$ such as "Scope 1 emissions in 2021 were ?? t", "Scope 1 emissions in 2020 were ?? t", and the test prompt x_{test} ="Provide the Scope 1 emissions in the year 2023 in the unit t".

The first step in in our framework provides an additional chunk $\mathcal{C}$ of text from the text corpus r_t which is appended to the training examples to obtain

$$\widetilde{\mathcal{S}}_n = (\mathcal{S}_n, \mathcal{C}).$$

Therefore, Equation 5 in Xie et al. (2022) becomes

$$\begin{aligned} p(y \mid \widetilde{\mathcal{S}}_n,\, x_{\text{test}}) \;\propto\; & \int_\theta \sum_{h \in \mathcal{H}} p(y \mid x_{\text{test}},\, h,\, \theta) \\ & \times p\big(\widetilde{\mathcal{S}}_n, x_{\text{test}} \mid \theta\big) \\ & \times p\big(h \mid \widetilde{\mathcal{S}}_n, x_{\text{test}}, \theta\big)\; p(\theta)\; \mathrm{d}\theta. \end{aligned}$$

In this setting, the prior $p(\theta)$ encodes the LLM's pretrained distribution. Including $\widehat{\mathcal{S}}$ in addition to the test prompt updates the model's posterior by answering the question which parts of the parameter space and which hidden states $h \in \mathcal{H}$ are most relevant with regard to the inputs, thus preventing the model from drifting to irrelevant states or modes.

B Similarity Metrics

Each similarity metric has inputs query and cell and returns a value between 0 and 1.

B.1 Regular Expression

To evaluate whether a query is contained within a string, we implemented five complementary approaches.

B.1.1 Complete Word Matching

The first approach converts both the query and cell string to lowercase and checks if the query is contained within the cell. It returns 1 for a match and 0 for no match.

B.1.2 Numerical Substring Matching

The second method extracts the numeric characters from both the query and cell string, then checks if the query's numbers appear in the cell. It returns 1 for a match and 0 for no match.

B.1.3 Word-Level Matching

The fourth method splits the query into words, converts them to lowercase, and calculates the fraction of words found in the target string. The result ranges from 0 (no matches) to 1 (all words matched).

B.2 Levenshtein Ratio

This method is based on the Levenshtein distance, which quantifies the minimum number of single-character edits – insertions, deletions, or substitutions – required to transform one string into another. To normalize this distance, the Levenshtein ratio divides it by the maximum possible string length, yielding a similarity score between 0 and 1, where 1 indicates identical strings and 0 denotes no similarity. This approach is particularly useful for handling minor spelling variations, typos, and fuzzy matching, making it a robust technique for evaluating approximate string containment.

B.3 Jaccard Similarity

The Jaccard-Coefficient is a statistic used for the similarity of two sets. In NLP this statistic is used to yield a similarity score between 0 and 1. During preprocessing, we create separate both the query and the cell into the sets A and B of strings with length n. Then we calculate the Jaccard similarity as so:

$$S(A, B) = \frac{|A \cap B|}{|A \cup B|}$$

B.4 Numerical Comparison

Here, we preprocess the strings such that we obtain the quantity and unit. (*"20,000 t CO2"* $\rightarrow$ *Quantity: 20000, Unit: "t CO2"*). If there is no unit, we directly compare the two quantities. To create a similarity score between the two quantities a and b, we use the following methods.

B.4.1 Binary Comparison

The binary comparison returns the score 1 if the absolute difference between a and b is smaller than 1. Else it returns 0.

$$S(a, b) = \begin{cases} 1, & |a - b| < 1 \\ 0, & \text{else} \end{cases}$$

B.4.2 Continuous Comparison

To allow minor differences between a and b we use a continuous function. It returns the relative difference between with respect to a.

$$S(a, b) = max\left(\frac{|a - b|}{a}, 0\right)$$

B.4.3 Step Function

The step function is an extension of the continuous function.

$$S(a, b) = \begin{cases} 0, & |a - b| < 1 \\ 0.9, & \frac{|a-b|}{a} < 0.1 \\ 0.8, & \frac{|a-b|}{a} < 0.2 \\ 0.6, & \frac{|a-b|}{a} < 0.4 \\ 0.4, & \frac{|a-b|}{a} < 0.6 \\ 0.2, & \frac{|a-b|}{a} < 0.8 \\ 1.0, & \text{else} \end{cases}$$

B.4.4 Word Embeddings and Cosine Similarity

To calculate the similarities between two words a and b, we first generate the word embeddings with the given model $\mathbf{e_a}$ and $\mathbf{e_b}$. Then we define the similarity of a and b as:

$$S(a, b) = \frac{\mathbf{e}_a \cdot \mathbf{e}_b}{\|\mathbf{e}_a\| \|\mathbf{e}_b\|},$$

where $e_a \cdot e_b$ denotes the scalar product.

C Measurements for Experiments

These are the formal definitions for the experiments to test the rule-based approach to create a candidate set:

$$\text{Average Precision} = \frac{1}{N} \sum_{i=1}^{N} \frac{TP_i}{TP_i + FP_i}$$

$$\text{Average Recall} = \frac{1}{N} \sum_{i=1}^{N} \frac{TP_i}{TP_i + FN_i}$$

$$\text{FPO} = \sum_{i=1}^{N} \mathbb{1}(TP_i = 0 \wedge FN_i = 0 \wedge FP_i > 0),$$

where N is the total number of test extractions, and TP_i, FP_i, FN_i correspond to the counts for extraction i.

These are the formal definitions for the experiments to test the full table IE:

$$\text{Precision} = \frac{TP}{TP + FP}$$

$$\text{Recall} = \frac{TP}{TP + FN}$$

D Prompting

We used the following prompt for our benchmark of table IE:

Benchmark Table IE Prompt

System: Help to extract the total Scope 3 emissions in the year t from a table given below.
Human: Therefore, choose the best answer for the given context. And fill in the json format: "Scope 3": <Scope 3 emissions>, where <Scope 3 emissions> is a string of the Scope 3 emission with unit.
Context: $\widehat{\mathbf{T}}(r_t)$
Question: What are the total scope 3 emissions in the year t given in the table?

We used the following prompt for our hybrid two-step table IE:

Table IE with candidate set

System: Help to extract the total Scope 3 emissions in the year t from a table given below.
Human: Help to extract the total Scope 3 emissions in the year t from a table given below from a preselection of possible answers. The previous year's emissions were y_{t-1}, and it is likely that this year's emissions do not deviate significantly from y_{t-1}. Therefore, choose the best answer for the given context out of the set of possible answers. And fill in the json format: {"Scope 3": <Scope 3 Emission>}, where <Scope 3 Emission> is a string of the Scope 3 Emission with unit.
Context: $\widehat{\mathbf{T}}(r_t)$
Question: What are the total Scope 3 emissions in the year t given in the table?
Select one of these possible answers $\{\widehat{y}^{(1)}, \ldots, \widehat{y}^{(l)}\}$ and make sure that it keeps the JSON format.

Listen to the Context: Towards Faithful Large Language Models for Retrieval Augmented Generation on Climate Questions

David Thulke[1,2] **Jakob Kemmler**[1,2] **Christian Dugast**[2] **Hermann Ney**[1,2]

[1]Machine Learning and Human Language Technology, RWTH Aachen University, Germany
[2]AppTek GmbH, Aachen, Germany
{thulke,jakob.kemmler,ney}@hltpr.rwth-aachen.de, cdugast@apptek.com

Abstract

Large language models that use retrieval augmented generation have the potential to unlock valuable knowledge for researchers, policymakers, and the public by making long and technical climate-related documents more accessible. While this approach can help alleviate factual hallucinations by relying on retrieved passages as additional context, its effectiveness depends on whether the model's output remains faithful to these passages. To address this, we explore the automatic assessment of faithfulness of different models in this setting. We then focus on ClimateGPT, a large language model specialised in climate science, to examine which factors in its instruction fine-tuning impact the model's faithfulness. By excluding unfaithful subsets of the model's training data, we develop ClimateGPT Faithful+, which achieves an improvement in faithfulness from 30% to 57% in supported atomic claims according to our automatic metric.

1 Introduction

As the urgency of climate action intensifies (Calvin et al., 2023), researchers, policymakers, and the public require efficient access to reliable climate information. Large language models (LLMs) have emerged as powerful tools for processing and summarizing vast amounts of climate-related literature. However, these models often suffer from factual hallucinations, where generated responses contain incorrect or misleading information. One promising approach to mitigating this issue is Retrieval Augmented Generation (RAG), which enables models to ground their responses in retrieved passages from trusted sources. Several climate-related RAG-based systems have been developed, including ChatClimate (Vaghefi et al., 2023), ClimateGPT (Thulke et al., 2024), and ClimateQ&A (Calzada et al., 2024). Despite these advancements, challenges remain in ensuring that generated outputs remain faithful to the retrieved content.

Prompt
What is the purpose of the Global Stocktake?

Context
[0] Recalling Article 14 of the Paris Agreement, the objective of the Global Stocktake is to assess the collective progress towards achieving [...]
[1] · Paragraph 34 of 19/CMA.1 decides that the outputs of this component of the global stocktake should: a) Identify opportunities for and [...]
[2] The SBSTA and the SBI invited submissions on views on the approach to the consideration of outputs component of the first global [...]

ClimateGPT 7B
The Global Stocktake is a key component of the Paris Agreement, scheduled to begin in 2023. It involves a comprehensive assessment of global progress in addressing climate change, considering scientific evidence, economic implications, and societal impacts. The Stocktake aims to inform the updates to countries' climate commitments, known as "nationally determined contributions" (NDCs), to ensure they are in line with the agreement's objectives and targets. [...]

ClimateGPT 7B Faithful+
The Global Stocktake's purpose, as outlined in Article 14 of the Paris Agreement, is to assess the collective progress made towards achieving the Agreement's purpose and long-term goals, which encompass mitigation, adaptation, and means of implementation. This assessment is conducted considering equity and the best available science. The Global Stocktake process spans two years, beginning with information collection and preparation and culminating in a consideration of [...]

Figure 1: Shortened example from the original ClimateGPT and the Faithful+ variant on one example from the Climate Policy Radar task. Text marked in red is not faithful, i.e. it is not supported by the given context. The full example is shown in Figure 3.

Proceedings of the 2nd Workshop on Natural Language Processing Meets Climate Change (ClimateNLP 2025), pages 245–259
July 31, 2025

Faithfulness, in this context, refers to the extent to which a model's response accurately reflects the information contained in the retrieved passages without introducing extraneous or misleading details. Importantly, factuality does not imply faithfulness. A response may be factually correct with respect to general world knowledge but still unfaithful if the information is not supported by the retrieved passages as shown in Figure 1. A lack of faithfulness undermines trust in these models, particularly in the climate domain, where misinformation has significant real-world consequences.

Moreover, we argue that faithfulness is even more important than general factuality in this setting, as large language models may inevitably hallucinate when faced with long-tail or rare knowledge. By requiring that all factual information in a response originates from the provided context, we can mitigate the risk of such hallucinations and ensure that model outputs are transparent, verifiable, and aligned with the available evidence. Thus, evaluating and improving faithfulness is a crucial step in enhancing the reliability of climate-focused LLMs.

In this work, we investigate methods for automatically assessing the faithfulness of RAG-based models in climate-related applications. We then focus on ClimateGPT (Thulke et al., 2024), a specialised open-weight LLM trained on climate-related texts to examine how different instruction fine-tuning (IFT) datasets influence faithfulness. By excluding parts of the training data with low faithfulness, we propose a new model ClimateGPT Faithful+ that on our main benchmark increases the percentage of supported claims from 30% to 57%.

2 Faithfulness and Factuality

Our definition of faithfulness and factuality follows the work of Dziri et al. (2022) and Huang et al. (2025). Given a question q, a set of N retrieved passages $K = (k_1, k_2, \ldots, k_N)$ from a knowledge base $\mathcal{KB}$, and a response r, we define faithfulness of r with respect to K as r should be supported by the information in K, i.e., r should not contain any information that contradicts the information in K or is not present in K. Factuality, on the other hand, refers to the correctness of the information in r with respect to general world knowledge. In our context, we assume that the relevant world knowledge is contained in $\mathcal{KB}$. Thus, we consider a response r to be factual if it is faithful to $\mathcal{KB}$.

2.1 Evaluation

To assess both the faithfulness and factuality of long-form responses, we build upon existing automated evaluation approaches, particularly RAGAs (Es et al., 2024) for faithfulness and FActScore (Min et al., 2023) and VeriScore (Song et al., 2024) for factuality. These methods share a common three-step pipeline: (1) claim decomposition, (2) evidence retrieval, and (3) claim verification. The main differences in evaluating for faithfulness versus factuality lie in the evidence retrieval step, as we describe below.

Claim Decomposition As long-form responses are typically composed of multiple claims, we first decompose the response into smaller and independent claims to simplify the subsequent steps. Given a response r, we decompose it into a set of claims $C = c_1, \ldots, c_I$. The definition of a claim and the granularity of the decomposition differs between different variants and use-cases. In this work, we use the claim decomposition method from RAGAs (Es et al., 2024) which prompts a large language model to decompose the full response into smaller claims in one step.

Evidence Retrieval The key distinction between evaluating faithfulness and factuality lies in this step. For faithfulness evaluation, we directly use the retrieved passages $K = k_1, k_2, \ldots, k_N$ from the RAG process as evidence. In contrast, for factuality evaluation, relevant evidence for each claim c_i is retrieved from a knowledge base $\mathcal{KB}$. In this work, we use the retrieval mechanism that is also used for RAG.

Claim Verification Finally, for each claim, we verify whether it is supported by the retrieved evidence. Therefore, we use an LLM to classify each claim c_i given the retrieved evidence (multiple retrieved evidence passages are concatenated into a single evidence). Similar to other work (Song et al., 2024), we do not differentiate between refuting and unrelated evidence. The overall faithfulness and factuality scores of r are then aggregated from these individual claim verifications by reporting the percentage of supported claims.

Implementation Details The exact prompts we used for each step are reported in Appendix B. GPT-4o (version `gpt-4o-2024-08-06`) is used as the large language model.

Model	#Tokens in Trillion	#Parameters in Billion	RAG	Avg. #Claims	Claim Support wrt. Ref. [%]	KB [%]
LLama 3.1 Instruct	15	8	-	22.7	-	59
			✓	17.3	**67**	**72**
LLama 2 Chat	2	7	-	23.3	-	60
			✓	21.2	48	65
ClimateGPT	2	7	-	21.6	-	59
			✓	21.1	30	61
ClimateGPT Faithful+ (ours)	2	7	-	20.2	-	57
			✓	19.2	<u>57</u>	<u>69</u>

Table 1: Results for claim support wrt. the reference, as a metric of faithfulness, and wrt. the knowledge base (KB) as a metric for factuality for different large language models with and without RAG.

3 ClimateGPT IFT Evaluation Task

We use the same evaluation dataset and RAG setup as Thulke et al. (2024) to evaluate the faithfulness and factuality of the generated responses. The test set is a held-out portion of the IFT data curated to train ClimateGPT. It was created in cooperation with domain experts and contains different open-ended tasks like QA, text generation, classification, chat, and brainstorming as well as closed-ended tasks like summarisation, extraction or rewrite. Our evaluation focuses on the subset of open-ended prompts of the held-out data (334 out of the 400 samples).

3.1 Information Retrieval

We use the dataset and retrieval pipeline as described by Thulke et al. (2024) for retrieving relevant contexts in our faithfulness evaluation. The dataset consists of climate-related documents from various sources, including IPCC reports and climate science related papers (see Appendix C for detailed statistics). For retrieval, we employ the `bge-large-en-v1.5` embedding model (Xiao et al., 2024) and a hierarchical retrieval strategy where we first retrieve the most relevant pages based on the query, selecting the top 5 ranked pages. Then, within these, we retrieve the top 5 most relevant 115-token snippets.

3.2 Large Language Models

We experiment with several language models in addition to ClimateGPT. As baselines, we include the 7B parameter variants of Llama 2 Chat (which shares the same foundation model as ClimateGPT) and Llama 3.1 Instruct. Further, we report results on the 70B parameters variants as well as on GPT-4o in Table 4 in the appendix. For all baseline models, we use a standardized RAG prompt that explicitly instructs the model to base its response solely on the provided references[1]. Both the user question and retrieved references are included within the user message to ensure a consistent evaluation setup. For ClimateGPT, we leverage its dedicated context role, which was introduced during training to optimize reference usage. We also use the model's default system prompt to align with its intended deployment configuration.

3.3 Results

We report the results with our faithfulness and factuality metrics for the small models in Table 1. Results of all models are reported in Appendix A. Overall, we observe that the more recent Llama 3.1 has significantly higher faithfulness than the predecessor Llama 2. For ClimateGPT, we observe that the faithfulness, as measured by claim support, is very low. Further, in contrast to the other models, using RAG with ClimateGPT does only slightly improve the claim support wrt. to the KB, i.e. the factuality. This is a strong indicator that the model does not make effective use of the provided paragraphs.

Factuality, i.e. claim support in the knowledge base might be underestimated. By looking at claims that are not supported by the knowledge base, we identify multiple instances of claims that are factual but where we fail to retrieve the relevant evidence. This either occurs due to the limited size of our knowledge base or due to a failure on

[1] Full prompt in Appendix D.

Source	Subset	Size	Avg. #Claims	Claim Support wrt. Ref. [%]
Senior Expert	Grounded	74	8.6	93
Expert	Grounded	403	13.1	52
Non-Expert	Open-Ended	8,503	19.1	-
	Closed-Ended	1,160	10.0	90
	(Open-Ended) Grounded	2,368	19.0	43
	(Closed-Ended) Grounded	1,024	9.6	91

Table 2: Climate-specific subsets of the ClimateGPT IFT data. For the closed-ended examples, claim support wrt. reference refers to the context given in the prompt and for grounded examples it refers to the given paragraphs.

retrieval. For an assessment of factuality, we therefore note that the reported metric should just be considered as a lower bound and more accurate results could be achieved.

4 Ablation of the IFT Data

Motivated by the suboptimal faithfulness of ClimateGPT, especially compared to Llama 2 Chat, we want to study the post-training of the model. We focus on the IFT step as we do not expect that the continued pre-training step has a significant impact on the faithfulness of the model. The IFT data of ClimateGPT consists of a general domain partition and a climate-specific partition that was specifically curated to train the model. The different subsets of the latter are listed in Table 2. A small portion of the data was generated in close cooperation with domain experts (Exp.), and the larger set generated by non-experts (Non-Exp.). In *closed-ended* questions, the model is given a reference text to perform its task, such as creating a summary of that text or extracting specific information from it. In contrast, for *open-ended* questions, no additional explicit references are given in the prompt, and the model is expected to use its parametric knowledge or to retrieve additional sources via RAG.

Grounded refers to examples where additional context is provided to the model as it would be the case when RAG is used during inference. For the expert and senior expert subsets, these references were directly provided during annotation. In the case of the non-expert subset, annotators only provided one or multiple URLs to sources the answer is based on. For a subset of the dataset, these URLs were crawled, chunked and Thulke et al. (2024) used a heuristic[2] to select the most relevant chunk as context for the response. Additionally, for each example up to four distractor paragraphs from other documents were selected to make the model more robust to noisy retrieval results. For closed-ended questions, only distractors were added as all the relevant content is already provided in the prompt.

We start our investigation by analysing the faithfulness of the gold responses in the IFT data with respect to their context. For closed-ended questions, we use the full prompt as context and for the grounded questions, the selected context paragraphs. The percentage of supported claims for each subset as well as the average number of claims per response are reported in Table 2. We notice that the Grounded Senior Expert and Closed-Ended Non-Expert are faithful to their context with 93% and 90% of claims being supported. The faithfulness of the Grounded Expert data is already much lower with only 52% claim support. Upon closer inspection, we found that the annotators only provided grounding passages for crucial claims in the response. Finally, we observe the lowest level of faithfulness for the Open-Ended Grounded Non-Expert data with only 43% claim support.

Next, we repeated the IFT step on different subsets of the data to observe the effect on the faithfulness on the final model. The results are reported in Table 3. As anticipated from our previous analysis of the IFT subsets, excluding the grounded non-expert data significantly increases the claim support from 30% to 57%. Furthermore, excluding the closed-ended but not grounded non-expert data reduces the claim support again to 49%. This indicates that closed-ended examples with high faithfulness seem to improve the faithfulness of the model despite the context being provided directly in the prompt, rather than via retrieval. Finally, removing the open-ended non-expert data without grounding does not have a significant effect on the claim support. For the final model, *ClimateGPT Faithful+*

[2]See Section 4.3 in Thulke et al. (2024) for more details.

	Other	Open-End.	Closed-End.	Grounded	Avg.	Claim Support
Size	65,000	8,503	1,160	3,328	**#Claims**	**wrt. Ref [%]**
ClimateGPT 7B	✓	✓	✓	✓	21.1	30
	✓	✓	✓	-	19.2	**57**
	✓	✓	-	-	18.9	49
	✓	-	✓	-	20.1	**58**
	✓	-	-	-	20.4	53

Table 3: Ablation study results showing test-time claim support for different training data combinations.

we still include the open-ended data as we expect it to improve other aspects.

4.1 Validation on Alternate Metrics and Tasks

To validate the generalizability and robustness of our improvements, we conducted additional experiments on a RAG dataset from Climate Policy Radar (Juhasz et al., 2024) focussing on questions on climate policy documents. For ClimateGPT Faithful+, we observe a similar improvement in faithfulness with an improvement in claim support from 44% to 58%. More details are discussed in Appendix F.

Further, we confirm the results on the ClimateGPT IFT Task by using an additional faithfulness metric (LettuceDetect, Ádám Kovács and Recski (2025)) and observe an improvement from 6% to 34% completely faithful responses with the Faithful+. Details are discussed in Appendix G.

5 Related Work

Similar to our work, Schimanski et al. (2024) study the faithfulness of a RAG system on climate questions. They restrict the output of the model so that one sentence always corresponds to exactly one reference passage and verify the faithfulness using an NLI model. This way they avoid the claim decomposition step. They also fine-tune the model on a synthetic dataset following these constraints to improve faithfulness. Our work focuses on improving faithfulness by fine-tuning on more complex human written responses.

In addition to the faithfulness evaluation approaches discussed in this work, there are other approaches to evaluate faithfulness of text generation. Early work on document-grounded dialog used simple overlap based metrics like unigram F1 scores between the response and retrieved passages as a proxy for faithfulness (Dinan et al., 2019; Thulke et al., 2023). Fadeeva et al. (2024) make use of uncertainty quantification to evaluate the factuality of generated responses. Other work does not consider the claim decomposition step and directly verify the full response against the reference (Honovich et al., 2022; Juhasz et al., 2024; Ádám Kovács and Recski, 2025).

6 Conclusion

Ensuring faithfulness of LLM outputs is crucial for improving the reliability of climate-related RAG setups. Our study evaluates automated faithfulness assessment methods. According to our metric, recent LLMs like Llama 3.1 Instruct and GPT-4o provide much higher faithfulness than Llama 2 Chat or the climate-specific ClimateGPT model. Based on our experiments, we assume that the main difference comes from the instruction fine-tuning and other post-training steps and not from the pre-training.

For ClimateGPT, we then do a detailed analysis, which subsets of the IFT data are most important for faithfulness. We show that faithful closed-ended prompts in training also improve the faithfulness in the context of RAG and that it is crucial to avoid unfaithful training examples in the IFT data. With these insights, we develop ClimateGPT Faithful+ which improves ClimateGPT's faithfulness from 30% to 57% according to our automatic metric. These results are confirmed by additional experiments on an additional task as well as by using an additional metric to measure faithfulness.

These initial findings point to promising directions for future work. Rather than discarding unfaithful training examples, one potential approach is to enrich them by retrieving supporting passages for each claim and using those passages as context during training. For cases where no suitable evidence can be retrieved, synthetic context could be generated using a LLM. This would keep a larger portion of the data while still encouraging faithful model behaviour.

Limitations

In this work, we discuss results from our ongoing work towards more faithful LLMs for RAG on climate questions. While our preliminary results are promising, there are still many open questions and limitations.

While RAGAs (Es et al., 2024) is a popular approach to evaluate faithfulness with RAG, we did not perform a systematic evaluation of its performance in the context of the task at hand. Spot-checking of results during the development progress indicated that the metric is reliable enough for our purposes. We tested the approach on two relevant climate datasets from the literature. The results are reported in Appendices F and I, but the results are inconclusive. Thus, a more thorough human evaluation is needed to fully verify the adequacy of the metric for the task and to validate the improvements reported in this work.

The claim verification step in our pipeline currently focuses on verifying direct support via a given evidence passage. This approach works well if the claim is directly expressed in the given passage and we can consider its content as truth. In the context of evaluating faithfulness one can argue that this is a valid assumption. But as soon as we want to also apply these methods to evaluate the factuality of more complex claims, this does not hold any more. Often claims are not directly stated in a retrieved passage and more complex reasoning is required to identify the support. Claims might express opinions or more holistic statements that require support from multiple sources to be considered as supported. Also, a binary decision between supporting and not supporting might not be adequate in many cases, or more nuance is needed. More complex claim verification approaches as proposed by Leippold et al. (2025) partially address many of these points but are also much more complex and computationally expensive than the approach we use.

The behavior of a RAG system is highly dependent on the relevancy and adequacy of the retrieved passages. In this work, we adopted the same knowledge base and retrieval method as used by Thulke et al. (2024). Thus, our results are also limited to this specific setting and generalization to other settings needs to be studied. Furthermore, the size of the knowledge base and the accuracy of the retrieval method limit the accuracy of the factuality evaluation during the evidence retrieval step. Additional analysis would be needed to study the impact of these factors on the claim support wrt. the knowledge base, we consider as a proxy for factuality.

The ablation experiments on the IFT data focus on the climate-specific subsets. We did not study the impact of the general domain IFT datasets included in IFT training, such as Open Assistant (Köpf et al., 2023), Dolly[3] and FLAN v2 (Longpre et al., 2023). Further, ClimateGPT is based on Llama 2. In our experiments, we observed higher faithfulness for Llama 3.1 Instruct than for Llama 2 Chat. The impact of the pre-training compared to different post-training steps on the faithfulness of the model remains unclear.

Finally, during our evaluation we only focused on claim support wrt. the reference and knowledge base which we consider as a proxy for faithfulness and factuality. We do not consider additional quality factors like the helpfulness or adequacy of generated responses. In some cases, a less faithful output can actually be more helpful or relevant. For example the unfaithful parts in Figure 1 like the information on the year might actually make the response more helpful for some users.

References

Katherine Calvin, Dipak Dasgupta, Gerhard Krinner, Aditi Mukherji, Peter W. Thorne, Christopher Trisos, José Romero, Paulina Aldunce, Ko Barrett, Gabriel Blanco, William W.L. Cheung, Sarah Connors, Fatima Denton, Aïda Diongue-Niang, David Dodman, Matthias Garschagen, Oliver Geden, Bronwyn Hayward, Christopher Jones, Frank Jotzo, Thelma Krug, Rodel Lasco, Yune-Yi Lee, Valérie Masson-Delmotte, Malte Meinshausen, Katja Mintenbeck, Abdalah Mokssit, Friederike E.L. Otto, Minal Pathak, Anna Pirani, Elvira Poloczanska, Hans-Otto Pörtner, Aromar Revi, Debra C. Roberts, Joyashree Roy, Alex C. Ruane, Jim Skea, Priyadarshi R. Shukla, Raphael Slade, Aimée Slangen, Youba Sokona, Anna A. Sörensson, Melinda Tignor, Detlef van Vuuren, Yi-Ming Wei, Harald Winkler, Panmao Zhai, Zinta Zommers, Jean-Charles Hourcade, Francis X. Johnson, Shonali Pachauri, Nicholas P. Simpson, Chandni Singh, Adelle Thomas, Edmond Totin, Andrés Alegría, Kyle Armour, Birgit Bednar-Friedl, Kornelis Blok, Guéladio Cissé, Frank Dentener, Siri Eriksen, Erich Fischer, Gregory Garner, Céline Guivarch, Marjolijn Haasnoot, Gerrit Hansen, Mathias Hauser, Ed Hawkins, Tim Hermans, Robert Kopp, Noëmie Leprince-Ringuet, Jared Lewis, Debora Ley, Chloé Ludden, Leila Niamir, Zebedee Nicholls, Shreya Some, Sophie Szopa, Blair Trewin, Kaj-Ivar van der Wijst, Gundula Winter, Maximilian Witting, Arlene

[3]https://huggingface.co/datasets/databricks/databricks-dolly-15k

Birt, and Meeyoung Ha. 2023. Ipcc, 2023: Climate change 2023: Synthesis report. contribution of working groups i, ii and iii to the sixth assessment report of the intergovernmental panel on climate change.

Natalia De La Calzada, Théo Alves Da Costa, Annabelle Blangero, and Nicolas Chesneau. 2024. Climateq&a: Bridging the gap between climate scientists and the general public. *Preprint*, arXiv:2403.14709.

Thomas Diggelmann, Jordan Boyd-Graber, Jannis Bulian, Massimiliano Ciaramita, and Markus Leippold. 2021. Climate-fever: A dataset for verification of real-world climate claims. *Preprint*, arXiv:2012.00614.

Emily Dinan, Stephen Roller, Kurt Shuster, Angela Fan, Michael Auli, and Jason Weston. 2019. Wizard of wikipedia: Knowledge-powered conversational agents. In *International Conference on Learning Representations*.

Nouha Dziri, Ehsan Kamalloo, Sivan Milton, Osmar Zaiane, Mo Yu, Edoardo M. Ponti, and Siva Reddy. 2022. FaithDial: A faithful benchmark for information-seeking dialogue. *Transactions of the Association for Computational Linguistics*, 10:1473–1490.

Shahul Es, Jithin James, Luis Espinosa Anke, and Steven Schockaert. 2024. RAGAs: Automated evaluation of retrieval augmented generation. In *Proceedings of the 18th Conference of the European Chapter of the Association for Computational Linguistics: System Demonstrations*, pages 150–158, St. Julians, Malta. Association for Computational Linguistics.

Ekaterina Fadeeva, Aleksandr Rubashevskii, Artem Shelmanov, Sergey Petrakov, Haonan Li, Hamdy Mubarak, Evgenii Tsymbalov, Gleb Kuzmin, Alexander Panchenko, Timothy Baldwin, Preslav Nakov, and Maxim Panov. 2024. Fact-checking the output of large language models via token-level uncertainty quantification. In *Findings of the Association for Computational Linguistics: ACL 2024*, pages 9367–9385, Bangkok, Thailand. Association for Computational Linguistics.

Or Honovich, Roee Aharoni, Jonathan Herzig, Hagai Taitelbaum, Doron Kukliansy, Vered Cohen, Thomas Scialom, Idan Szpektor, Avinatan Hassidim, and Yossi Matias. 2022. TRUE: Re-evaluating factual consistency evaluation. In *Proceedings of the 2022 Conference of the North American Chapter of the Association for Computational Linguistics: Human Language Technologies*, pages 3905–3920, Seattle, United States. Association for Computational Linguistics.

Lei Huang, Weijiang Yu, Weitao Ma, Weihong Zhong, Zhangyin Feng, Haotian Wang, Qianglong Chen, Weihua Peng, Xiaocheng Feng, Bing Qin, and Ting Liu. 2025. A survey on hallucination in large language models: Principles, taxonomy, challenges, and open questions. *ACM Trans. Inf. Syst.*, 43(2).

Matyas Juhasz, Kalyan Dutia, Henry Franks, Conor Delahunty, Patrick Fawbert Mills, and Harrison Pim. 2024. Responsible retrieval augmented generation for climate decision making from documents. *Preprint*, arXiv:2410.23902.

Andreas Köpf, Yannic Kilcher, Dimitri von Rütte, Sotiris Anagnostidis, Zhi Rui Tam, Keith Stevens, Abdullah Barhoum, Duc Minh Nguyen, Oliver Stanley, Richárd Nagyfi, Shahul ES, Sameer Suri, David Alexandrovich Glushkov, Arnav Varma Dantuluri, Andrew Maguire, Christoph Schuhmann, Huu Nguyen, and Alexander Julian Mattick. 2023. Openassistant conversations - democratizing large language model alignment. In *Thirty-seventh Conference on Neural Information Processing Systems Datasets and Benchmarks Track*.

Markus Leippold, Saeid Ashraf Vaghefi, Dominik Stammbach, Veruska Muccione, Julia Bingler, Jingwei Ni, Chiara Colesanti Senni, Tobias Wekhof, Tobias Schimanski, Glen Gostlow, Tingyu Yu, Juerg Luterbacher, and Christian Huggel. 2025. Automated fact-checking of climate claims with large language models. *npj Climate Action*, 4(1).

Shayne Longpre, Le Hou, Tu Vu, Albert Webson, Hyung Won Chung, Yi Tay, Denny Zhou, Quoc V Le, Barret Zoph, Jason Wei, and Adam Roberts. 2023. The flan collection: Designing data and methods for effective instruction tuning. In *Proceedings of the 40th International Conference on Machine Learning*, volume 202 of *Proceedings of Machine Learning Research*, pages 22631–22648. PMLR.

Sewon Min, Kalpesh Krishna, Xinxi Lyu, Mike Lewis, Wen-tau Yih, Pang Koh, Mohit Iyyer, Luke Zettlemoyer, and Hannaneh Hajishirzi. 2023. FActScore: Fine-grained atomic evaluation of factual precision in long form text generation. In *Proceedings of the 2023 Conference on Empirical Methods in Natural Language Processing*, pages 12076–12100, Singapore. Association for Computational Linguistics.

Cheng Niu, Yuanhao Wu, Juno Zhu, Siliang Xu, KaShun Shum, Randy Zhong, Juntong Song, and Tong Zhang. 2024. RAGTruth: A hallucination corpus for developing trustworthy retrieval-augmented language models. In *Proceedings of the 62nd Annual Meeting of the Association for Computational Linguistics (Volume 1: Long Papers)*, pages 10862–10878, Bangkok, Thailand. Association for Computational Linguistics.

Tobias Schimanski, Jingwei Ni, Mathias Kraus, Elliott Ash, and Markus Leippold. 2024. Towards faithful and robust LLM specialists for evidence-based question-answering. In *Proceedings of the 62nd Annual Meeting of the Association for Computational Linguistics (Volume 1: Long Papers)*, pages 1913–1931, Bangkok, Thailand. Association for Computational Linguistics.

Yixiao Song, Yekyung Kim, and Mohit Iyyer. 2024. VeriScore: Evaluating the factuality of verifiable claims in long-form text generation. In *Findings of the Association for Computational Linguistics: EMNLP 2024*, pages 9447–9474, Miami, Florida, USA. Association for Computational Linguistics.

David Thulke, Nico Daheim, Christian Dugast, and Hermann Ney. 2023. Task-oriented document-grounded dialog systems by hltpr@rwth for dstc9 and dstc10. *IEEE/ACM Transactions on Audio, Speech, and Language Processing*, 32:733–741.

David Thulke, Yingbo Gao, Petrus Pelser, Rein Brune, Rricha Jalota, Floris Fok, Michael Ramos, Ian van Wyk, Abdallah Nasir, Hayden Goldstein, Taylor Tragemann, Katie Nguyen, Ariana Fowler, Andrew Stanco, Jon Gabriel, Jordan Taylor, Dean Moro, Evgenii Tsymbalov, Juliette de Waal, Evgeny Matusov, Mudar Yaghi, Mohammad Shihadah, Hermann Ney, Christian Dugast, Jonathan Dotan, and Daniel Erasmus. 2024. Climategpt: Towards ai synthesizing interdisciplinary research on climate change. *Preprint*, arXiv:2401.09646.

Saeid Ashraf Vaghefi, Dominik Stammbach, Veruska Muccione, Julia Bingler, Jingwei Ni, Mathias Kraus, Simon Allen, Chiara Colesanti-Senni, Tobias Wekhof, Tobias Schimanski, Glen Gostlow, Tingyu Yu, Qian Wang, Nicolas Webersinke, Christian Huggel, and Markus Leippold. 2023. Chatclimate: Grounding conversational ai in climate science. *Communications Earth & Environment*, 4(1).

Benjamin Warner, Antoine Chaffin, Benjamin Clavié, Orion Weller, Oskar Hallström, Said Taghadouini, Alexis Gallagher, Raja Biswas, Faisal Ladhak, Tom Aarsen, Nathan Cooper, Griffin Adams, Jeremy Howard, and Iacopo Poli. 2024. Smarter, better, faster, longer: A modern bidirectional encoder for fast, memory efficient, and long context finetuning and inference. *Preprint*, arXiv:2412.13663.

Shitao Xiao, Zheng Liu, Peitian Zhang, Niklas Muennighoff, Defu Lian, and Jian-Yun Nie. 2024. C-pack: Packed resources for general chinese embeddings. In *Proceedings of the 47th International ACM SIGIR Conference on Research and Development in Information Retrieval*, SIGIR '24, page 641–649, New York, NY, USA. Association for Computing Machinery.

Ádám Kovács and Gábor Recski. 2025. Lettucedetect: A hallucination detection framework for rag applications. *Preprint*, arXiv:2502.17125.

A Full Results on the ClimateGPT IFT Task

Table 4 shows the claim support of all models that we tested on the ClimateGPT IFT task. In contrast to the table in the main part of the paper, here we also report the claim support wrt. the reference for the case that no RAG was used. As the reference is not given to the model as additional input, we do expect low claim support. The value is interesting as an indication for the percentage of claims that are faithful to the reference by chance. Interestingly, we observe that the claim support of the original ClimateGPT models is close to this value. This further supports the interpretation that these models do not make effective use of the provided context. We omitted these results in the main part of the paper for better clarity as they are not directly relevant to the main claims of the paper.

B Evaluation Prompts

Listing 1 and Listing 2 show the prompts that were used for the claim extraction and verification steps in the evaluation pipeline. Both prompts are based on the implementation of RAGAs[4] (Es et al., 2024).

C Knowledge Base Details

Table 5 shows the statistics of the ClimateGPT knowledge base.

D RAG Prompts

Listing 3 shows the prompt used in RAG for inference for all models except ClimateGPT.

Listing 3: Prompt used in RAG for inference for all models except ClimateGPT.

```
You're a helpful assistant supporting users
with their questions on climate change. Answer
the question based on the given contexts. Make
sure to only use information that is fully
grounded in the contexts.

Context:
[[0]] "{passage[0].title}", {passage[0].year}
{passage[0].content}
{...}
[[4]] "{passage[4].title}", {passage[4].year}
{passage[4].content}

Question:
{question}
```

E Training Details

In our training pipeline, we follow the setup from Thulke et al. (2024). The models are trained using Megatron-LLM[5] a fork of NVIDIA's Megatron-LM [6] by the EPFL LLM team. A cosine learning rate schedule with a peak LR of 10^{-5} and 100 warmup steps are used. The batch size is 64 and the sequence length is 4096. Additionally, a weight decay of 10^{-2} and dropout are used.

All 7B parameter models are trained with full parameter fine-tuning on 4xA100 80GB GPUs. One training run takes approximately 4 hours, so in total 64 GPU hours were needed to train the models reported in this paper.

F CPR's RAG Dataset Evaluation

To further evaluate the generalization of ClimateGPT 7B Faithful+ to other datasets, we tested it on a set of question–passage pairs published by the Climate Policy Radar team (Juhasz et al., 2024). This dataset contains 1,013 examples, with the retrieved passages taken from Climate Policy Radar's internal database. We generated responses using both ClimateGPT 7B and ClimateGPT 7B Faithful+, and evaluated their faithfulness to the provided reference passages using our RAGAs-based metric. On this dataset, ClimateGPT 7B Faithful+ achieved a claim support of 58%, substantially outperforming the base ClimateGPT 7B model, which achieved 44%. These results demonstrate that the improvements made in the refined model generalize effectively to other climate-domain datasets.

In addition, Juhasz et al. (2024) also collected expert annotations for model outputs from GPT-4o, GPT-3.5, Gemini 1.0 and 1.5, and Mistral 7B v0.2. Each response was evaluated for faithfulness using a definition closely aligned with ours. Expert annotators labeled responses as either faithful (58.9%), not faithful (9.6%), not applicable (28%), or don't know (3.5%). We used this data to evaluate how well our RAGAs-based metric aligns with human judgments. For the analysis, we focused only on examples that were labeled as either faithful or not faithful, excluding cases where the model refused to answer. This resulted in a total of 1,367 samples. To convert the claim support from our metric into a binary label for each example, we classify an output as faithful if the claim support exceeds 50%. On this test set, our metric achieved an overall agreement of 86.7% with the human annotations. However, accuracy varied between label categories: it reached 93.7% for human-labeled

[4]https://github.com/explodinggradients/ragas

[5]https://github.com/epfLLM/Megatron-LLM

[6]https://github.com/nvidia/megatron-lm

Model	#Tokens in Trillion	#Parameters in Billion	RAG	Avg. #Claims	Claim Support wrt. Ref. [%]	KB [%]
GPT-4o	n/a	n/a	-	17.4	33	68
			✓	16.2	**72**	**74**
LLama 3.1 Instruct	15	8	-	22.7	24	59
			✓	17.3	67	<u>72</u>
		70	-	21.8	25	60
			✓	16.1	<u>70</u>	**74**
LLama 2 Chat	2	7	-	23.3	25	60
			✓	21.2	48	65
		70	-	25.1	24	60
			✓	21.6	54	68
ClimateGPT	2	7	-	21.6	25	59
			✓	21.1	30	61
		70	-	21.8	27	61
			✓	22.2	30	62
ClimateGPT Faithful+ (ours)	2	7	-	20.2	27	57
			✓	19.2	57	69

Table 4: Results of all tested models for claim support wrt. the reference, as a metric of faithfulness, and wrt. the knowledge base (KB) as a metric for factuality for different large language models with and without RAG. The best values are in bold and the second best values underlined.

Listing 1: Prompt template used for Claim Extraction adapted from RAGAs.

```
Given a question, an answer, and sentences from the answer, analyze the complexity of
each sentence and break it down into one or more fully understandable statements.
Ensure that no pronouns are used in each statement and that every claim is explicit
and self-contained. Format the output as a structured JSON response.

EXAMPLE
Question: Who was Albert Einstein and what is he best known for?
Answer: He was a German-born theoretical physicist, widely acknowledged to be one of
the greatest and most influential physicists of all time. He was best known for
developing the theory of relativity. He also made important contributions to the
development of quantum mechanics.
Statements:
{
    "statements": [
        "Albert Einstein was a German-born theoretical physicist.",
        "Albert Einstein is recognized as one of the greatest and most influential physicists of all time.",
        "Albert Einstein was best known for developing the theory of relativity.",
        "Albert Einstein also made important contributions to the development of quantum mechanics."
    ]
}

YOUR TURN
Question: {{question}}
Answer: {{sentences}}
Statements:
```

Listing 2: Prompt template used for Claim Verification adapted from RAGAs.

```
Your task is to judge the faithfulness of a series of claims based on a given context. For each claim you
must return verdict as 1 if the claim can be directly inferred based on the context or 0 if the claim can
not be directly inferred based on the context.

EXAMPLE 1:
Context: John is a student at XYZ University. He is pursuing a degree in Computer Science. He is enrolled in
 several courses this semester, including Data Structures, Algorithms, and Database Management. John is a
diligent student and spends a significant amount of time studying and completing assignments. He often stays
 late in the library to work on his projects.

Claims:
1. John is majoring in Biology.
2. John is taking a course on Artificial Intelligence.
3. John is a dedicated student.
4. John has a part-time job.

Analysis:
{"analysis": [
{
    "claim": "John is majoring in Biology.",
    "reason": "John's major is explicitly mentioned as Computer Science. There is no information suggesting
    he is majoring in Biology.",
    "verdict": 0
},
{
    "claim": "John is taking a course on Artificial Intelligence.",
    "reason": "The context mentions the courses John is currently enrolled in, and Artificial Intelligence
    is not mentioned. Therefore, it cannot be deduced that John is taking a course on AI.",
    "verdict": 0
},
{
    "claim": "John is a dedicated student.",
    "reason": "The context states that he spends a significant amount of time studying and completing
    assignments. Additionally, it mentions that he often stays late in the library to work on his projects,
    which implies dedication.",
    "verdict": 1
},
{
    "claim": "John has a part-time job.",
    "reason": "There is no information given in the context about John having a part-time job.",
    "verdict": 0
}
]}

EXAMPLE 2:
Context: Photosynthesis is a process used by plants, algae, and certain bacteria to convert light energy
into chemical energy.

Claims:
1. Albert Einstein was a genius.

Analysis:
{"analysis": [
{
    "claim": "Albert Einstein was a genius.",
    "reason": "The context and claim are unrelated.",
    "verdict": 0
}
]}

YOUR TURN:
Context: {{context}}
Claims:
{{claims}}
Analysis:
```

Source	# Docs	# 512 Chunks
IPCC Reports	16	17,897
Potsdam Papers	390	8,539
Earth4All	14	235
Other	336	8,648
Total	756	35,319

Table 5: Statistics of the different data sources of the ClimateGPT knowledge base.

faithful responses, but only 29.5% for not faithful ones. Notably, Juhasz et al. (2024) themself acknowledged that their annotations were sometimes "too noisy along the faithfulness dimension". In addition, limited spot-checking on our part more frequently agreed with our metric's assessments than with the human annotations.

G Evaluation with LettuceDetect

In addition to our primary faithfulness evaluation using RAGAs, we include results using **LettuceDetect** (Ádám Kovács and Recski, 2025), a recent hallucination detection framework designed for RAG systems. LettuceDetect is a token-level classifier based on ModernBERT (Warner et al., 2024), trained on the RAGTruth dataset (Niu et al., 2024) to identify hallucinated spans in LLM responses given the input question and context. As LettuceDetect's definition of hallucination closely aligns with our notion of faithfulness, we use it to validate the results obtained with RAGAs.

For our evaluation, we convert LettuceDetect's span-level predictions into a binary faithfulness score by marking a generation as faithful if no hallucinated spans are detected. Results, using the `lettucedetect-large-v1` variant of the model, are reported in Table 6.

The LettuceDetect results support the conclusions drawn from our RAGAs-based evaluation. Without RAG, both ClimateGPT and ClimateGPT Faithful+ achieve low scores (6% and 2% respectively), providing a baseline for how often generations align with the reference context by chance. With RAG, ClimateGPT Faithful+ shows a substantial improvement, reaching 34% hallucination-free responses compared to only 6% for the original ClimateGPT. This underpins the claim support results obtained with RAGAs (57% vs. 30%), reinforcing the effectiveness of our instruction fine-tuning strategy in improving the model's ability to ground its generations in the retrieved context.

Listing 4: Climate-FEVER example where our claim verifier disagrees with the gold label.

```
Claim:
Global warming is driving polar bears toward
extinction

Evidence:
[Global Warming] Environmental impacts include
the extinction or relocation of many species as
 their ecosystems change, most immediately the
environments of coral reefs, mountains, and the
 Arctic.

Gold label: supported
Predicted label: not_supported
```

H ClimateGPT IFT Example Output

Figure 2 shows an example output of ClimateGPT 7B and the ClimateGPT 7B Faithful+ model proposed in this work. While both responses address the question in the prompt and are generally correct, the refined model's response demonstrates a significantly closer alignment with the reference passages.

I Climate FEVER Claim Verification

To evaluate our claim verification method, we applied it to the Climate-FEVER dataset (Diggelmann et al., 2021). The dataset consists of 1,535 claims, each paired with 5 corresponding evidence passages. Each claim-evidence pair is annotated by at least two annotators as either supported, refuted, disputed, or lacking sufficient information. For our analysis, we classify refuted and lacking sufficient information as not supported, and exclude all pairs labelled as disputed. In addition, we only include examples where all annotators agree on the label, resulting in a total of 1,146 claims and 3,348 claim-evidence pairs. On this subset, the RAGAs-based claim verifier achieves an overall accuracy of 67.1%. For pairs with the gold label "not supported", the accuracy is 99.7%, while for supported pairs, it drops to 20.6%. Qualitatively, we observe that our claim verifier is relatively strict, requiring the claim to be explicitly stated in the evidence. An example of this is given in Listing 4. In this instance, the evidence does not explicitly state that polar bears are one of the affected species. However, it could be argued that it is plausible to infer that polar bears are included among the "many species" mentioned in the evidence.

Model	#Tokens in Trillion	#Parameters in Billion	Hallucination-Free Responses [%] w/o RAG	w/ RAG
GPT-4o	n/a	n/a	9	**73**
LLaMA 3.1 Instruct	15	8	4	<u>65</u>
		70	2	<u>65</u>
LLaMA 2 Chat	2	7	2	18
		70	2	29
ClimateGPT	2	7	4	6
		70	4	4
ClimateGPT Faithful+ (ours)	2	7	2	34

Table 6: Percentage of hallucination-free responses according to `lettucedetect-large-v1` (Ádám Kovács and Recski, 2025) for various models with and without RAG. The best values are in bold and the second best values underlined.

Prompt
How do ocean ecosystems, ranging from coral reefs to the deep sea, support biodiversity and provide essential services? Answer in a single detailed paragraph.
Context
[0] "Understanding the Effectiveness of Coastal Nature-based Solutions: Practitioner-based Learning", 2023
despite the high cost and failure rate. oyster reefs are effective for wave attenuation, and research suggests that focusing on positive species interactions can provide a framework for restoration. biodiversity enhancement supports a shellfish reef's ability to provide ecosystem services. ecosystem services research highlights that healthy ecosystems provide high - quality services, while stressed ecosystems produce degraded services and may harm human well - being. most ecosystem services are supported by biodiversity. biodiversity in coastal systems contributes to providing ecosystem services such as fish habitat, nutrient cycling, and various cultural services. additionally, biodiversity may play a role in the
[1] "The Role of Blue Carbon in Climate Change Mitigation and Carbon Stock Conservation", 2021
is an essential to sustainably manage and develop marine resources to their maximum potential. environmental measures should tackle both terrestrial and marine ecosystems, with one as a continuum of the other. coral reef restoration can increase coastal resilience to sea level rise and flooding and provide valuable environmental services for local populations. water pollution in rivers contributes to ocean ecosystem degradation, via eutrophication and the formation of
[2] "The Ocean and Cryosphere in a Changing Climate", 2020
level of risk than the high emission scenario. changing marine ecosystem services and human well - being ecosystem services are environmental processes and functions that provide benefits to people and support human well - being. marine ecosystem services are generated throughout the ocean, from shallow water to the deep sea. these services can be divided into provisioning services, regulating services, supporting services, and cultural services. they are also described as'nature's contribution to people '. changes in key ecosystem
[3] "Climate Change 2022: Impacts,Adaptation and Vulnerability", 2022
logical shifts and trophic mismatches, changes in communities and biodiversity, time of emergence of climate - impact signals in ecological systems from background natural variability, and changes in biomass, primary productivity, and carbon export. coastal ecosystems and seas warm - water coral reefs warm - water coral reef ecosystems, which house a significant portion of marine biodiversity, provide food, income, and shoreline protection to coastal communities. these ecosystems are threatened by climate - induced and non - climate drivers, including ocean warming, marine heatwaves, ocean acidification, sea - level rise, tropical cyclones
[4] "Understanding the Effectiveness of Coastal Nature-based Solutions: Practitioner-based Learning", 2023
research indicates that biodiversity plays a crucial role in the functioning of ecosystem services provided by habitats. recent studies have focused on the climate mitigation capacity of forests in the context of biodiversity and nature - based solutions (nbs). challenges arise in prioritizing monoculture plantations for carbon storage over intact and diverse ecosystems. the identity and richness of species, as well as the concept of'niche complementary ', contribute to boosting ecosystem function and their services. significant restoration efforts are underway for shellfish reefs, despite the high cost and failure rate. oyster reefs"

ClimateGPT 7B (Faithfulness: 19%, Factuality: 67%)
Ocean ecosystems, encompassing a diverse range from coral reefs to the deep sea, demonstrate unparalleled biodiversity and provide a multitude of essential services. In the coral reef ecosystem, vibrant arrays of colorful fish and intricate coral structures serve as vital breeding grounds and habitats for countless marine organisms. These ecosystems act as natural breakwaters, reducing coastal erosion and protecting shorelines. Additionally, they store vast amounts of carbon, contributing to climate change mitigation. Moving out from the coast, the deep sea presents a different world, characterized by complete darkness, extreme pressure, and cold temperatures. Despite the harsh conditions, it sustains unique species adapted to these environments, many of which are still undiscovered. The deep sea plays a crucial role in regulating climate and weather patterns, driven by its immense biological pump, which captures carbon from the surface and transports it to the deep oceans. In essence, ocean ecosystems are not only a source of aesthetic beauty and recreation but also linchpins of the Earth's ecological balance and a treasure trove of biodiversity.
ClimateGPT 7B Faithful+ (Faithfulness: 92%, Factuality: 96%)
Ocean ecosystems, including coral reefs and the deep sea, play a vital role in biodiversity support and providing essential services. They provide provisional services, regulating services, supporting services, and cultural services. Coral reefs contribute to coastal resilience to sea level rise and flooding and provide valuable environmental services for local populations. However, they are threatened by climate-induced and non-climate drivers, impacting coastal ecosystems and warm-water coral reefs. These ecosystems provide food, income, and shoreline protection to coastal communities. Restoration efforts, such as oyster reefs, are essential for wave attenuation, but their failure and high cost are challenges. Recent research has focused on the climate mitigation capacity of forests in the context of biodiversity and nature-based solutions. Prioritizing monoculture plantations for carbon storage over intact and diverse ecosystems is a challenge. The identity and richness of species and the concept of niche complementarity contribute to boosting ecosystem function and services.

Figure 2: Example comparing the outputs of ClimateGPT and ClimateGPT 7B Faithful+ on one example of the held-out test set. Parts marked in red correspond to claim that are not supported in the given context according to our automatic evaluation.

Prompt
What is the purpose of the Global Stocktake?
Context
[0] Recalling Article 14 of the Paris Agreement, the objective of the Global Stocktake is to assess the collective progress towards achieving the purpose of the Agreement and its long-term goals, considering mitigation, adaptation and the means of implementation and support, in light of equity and the best available science. The outcome of the global stocktake shall inform Parties in updating and enhancing, in a nationally determined manner, their actions and support in accordance with the relevant provisions of this Agreement, as well as in enhancing international cooperation for climate action. The GST two-year process, begins with an information collection and preparation component, followed by a technical assessment component and culminates in a consideration of outputs component, which includes the outcomes of the GST. The SBSTA and the SBI invited submissions on views on the approach to the consideration of outputs component of the first global stocktake. This submission responds to that invitation and sets out to identify important characteristics for the outputs phase and the shaping of the outcomes of the GST.
[1] · Paragraph 34 of 19/CMA.1 decides that the outputs of this component of the global stocktake should: a) Identify opportunities for and challenges in enhancing action and support for collective progress in relation to the thematic areas of the global stocktake referred to in paragraph 6(b) above, as well as possible measures and good practices and international cooperation and related good practices;
[2] The SBSTA and the SBI invited submissions on views on the approach to the consideration of outputs component of the first global stocktake. This submission responds to that invitation and sets out to identify important characteristics for the outputs phase and the shaping of the outcomes of the GST. Approach to the Consideration of Outputs:
ClimateGPT 7B
The Global Stocktake is a key component of the Paris Agreement, scheduled to begin in 2023. It involves a comprehensive assessment of global progress in addressing climate change, considering scientific evidence, economic implications, and societal impacts. The Stocktake aims to inform the updates to countries' climate commitments, known as "nationally determined contributions" (NDCs), to ensure they are in line with the agreement's objectives and targets. This process seeks to enhance international cooperation and commitment to mitigating climate change on a global scale.
ClimateGPT 7B Faithful+
The Global Stocktake's purpose, as outlined in Article 14 of the Paris Agreement, is to assess the collective progress made towards achieving the Agreement's purpose and long-term goals, which encompass mitigation, adaptation, and means of implementation. This assessment is conducted considering equity and the best available science. The Global Stocktake process spans two years, beginning with information collection and preparation and culminating in a consideration of outputs. The outputs of the latter component are decided upon in paragraph 34 of 19/CMA.1, which states that they should identify opportunities for enhancing action and support for collective progress in relation to thematic areas of the global stocktake and possible measures and good practices.

Figure 3: Example comparing the outputs of ClimateGPT and ClimateGPT 7B Faithful+ on one example from the Climate Policy Radar data. Parts marked in red correspond to claim that are not supported in the given context according to our automatic evaluation.

Interactive platform for the exploration of large-scale 'living' systematic maps

Tim Repke
Potsdam Institute for Climate Impacts Research / Potsdam, Germany
tim.repke@pik-potsdam.de

Abstract

Research syntheses, such as systematic maps or evidence and gap maps, provide valuable overviews of the coverage of research in a particular field. They serve as pointers for funders and researchers to identify important gaps in the literature where more research is needed but also to find relevant work for more in-depth systematic reviews or meta-analyses. However, systematic maps become outdated quickly, sometimes even after they are released due to the time it takes to screen and code the available literature and long publication processes. Furthermore, the write-up of the synthesis (in form of a peer-reviewed article) can only serve as a high-level summary—for detailed questions one would need full access to the underlying data. To this end, we developed an interactive web-based platform to share annotated datasets. For some datasets, where automated categorisation passes the necessary scientific quality standards, we also update the data as new research becomes available and thus make them 'living'.

1 Introduction

The number of scientific publications is continually growing at an exponential rate. For example, more articles on climate change were published during the sixth assessment cycle of the IPCC than during all previous cycles since 1985 combined (Callaghan et al., 2020). Systematic maps of timely topics that are up-to-date are crucial tools to get an overview of a specific field, to identify research gaps, or to identify articles that are relevant for a particular meta-study or review (JPT et al., 2024; Kastner et al., 2016). The sheer amount of potentially relevant literature to consider and the rapid growth make it increasingly prohibitive to conduct systematic maps by hand. Digital evidence synthesis tools can speed up the most time-consuming of a synthesis, particularly screening abstracts in search for relevant articles (Haddaway et al., 2020; Tsafnat et al., 2014). The Covid-19 pandemic has shown the value of so-called 'living evidence syntheses' that are continually updated to capture findings from the latest clinical trials and other research strands (Chakraborty et al., 2024; Elliott et al., 2014). The crises of our time require similar up-to-date repositories of evidence to support evidence-based policy-making.

Traditional publication models, however, are not able accommodate requirements of regular and frequent updates (Thomas et al., 2017). By the time a research team submit their initial draft, especially until the final publication, the synthesis might already be outdated. The publication is by definition a high-level overview of the underlying data, for which the authors and contributors have spent a lot of time to compile and annotate. However, this raw data is often not published alongside the article or is no longer available. Even where data is available, it might only be available in a proprietary format or the schema used in a csv file might not be self-explanatory. This makes it hard for other researchers or policy analysts to utilise existing categorisations to find relevant literature of their particular questions.

To this end, we developed an interactive web-platform for sharing the underlying data of systematic maps. The initial prototype[1] hosts four projects: A systematic map of literature on climate policy instruments (Callaghan et al., 2025), a systematic evidence and gap map of literature on carbon pricing (Döbbeling-Hildebrandt et al., 2024), a systematic map of literature on carbon dioxide removal (Lück et al., 2024; Smith et al., 2023, 2024), and a systematic map of literature on climate and health (Berrang-Ford et al., 2021; Romanello et al., 2023, 2024). We are also working on adapting data from past publications and making all maps 'living' by building pipelines that automatically retrieve new publications and classify them. Depending on

[1] https://climateliterature.org/

Proceedings of the 2nd Workshop on Natural Language Processing Meets Climate Change (ClimateNLP 2025), pages 260–265
July 31, 2025 ©2025 Association for Computational Linguistics

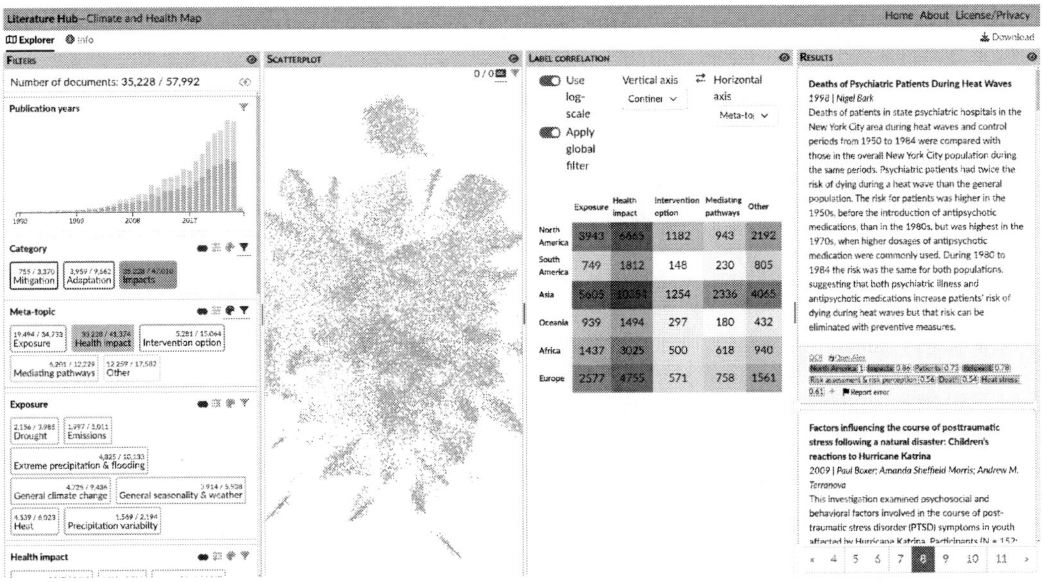

Figure 1: Screenshot of the interactive climate literature hub; Left panel contains all available filters, the second panel shows a scatterplot of the semantic landscape with a descriptive keyword overlay (hidden in the screenshot), the third panel provides a heat-map (or gap-map) to show how different filters or labels correlate, the right panel shows the abstracts that match the current set of filters. Not shown is a panel of a geographic map to see which places are mentioned in abstracts or where authors' affiliations are located.

the purpose of the map and the quality of machine-learning classifiers, this entire process can—with several caveats—be fully automated. With this platform we want to foster open research, transparency, and reusability and make up-to-date evidence easily accessible for anyone. The platform itself is also open-source and available for anyone to adapt or host.[2] The system can easily be adapted to include additional datasets by adding a meta-data and database file using a very basic format (see project repository for details).

Figure 1 shows the main screen of the latest prototype (March 2025) for our map of literature at the nexus of climate and health. The interface is modular, so authors can decide how to best showcase their data. Current components feature various filters for publication year (with a histogram), normal labels (boxes that can be selected), as well as full-text and author search. Furthermore, the dataset can be explored on a 'semantic landscape', a scatter-plot where each dot represents an article and their close proximity indicates high similarity. This explorative visualisation, inspired by Nomic AI (González-Márquez et al., 2024), may be useful to quickly identify clusters of similar works or to see how specific filters cover the topical spectrum. The lasso-selection tool on the landscape also acts as an additional filter. Aside from the semantic landscape, there is also a component to display geographical locations associated with the articles, for example by mentions in the abstract or author affiliations. Regions and location on this map can also be used as filters. The heat-map component provides a quick overview of how labels correlate and is inspired by gap-maps (Snilstveit et al., 2016). Lastly, we implemented a list component to show the most relevant records based on the current global filter.

In the remainder of this article, we describe some of the technical aspects that enable this web-platform to be highly interactive—even for large datasets and many filters. Furthermore, we discuss some challenges of automated updates of datasets shared on this platform and considerations for future work.

2 Real-time filtering

One of the key features of the web-platform is the ability to combine arbitrary filters and receive real-time feedback of how they influence the overall statistics. The platform is built to handle large datasets—tested for a million records. This poses

[2] `https://gitlab.pik-potsdam.de/mcc-apsis/living-evidence-maps/literature-hub`

a particular challenge for optimising the traffic between the front-end and the server. We address this challenge by only transmitting binary bit-masks for each label during the initial page-load. This allows us to do all computations in the client's browser very efficiently, including updating the visualisations. The required traffic for each bit-mask is around 1.2kb per 10,000 records per filter before additional gzip compression on the transport layer. In the example shown in Figure 1, there are three filters grouped under 'Category'. Publication years and x/y coordinates for the scatter-plot are transmitted as uint16 and float16 in a light-weight batched Apache arrow file amounting to around 120kb per 10,000 records.

While we have not conducted dedicated performance experiments, we have not encountered any noteworthy lags to hinder any interactive exploration of the data. For example, on a basic laptop from 2021,[3] a dataset with 78k records and 20 filters only takes a few milliseconds to update all counts and has rendered the scatterplot component in under 200ms after a click event on one of the filters is triggered.

This design has the added benefit, that server requirements are very limited. The raw data is stored by sqlite files that contain raw classifier or topic-model scores. Where human annotations are available, they supersede automatically assigned labels and are set as explicit zero or one scores, whereas all others are limited to the range 0.01–0.99. Alternative bit-masks are transmitted if the user sets different thresholds. By setting the thresholds accordingly, users can decide to only show human annotations. For full-text and author search, the same mechanism is used and generates specific bit-masks.

The result set is loaded in the frontend by sending the bit-mask of the current global filter—the combination of all active filters—to the server, which then responds with the records ranked by the sum of the stored classifier scores.

Overall, users have given very positive feedback and were excited to explore our systematic maps in real-time, filter for what they need, and download records including all labels as a csv for the selection they made. At this point, we have not conducted systematic user studies, but plan to do so to inform future developments.

[3]ThinkPad T14s, no dedicated GPU

3 Considerations for 'living' maps

The conventional process for a systematic map follows a linear and very labour-intensive structure. Once an appropriate (boolean) query is developed, the author team would retrieve bibliographic metadata from a search engine like the Web of Science or Scopus and then screen all records by hand for inclusion. Included records are then annotated further, in case of a review additional information is extracted from the full-text. With the help of automation (Thomas et al., 2017), we can speed up the process by prioritising which records to screen and stop early without having to look at all records (Callaghan and Müller-Hansen, 2022). This also means, that we can design more inclusive queries may lead to more complete systematic maps overall. Once enough labels are collected, other categorisations might also be done with machine-learning classifiers to automate future updates once the first version of a systematic map is published. This means, that we can build fully-automated pipelines to reproduce the original study and run this pipeline regularly (for example daily) to also include newly published research. At this point, we did not develop a standardised framework for machine-learning-based classifications and refer to the original publications the respective datasets came from for how automation was developed and how well it performs.

However, such an automated update should also come with a protocol for how the quality is monitored over time. As a research area evolves, keywords that are relevant for the topic of the systematic map may change, which requires updates to the boolean query at the first step of the pipeline. This then may also require additional annotations as the scope changes to ensure a high-quality scientific standard. Depending on the use-case, a systematic map might prioritise inclusiveness over precision. However, that trade-off might be hard to communicate to users, especially with varying levels of classifier performance. The versions of the dataset also need to be clearly marked, for example to distinguish if only new data was added or whether classifiers were re-trained or the process changed in any way.

Furthermore, the research community needs to develop guidelines for the safe and responsible use of automation. Particularly the rise of generative large language models has already found early adopters in research synthesis. The performance of such models are very hard to validate, as they are also shown to suddenly fail. However, more conventional supervised classification models are also rarely perfect, especially with limited data available for training. In these cases, we need clear guidelines when it is acceptable to still use automation or in which use-cases some categorisation have to be omitted from automated updates and rather need to be reviewed by experts before including them in the published dataset.

Another aspect to consider is continuity of a living systematic map that users return to. Visualisations, such as semantic landscapes need to remain relatively stable over time and should not suddenly use a completely different layout, which, to an extent, can be ensured by adding new data to existing layouts and some additional fine-tuning (Poličar et al., 2024; Repke and Krestel, 2021). As mentioned before, classifiers may be updated during the life-time of a living map. In this case, labels for records that were in earlier versions might change, which could have an impact on downstream users.

Initially, the update iterations are just available via additional fine-grained publication date filters. In future work, we are planning to develop newsletters that interested parties can subscribe to that sends them a list of latest publications that apply to their filter settings. Furthermore, we plan to include a dashboard that showcases the latest trends, ideally highlighting semantic shifts.

One major impact on the continuity across versions of a living systematic map is the data availability in academic search engines. Proprietary databases, such as the Web of Science or Scopus are not accessible to all researchers and the number of results may vary based on the institutional subscription. Open repositories such as OpenAlex or SemanticScholar have shown very good interoperability and coverage (Priem et al., 2022; Culbert et al., 2024), but are increasingly sabotaged by large publishers who enforce the deletion of abstracts from these public indices, which renders the database effectively useless for use in automated pipelines and prevents many researchers from doing their work effectively.

On a similar note, the data shared on such a platform should also be subject to clear licenses where all authors and contributors need to agree to. The license should ideally be very permissive so that other researchers are free to use the annotations for their own work—be it to improve tools for digital evidence synthesis or as a starting point for a systematic review, meta-study, or even a companion map with additional labels or adjusted scope. Finally, as the user base of such a platform grows, they might also identify errors and provide feedback. The maintainers of the living map should to consider how they might want to incorporate the support by a (potentially) global community of experts and laypeople alike. Incorporating feedback and improving the classifications may also mean that annotations for historic data changes. This requires a clear way to reference specific versions of the underlying database for reproducibility, for example by providing daily changelogs.

At this point, we have automated updates for two of the publicly accessible datasets orchestrated by running modified versions of the original studies' scripts and models using scheduled GitLab runners.

4 Limitations

This platform—in its current form—certainly qualifies as an 'expert system'. That means, that some functionality may not be intuitive, especially the combination of filters. We deliberately opted to provide very fine-grained control of how filters can be combined or choosing custom thresholds for classifier scores to adjust the precision/recall trade-offs. Since the development is not directly funded, prioritising usability improvements over additional features or bug fixes is challenging. This highlights another consideration about the sustainability of providing such a platform. The original data should additionally always be published through conventional channels such as zenodo, the OSF, or companion platforms of the publisher.

That said, a systematic evaluation of the usability of the platform should be conducted as part of future work. In particular, such user studies should focus on how each component contributes to a better understanding to contextualise the available evidence and how it can reduce the time required to find relevant evidence for a user's information needs.

5 Conclusion

We have released the climate literature hub, a prototype of a web-platform for sharing data from large-scale systematic maps as we believe in open and transparent research that serves the wider community of researchers and policy-makers. We hope that in exploring our datasets, we inspire 'ecosystems of reviews' in which expert teams can use our maps as starting points for their in-depth analyses on more specific questions or offer a tool for policy-makers to identify relevant research more quickly. Feeding into the platform, we built (semi-)automated pipelines to update the data as new relevant research is published. This submission should serve as a starting point for the wider community how digital evidence synthesis tools can be used responsibly in the future. Future work is needed to develop guidelines for safe use of AI and automation systematic maps feeding into platforms like this, how results can be presented in such a way, that users can make informed decisions if the (possibly imperfect) automated classifications meet their quality needs, and how to address challenges around continuity.

Acknowledgements

Over the course of this project, the authors of this article were supported by the European Research Council (ERC-2020-SyG 'GENIE', grant no. ID951542), the German Federal Ministry for Education and Research ('CDR-SynTra', grant no 01LS2101F), and the Wellcome Trust (PATHFINDER2, grant no. 313586/Z/24/Z and DESTinY, grant no. 227165/Z/23/Z).

References

Lea Berrang-Ford, Anne J Sietsma, Max Callaghan, Jan C Minx, Pauline FD Scheelbeek, Neal R Haddaway, Andy Haines, and Alan D Dangour. 2021. Systematic mapping of global research on climate and health: a machine learning review. *The Lancet Planetary Health*, 5(8):e514–e525.

Max Callaghan, Lucy Banisch, Niklas Doebbeling-Hildebrandt, Duncan Edmondson, Christian Flachsland, William F Lamb, Sebastian Levi, Finn Müller-Hansen, Eduardo Posada, Shraddha Vasudevan, et al. 2025. Machine learning map of climate policy literature reveals disparities between scientific attention, policy density, and emissions. *npj Climate Action*, 4(1):7.

Max W Callaghan, Jan C Minx, and Piers M Forster. 2020. A topography of climate change research. *Nature Climate Change*, 10(2):118–123.

Max W. Callaghan and Finn Müller-Hansen. 2022. Statistical stopping criteria for automated screening in systematic reviews. *Systematic Reviews*, 9.

Samantha Chakraborty, Tanja Kuchenmüller, John Lavis, Fadi El-Jardali, Laurenz Mahlanza-Langer, Sally Green, Ludovic Reveiz, Victoria Carter, Emma McFarlane, Cheryl Pace, Lisa Askie, Fiona Glen, and Tari Turner. 2024. Implications of living evidence syntheses in health policy. *Bulletin of the World Health Organization*, 102:757 – 759.

Jack H. Culbert, Anne Hobert, Najko Jahn, Nick Haupka, Marion Schmidt, Paul Donner, and Philipp Mayr. 2024. Reference coverage analysis of openalex compared to web of science and scopus. *ArXiv*, abs/2401.16359.

Niklas Döbbeling-Hildebrandt, Jan Minx, and Tim Repke. 2024. Protocol: Carbon pricing outcomes – an evidence and gap map of the ex-post literature.

Julian H. Elliott, Tari Turner, Ornella Clavisi, James Thomas, Julian P. T. Higgins, Chris Mavergames, and Russell Lindsay Gruen. 2014. Living systematic reviews: An emerging opportunity to narrow the evidence-practice gap. *PLoS Medicine*, 11.

Rita González-Márquez, Luca Schmidt, Benjamin M. Schmidt, Philipp Berens, and Dmitry Kobak. 2024. The landscape of biomedical research. *Patterns*, 5.

Neal R Haddaway, Neal R Haddaway, Max W. Callaghan, Alexandra Mary Collins, William F. Lamb, Jan C. Minx, James Thomas, and D. John. 2020. On the use of computer- assistance to facilitate systematic mapping. *Campbell Systematic Reviews*, 16.

Higgins JPT, Thomas J, Chandler J, Cumpston M, Li T, Page MJ, and Welch VA. 2024. Cochrane handbook for systematic reviews of interventions version 6.5.

Monika Kastner, Jesmin Antony, Charlene Soobiah, Sharon E Straus, and Andrea C Tricco. 2016. Conceptual recommendations for selecting the most appropriate knowledge synthesis method to answer research questions related to complex evidence. *Journal of clinical epidemiology*, 73:43–49.

Sarah Lück, Max Callaghan, Malgorzata Borchers, Annette Cowie, Sabine Fuss, Oliver Geden, Matthew Gidden, Jens Hartmann, Claudia Kammann, David P Keller, et al. 2024. Scientific literature on carbon dioxide removal much larger than previously suggested: insights from an ai-enhanced systematic map. *ResearchSquare*.

Pavlin G Poličar, Martin Stražar, and Blaž Zupan. 2024. opentsne: a modular python library for t-sne dimensionality reduction and embedding. *Journal of Statistical Software*, 109:1–30.

Jason Priem, Heather A. Piwowar, and Richard Orr. 2022. Openalex: A fully-open index of scholarly works, authors, venues, institutions, and concepts. *ArXiv*, abs/2205.01833.

Tim Repke and Ralf Krestel. 2021. Robust visualisation of dynamic text collections: measuring and comparing dimensionality reduction algorithms. In *Proceedings of the 2021 Conference on Human Information Interaction and Retrieval*, pages 255–259.

Marina Romanello, Claudia Di Napoli, Carole Green, Harry Kennard, Pete Lampard, Daniel Scamman, Maria Walawender, Zakari Ali, Nadia Ameli, Sonja Ayeb-Karlsson, et al. 2023. The 2023 report of the lancet countdown on health and climate change: the imperative for a health-centred response in a world facing irreversible harms. *The Lancet*, 402(10419):2346–2394.

Marina Romanello, Maria Walawender, Shih-Che Hsu, Annalyse Moskeland, Yasna Palmeiro-Silva, Daniel Scamman, Zakari Ali, Nadia Ameli, Denitsa Angelova, Sonja Ayeb-Karlsson, et al. 2024. The 2024 report of the lancet countdown on health and climate change: facing record-breaking threats from delayed action. *The Lancet*, 404(10465):1847–1896.

Stephen Smith, Oliver Geden, Matthew Gidden, W Lamb, G Nemet, Jan Minx, Holly Buck, Josh Burke, Emily Cox, Morgan Edwards, et al. 2024. The state of carbon dioxide removal.

Stephen M Smith, Oliver Geden, Gregory F Nemet, M Gidden, WF Lamb, C Powis, R Bellamy, M Callaghan, A Cowie, E Cox, et al. 2023. The state of carbon dioxide removal–the state of carbon dioxide removal.

Birte Snilstveit, Martina Vojtkova, Ami Bhavsar, Jennifer L. Stevenson, and Marie Moland Gaarder. 2016. Evidence & gap maps: A tool for promoting evidence informed policy and strategic research agendas. *Journal of clinical epidemiology*, 79:120–129.

James Thomas, Anna Noel-Storr, Iain James Marshall, Byron C. Wallace, Steve McDonald, Chris Mavergames, Paul P. Glasziou, Ian Shemilt, Anneliese J Synnot, Tari Turner, and Julian H. Elliott. 2017. Living systematic reviews: 2. combining human and machine effort. *Journal of clinical epidemiology*, 91:31–37.

Guy Tsafnat, Paul P. Glasziou, Miew Keen Choong, Adam G. Dunn, Filippo Galgani, and Enrico W. Coiera. 2014. Systematic review automation technologies. *Systematic Reviews*, 3:74 – 74.

Transforming adaptation tracking: Benchmarking Transformer-based NLP approaches to retrieve adaptation-relevant information from climate policy text

Jetske Bonenkamp, Robbert Biesbroek, Ioannis Athanasiadis
Wageningen University and Research, Wageningen, The Netherlands
(jetske.bonenkamp, robbert.biesbroek, ioannis.athanasiadis)@wur.nl

Abstract

The voluminous, highly unstructured, and intersectoral nature of climate policy data resulted in increased calls for automated methods to retrieve information relevant to climate change adaptation. Collecting such information is crucial to establish a large-scale evidence base to monitor and evaluate current adaptation practices. Using a novel, hand-labelled dataset, we explored the potential of state-of-the-art Natural Language Processing methods and compared the performance of various Transformer-based solutions to classify text based on adaptation-relevance in both zero-shot and fine-tuned settings. We find that fine-tuned, encoder-only models, particularly those pre-trained on data from a related domain, are best suited to the task, outscoring zero-shot and rule-based approaches. Furthermore, our results show that text granularity played a crucial role in performance, with shorter text splits leading to decreased performance. Finally, we find that excluding records with below-moderate annotator confidence enhances model performance. These findings reveal key methodological considerations for automating and upscaling text classification in the climate change (adaptation) policy domain.

1 Introduction

The urgent need for climate change adaptation (referred to as 'adaptation' hereafter) has driven governments to formulate and implement ambitious policies and actions (Orlove, 2022). A comprehensive understanding of global adaptation progress, however, has remained absent. Despite conceptual proposals, no consistent, large-scale framework for tracking progress has been implemented to date (Magnan & Chalastani, 2019).

A key factor in this challenge is the abundance and unstructured nature of the relevant evidence, with adaptation information often being embedded in long climate policy documents. This hinders accessibility of relevant information to inform monitoring and evaluation, making identification of adaptation-relevant text essential for a tracking framework. The sheer volume of the text available, however, makes manual analysis infeasible, thus requiring an automated text classification approach. The field of Natural Language Processing (NLP) has shown great promise to contribute to adaptation tracking (Ford et al., 2016; Sietsma et al., 2024), but the multitude of approaches, setups, and data strategies that can potentially influence performance makes selecting the most suitable method challenging.

Rule-based approaches (e.g., keyword search) are most transparent and may achieve satisfactory results for non-complex topics, but their statistics-based successors are typically more accurate and stable (Li et al., 2022). For classification of short texts, early Deep Learning-based approaches continued this rising trend in accuracy, albeit with small margins – particularly when the dataset gets more imbalanced – and at the cost of computational efficiency (Shyrokykh et al., 2023).

More recently, the NLP field has shifted to the use of pre-trained models based on the Transformer architecture (Fields et al., 2024; Vaswani et al., 2017), of which encoder-only language models (ELMs) like BERT (Devlin et al., 2019) and large language models (LLMs) like GPT (Radford et al., 2019) are examples. For text classification tasks, state-of-the-art (SOTA) models have shown potential through three main approaches: (1) supervised fine-tuning of an ELM on a labelled dataset; (2) using an existing ELM fine-tuned on Natural Language Inference (NLI) for zero-shot classification, and; (3) prompting an advanced, general-purpose LLM to classify in a zero- or few-shot setting.

Proceedings of the 2nd Workshop on Natural Language Processing Meets Climate Change (ClimateNLP 2025), pages 266–277
July 31, 2025 ©2025 Association for Computational Linguistics

The choice of approach and model, and their performance relative to more traditional NLP approaches, depends on numerous aspects. Prior research has shown that fine-tuned ELMs tend to outperform general-purpose LLMs on classification tasks when sufficient training data is available (Bucher & Martini, 2024), when the model is pre-trained on domain-relevant data (Dimitar et al., 2023), or when the task is of limited complexity (Yu et al., 2023). However, when training data is scarce or the text complexity requires advanced language understanding, LLMs may outperform fine-tuned ELMs (Yu et al., 2023), as well as traditional and NLI-based models (Z. Wang et al., 2023).

Other influential factors include text splitting strategy and inclusion threshold. Longer texts preserve context but pose challenges for SOTA models, as: (1) these are typically pre-trained on shorter texts (Fiok et al., 2021); (2) the models have difficulties with identifying information when text becomes more sparse (D'Cruz et al., 2024), and; (3) computation of Transformers scales quadratically with input length (Beltagy et al., 2020), making it challenging to determine the right text granularity and splitting strategy. Inclusion threshold refers to the extent to which a given text block must align with a label to belong to that class, potentially affecting model adaptability, and, therefore, performance. For zero-shot classification, setup choices like labels, task type (e.g., binary or multi-class), and prompt design (for LLMs) may also impact results.

To address the uncertainties discussed above, SOTA and traditional, automated text classification approaches are benchmarked against manually labelled climate policy texts. In addition, the impact of text granularity and inclusion thresholds is assessed. The aim is to identify the best method – i.e., the combination of approach, setup, and dataset variant (see Appendix A for the nomenclature) getting closest to human-labelled, 'ground truth' examples – for extracting adaptation-relevant information from climate policy texts. By doing so, this work supports the creation of a global evidence base of adaptation progress.

2 Data

The main dataset comprises text extracted from national policy documents in the Climate Policy Radar (CPR) database[1] (Climate Policy Radar, n.d.), filtered to include only documents pre-labelled as adaptation-relevant and UNFCCC submissions, excluding mitigation-focused NIRs. A sample of 14 countries[2] (243 documents) was carefully selected to represent variety in climatic zones (WorldAtlas, 2023), developmental levels, number of available documents, and administrative language. All text was parsed from publicly available PDFs, transformed into Markdown format based on PDF layout, and non-English texts were translated via the Google Translate API using the default API settings (Han, n.d.). Subsets were created to evaluate the effects of text splitting and data cleaning strategies, as detailed in the following sections.

2.1 Chunking strategy

For assessing the effect of text granularity and context, three subsets of the main dataset were created. Each subset, referred to as 'dataset', uses a different strategy for splitting the texts into smaller blocks (i.e., chunking), as introduced below.

Dataset 1: Full chunks

First, the documents were split into text chunks of, on average, 3,186 characters and 10 paragraphs, using a Markdown-aware semantic splitter (*Semantic Text Splitter (API Documentation)*, n.d.). The chunks were sampled by document type, resulting in a set of 3,159 chunks, which were manually labelled by trained, graduate-level students and the authors of this paper. 24% of the dataset was labelled as relevant to adaptation. The inter-annotator agreement is 83%, which is considered acceptable. For the cases of disagreement between two annotators, the label with the highest confidence score was taken as the ground truth label. These confidence scores are further explained in section 2.2.

Dataset 2: Sub-chunks

To facilitate experimenting with variation in text splitting strategies, the text of dataset 1 was further split into sub-chunks of 500 to 800 characters so

[1] The data was retrieved from the database in June 2024. Documents added after this data are, therefore, not included in the dataset.

[2] Australia, Azerbaijan, Canada, Cyprus, Ecuador, Finland, Haiti, Iceland, Mexico, Nigeria, Sierra Leone, United Kingdom, Vanuatu, and Vietnam

that the chunks average approximately one paragraph and stay within the common NLP model limit of 512 tokens. For the full chunks previously labelled as 'not adaptation', the corresponding sub-chunks, totalling 13,132 items, were automatically assigned the same label. The remaining 5,356 'adaptation' sub-chunks were re-labelled.

To address missing context caused by unclear coreferences, the sub-chunk experiments are conducted in two settings: one using the original sub-chunks without preprocessing (dataset 2a), and another applying coreference resolution (Elango, 2005) to replace unclear noun phrases (e.g., 'the country') with their parent entity (e.g., 'Vietnam') from outside the sub-chunk (dataset 2b). 71% of the sub-chunks retrieved from the relevant full chunks were re-labelled as adaptation-relevant, representing 20% of the full sample of 18,488 sub-chunks.

Dataset 3: Summarized chunks

To balance the advantages of shorter text but retaining crucial context, a third experimental dataset was created, in which the full chunks were summarized to single paragraphs using `bart-large-cnn`, a Transformer-based summarization model (Lewis et al., 2020). The automatically generated summaries were not evaluated at scale and thus may contain errors or inaccurate information. They are, therefore, solely used for the classification stage and not as actual adaptation evidence: the predicted labels are connected to the original, full chunks.

2.2 Data cleaning strategy

Besides the datasets resulting from applying the different chunking strategies, more dataset variants were created to assess differences in performance when applying different data cleaning strategies. Below, these strategies, each resulting in additional dataset variants*, are introduced.

Confidence score threshold

A distinctive step was added in the annotation process. While labelling the chunks, the annotators specified a confidence score, indicating how sure they were about the assigned label (i.e., 'adaptation' or 'not adaptation') on a 0-100 scale. This score is used during evaluation to allow for assessing to what extent exclusion of chunks below a certain confidence score threshold (CST) affects performance. Figure 1 shows the distribution of the scores across the text chunks, indicating that for the majority of the chunks (i.e., >80%), the annotators were very confident (i.e., 80-100% certain).

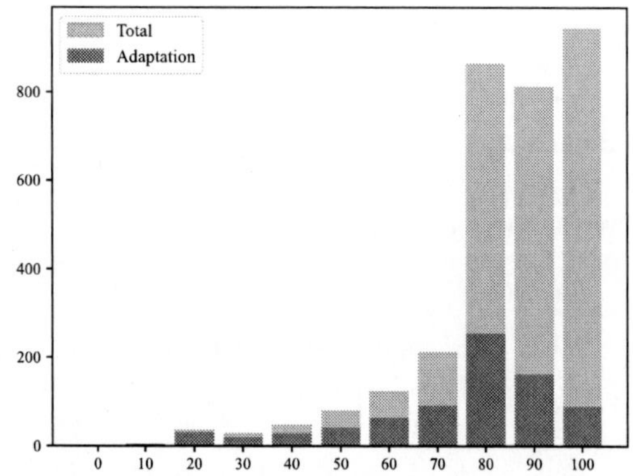

Figure 1: Distribution of confidence scores among 3,159 hand-labelled text chunks. The darker bars show the ratio of chunks labelled as 'adaptation'.

Document type filter

When training classifiers, imbalanced data (i.e., uneven distribution of the classes) can cause difficulties for these models to correctly predict the right label, particularly for the under-represented class (Paduraru & Breaban, 2019). Since the document types of the full documents retrieved from CPR are known, an analysis of the class distribution per document type, based on the labels of dataset 2, revealed that there were multiple document types with a sub-chunk relevance ratio of 4% or lower[3]. Removing all chunks of these low-relevance document types would increase the initial (i.e., with no CST applied) ratio of adaptation-relevant chunks from 24% to 29% (dataset 1 and 3) and from 20% to 28% (dataset 2). The size of the datasets reduces from 3,159 to 2,572 (dataset 1 and 3) and from 18,488 to 13,948 (dataset 2) when applying this document type filter. To assess whether this increased balance, despite the decreased size of the training data, leads to improved performance that compensates for potentially missed relevant data, both strategies are applied and evaluated in combination with all datasets introduced in section 2.1. The resulting

[3] Decision and Plan, Regulation, Vision, Roadmap, Constitution, Act, Long-Term Low-Emission Development Strategy, and Biennial Update Report

* See Appendix A. for the nomenclature

dataset variants are referred to as *unfiltered* (i.e., all document types included) and *filtered* (i.e., low-relevance document types removed).

3 Methodology

Four main approaches are benchmarked against the dataset variants presented in section 2. Each approach and the corresponding sub-methods (i.e., models, queries, tasks, and/or prompts) are introduced in below sections. The confidence scores of the labels and the two filtering strategies (see section 2.2) are used to prepare ten variants of each dataset, each corresponding to a CST value of 0 (i.e., original labels maintained), 50, 60, 70, or 80 combined with *filtered* or *unfiltered* as the data cleaning strategy. For each CST iteratively, the items with label 'adaptation' but a score lower than the CST are excluded from the dataset. For each dataset variant, random splits are created, where 15% is used for evaluation, and, where applicable, 70% for training and 15% for validation. Performance is evaluated by computing precision, recall, and F1-score compared to the human-coded labels. Additional criteria, such as computational cost, are also noted during the final evaluation.

3.1 Rule-based classification (RBC)

The rule-based pattern matching technique is arguably the simplest approach evaluated, querying for (sets of) keywords to classify the chunks. Three different queries are applied. The first is a baseline query, focusing on the text sequence 'adapt' only. The second is theory-based, following the concept of adaptation (Orlove, 2022). The third query is data-driven, following the prominent topics in the data labelled as relevant, determined by applying a topic model. The queried topics were additionally filtered to exclude the topics that occurred in more than 200 of the chunks that were labelled as 'not adaptation' (e.g., 'Paris Agreement'). The resulting queries can be found in table 1.

3.2 Natural Language Inference (NLI)

Four NLI-based zero-shot classifiers are evaluated for identifying adaptation-relevant text: `deberta-small-long-nli` (Sileo, 2024), `bart-large-mnli` (Facebook, 2024), `deberta-v3-large-zero-shot-v2.0` (Laurer et al., 2024), and `nli-MiniLM2-L6-H768` (W. Wang et al., 2021). NLI models leverage their understanding of language obtained through pre-training to determine whether a hypothesis (label) is true given a premise (text) (Laurer et al., 2024). The model selection is based on compatibility with longer texts, model transparency, and reported performance in prior work. For each model, different tasks were evaluated, adding variety in used labels and task type (i.e., binary versus multi-class). An overview of the different tasks can be found in Appendix B. . For the 'multi-class' task type, where the model is asked to assign scores to multiple labels rather than a binary judgment about only the presence of an adaptation-related label, the additional label (i.e., 'mitigation') was ignored during evaluation, and the experiments were repeated with different thresholds for the adaptation label.

3.3 Fine-tuned encoder-only models (FEM)

Four models were selected to be fine-tuned for the classification task. They were chosen to include both general-purpose and domain-specific models, taking into account important criteria such as context length compatibility, pre-training data characteristics, and model parameters. The first one is the general-purpose model `distilroberta-`

Table 1: Overview of search queries

Title	Simplified expression
Baseline query	adapt[a-z]*
Theory-based query	adapt[a-z]* OR ((decreas[a-z]+ OR reduc[a-z]+ OR mitigat[a-z]+ OR avoid[a-z]*) NEAR (impact OR vulnerab[a-z]+ OR hazard OR exposure OR risk)) OR ((increas[a-z]+ OR improv[a-z]+ OR enhanc[a-z]+ OR build) NEAR resilien[a-z]+)
Data-driven query	((climat[a-z]+)? (change)? adapt[a-z]+) OR ((natural)? disaster NEAR (prevent[a-z]* OR control OR respons[a-z]+)) OR (risk NEAR (reduc[a-z]+ OR manag[a-z]+)) OR ((negative)? climat[a-z]+ NEAR impacts?) OR (climat[a-z]+ NEAR respons[a-z]+) OR ((sea level) NEAR rise) OR (capacit[a-z]+ NEAR build[a-z]*) OR ((climat[a-z]+ OR (fast start)) NEAR financ[a-z]+) OR ((early) warning NEAR system) OR (environment[a-z]* NEAR protect[a-z]*) OR (natural NEAR resource[a-z]+ NEAR manag[a-z]+)

Table 2: Overview of prompts used for LLM-based classification

ID	Prompt
P1 (concise)	Classify the following climate policy text chunk as "Adaptation" or "Not adaptation". Do not include any text other than the label.
P2 (specific)	Your task is to categorize text chunks as "Adaptation" or "Not adaptation". If the text contains any information about climate change adaptation policy, categorize it as "Adaptation". If not, for example when it only contains information about mitigation, categorize it as "Not adaptation". Do not include any text other than the label.

`base` (Sanh et al., 2019). The second is a climate domain-specific model from the ClimateBERT family, namely `distilroberta-base-climate-f` (Webersinke et al., 2022). The third model evaluated is `legal-bert-small-uncased` (Chalkidis et al., 2020), a model tuned to the legal domain, and the final one involves a model trained for understanding of environmental texts, namely `EnvironmentalBERT-base` (Schimanski et al., 2024).

All listed models are iteratively fine-tuned on the different dataset variants to assess the performance of the models themselves, as well as the impact of tuning the training data on the results.

3.4 Large Language Models (LLMs)

State-of-the-art LLMs are prompted to assign a binary label (i.e., 'adaptation' or 'not adaptation') to the text chunks they are provided with. Here, the experiments are conducted with OpenAI's `GPT-4o` (OpenAI et al., 2024) and the 8 billion parameter version of `Llama 3.1` (Touvron et al., 2023). Table 2 provides an overview of the two prompts used. Given the length of the chunks combined with the abundance of the dataset, only zero-shot prompting techniques were included in the experiments: providing examples in the prompt (i.e., few-shot learning) would require excessive computational resources (Sahoo et al., 2025). Two prompt variations were applied, were the first one (P1) only provides the task to the model, and the second (P2) elaborates further on the label definitions.

The prompts were carefully composed to vary in conciseness (P1) and specificity (P2), following common prompt engineering principles (Geroimenko, 2025). This experiment is intended to bring insights into how extending the prompt with additional context information and elaborated instructions, thereby limiting conciseness, affects performance.

4 Results

The variations in approaches, setups, and dataset variants resulted in 791 different methods*. A complete overview of the evaluation scores of each method are accessible via the Git repository of this paper[4]. In the following sections, a selection of the most noteworthy results is presented.

4.1 Approaches

For each approach introduced in section 3, the two best methods based on F1 score and the best method based on recall are plotted in figure 2. The bars show the distribution of true positives (darker green), false positives (pastel green), and false negatives (orange). In below subsections, the results of each approach are discussed.

Rule-based classification (RBC)

The results of the RBC experiments reveal, as can be obtained from figure 2, that both the theory-based and data-driven queries outscored the baseline query on the recall metric. This indicates that these setups* excel at correctly identifying the largest ratio of relevant chunks. This increase, however, negatively affects precision, as the baseline query (i.e., only searching for the word 'adaptation') shows better results at limiting the number of irrelevant items being predicted as relevant. Overall, the data-driven query mainly outperforms the theory-driven one on recall.

Zero-shot classification (NLI/LLM)

The results of the two zero-shot approaches (see the bars of NLI and LLM in figure 2) show that the instructed LLMs provide better scores compared to the NLI-based models. Although the *BART-large*

[4] git.wur.nl/bonen003/transforming-adaptation-tracking

* See Appendix A. for the nomenclature

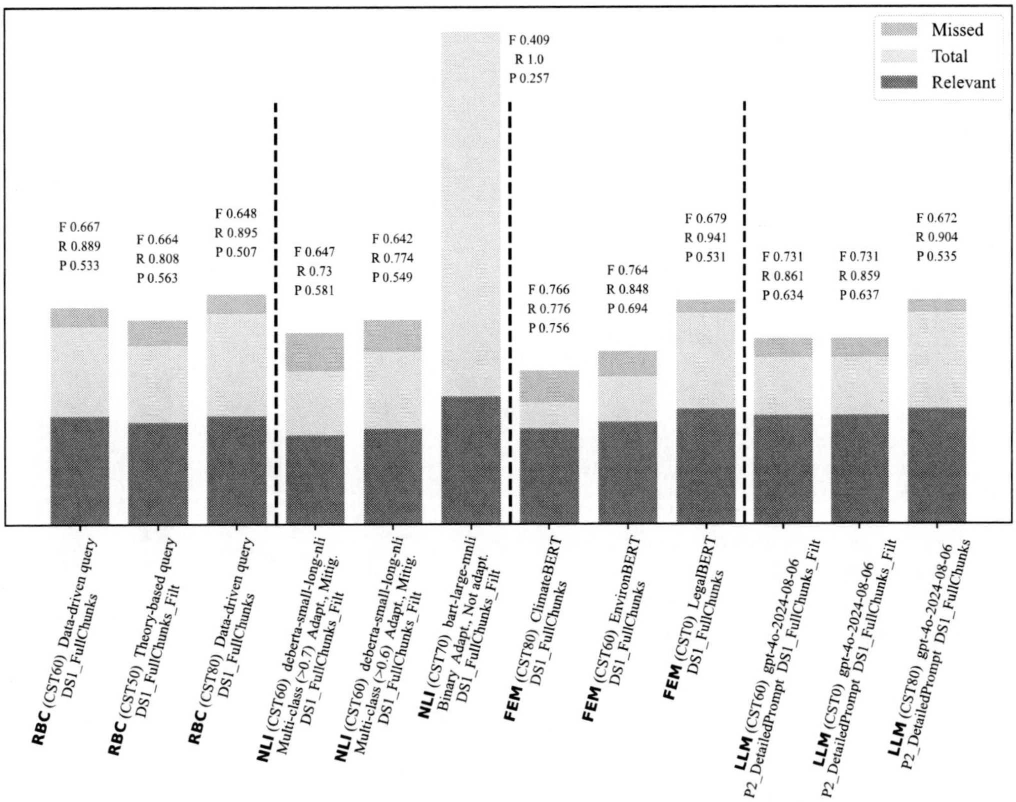

Figure 2: Selection of results for the four approaches, namely Rule-Based Classification (RBC), Natural Language Inference (NLI), Large Language Models (LLM) and Fine-tuned Encoder Models (FEM). For each approach, the two top-performing models based on F1 score (F) and the top one according to recall (R) were selected. Precision is also reported (P).

model achieves a perfect recall, it incorrectly classifies most of the irrelevant items as adaptation-relevant. Among the two LLMs evaluated, *GPT-4o* outperforms *Llama* on all occasions, being particularly well-capable of identifying relevant chunks. For *GPT-4o*, the specific prompt (P2) shows an increase in recall, although at the cost of precision.

Fine-tuned models (FEM)

The FEM experiment results show that three different domain-specific models occur among the three top scoring methods (see figure 2). The methods using *ClimateBERT* and *Environmental-BERT* achieve the best F1-score, indicating good capability of balancing inclusion of relevant items and exclusion of irrelevant items, with the differences mainly found in the balance between recall and precision. The *LegalBERT*-based method excels at recall, predicting 94.1% of the adaptation-relevant items as such.

4.2 Annotator confidence

For analysis of the effects of applying CSTs on the training and evaluation data, the CSTs were clustered into low (i.e., all labelled data included), medium (i.e., all items labelled with a score of 50% or lower excluded), and high (i.e., all items with a score of 70% or lower excluded). The bars in figure 3 show the mean evaluation scores of the best five methods per approach based on F1 score. Here, it becomes clear that applying a CST affects performance for all approaches, as a low CST yields the lowest scores in all four cases. For the traditional and zero-shot approaches, a higher CST positively affects the ratio of items correctly identified, whereas it mainly results in increased precision (i.e., the ratio of non-relevant items incorrectly predicted as relevant) for the fine-tuned models.

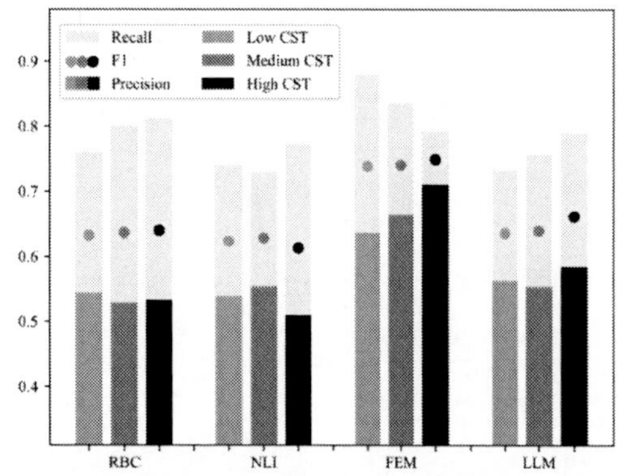

Figure 3: Results of applying a low, medium, or high confidence score threshold (**CST**) on the dataset. The average performance metrics of the top **five** methods per approach, ranked by F1-score, are plotted.

In the prompt variations used for classification with *GPT-4o*, it is observed that the prompt resulting in the best scores depends on what CST is applied. Although recall increases in all cases, the specific prompt (P2) lead to degraded performance when no CST was applied (i.e., CST0) or when a high threshold was used (i.e., CST80). For this zero-shot approach, a medium CST combined with a specific prompt (P2) results in the best scores, excelling on both recall and precision (see figure 4).

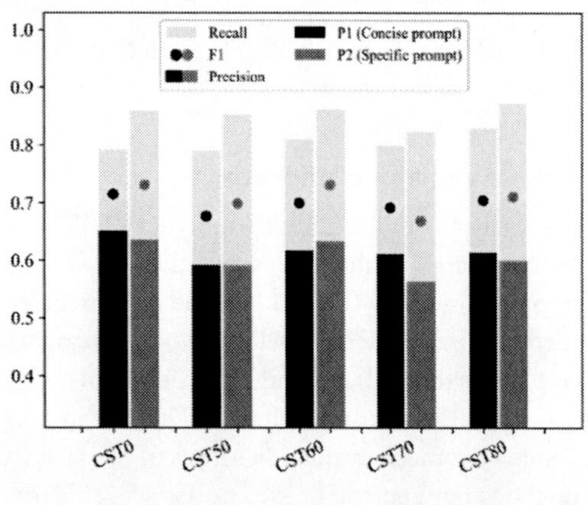

Figure 4: Results of classification with GPT-4o. The bars show F1-score, recall, and precision for each confidence score threshold (CST) and compares the scores of using a **concise** versus a **specific** prompt.

4.3 Chunking strategy

A comparison of the overall performance of each approach on the different datasets (see figure 5) shows that the best balance between maximized true positives and minimized false positives is achieved with the dataset of full-length text chunks (dataset 1). The models fine-tuned and/or evaluated on this dataset particularly excel on recall. Although the margins vary, the summarized dataset (dataset 3) resulted in the lowest F1 score for all approaches. The methods in non-fine-tuned settings do show an increase in precision for this dataset compared to the others.

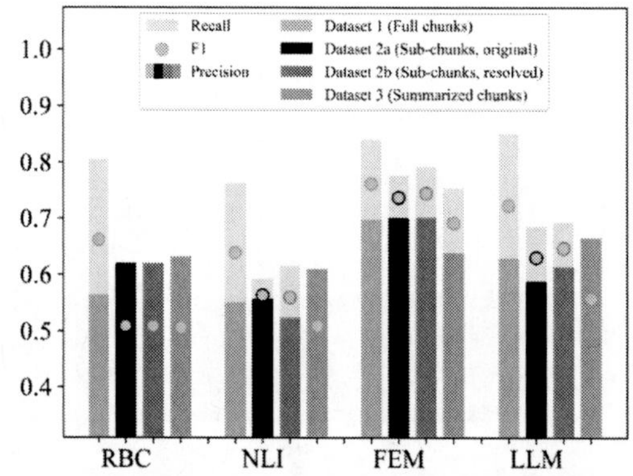

Figure 5: Results of evaluating classification with four different **datasets**: dataset 1 (full chunks), dataset 2a (sub- chunks), dataset 2b (sub-chunks with resolved coreferences), and dataset 3 (summarized chunks). The average performance metrics of the top **five** sub-methods per approach, ranked by F1-score, are plotted.

Comparing the two versions of the sub-chunk dataset, no major differences in performance between the original dataset (2a) and the one with resolved coreferences (2b) are observed. For the FEM approach, coreference resolution shows a slight increase in evaluation scores. However, in most other cases, as figure 5 also indicates, the models become less capable of identifying relevant items, hence a decrease in recall.

4.4 Data cleaning strategy

The results plotted in figure 6 reveal that adding a document type filter positively affects the F1 score for all approaches, except for the fine-tuned models. For this FEM approach, the results show that the data strategy (i.e. filtered versus unfiltered on document type) that outscores the other varies per model and CST. This is expected, as applying the filter lead to a more balanced dataset, typically improving classification performance, but the size of the training dataset decreases, meaning the model has less examples to learn from. Of all experiments conducted overall, both strategies occur in the top 20 (sorted by F1 score). The absolute numbers of a confusion matrix of the best-scoring variant of each filtering strategy[5] (see Appendix C.) also indicate that there is no clear outperforming data strategy*. The most suitable choice depends on various design choices, as further explored in the following section.

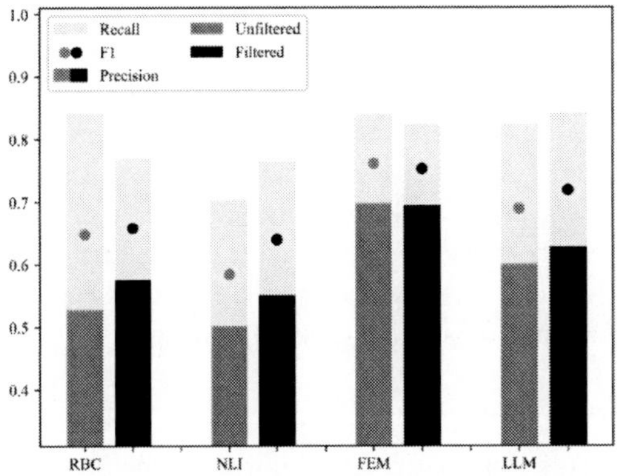

Figure 6: Results of evaluating classification with two different data strategies: with and without an applied **document type filter**. The average performance metrics of the top **five** methods per approach, ranked by F1-score, are plotted.

4.5 Overall comparison

In this research, recall is prioritized over precision, meaning that the 'best' method is not purely determined based on F1 score. Setting a precision threshold of 0.66 and sorting the results on recall leads to a set of four FEM methods considered most suitable to the task, each with its own strength. An overview of these methods, including their results, is provided in table 3. All models in this selection are fine-tuned and applied on/to the dataset of full chunks (dataset 1).

Table 3: Overview of four selected methods, referring to models fine-tuned on specific dataset variants

Model	**Var.**	**Prec.**	**Rec.**	**Comp. Cost**
ClimateBERT	CST70, Filtered	0.673	0.871	-
ClimateBERT	CST60, Unfilt.	0.672	0.869	-/o
LegalBERT	CST0, Filtered	0.671	0.855	-
EnvironBERT	CST60, Unfilt.	0.694	0.848	o

[5] Determined by setting a minimum precision of 0.66, then sorting by recall (descending)

This selection shows that there are multiple methods* that lead to satisfactory results. The selected methods show comparable performance, from where it is obtained that the two models fine-tuned and evaluated on a filtered dataset variant are most computationally efficient (measured by duration of the fine-tuning process), the *EnvironmentalBERT*-based method excels on precision and F1 score, and the first *ClimateBERT* model achieves the best recall.

5 Discussion

The main objective of this paper was to determine the best classification method to identify adaptation-relevant text chunks in large and unstructured climate policy documents. The results reveal that each approach comes with its own strengths and weaknesses, but domain-specific models fine-tuned on a labelled dataset showed the best balance between ratio of correctly identified, relevant items and minimized presence of irrelevant items among those predicted as being relevant. With F1 scores of 0.759, 0.758, 0.752, and 0.764 respectively, four fine-tuned models (listed in table 3), including three different base models and multiple dataset variants, have proven their potential for identifying relevant information needed to track adaptation globally. These findings align with those of Bucher & Martini (2024), i.e., that fine-tuned models outperform LLMs when sufficient training data is available, and those of Dimitar et al. (2023), i.e., that better scores are achieved when such models have been pre-trained and/or previously fine-tuned on domain-specific data. In this research, where the labelled data has been created, this supervised FEM approach is considered most suitable, as it outperforms the benchmarked RBC and NLI approaches by large margins and shows small performance advancements over the top-scoring zero-shot method with *GPT-4o*. As the differences with the latter are relatively minor, however (i.e., an F1 difference of 0.07), LLM-based zero-shot classification has also demonstrated its potential. Especially in future cases, when the resources to (re-)create a labelled dataset are limited, this approach may be a valid alternative. The results have shown, however, that the chosen CST and

* See Appendix A. for the nomenclature

prompt design can majorly affect the performance of LLMs as classifiers, making this approach less reliable and robust when there is no labelled dataset available for validation.

What specific FEM and dataset variant should be selected, however, depends on prioritization of trade-offs. First, what CST is applied should be taken into account for the final choice of method. A low CST means that even chunks that are somewhat relevant will be included in the eventual dataset, which limits the possibility of missing out on potentially relevant information. The relatedness of the text in the final dataset, however, likely improves when its content is relevant to adaptation with high confidence. Using a medium CST (i.e., 50-70) is, therefore, preferred, as this balances out these (dis)advantages.

In contrast to prior studies which suggest that performance of Transformer-based models typically improves when text is relatively short and consistent, the results show that for all approaches, the best scores were achieved using the full chunk dataset. Applying coreference resolution to the short text splits did not solve the 'missing context' problem, showing negligible differences, nor did automated summarization overcome the challenge of dealing with longer text lengths. This emphasizes the importance of context information, which likely connects to adaptation's conceptual indistinctness described by Dupuis & Biesbroek (2013). Using the full chunks (dataset 1) is, therefore, preferred here. Determining whether a document type filter should be applied, however, turned out a greater challenge. The main advantage of filtering the dataset on document type is that it limits the size (by more than 20%) of the dataset, positively affecting computational cost, and that it improves class balance. This, depending on the sub-method, results in an increased recall, compensating for the small ratio of relevant chunks that are missed out on by applying the filter.

Determining the overall best method requires an optimal trade-off between precision and recall. Although capturing all relevant information is crucial, ensuring sufficient precision to minimize the presence of irrelevant information and, with that, improve the quality of the evidence base, should not be ignored. Therefore, a precision minimum of 0.66 was set, after which the results were ranked by recall. In addition, computational cost also plays a role in determining the optimal method. For establishing the adaptation evidence base, therefore, the *ClimateBERT* model fine-tuned on the filtered dataset with a CST of 70 (see table 3) is considered the most appropriate choice.

These findings underscore the potential of state-of-the-art NLP methods to narrow down relevant policy information at large scale, which may also be interesting to explore in other (policy) domains. Other suggested future research directions involve successive steps in establishing an NLP-driven adaptation tracking framework by, e.g., further unpacking and structuring the unstructured climate policy texts by identifying and categorizing adaptation-specific (policy) elements.

6 Conclusion

This work has revealed important methodological considerations for classification of adaptation policy texts. For an automated framework for identifying relevant information, with the aim of creating a dataset of adaptation policy and, with that, increasing accessibility of information needed to track progress, a fine-tuned *ClimateBERT* model has shown optimal performance. This method ensures a sufficient balance between correctly identified text, minimized missed items, and maximization of irrelevant items filtered out. To boost performance, label confidence should be taken into account during manual labelling. Following, items labelled with a confidence score of less than 60% should be excluded. Also, documents should be filtered to include only those that are known to contain adaptation-relevant information and should be split based on Markdown structure and semantic meaning, with an average of 10 paragraphs per splits. The exact length is determined by the semantic splitter, ideally with a range of 2,000-8,000 characters.

Limitations

The discussed work comes with several limitations. First, the text chunks were automatically parsed from the original PDFs and non-English text was machine-translated. The data may, therefore, contain parsing and/or translation errors, potentially affecting the results. Second, relevance labels and confidence scores were assigned by human annotators, making them exposed to subjectivity. This was also observed in the inter-annotator agreement, where the annotators disagreed in 17% of the cases. Considering this, despite extensive training, the labels and scores may not always reflect true certainty, highlighting the ambiguity of the classification task and the challenge of aligning AI predictions with human judgment. Third, the automatically generated summaries (dataset 3) were not extensively reviewed and no alternative methods or models for summarization were explored, limiting comprehensive assessment of the potential of this approach. Fourth, only two prompt variations were evaluated, which were based on prompt engineering principles (Geroimenko, 2025) and may not reflect the full potential of the zero-shot LLM approach.

Following these limitations, future work should enhance validity by, e.g., delving further into annotation consistency, evaluating alternative summarization models, and full-scale evaluation of more than these two prompt variations to assess whether the practical results align with prompt design theory.

Acknowledgements

We thank Demi Ursem, Denatalie Chrisdameria, Suze Zander, and Robin Blersch for their valuable assistance in hand-labelling the datasets used in this research. In addition, we thank the anonymous reviewers for their useful feedback on an earlier version of this manuscript. Finally, JB and RB would like to acknowledge funding support from the Dutch Research Council (NWO VI.Vidi.211.132).

References

Beltagy, I., Peters, M. E., & Cohan, A. (2020). *Longformer: The Long-Document Transformer* (arXiv:2004.05150). arXiv. https://doi.org/10.48550/arXiv.2004.05150

Bucher, M. J. J., & Martini, M. (2024). *Fine-Tuned 'Small' LLMs (Still) Significantly Outperform Zero-Shot Generative AI Models in Text Classification* (arXiv:2406.08660). arXiv. https://doi.org/10.48550/arXiv.2406.08660

Chalkidis, I., Fergadiotis, M., Malakasiotis, P., Aletras, N., & Androutsopoulos, I. (2020). *LEGAL-BERT: The Muppets straight out of Law School* (arXiv:2010.02559). arXiv. https://doi.org/10.48550/arXiv.2010.02559

Climate Policy Radar. (n.d.). *Climate Policy Radar | AI for climate law and policy research*. Climate Policy Radar. Retrieved 30 May 2024, from https://www.climatepolicyradar.org/

D'Cruz, C., Bereder, J.-M., Precioso, F., & Riveill, M. (2024). *Domain-specific long text classification from sparse relevant information* (arXiv:2408.13253). arXiv. https://doi.org/10.48550/arXiv.2408.13253

Devlin, J., Chang, M.-W., Lee, K., & Toutanova, K. (2019). *BERT: Pre-training of Deep Bidirectional Transformers for Language Understanding* (arXiv:1810.04805). arXiv. https://doi.org/10.48550/arXiv.1810.04805

Dimitar, T., Gorgi, L., Ljubomir, C., & Irena, V. (2023). Comparing the performance of ChatGPT and state-of-the-art climate NLP models on climate-related text classification tasks. *E3S Web of Conferences*, *436*, 02004. https://doi.org/10.1051/e3sconf/202343602004

Dupuis, J., & Biesbroek, R. (2013). Comparing apples and oranges: The dependent variable problem in comparing and evaluating climate change adaptation policies. *Global Environmental Change*, *23*(6), 1476–1487. https://doi.org/10.1016/j.gloenvcha.2013.07.022

Elango, P. (2005). *Coreference Resolution: A Survey*.

Facebook. (2024, January 4). *Model card for bart-large-mnli*. Hugging Face. https://huggingface.co/facebook/bart-large-mnli

Fields, J., Chovanec, K., & Madiraju, P. (2024). A Survey of Text Classification With Transformers: How Wide? How Large? How Long? How Accurate? How Expensive? How Safe? *IEEE Access*, *12*, 6518–6531. IEEE Access. https://doi.org/10.1109/ACCESS.2024.3349952

Fiok, K., Karwowski, W., Gutierrez-Franco, E., Davahli, M. R., Wilamowski, M., Ahram, T., Al-Juaid, A., & Zurada, J. (2021). Text Guide: Improving the Quality of Long Text Classification by a Text Selection Method Based on Feature Importance. *IEEE Access*, *9*, 105439–105450. IEEE Access. https://doi.org/10.1109/ACCESS.2021.3099758

Ford, J. D., Tilleard, S. E., Berrang-Ford, L., Araos, M., Biesbroek, R., Lesnikowski, A. C., MacDonald, G. K., Hsu, A., Chen, C., & Bizikova, L. (2016). Big data has big potential for applications to climate change adaptation. *Proceedings of the National Academy of Sciences*, *113*(39), 10729–10732. https://doi.org/10.1073/pnas.1614023113

Geroimenko, V. (2025). Key Principles of Good Prompt Design. In V. Geroimenko (Ed.), *The Essential Guide to Prompt Engineering: Key Principles, Techniques, Challenges, and Security Risks* (pp. 17–36). Springer Nature

Switzerland. https://doi.org/10.1007/978-3-031-86206-9_2

Han, S. (n.d.). *googletrans: Free Google Translate API for Python. Translates totally free of charge.* (Version 3.0.0) [Python; MacOS :: MacOS X, Microsoft :: Windows, POSIX].

Laurer, M., Atteveldt, W. van, Casas, A., & Welbers, K. (2024). *Building Efficient Universal Classifiers with Natural Language Inference* (arXiv:2312.17543). arXiv. https://doi.org/10.48550/arXiv.2312.17543

Lewis, M., Liu, Y., Goyal, N., Ghazvininejad, M., Mohamed, A., Levy, O., Stoyanov, V., & Zettlemoyer, L. (2020). BART: Denoising Sequence-to-Sequence Pre-training for Natural Language Generation, Translation, and Comprehension. *Proceedings of the 58th Annual Meeting of the Association for Computational Linguistics*, 7871–7880. https://doi.org/10.18653/v1/2020.acl-main.703

Li, Q., Peng, H., Li, J., Xia, C., Yang, R., Sun, L., Yu, P. S., & He, L. (2022). A Survey on Text Classification: From Traditional to Deep Learning. *ACM Trans. Intell. Syst. Technol.*, *13*(2), 31:1-31:41. https://doi.org/10.1145/3495162

Magnan, A. K., & Chalastani, V. I. (2019). *Towards a Global Adaptation Progress Tracker: First thoughts.* Institute for Sustainable Development and International Relations (IDDRI). https://www.jstor.org/stable/resrep52383

OpenAI, Hurst, A., Lerer, A., Goucher, A. P., Perelman, A., Ramesh, A., Clark, A., Ostrow, A. J., Welihinda, A., Hayes, A., Radford, A., Mądry, A., Baker-Whitcomb, A., Beutel, A., Borzunov, A., Carney, A., Chow, A., Kirillov, A., Nichol, A., … Malkov, Y. (2024). *GPT-4o System Card* (arXiv:2410.21276). arXiv. https://doi.org/10.48550/arXiv.2410.21276

Orlove, B. (2022). The Concept of Adaptation. *Annual Review of Environment and Resources*, *47*(1), 535–581. https://doi.org/10.1146/annurev-environ-112320-095719

Padurariu, C., & Breaban, M. E. (2019). Dealing with Data Imbalance in Text Classification. *Procedia Computer Science*, *159*, 736–745. https://doi.org/10.1016/j.procs.2019.09.229

Radford, A., Wu, J., Child, R., Luan, D., Amodei, D., & Sutskever, I. (2019). *Language Models are Unsupervised Multitask Learners.*

Sahoo, P., Singh, A. K., Saha, S., Jain, V., Mondal, S., & Chadha, A. (2025). *A Systematic Survey of Prompt Engineering in Large Language Models: Techniques and Applications* (arXiv:2402.07927). arXiv. https://doi.org/10.48550/arXiv.2402.07927

Sanh, V., Debut, L., Chaumond, J., & Wolf, T. (2019, October 2). *DistilBERT, a distilled version of BERT: Smaller, faster, cheaper and lighter*. arXiv.Org. https://arxiv.org/abs/1910.01108v4

Schimanski, T., Reding, A., Reding, N., Bingler, J., Kraus, M., & Leippold, M. (2024). Bridging the gap in ESG measurement: Using NLP to quantify environmental, social, and governance communication. *Finance Research Letters*, *61*, 104979. https://doi.org/10.1016/j.frl.2024.104979

Semantic text splitter (API documentation). (n.d.). Retrieved 19 December 2024, from https://semantic-text-splitter.readthedocs.io/en/stable/semantic_text_splitter.html

Shyrokykh, K., Girnyk, M., & Dellmuth, L. (2023). Short text classification with machine learning in the social sciences: The case of climate change on Twitter. *PLOS ONE*, *18*(9), e0290762. https://doi.org/10.1371/journal.pone.0290762

Sietsma, A. J., Ford, J. D., & Minx, J. C. (2024). The next generation of machine learning for tracking adaptation texts. *Nature Climate Change*, *14*(1), 31–39. https://doi.org/10.1038/s41558-023-01890-3

Sileo, D. (2024). tasksource: A Large Collection of NLP tasks with a Structured Dataset Preprocessing Framework. In N. Calzolari, M.-Y. Kan, V. Hoste, A. Lenci, S. Sakti, & N. Xue (Eds.), *Proceedings of the 2024 Joint International Conference on Computational Linguistics, Language Resources and Evaluation (LREC-COLING 2024)* (pp. 15655–15684). ELRA and ICCL. https://aclanthology.org/2024.lrec-main.1361

Touvron, H., Lavril, T., Izacard, G., Martinet, X., Lachaux, M.-A., Lacroix, T., Rozière, B., Goyal, N., Hambro, E., Azhar, F., Rodriguez, A., Joulin, A., Grave, E., & Lample, G. (2023). *LLaMA: Open and Efficient Foundation Language Models* (arXiv:2302.13971). arXiv. https://doi.org/10.48550/arXiv.2302.13971

Vaswani, A., Shazeer, N., Parmar, N., Uszkoreit, J., Jones, L., Gomez, A. N., Kaiser, Ł. ukasz, & Polosukhin, I. (2017). Attention is All you Need. *Advances in Neural Information Processing Systems*, *30*. https://proceedings.neurips.cc/paper_files/paper/2017/hash/3f5ee243547dee91fbd053c1c4a845aa-Abstract.html

Wang, W., Bao, H., Huang, S., Dong, L., & Wei, F. (2021). *MiniLMv2: Multi-Head Self-Attention Relation Distillation for Compressing Pretrained Transformers* (arXiv:2012.15828). arXiv. https://doi.org/10.48550/arXiv.2012.15828

Wang, Z., Pang, Y., & Lin, Y. (2023). *Large Language Models Are Zero-Shot Text Classifiers* (arXiv:2312.01044). arXiv. https://doi.org/10.48550/arXiv.2312.01044

Webersinke, N., Kraus, M., Bingler, J. A., & Leippold, M. (2022). *ClimateBert: A Pretrained Language Model for Climate-Related Text* (arXiv:2110.12010). arXiv. http://arxiv.org/abs/2110.12010

WorldAtlas. (2023, April 13). *The Climate Zones Of The World*. WorldAtlas. https://www.worldatlas.com/climate/the-climate-zones-of-the-world.html

Yu, H., Yang, Z., Pelrine, K., Godbout, J. F., & Rabbany, R. (2023). *Open, Closed, or Small Language Models for Text Classification?* (arXiv:2308.10092). arXiv. https://doi.org/10.48550/arXiv.2308.10092

Appendices

Appendix A. Table of nomenclature

For a clear overview of the terms used throughout the paper, one can refer to table 4. Any combination of the different levels (e.g., LLM-based classification with GPT-4o on the filtered variant of dataset 1) is referred to as a *method.* Any combination of levels 2 up to and including 5, for a given approach, is called a *sub-method*.

Appendix B. NLI task types

Table 5: Overview of tasks included in the NLI-based classification experiments

Labels	Task type
Adaptation	Binary
Adaptation policy	Binary
Climate change adaptation	Binary
Adaptation, Mitigation	Multi-class (>0.5) Multi-class (>0.6) Multi-class (>0.7)

Appendix C. Confusion matrix data strategy

Table 6: Confusion matrix of classification results on the test set in absolute numbers. Each cell shows the results of the filtered (L) versus unfiltered (R) data strategy.

True label	Predicted label: A	Predicted label: NA
A	**83 / 87**	**11** / 13
NA	**40** / 42	**188** / 301

Table 4: Table of nomenclature

Level	Name	Example(s)	Applies to...
Method (any combination of the different levels)			
1	Approach	RBC, NLI, FEM, LLM	*n/a*
Sub-method (any combination of a *setup* and *dataset variant;* a method for a given *approach*) **Setup** (levels 2 and 3; any combination of a *model, query, task (type)*, and/or *prompt*)			
2	Model	BART-large-mnli, ClimateBERT, GPT-4o	NLI, FEM, LLM
	or Query	Baseline query, data-driven query	RBC
3	Task (type)	Labels ('adaptation', 'mitigation'), task type (multi-class)	NLI
	or Prompt	P1 (concise), P2 (specific)	LLM
Dataset variant (levels 4 and 5; any combination of a *dataset* and *data strategy*)			
4	Dataset	Dataset 1 (full chunks), dataset 2b (sub-chunks, resolved)	*all*
5	Data strategy	CST0, CST70, filtered, unfiltered	*all*

LLM-Driven Estimation of Personal Carbon Footprint from Dialogues

Shuqin Li[1] **Huifang Du**[2] **Haofen Wang**[2*]
[1]Hangzhou Dianzi University Information Engineering College, Hangzhou, China
[2]Tongji University, Shanghai, China
shuqinlee9683@gmail.com, duhuifang@tongji.edu.cn, carter.whfcarter@gmail.com

Abstract

Personal Carbon Footprint (PCF) Estimation is crucial for raising individual environmental awareness by linking daily activities to their environmental impact. However, existing tools are limited by fragmented scenarios and labor-intensive manual data entry. We present PCCT, an LLM-powered system that combines conversational understanding with emission knowledge grounding for PCF Estimation. We address two key challenges: (1) resolving incomplete activity information across turns through knowledge-guided and context-aware tracking, and (2) accurately mapping emission factors using multi-step LLM inference and vector-based similarity search. The system dynamically combines knowledge-guided activity extraction, and context-aware memory management, generating accurate carbon footprint estimates. We validate the effectiveness with the *CarbonDialog-1K* benchmark, comprising 1,028 annotated user activity narratives. Experimental results demonstrate that our method outperforms baseline systems in accuracy, while subjective evaluations show superior appropriateness, usability, efficiency, and naturalness.

1 Introduction

Personal Carbon Footprint (PCF) estimation plays a pivotal role in fostering individual environmental awareness by translating daily activities into environmental impacts (Mancini et al., 2016; Lannelongue et al., 2021). Current tools, though effective in narrow domains like transportation tracking, face systemic limitations: (1) they rely on fragmented scenario definitions that hinder cross-domain analysis and (2) demand laborious manual data entry, which prevents long-term user engagement (Scrucca et al., 2021; Dreijerink and Paradies, 2020; Chen et al., 2016).

Recently, large language models (LLMs) have demonstrated impressive semantic understanding and reasoning capabilities (Radford et al., 2019; Mann et al., 2020; Bi et al., 2024). These advancements offer a promising path toward passive carbon footprint estimation through conversational interfaces. We inquire *how to accurately calculate the PCF from the casual conversations by leveraging the LLM's capabilities?*

In order to calculate the PCF, we need two parts: the user's activity parameters and the emission factor accordingly (Mariette et al., 2022). The first challenge arises from the progressive nature of conversational data. Users tend to describe their activities across multiple conversation turns with varying levels of detail and precision. For example, a user might initially mention "driving to work" and later specify "in an electric car for 15 kilometers." This requires the system to reason over the conversation history and Emission Factor database and infer the missing information or ask the user for clarification when necessary (Tu et al., 2024). Even with complete parameters, matching activities to emission factors (EFs) (Solazzo et al., 2021) is a challenging task (Balaji et al., 2023; Wood et al., 2017; Oehlert et al., 2022). Traditional practitioners either rely on manually curated mappings from millions of entries of EF database, which are time-consuming and prone to errors or use coarse-grained rule-based systems that are not scalable and accurate (Scrucca et al., 2021). This demands precise integration of domain knowledge and context-aware reasoning (Deng et al., 2023).

In this work, we propose a Progressive Contextual Carbon Tracking (PCCT) framework. Figure 2 presents our framework architecture. The PCCT framework consists of three components that work together to process multi-turn conversations about daily activities: (1) Knowledge-Guided Progressive Activity Extraction, which identifies user activities while leveraging emission factor knowledge to guide the extraction process; (2) Context-Aware Memory Management, which main-

* Corresponding author

Proceedings of the 2nd Workshop on Natural Language Processing Meets Climate Change (ClimateNLP 2025), pages 278–287
July 31, 2025

tains structured representations of activities and parameters across conversation turns; and (3) Carbon Footprint Calculation. To validate performance, we construct the *CarbonDialog-1K* benchmark containing 1,028 annotated user activity narratives with ground-truth emission values. Experimental comparisons against a rule-based expert system demonstrate our method's superior accuracy across all metrics. Complementary user studies reveal 30% higher engagement rates compared to conventional carbon tracking tools.

Our contributions are as follows:

- We propose a Progressive Contextual Carbon Tracking (PCCT) framework that calculates carbon footprint progressively in dialogues guided by knowledge. To our knowledge, this work represents the first use of LLMs for PCF estimation from conversations.
- We introduce CarbonDialog-1K to facilitate the research on carbon footprint estimation from natural language.
- Extensive experimental results demonstrate our system's superior performance in accuracy and usage effectiveness.

2 Related work

2.1 Carbon Footprint Calculation

Carbon footprint calculation has been a cornerstone of environmental research (Wiedmann and Minx, 2008), focusing on quantifying the greenhouse gas emissions associated with various activities, products, and organizations. Early methodologies, such as life cycle assessment (LCA), provided frameworks for calculating emissions across supply chains and industrial processes (Agyei Boakye et al., 2023; Chen et al., 2021). These approaches were later adapted to assess the environmental impact of individual actions, giving rise to the concept of Personal Carbon Footprint (PCF). The calculation of PCF involves translating daily activities, such as transportation, energy consumption, and dietary habits, into measurable emissions. Traditional methods rely on structured data inputs, often requiring users to manually provide detailed information such as vehicle type, fuel consumption, and travel distance (Scrucca et al., 2021). Some work is proposed to facilitate this process, enabling emissions tracking in specific domains such as commuting and household energy use (Dreijerink and Paradies, 2020; Balaji et al., 2023). However, these approaches are limited by their fragmented nature, as they often fail to account for cross-domain activities or incomplete user inputs. Recent advancements have attempted to address these limitations by integrating automated data collection methods, such as GPS tracking for transportation emissions (Chen et al., 2016; Gately et al., 2017) and smart meter integration for energy usage (Miao et al., 2024). Despite these improvements, existing tools remain labor-intensive and lack the flexibility to handle diverse and complex user activities, leading to low user engagement and long-term abandonment.

2.2 NLP Techniques for Carbon Footprint Estimation

The application of Natural Language Processing (NLP) techniques to estimate carbon footprints has gained significant attention in recent years, driven by the need for scalable and automated solutions to address the challenges of environmental impact assessment (Murphy, 1998). Traditional methods rely on manual annotation or supervised classification approaches (Sousa and Wallace, 2006), which require large labeled datasets. Recent works have explored to automate the estimation process. For instance, term frequency-inverse document frequency (TF-IDF) features are employed to classify companies into American Industry Classification System (NAICS) codes based on web-scraped text data (Wood et al., 2017; Oehlert et al., 2022). While effective, these methods are limited by their reliance on large labeled datasets and their inability to generalize to unseen or zero-shot scenarios. CaML automates the Economic Input-Output based Life Cycle Assessment (EIO-LCA) (Hendrickson et al., 1998) process by using semantic text similarity matching with the pre-trained model SBERT (Reimers and Gurevych, 2019) to screen and rank product and industry sector matches (Balaji et al., 2023). However, despite the progress made, current NLP-based approaches still struggle to perform accurate carbon emission calculations due to their inadequate context-aware reasoning capabilities, which fail to fully capture the complex relationships in environmental data.

3 Dataset Construction

To enable research on carbon footprint estimation through natural conversations, we construct a comprehensive dataset containing 1,028 multi-turn dialogues (*CarbonDialog-1K*) through a two-stage process: (1) building a daily activity emission factor database and (2) generating natural multi-turn dialogues. Rather than collecting real-world conversations and manually annotating them, we adopt a reverse engineering approach (Wang et al., 2015) to ensure comprehensive coverage and calculation accuracy.

3.1 Activity-EF Database Construction

The first stage focuses on building a comprehensive database of emission factors (EFs) for daily activities. We first collect a diverse set of daily activities that potentially contribute to personal carbon footprints. To ensure comprehensive coverage, we leverage large language models to generate a wide range of activities in 6 categories: transportation, energy consumption, food and beverages, consumer goods, services, and entertainment. For each activity, we query the Climatiq API[1] to obtain standardized emission factors.

Given the significant regional variations in emission factors due to differences in energy mix, infrastructure, and economic development, we focus on activities with well-documented emission factors in selected regions (New Zealand, United Kingdom, and Germany, China, and United States).

3.2 Dialogue Generation

We then adopt the Activity-EF Database to generate dialogues through a reverse engineering approach. We first construct the ground truth activities and their carbon footprints, then generate dialogues that would lead to these calculations. The process consists of three main phases: (1) activity set construction, (2) narrative and parameter generation, and (3) progressive dialogue synthesis. Algorithm 1 formalizes this process.

For each dialogue, we first sample 3-5 activities from the database that are compatible both logically (e.g., avoiding conflicting transportation modes) and regionally (sharing the same region for consistent emission factors). Each activity is then assigned temporal context and enriched with necessary parameters required by its emission factor calculation:

[1] https://www.climatiq.io, Climatiq maintains a database of emission factors from authoritative sources such as government environmental agencies and research institutions.

Algorithm 1 Dialogue Generation

Input: Activity-EF database $\mathcal{E}$, Sample size N
Output: Dialogue dataset $\mathcal{D}$
for $i = 1$ to N **do**
  Select region r randomly
  $n \leftarrow$ RandActivityNum()
  $A_r \leftarrow$ SampleActivities(n, $\mathcal{E}$, r)
  for each activity $a \in A_r$ **do**
    Assign temporal context to a
    $D_a \leftarrow$ GenerateDescription(a)
    $P_a \leftarrow$ GenerateParameter(a)
    $C_a \leftarrow$ CalculateCarbonFootprint(a)
  end for
  $C_r \leftarrow$ SumCarbonFootprint($\{C_a\}$)
  $A_r \leftarrow$ GroupActivities(A_r)
  $T \leftarrow$ ToDialogue(A_r, $\{D_a, P_a, C_a\}$, C_r)
  $\mathcal{D} \leftarrow \mathcal{D} \cup \{T, C_r\}$
end for
return $\mathcal{D}$

$$CO_2e = \sum_{i=1}^{n} P_i \times EF_i \tag{1}$$

where P_i represents the activity parameter (e.g., distance traveled, fuel consumed) and EF_i is the corresponding emission factor from our database.

The selected activities are then transformed into natural narratives using LLMs, with parameters strategically distributed across multiple dialogue turns. Then, we break an activity's description into multiple turns. This design mirrors typical human conversation patterns, where users often begin with basic activity descriptions before gradually adding specific details. Table 1 shows an example dialogue constructed through this process.

3.3 Quality Control

We implement a comprehensive quality control pipeline integrating automated filtering, manual review, and LLM-assisted validation. Initially, automatic filters exclude anomalous data by removing: 1) activities with emission factors beyond three standard deviations from category means, indicative of industrial rather than personal behavior; 2) parameter outliers exceeding typical personal usage scenarios (e.g., flights over 20,000 km, meals exceeding \$500); and 3) activities incompatible with personal carbon footprints (e.g., industrial production, commercial shipping).

Subsequently, we validate remaining activities via a hybrid approach, combining manual verifi-

```
User: Hi, I'd like to track my carbon footprint for today.
System: Hello! I'll help you track your carbon footprint. Please tell me about your activities
        for today.
User: I swung by a cafe on my way to work and picked up a coffee and pastry for $5. Then I took
      a taxi to the city centre. That's all.
System: Thank you. I've recorded the cafe purchase. For the taxi ride, I'll need to know the
        distance traveled. Could you tell me how long was your taxi ride?
User: The taxi ride was about 5km.
System: Thank you. Here's your carbon footprint breakdown:
        1. Cafe purchase: $5 × 1.1224 kg/NZD = 5.612 kg COe
        2. Taxi ride: 5km × 0.018894 kg/km = 0.09447 kg COe
        Total Carbon Footprint: 5.71 kg COe
        Would you like to add any other activities?
```

Table 1: Example dialogue from our dataset showing users tend to describe their activities progressively across multiple turns and the system ask for missing information incrementally and calculate the carbon footprint when all information is provided.

Metric	Value
Total Dialogues	1,028
Unique Activities	196
Avg. Turns / Dialogue	6.96 (±3.47)
Avg. Activities / Dialogue	2.64 (±0.82)
Avg. COe / Dialogue	16.65 (±13.82) kg

Table 2: Main statistics of the dataset

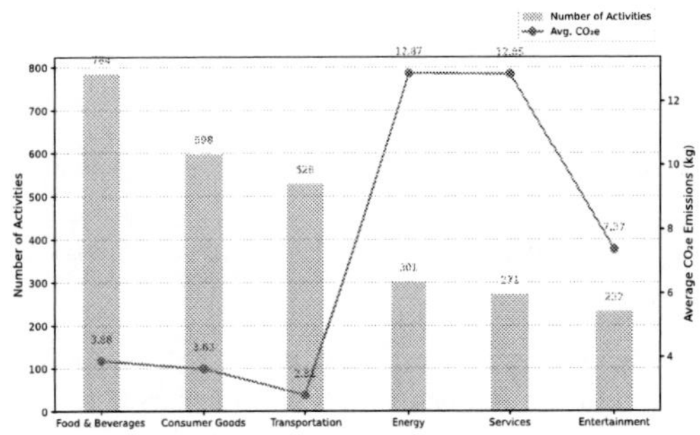

Figure 1: Category-wise distribution of activities and their carbon footprint. The bars (blue) show the number of activities in each category, while the line (red) shows the average CO_2e emissions per activity.

cation (20% of the dataset) with LLM-based validation (80%), examining activity compatibility, parameter reasonableness, and calculation accuracy. The LLM-based validation achieves 89% consistency with human evaluators on a test set of 50 dialogues. Overall, our process filters out approximately 10% of initial dialogues, significantly enhancing dataset quality.

3.4 Dataset Statistics

Table 2 presents the key statistics of our dataset. The final dataset comprises 1,028 multi-turn conversations distributed across seven regions (New Zealand, United Kingdom, Germany, France, United States, China, and Global). Figure 1 shows the distribution of activities across six major categories. Food & Beverages represents the largest category with 784 activities, followed by Consumer Goods (598) and Transportation (528). For average emissions, Energy and Services categories contribute the highest average emissions.

The comprehensive annotation and diverse activity coverage make this dataset valuable for advancing research in conversational carbon footprint estimation. The dataset will be made publicly available to support further research in this important area[2].

4 Method

Our approach addresses the challenges of carbon footprint calculation through natural conversations by introducing a Progressive Contextual Carbon Tracking (PCCT) framework. Figure 2 presents our framework architecture.

The PCCT framework consists of three components that work together to process multi-turn conversations about daily activities: (1) Knowledge-Guided Progressive Activity Extraction, which identifies user activities while leveraging emission factor knowledge to guide the extraction process; (2) Context-Aware Memory Management, which maintains structured representations of activities and parameters across conversation turns; and (3)

[2]Dataset available at: `https://github.com/shuqinlee/Chat2CarbonFootprint.git`

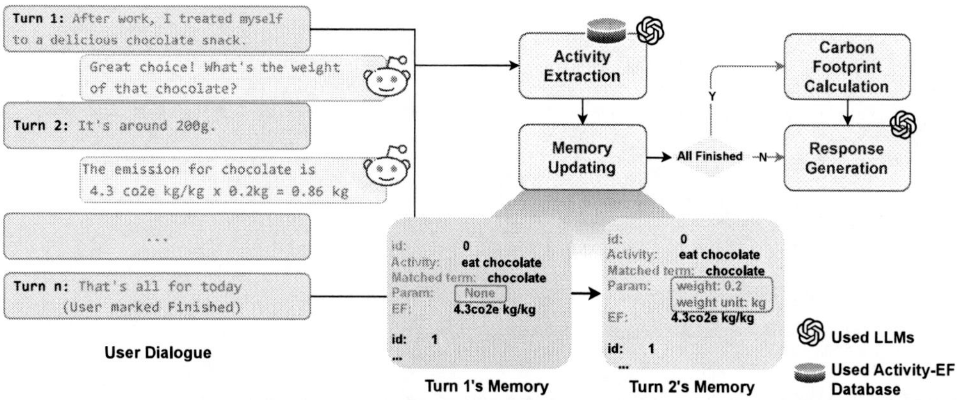

Figure 2: Overview of our Progressive Contextual Carbon Tracking (PCCT) framework. The system processes user inputs through three main components: (1) Activity Extraction, which identifies activities and parameters guided by Activity-EF Database, (2) Memory Updating, which maintains a context memory of previously extracted activities and their parameters, enabling incremental updates and refinements as new information becomes available, (3) Carbon Footprint Calculation.

Carbon Footprint Calculation

4.1 Knowledge-Guided Progressive Activity Extraction

A key challenge in conversational carbon footprint calculation is that users tend to describe their activities progressively across multiple turns, often with varying levels of detail and precision. For example, a user might initially mention "driving to work" and later specify "in an electric car for 15 kilometers." This natural communication pattern creates the challenge of correctly identifying and updating activities with their parameters across multiple turns. Our knowledge-Guided approach addresses these challenges through a bidirectional interaction between LLMs and activity-emission factor database.

We develop a progressive extraction process where emission factor knowledge guides parameter identification (Figure 3). The system first uses LLMs to recognize activity descriptions from user utterances. These descriptions are then matched against our emission factor database, which identifies the most relevant emission factors along with their required parameters. For example, if a user mentions "driving to work," the system identifies potential emission factors that might require parameters such as distance and passengers.

4.1.1 Efficient Emission Factor Retrieval

A critical support component for our knowledge-guided approach is an efficient retrieval system that

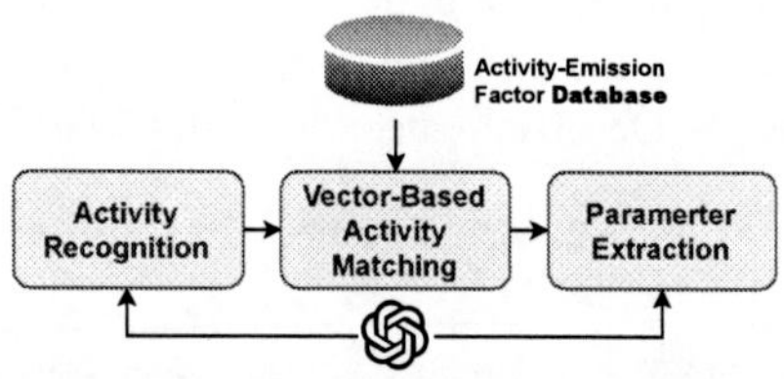

Figure 3: Knowledge-Guided Progressive activity extraction process. The system maintains awareness of previously extracted activities and their parameters, enabling it to identify both new activities and updates to existing ones.

can quickly match described activities with appropriate emission factors. Matching activities with emission factors from large databases is typically time-consuming and requires professional expertise(Balaji et al., 2023). To address this challenge, we implement a vector-based retrieval system using FAISS library (Douze et al., 2024).

The retrieval system operates in two stages. Firstly, we construct offline index by precomputing dense vector representations for all emission factors in our database, incorporating both activity descriptions and their associated parameters. Next, during conversation, extracted activities are encoded into the same vector space and matched against the pre-built indices. This enables sub-second retrieval of relevant emission factors, even from databases containing thousands of entries which guides the subsequent parameter extraction stage.

4.2 Context-Aware Memory Management

To maintain a structured representation of all extracted information across turns, our memory module maintains two interconnected components: 1) Activity Registry maintains all identified activities with their emission factor mappings, tracking the state of each activity (complete or partially specified); 2) Parameter Registry tracks all extracted parameters for each activity and maintains all missing information. This including their values, units, extraction confidence, and source turn information.

In each turn, the activity extraction module identifies incremental activity information and updates the memory rather than creating duplicates. It provides a complete view of missing parameters, enabling targeted follow-up questions; and it maintains extraction confidence scores that help prioritize which missing information is most critical to obtain.

4.3 Carbon Footprint Calculation

The final component of our framework transforms the structured activity and parameter information into accurate carbon footprint calculations. Once the system has collected sufficient information about the user's activities, it applies the appropriate emission factors to calculate the carbon footprint. For each activity, the system applies the appropriate emission factor to the validated parameters, calculating the carbon footprint using the formula: $CO_2e = \sum_{i=1}^{n} P_i \times EF_i$ where P_i represents the activity parameter (e.g., distance traveled) and EF_i is the corresponding emission factor.

After the calculation for each activity, the system generates detailed explanations of the calculation process, breaking down the contribution of each activity to the total carbon footprint and gives advice.

5 Objective Experiments

We conduct objective experiments to evaluate the effectiveness of our PCCT framework against baseline approaches. Our evaluation focuses on three key aspects: activity recognition accuracy, parameter extraction completeness, and carbon footprint calculation accuracy.

5.1 Experimental Setup

Dataset. We evaluate all systems using our CarbonDialog-1K dataset. For testing, we use a randomly selected subset of 560 dialogues, ensuring balanced coverage across activity categories and regions. The remaining dialogues are used for system development and parameter tuning.

Metrics. We assess system performance using several key metrics. For activity recognition, we measure the F1-score for correctly identifying activities from user descriptions. Parameter extraction performance is evaluated using both the F1-score for parameter identification and the Missing Critical Parameter Rate (MCPR) for missing critical parameters. Finally, to assess emission calculation accuracy, we compute the Mean Absolute Error (MAE) in kg CO_2e and Mean Absolute Percentage Error (MAPE).

Baseline System. Our goal is to estimate personal carbon footprint (PCF) from open-ended, multi-domain text-based dialogues. To evaluate this, we compare PCCT with a rule-based calculator (**RBC**) that uses keyword matching and fixed templates to compute emissions from structured inputs. This reflects conventional carbon tools based on forms or explicit prompts. GPS-based trackers focus mainly on transport, require continuous sensing and device integration, and cover a limited range of activities. As they are not comparable to general-purpose, text-based dialogue systems, we exclude them from our evaluation.

Implementation Details. Our PCCT implementation uses a combination of LLMs and vector similarity search. For the LLM component, we employ the DeepSeek-V3 model (Liu et al., 2024) for activity extraction and parameter identification. The vector similarity search uses FAISS (Douze et al., 2024) with 768-dimensional embeddings from sentence transformer (Reimers and Gurevych, 2019) for emission factor retrieval.

5.2 Results and Analysis

5.2.1 Overall Performance

Table 3 presents the overall performance comparison between our PCCT system and the rule-based baseline.

The results demonstrate that PCCT significantly outperforms the rule-based approach across all metrics. Our system achieves a substantial improvement in activity recognition and parameter extraction. This can be attributed to the knowledge-augmented activity extraction that effectively matches user descriptions with standard-

System	Activity F1 (%)	Parameter F1 (%)	MCPR (%)	MAE (kg CO_2e)	MAPE (%)
Rule-Based Calculator	30.8	18.5	45.0	9.7	74.3
PCCT (Ours)	**74.2**	**75.5**	**15.5**	**4.8**	**39.6**

Table 3: Overall performance comparison of PCCT vs. Rule-based Calculator. Our system outperforms the rule-based calculator across all metrics. MCPR reveals the missing rate of critical parameter. MAPE reveals relative error of carbon footprint.

Category	MAPE (%)	
	RBC	PCCT
Transportation	55.8	**28.5**
Food & Beverages	68.5	**42.3**
Consumer Goods	95.8	**45.8**
Entertainment	85.5	**38.5**
Services	115.6	**58.6**
Energy	95.4	**65.4**

Table 4: Category-level performance comparison sorted by MAPE (lower values indicate better performance)

ized emission factors, (2) the context-aware memory management that maintains coherent information across turns, and (3) the progressive parameter-gathering strategy that ensures the completeness of critical information.

The improved activity recognition and parameter extraction directly translate to calculation accuracy, with PCCT achieving a 50% lower Mean Absolute Error (MAE: 4.8 vs 9.7 kg CO_2e) and a reduction of 34.7 percentage points in Mean Absolute Percentage Error (MAPE: 39.6% vs 74.3%). Importantly, while the rule-based system can only attempt calculations for the small subset of activities where it successfully extracts all parameters (18.5% of cases), PCCT maintains high accuracy while handling a much broader range of activities and conversation patterns.

5.2.2 Category-level Performance

A detailed analysis of performance across different activity categories reveals significant variations in calculation accuracy, as shown in Table 4.

Our analysis reveals several important patterns that highlight the effectiveness of our knowledge-augmented approach:

Transportation activities show the highest accuracy for both systems, with PCCT achieving the lowest MAPE (28.5%). The well-defined parameter requirements (primarily distance) benefit from our progressive extraction approach, allowing the system to focus on gathering specific, critical information.

Food & Beverages and Consumer Goods show moderate error rates with PCCT (MAPE: 42-46%). These categories represent the most frequent activities in our dataset, they demonstrate the value of our retrieval system.

Energy and Services present the greatest challenges for both systems, with the highest error rates even with PCCT (MAPE: 58-65%). These categories involve inducing parameters that the users may not themselves know, and the system may not have access to. For example, the amount of electricity consumed by a service may not be known, therefore this brings in more uncertainty.

The performance gap between PCCT and RBC is consistent across all categories, with PCCT reducing MAPE by 48-57%. Note that the rule-based system's MAPE values are only applicable to the small subset of activities where it successfully extracted all critical parameters (Activity F1: 30.8%, Parameter F1: 18.5%). Therefore, the performance gap between PCCT and RBC is even larger for these categories.

These results demonstrate that while certain activity types remain challenging for carbon footprint calculation, PCCT's integrated approach of knowledge-augmented extraction, context-aware memory, and progressive parameter gathering significantly improves accuracy across all categories.

6 Subjective Experiments

To further evaluate the effectiveness of our system in comparison to the baseline system carbon footprint calculation methods, we conducted a controlled subjective experiment.

6.1 Experiment Design

We recruited 20 participants, including environmental enthusiasts, general users, and sustainability researchers, to calculate their carbon footprint for 10 predefined daily activities of varying complexity. Participants used our system, a conversational interface that guided them through activity details, resolved incomplete information, and provided carbon footprint estimates with confidence bounds.

They also used traditional tools like spreadsheets or online calculators requiring manual data entry. Participants rated their experience on a 7-point Likert scale for usability (Zwakman et al., 2020), appropriateness (Torrey et al., 2013; Peng et al., 2019), efficiency (Siro et al., 2022), and naturalness (CAO et al., 2023), and provided qualitative feedback through semi-structured interviews. Task completion time, result accuracy, and user engagement metrics (e.g., errors, frustration) were recorded.

6.2 Results

Usability. Our system received higher usability ratings (mean = 4.6, SD = 0.5) than the baseline system (mean = 3.2, SD = 0.8). Users praised the intuitive interface of our syetem, while the baseline system users found data entry frustrating.

Appropriateness. Both methods scored similarly (Our system: mean = 4.7, SD = 0.4; The baseline system: mean = 4.5, SD = 0.6), though our system's context-aware explanations were appreciated.

Efficiency. Our system was more efficient (mean = 4.5, SD = 0.5) than the baseline system (mean = 2.8, SD = 0.9), with task completion times of 12 minutes vs. 28 minutes.

Naturalness. Our system scored higher on naturalness (mean = 4.8, SD = 0.3) compared to the baseline system (mean = 2.5, SD = 0.7), with its conversational interface being a key factor.

6.3 Statistical Analysis

A two-way ANOVA confirmed significant differences between methods ($F(1, 236) = 45.3$, $p < 0.001$) and dimensions ($F(3, 236) = 12.7$, $p < 0.001$), with an interaction effect ($F(3, 236) = 8.2$, $p < 0.001$). Post-hoc tests showed our system outperformed the baseline system in usability, efficiency, and naturalness ($p < 0.001$), but not in appropriateness ($p > 0.05$).

6.4 Qualitative Feedback

Participants praised our system for its intuitive, conversational interface and transparency, with one noting, "It was so easy to use—I didn't need prior knowledge of carbon emissions," and another appreciating its guidance and confidence-bound estimates. In contrast, the baseline system users found data entry and emission factor lookup tedious and error-prone, with comments like, "I spent more time looking up emission factors than calculating," and frustration over the lack of guidance. Many wished for explanatory features similar to our system. Feedback highlighted our system's strengths in usability, transparency, and engagement while revealing inefficiencies in rule-based system.

7 Conclusion

We introduce PCCT, a novel framework for calculating personal carbon footprints through natural conversations. By integrating knowledge-guided activity extraction, context-aware memory management, and progressive parameter gathering, PCCT bridges the gap between casual dialogue and precise carbon footprint estimation. Experimental results demonstrate PCCT's significant advantages over traditional approaches, achieving substantial improvements in both automated understanding and calculation accuracy. Our subjective experiments further validate PCCT's practical value, with users particularly appreciating its intuitive interface and conversational guidance. Our work establishes a promising direction for making carbon footprint estimation more accessible through natural interactions.

Limitations

While PCCT demonstrates promising results in personal carbon footprint estimation, several limitations should be acknowledged:

First, the accuracy of our system heavily depends on the quality and coverage of the emission factor database. For emerging or region-specific activities, the system may struggle to find appropriate emission factors, potentially leading to less accurate calculations. This limitation is particularly evident in service-related activities where standardized emission factors are often lacking. Second, our evaluation, while comprehensive, is limited to a curated benchmark dataset and controlled user studies. Real-world deployment may present additional challenges, such as handling extremely rare activities and adapting to regional variations in emission factors, and maintaining consistent performance across different user demographics. Third, the PCCT pipeline relies on LLM inference, which itself incurs a non-negligible carbon cost. Future work will fine-tune an expert model to a 7 B parameter variant, further reducing inference emissions.

Acknowledgements

This work was supported by the National Natural Science Foundation of China (U23B2057, 62176185), and the Youth Fund Special Project at Hangzhou Dianzi University Information Engineering College (KYP0323010).

References

Amma Asantewaa Agyei Boakye, Terrie Boguski, Sarah Cashman, Christoph Koffler, Ashley Kreuder, Manish Kumar, Naveen Kumar Vipparla, and Lisa Peterson. 2023. At the intersection of life cycle assessment and indirect greenhouse gas emissions accounting. *The International Journal of Life Cycle Assessment*, 28(4):321–335.

Bharathan Balaji, Venkata Sai Gargeya Vunnava, Geoffrey Guest, and Jared Kramer. 2023. Caml: Carbon footprinting of household products with zero-shot semantic text similarity. In *Proceedings of the ACM Web Conference 2023*, pages 4004–4014.

Xiao Bi, Deli Chen, Guanting Chen, Shanhuang Chen, Damai Dai, Chengqi Deng, Honghui Ding, Kai Dong, Qiushi Du, Zhe Fu, et al. 2024. Deepseek llm: Scaling open-source language models with longtermism. *arXiv preprint arXiv:2401.02954*.

Jianqin CAO, Jingyu ZHANG, Liang ZHANG, and Xiaoyu WANG. 2023. The psychological structure and influence of interactive naturalness. *Acta Psychologica Sinica*, 55(1):55.

Chen Chen, Zengfeng Zhao, Jianzhuang Xiao, and Robert Tiong. 2021. A conceptual framework for estimating building embodied carbon based on digital twin technology and life cycle assessment. *Sustainability*, 13(24):13875.

Shimon Chen, Shlomo Bekhor, David M Broday, et al. 2016. Aggregated gps tracking of vehicles and its use as a proxy of traffic-related air pollution emissions. *Atmospheric Environment*, 142:351–359.

Zhu Deng, Jinjie Liu, Biao Luo, Can Yuan, Qingrun Yang, Lei Xiao, Wenwen Zhou, and Zhu Liu. 2023. Autopcf: Efficient product carbon footprint accounting with large language models. *arXiv preprint arXiv:2308.04241*.

Matthijs Douze, Alexandr Guzhva, Chengqi Deng, Jeff Johnson, Gergely Szilvasy, Pierre-Emmanuel Mazaré, Maria Lomeli, Lucas Hosseini, and Hervé Jégou. 2024. The faiss library. *arXiv preprint arXiv:2401.08281*.

LJM Dreijerink and GL Paradies. 2020. How to reduce individual environmental impact? a literature review into the effects and behavioral change potential of carbon footprint calculators. *TNO Report*, 11148.

Conor K Gately, Lucy R Hutyra, Scott Peterson, and Ian Sue Wing. 2017. Urban emissions hotspots: Quantifying vehicle congestion and air pollution using mobile phone gps data. *Environmental pollution*, 229:496–504.

Chris Hendrickson, Arpad Horvath, Satish Joshi, Octavio Juarez, Lester Lave, H Scott Matthews, Francis C McMichael, and Elisa Cobas-Flores. 1998. Economic input-output-based life cycle assessment (eiolca). *mental*.

Loïc Lannelongue, Jason Grealey, and Michael Inouye. 2021. Green algorithms: quantifying the carbon footprint of computation. *Advanced science*, 8(12):2100707.

Aixin Liu, Bei Feng, Bing Xue, Bingxuan Wang, Bochao Wu, Chengda Lu, Chenggang Zhao, Chengqi Deng, Chenyu Zhang, Chong Ruan, et al. 2024. Deepseek-v3 technical report. *arXiv preprint arXiv:2412.19437*.

Maria Serena Mancini, Alessandro Galli, Valentina Niccolucci, David Lin, Simone Bastianoni, Mathis Wackernagel, and Nadia Marchettini. 2016. Ecological footprint: refining the carbon footprint calculation. *Ecological indicators*, 61:390–403.

Ben Mann, N Ryder, M Subbiah, J Kaplan, P Dhariwal, A Neelakantan, P Shyam, G Sastry, A Askell, S Agarwal, et al. 2020. Language models are few-shot learners. *arXiv preprint arXiv:2005.14165*, 1:3.

Jérôme Mariette, Odile Blanchard, Olivier Berné, Olivier Aumont, Julian Carrey, AnneLaure Ligozat, Emmanuel Lellouch, Philippe-Emmanuel Roche, Gaël Guennebaud, Joel Thanwerdas, et al. 2022. An open-source tool to assess the carbon footprint of research. *Environmental Research: Infrastructure and Sustainability*, 2(3):035008.

Bo Miao, Jie Wen, Qiujie Yuan, Kaisaier Aisikaer, Daming Zhao, and Zhao Zhao. 2024. Research on carbon footprint analysis method of a-class single-phase cost-controlled intelligent energy meter product. In *2024 8th International Conference on Electrical, Mechanical and Computer Engineering (ICEMCE)*, pages 558–563. IEEE.

John B Murphy. 1998. Introducing the north american industry classification system. *Monthly Lab. Rev.*, 121:43.

Christine Oehlert, Evan Schulz, and Anne Parker. 2022. Naics code prediction using supervised methods. *Statistics and Public Policy*, 9(1):58–66.

Zhenhui Peng, Yunhwan Kwon, Jiaan Lu, Ziming Wu, and Xiaojuan Ma. 2019. Design and evaluation of service robot's proactivity in decision-making support process. In *Proceedings of the 2019 CHI Conference on Human Factors in Computing Systems*, pages 1–13.

Alec Radford, Jeffrey Wu, Rewon Child, David Luan, Dario Amodei, Ilya Sutskever, et al. 2019. Language models are unsupervised multitask learners. *OpenAI blog*, 1(8):9.

Nils Reimers and Iryna Gurevych. 2019. Sentence-bert: Sentence embeddings using siamese bert-networks. *arXiv preprint arXiv:1908.10084*.

Flavio Scrucca, Grazia Barberio, Valentina Fantin, Pier Luigi Porta, and Marco Barbanera. 2021. Carbon footprint: Concept, methodology and calculation. *Carbon Footprint Case Studies: Municipal Solid Waste Management, Sustainable Road Transport and Carbon Sequestration*, pages 1–31.

Clemencia Siro, Mohammad Aliannejadi, and Maarten de Rijke. 2022. Understanding user satisfaction with task-oriented dialogue systems. In *Proceedings of the 45th International ACM SIGIR conference on research and development in information retrieval*, pages 2018–2023.

Efisio Solazzo, Monica Crippa, Diego Guizzardi, Marilena Muntean, Margarita Choulga, and Greet Janssens-Maenhout. 2021. Uncertainties in the emissions database for global atmospheric research (edgar) emission inventory of greenhouse gases. *Atmospheric Chemistry and Physics*, 21(7):5655–5683.

Ines Sousa and David Wallace. 2006. Product classification to support approximate life-cycle assessment of design concepts. *Technological Forecasting and Social Change*, 73(3):228–249.

Cristen Torrey, Susan R Fussell, and Sara Kiesler. 2013. How a robot should give advice. In *2013 8th ACM/IEEE International Conference on Human-Robot Interaction (HRI)*, pages 275–282. IEEE.

Qingshi Tu, Jing Guo, Nan Li, Jianchuan Qi, and Ming Xu. 2024. Mitigating grand challenges in life cycle inventory modeling through the applications of large language models. *Environmental Science & Technology*, 58(44):19595–19603.

Yushi Wang, Jonathan Berant, and Percy Liang. 2015. Building a semantic parser overnight. In *Proceedings of the 53rd Annual Meeting of the Association for Computational Linguistics and the 7th International Joint Conference on Natural Language Processing (Volume 1: Long Papers)*, pages 1332–1342, Beijing, China. Association for Computational Linguistics.

Thomas Wiedmann and Jan Minx. 2008. A definition of 'carbon footprint'. *Ecological economics research trends*, 1(2008):1–11.

Sam Wood, Rohit Muthyala, Yi Jin, Yixing Qin, Nilaj Rukadikar, Amit Rai, and Hua Gao. 2017. Automated industry classification with deep learning. In *2017 IEEE International Conference on Big Data (Big Data)*, pages 122–129. IEEE.

Dilawar Shah Zwakman, Debajyoti Pal, Tuul Triyason, and Chonlameth Arpnikanondt. 2020. Voice usability scale: measuring the user experience with voice assistants. In *2020 IEEE International Symposium on Smart Electronic Systems (iSES)(Formerly iNiS)*, pages 308–311. IEEE.

Can Reasoning LLMs Synthesize Complex Climate Statements?

Yucheng Lu
New York University, New York, USA
yuchenglu@nyu.edu

Abstract

Accurately synthesizing climate evidence into concise statements is crucial for policy making and fostering public trust in climate science. Recent advancements in Large Language Models (LLMs), particularly the emergence of reasoning-optimized variants, which excel at mathematical and logical tasks, present a promising yet untested opportunity for scientific evidence synthesis. We evaluate state-of-the-art reasoning LLMs on two key tasks: (1) *contextual confidence classification*, assigning appropriate confidence levels to climate statements based on evidence, and (2) *factual summarization of climate evidence*, generating concise summaries evaluated for coherence, faithfulness, and similarity to expert-written versions. Using a novel dataset of 612 structured examples constructed from the Sixth Assessment Report (AR6) of the Intergovernmental Panel on Climate Change (IPCC), we find reasoning LLMs outperform general-purpose models in confidence classification by 8 percentage points in accuracy and macro-F1 scores. However, for summarization tasks, performance differences between model types are mixed. Our findings demonstrate that reasoning LLMs show promise as auxiliary tools for confidence assessment in climate evidence synthesis, while highlighting significant limitations in their direct application to climate evidence summarization. This work establishes a foundation for future research on the targeted integration of LLMs into scientific assessment workflows. Code and data are publicly available at `https://github.com/YuchengLu-NYU/LLMClimateSynthesis`.

1 Introduction

Climate science involves complex systems, intricate modeling approaches, and specialized terminology that create significant barriers to public understanding (Sterman, 2011; Somerville and Hassol, 2011; Bernauer and McGrath, 2016). Despite overwhelming scientific consensus on climate change, this complexity hinders widespread awareness and informed decision-making, even among policymakers responsible for addressing this global challenge (Pidgeon and Fischhoff, 2011). The extensive body of scientific evidence, while providing nuanced understanding of the systems and causal mechanisms driving climate change, simultaneously complicates efforts to communicate clear, actionable information—a fundamental challenge at the intersection of science, policy, and public engagement (van Eck, 2023). Large Language Models (LLMs) offer promising capabilities for addressing this communication gap. With their ability to process and synthesize vast amounts of text data, LLMs could potentially serve as powerful tools for distilling complex climate science into accessible formats (To et al., 2024; Bulian et al., 2024a). However, the nuanced nature of scientific evidence in climate research, with its inherent uncertainties and complex causal relationships, presents challenges that may exceed the capabilities of general-purpose LLMs (Bulian et al., 2024b). Recent developments in AI have produced specialized reasoning-optimized LLMs, which are explicitly designed to perform multi-step logical analysis and incorporate chain-of-thought processes that mirror analytical reasoning. These models are trained using reinforcement learning techniques to improve their ability to handle complex logical and mathematical problems (Cheng et al., 2025). In this study, we evaluate two state-of-the-art (SOTA) reasoning LLMs: DeepSeek-R1 (DeepSeek-AI et al., 2025) and OpenAI's o3-mini (OpenAI, 2025).[1] As a baseline, we also perform the same two tasks on

[1]We selected o3-mini over OpenAI's flagship reasoning model o1-pro and o1 due to availability and cost considerations. At the time of writing, o1-pro is not available as an API, whereas o1 costs $60.00 per million output tokens, including reasoning tokens, compared to o3-mini's $4.40 and DeepSeek-R1's regular price of $2.19 (discount price $0.55). These cost differences have significant implications for practical applications in research and deployment settings.

Proceedings of the 2nd Workshop on Natural Language Processing Meets Climate Change (ClimateNLP 2025), pages 288–303
July 31, 2025

GPT-4o, one of the most widely used and capable general-purpose LLMs available. Our research makes three key contributions:

- We develop a focused dataset of 612 structured examples derived from the IPCC AR6, specifically designed for evaluating climate science evidence synthesis. Though modest in size, this curated resource offers high-quality pairs of scientific evidence bases with expert-written summaries and standardized confidence levels, providing a specialized benchmark for both classification and generative tasks in climate communication.

- To our knowledge, we conduct the first evaluation of reasoning LLMs for climate evidence synthesis, assessing their ability to assign appropriate confidence levels to climate statements based on presented evidence. Moreover, we show that the strong performance of LLMs is not the result of pure memorization by benchmarking against "no evidence" prompts, where we provide reference to specific sections in IPCC AR6 but withhold actual evidences in context.

- We evaluate these models' summarization abilities on complex climate evidence, revealing important insights about the distinct skills required for effective scientific communication versus classification tasks. This analysis highlights the specific capabilities needed for translating scientific evidence into accessible formats for policymakers and the public.

These contributions collectively advance our understanding of how AI systems might address the critical challenge of communicating climate science more effectively, potentially facilitating greater public understanding and more informed policymaking in this crucial domain.

2 Related Work

Climate Science and NLP The application of NLP techniques to climate science has gained increasing popularity in recent years (Stammbach et al., 2024). Incorporating artificial intelligence in the assessment and communication of climate statements is among the most important research directions within the Climate NLP research program. Costa et al. (2024) introduced ClimateQ&A, a dataset and LLM-based assistant that answers climate and biodiversity-related questions grounded in scientific reports from the IPCC and IPBES, which builds upon previous related works (Morio and Manning, 2023; De-Gol et al., 2023; Muccione et al., 2024; Schimanski et al., 2024; Mullappilly et al., 2023).

However, research specifically focusing on climate evidence synthesis and assessment remains nascent. Joe et al. (2024) conducted a preliminary evaluation of GPT-4o's capabilities for climate change evidence synthesis and systematic assessments, but primarily focused on information extraction rather than comprehensive evidence evaluation. Similarly, Li et al. (2024b) extracted climate change statements in IPCC reports to understand patterns of confidence levels and evidence types, while Lacombe et al. (2023) developed CLIMATEX, which assessed statements from IPCC AR6 reports without their supporting evidence bases. These works emphasized information retrieval capabilities of general-purpose LLMs rather than evidence synthesis or confidence attribution.

Our work differs significantly by evaluating models' abilities to not only extract climate knowledge but to synthesize evidence and assign appropriate confidence levels—tasks more directly aligned with scientific communication needs. Furthermore, we specifically examine reasoning-optimized LLMs, which have not previously been evaluated for climate evidence synthesis tasks.

Reasoning LLMs Recent advancements in LLMs have led to specialized variants designed specifically for reasoning tasks. These models incorporate architectural innovations and targeted training methodologies to enhance their logical and multi-step reasoning capabilities. DeepSeek-R1 and OpenAI's o3-mini represent SOTA examples in this class of models, balancing exceptional performance with computational efficiency.

The broader landscape of reasoning in LLMs has been extensively studied. Huang and Chang (2023) provides a comprehensive survey of reasoning capabilities in LLMs, identifying key methodologies that enable more sophisticated logical analysis. Notably, Wei et al. (2022) demonstrated that chain-of-thought prompting significantly enhances reasoning performance across various benchmarks. Both DeepSeek-R1 and OpenAI's o3-mini incorporates explicit chain-of-thought in their reasoning. Additionally, Sun et al. (2024) categorizes various reasoning frameworks in foundation models, emphasizing the unique strengths of models optimized for reasoning tasks. Xu et al. (2025) surveyed the application of reinforcement learning (RL) in improving LLMs' reasoning capacity, a training technique employed by both DeepSeek-R1 and o3-mini.

Despite these advances, the application of reasoning LLMs to scientific evidence synthesis remains relatively unexplored, particularly in domains like climate science where uncertainty quantification and nuanced interpretation are essential for effective communication and policy guidance.

Evidence Synthesis with LLMs The task of synthesizing scientific evidence and assigning appropriate confidence levels has traditionally been performed by human experts following established protocols (IPCC, 2010; Mastrandrea et al., 2011). Recent work by (Van Veen et al., 2023; Peng et al., 2023; Delgado-Chaves et al., 2025) explored the use of LLMs for evidence synthesis in medical contexts, finding promising capabilities while acknowledging significant challenges remain, especially regarding trust and robustness. However, evidence synthesis in the climate domain remains largely unexplored. Reasoning LLMs, with their enhanced capabilities for logical analysis, represent a particularly promising approach for addressing the unique challenges of climate evidence synthesis, where nuanced interpretation of evidence is essential for effective science communication and policy guidance.

3 Dataset

The Intergovernmental Panel on Climate Change (IPCC) Sixth Assessment Report (AR6) IPCC AR6 represents the most comprehensive synthesis of climate science to date, compiled by hundreds of leading scientists and approved by 195 member governments. Published between 2021 and 2023, AR6 consists of contributions from three Working Groups covering the physical science basis (IPCC AR6 WGI Masson-Delmotte et al. (2021)), impacts and adaptation (IPCC AR6 WGII Pörtner et al. (2022)), and mitigation of climate change (IPCC AR6 WGIII Shukla et al. (2022)), along with a Synthesis Report that integrates findings across all components. A distinguishing feature of the IPCC AR6 is its rigorously structured format that follows a systematic evidence-to-conclusion framework. Each section presents detailed evidence bases drawn from peer-reviewed literature, followed by carefully crafted summary statements with explicitly assigned confidence levels. These confidence assessments follow a standardized methodology (Mastrandrea et al., 2011) that combines scientific agreement and evidence quality, producing calibrated language that expresses varying degrees of certainty (see Figure 5 in Appendix B for details). This structured approach makes AR6 an ideal source for systematic extraction of evidence-conclusion pairs with associated confidence assessments. Figure 1 illustrates

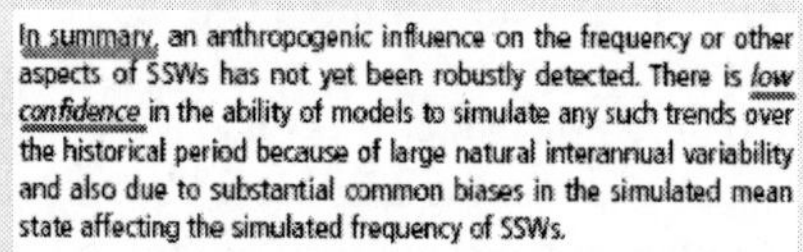
In summary, an anthropogenic influence on the frequency or other aspects of SSWs has not yet been robustly detected. There is *low confidence* in the ability of models to simulate any such trends over the historical period because of large natural interannual variability and also due to substantial common biases in the simulated mean state affecting the simulated frequency of SSWs.

Figure 1: Example conclusion from IPCC AR6 WGI

a sample conclusion from the the *Sudden Stratospheric Warming Activity* subsection from Chapter 3 *Human Influence on the Climate System* from IPCC AR6 WGI Masson-Delmotte et al. (2021). Figure 7 in the Appendix shows the subsection, which includes section header, evidence bases, and conclusion in its original layout.

The report's consistent organization enables reliable parsing of the relationship between supporting evidence and resulting conclusions. Each finding is traceable to its underlying evidence base [2], with transparent reasoning that connects specific climate observations, model outputs, and scientific theories to summary statements. This evidence-conclusion structure, combined with standardized confidence metrics, provides a gold-standard dataset for evaluating how effectively LLMs can process complex scientific information, determine appropriate confidence levels, and generate accurate summaries that preserve key scientific content while maintaining appropriate expressions of certainty.

Data Extraction Process We follow a three-step procedure to extract evidence-conclusion data pairs.

[2]Note that evidences presented in these subsections are already summaries with interpretations produced by climate experts, much like the exposition of literature in the related work or literature review sections of any scientific publication. That being said, for future research, one might be interested in retrieving the original, source research articles and having LLMs synthesizing from ground up.

1. **Document Preprocessing:** We converted PDF files to Markdown format using MinerU (Wang et al., 2024), a SOTA open-source PDF information extraction tool.[3] Given the extensive length of AR6 reports, we segmented them into manageable chunks based on the reports' table of contents. We incorporated one-page overlaps between segments to prevent information loss at section boundaries, as often one section begins on the same page where the previous section ends.

2. **Argument Identification:** We parsed each Markdown file using header tags (*#*) to identify distinct sections. To ensure the extraction of genuine evidence-conclusion pairs, we applied filtering criteria to identify argumentative sections. A section was classified as containing an argument if it: (1) consisted of three or more paragraphs, and (2) concluded with a paragraph containing one of the following concluding phrases: "in summary", "to summarize", "in conclusion", "overall,"[4], "to conclude", "in short", or "to sum up". While this approach may have excluded some valid evidence-conclusion pairs, it prioritized data quality over quantity.

3. **Confidence Level Extraction:** We identified and extracted the confidence levels associated with each conclusion. For conclusions containing multiple assessments with distinct confidence levels, we segmented the conclusion paragraph into individual statements. For example, the statement "To conclude, atmospheric aerosols sampled by ice cores, influenced by northern mid-latitude emissions, show positive trends from 1700 until the last quarter of the 20th century and decreases thereafter (*high* confidence), but there is *low* confidence in observations of systematic changes in other parts of the world in these periods" was divided into two separate conclusions with their respective confidence levels. Since there are too few "very low" and "very high" confidence conclusions at the end of the process, we keep only conclusions with "low", "medium", or "high" confidence.

[3]MinerU allows the extraction of pictures. However, we choose to disregard these pictures for the sake of fairness in comparison. While GPT-4o allows pictures as inputs, reasoning LLM APIs do not currently take pictures as input.

[4]The "," comma after overall is important to reduce false positives.

We deliberately employed a rule-based parsing strategy rather than relying on LLMs for data extraction to avoid potential issues of content hallucination or misrepresentation. Previous research by (Huang et al., 2023; Mohamed et al., 2025) has demonstrated that LLMs can inadvertently introduce factual distortions or fabricate content when processing scientific text, which could compromise dataset integrity. Our rule-based approach ensures reproducibility and maintains the original scientific meaning of the extracted evidence-conclusion pairs. After all, part of the purpose this paper is to evaluate LLMs's capacity to digest scientific text. Below is an example evidence excerpt extracted from this process (from WGI 3.3.3.4 *Sudden Stratospheric Warming Activity*, excerpt in support of the conclusion shown in Figure 1):

> Sudden stratospheric warmings (SSWs) are stratospheric weather events associated with anomalously high temperatures at high latitudes persisting from days to weeks .....
>
> Seviour et al. (2016) found that stratosphere-resolving CMIP5 models, on average, reproduce the observed frequency of vortex splits (one form of SSWs) but with a wide range of model-specific biases ......
>
> Some studies find an increase in the frequency of SSWs under increasing greenhouse gases ......

Dataset Characteristics Our extraction process yielded a compact dataset of 612 distinct "arguments" (evidence-conclusion pairs) from the IPCC AR6 reports. Each data point in our dataset contains the following features: (1) source information (Working Group report identifier and subsection header), (2) evidence bases (the supporting scientific content preceding the conclusion), (3) full conclusion paragraph, (4) individual conclusion statements (when a conclusion paragraph contains multiple assessments), and (5) the confidence level explicitly assigned to each individual conclusion statement (ranging from "low" to "high"). For the confidence classification task, we additionally created a field called "masked conclusion" where the original confidence level expressions were replaced with `<MASKED>`, allowing for evaluation of models' ability to assign appropriate confidence levels without worrying about the potential bias paraphrasing introduces.

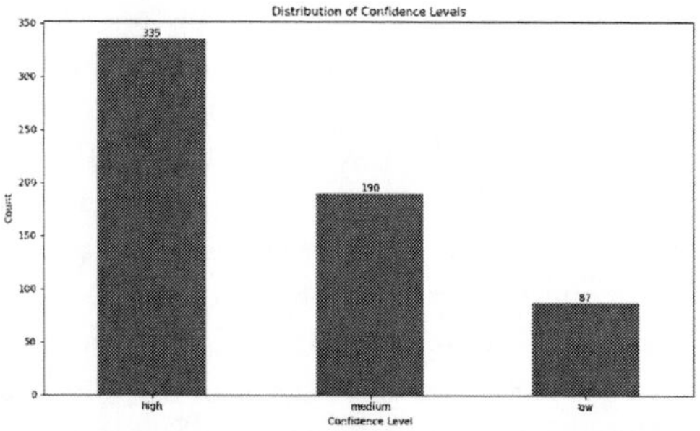

Figure 2: Confidence Level Distribution

Figure 2 shows that most conclusions have confidence, reflecting the scientific rigor of IPCC reports and the growing consensus in climate science (Cook et al., 2016). The distribution of confidence levels in our dataset is in line with what Lacombe et al. (2023); Li et al. (2024b) have observed in their climate statements datasets.

Figure 3 plots the distribution of the length of evidence texts, measured in tokens using the `cl100k base` tokenizer, where the average length is 1654 tokens. In contrast, the average length of individual conclusion statements is only 62 tokens. This substantial difference (approximately 27:1 ratio) highlights the condensation of information required when synthesizing evidence into concise conclusions, making this a challenging task for LLMs.

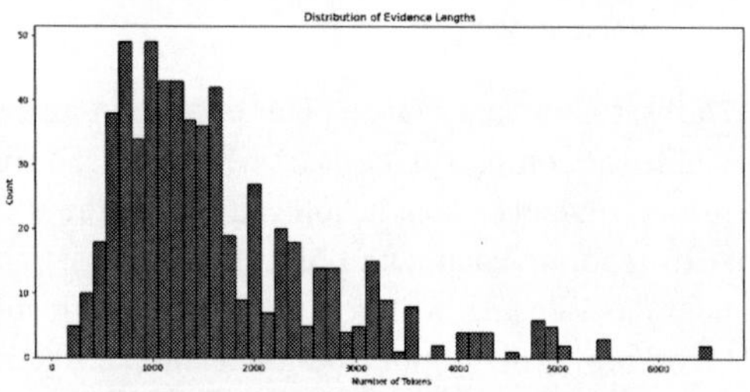

Figure 3: Evidence Length Distribution

4 Methods

Contextual confidence classification To rigorously evaluate the performance differences between reasoning-optimized LLMs and general-purpose LLMs, while also controlling for potential memorization effects, we developed three distinct prompting strategies:

1. **Zeroshot Contextual**: Models are provided with evidence bases and conclusion statements (with confidence levels masked), then asked to classify the appropriate confidence level according to IPCC standards without any examples.

2. **Fewshot Contextual**: Similar to the zero-shot approach, but with three randomly selected examples demonstrating low, medium, and high confidence classifications to provide models with context on the task.

3. **Reference Only**: Models are given only the conclusion statements, source metadata (i.e., the working group report and subsection), and standard definitions of the confidence levels- without any supporting evidence or examples. This setup serves as a control condition to test whether models are relying on memorization of the IPCC reports rather than reasoning over evidence.[5]

For all prompting strategies, we instructed models to select from three confidence levels ("low," "medium," or "high") based on the IPCC's standardized confidence assessment framework (Mastrandrea et al., 2011). Details about prompts are found in Appendix A.

Factual Summarization In the summarization task, models were given evidence bases and one example evidence-conclusion pair and then asked to generate concise summary statements that faithfully reflect the evidence while assigning appropriate confidence levels. Summaries are compared against the full conclusion, not the individual conclusions. This task evaluates models' ability to both synthesize complex scientific information and accurately represent uncertainty—two critical components of scientific communication.

Evaluation Metrics For the confidence classification task, we used **accuracy** and **macro-averaged F1** score as our primary metrics. Macro-F1 is the primay metric to look at since confidence levels are somewhat imbalanced in our dataset (as shown in Figure 2).

[5]In the absence of a custom-trained LLM explicitly excluding IPCC AR6 materials, we concede that we cannot definitively rule out memorization. Our evaluation design instead aims to approximate this distinction by comparing performance across content-based and reference-only conditions

For factual summarization task, we adopt three commonly used metrics: [6]

1. **ROUGE** (Lin, 2004). ROUGE computes the overlap of n-grams between model-generated summaries and expert-written conclusions from the IPCC, providing a basic measure of content coverage.We report ROUGE-1 (unigram overlap) and ROUGE-L (longest common subsequence), using the F1 variant, which is the harmonic mean of precision (how much of the candidate matches the reference) and recall (how much of the reference is covered by the candidate).

2. **BERTScore** (Zhang et al., 2020). BERTScore improves upon ROGUE by measuring semantic similarity between generated and expert-written conclusions beyond exact word matches, using contextual embeddings from pretrained language models. We use the version based on RoBERTa-Large (Liu et al., 2019) and report the F1 score, which is standard practice in BERTScore evaluations.

3. **G-Eval** (Liu et al., 2023). with GPT-4o. G-Eval leverages LLMs with structured prompts and promises to provide human-like assessment of summary quality. We use a customized prompt tailored to our scientific evidence synthesis context to focus on relevance, faithfulness, and appropriateness of confidence levels of LLM-generated conclusions.

Unlike the evaluation of classification tasks, which benefits from clear-cut ground truth, reliable evaluation of summarization task remains an ongoing area of research (Zhang et al., 2025). We choose our evaluation metrics to balance surface-level coverage (ROUGE), semantic similarity (BERTScore), and more human-aligned quality judgments (G-Eval), given the lack of climate-specific summarization evaluation metrics. While it would be valuable to adapt existing metrics, such as BERTScore or FACTCC, using domain-specific models like ClimateBERT (Webersinke et al., 2022), we leave this to future work.

5 Classification Results

Table 1 presents the performance of both reasoning-optimized LLMs (DeepSeek-R1 and o3-mini) and a general-purpose LLM (GPT-4o) on the confidence classification task across different prompting strategies. For context, random guessing on this three-class problem would yield an expected accuracy of 33.3%, while majority class guessing (predicting "high" confidence for all examples, which constitutes approximately 55% of our dataset) would result in an accuracy of 55% with a macro-F1 score of 0.24.

Reasoning LLMs Outperform General-Purpose Models Both reasoning-optimized LLMs consistently outperform GPT-4o across all prompting strategies. In the zero-shot contextual setting, DeepSeek-R1 and o3-mini achieve macro-F1 scores of 65% and 63% respectively, compared to 57% for GPT-4o, representing a performance gap of 8 percentage points between DeepSeek-R1 and GPT-4o. This advantage persists in the few-shot contextual setting, where reasoning models maintain a 7 percentage point lead. The accuracy scores follow a similar pattern.

Interestingly, the few-shot approach did not consistently improve performance over zero-shot for any of the models. While o3-mini increased its F1 score from 63% to 68%, DeepSeek-R1 decreased from 65% to 63%. One potential explanation is context length limitations. Including three complete evidence-conclusion pairs in addition to the task instructions may have caused information overload, making it difficult for the models to effectively process the lengthy context.

Memorization Is Not the Primary Driver of Performance Given that the IPCC AR6 was published in 2023, and the knowledge cutoff dates for all tested models extend beyond this date, a natural concern is whether models are simply retrieving memorized content rather than performing genuine reasoning. The reference-only condition allows us to investigate this possibility by providing models with only the conclusion statement and retrieval-relevant information (working group and section reference) without the actual text of supporting evi-

[6]We included FACTCC (Kryscinski et al., 2020) in earlier versions but removed it in the final version for two reasons. First, FACTCC was trained on news-style summarization datasets and may not generalize well to scientific domains like climate synthesis, where factual consistency involves nuanced reasoning and domain-specific terminology. Second, we observed potential implementation issues where FACTCC returned nearly identical scores across model outputs (up to the 4th decimal point), whereas other evaluation metrics, though close, showed more meaningful variance. This suggests that FACTCC was not a reliable discriminator in our setting.

dence.

The results reveal several important patterns. First, all models experience a performance drop in the reference-only condition compared to the contextual conditions, with GPT-4o showing the steepest decline (17 percentage points from zero-shot to reference-only). This suggests that access to evidence is indeed crucial for the task for all models. Second, even in the reference-only condition, reasoning models maintain accuracies of 57-58%, substantially above both random and majority-class baselines, while GPT-4o's performance drops to 41%, only marginally better than a random classifier and below the majority class baseline.

The relatively strong performance of reasoning models even without evidence suggests they may be better at leveraging minimal contextual cues to retrieve information or perhaps applying general reasoning principles to scientific uncertainty assessment. However, the significant performance gap between contextual and reference-only conditions across all models indicates that genuine evidence evaluation, rather than pure memorization, drives the superior performance observed in the contextual settings.

Performance Inference Cost Trade-off While reasoning LLMs demonstrate superior performance, this advantage comes with significant computational costs. DeepSeek-R1 and o3-mini consume substantially more tokens during inference compared to GPT-4o, as shown in Figure 4. This difference stems from reasoning models' explicit chain-of-thought inference-time scaling processes, where they generate extensive internal reasoning before producing a final answer.[7] In contrast, GPT-4o produces just 2 tokens: the prediction token and the EOS token. In practice, however, the more pressing concern is latency. Inference on DeepSeek-R1 took significantly longer than any other model, requiring over 12 hours to complete 612 requests sent asynchronously. While this largely reflects DeepSeek server's capacity and load constraints, the pattern holds even among OpenAI models. o3-mini required approximately four times longer to complete identical tasks compared to GPT-4o.

[7]Interestingly, performance appears to correlate with tokens consumed during inference. In fewshot settings, models actually spend fewer tokens on reasoning, as if the additional input tokens from demonstrations crowded out the model's chain-of-thoughts.

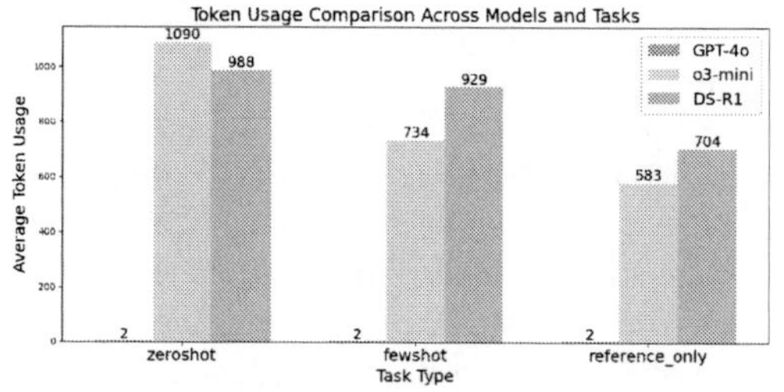

Figure 4: Token Cost Comparison

Note: The bars for GPT-4o are barely visible as it uses only 2 tokens per classification.

6 Summarization Results

DeepSeek-R1 seems to have a slight edge but reasoning LLMs in general do not. As shown in Table 2, DeepSeek-R1 slightly outperforms other models on lexical and semantic similarity metrics, achieving higher scores on ROUGE-1 (0.22), ROUGE-L (0.19), and BERTScore (0.84) compared to o3-mini and GPT4o. Similarly, the differences in G-Eval are minimal. Notably, the other reasoning LLM o3-mini, while clearly outperforming GPT-4o in classification tasks, shows negligible differences in summarization performance. We are inclined to believe that reasoning LLMs may not hold a general advantage in summarization tasks, and DeepSeek-R1's better performance may be idiosyncratic. One possible explanation for this phenomenon is that reasoning LLMs are primarily

Model	Zeroshot Contextual		Fewshot Contextual		Reference Only	
	ACC	F1	ACC	F1	ACC	F1
DS-R1	0.66	0.65	0.65	0.63	0.57	0.54
o3-mini	0.65	0.63	0.63	0.68	0.58	0.60
GPT-4o	0.58	0.57	0.57	0.56	0.41	0.41

Table 1: Classification Results. The table shows accuracy (ACC) and macro-averaged F1 (F1) scores for DeepSeek-R1, o3-mini, and GPT-4o in Zeroshot Contextual, Fewshot Contextual, and Reference only prompting settings.

Model F1	ROGUE-1 F1	ROGUE-L F1	BERTScore F1	G-Eval Faithfulness	G-Eval Relevance	G-Eval Confidence
DS-R1	0.22	0.19	0.84	4.80	4.90	4.94
o3-mini	0.14	0.12	0.82	4.74	4.86	4.98
GPT-4o	0.14	0.13	0.82	4.76	4.88	4.87

Table 2: Performance comparison of models on climate evidence summarization tasks. ROGUE-1 and ROUGE-L measures lexical overlap, BERTScore captures semantic similarity, and G-Eval metrics assess human-aligned quality dimensions including faithfulness, relevance, and appropriateness of confidence assessment. Higher scores indicate better performance across all metrics. Detailed evaluation prompts are provided in Appendix A.

trained to solve mathematical and logical tasks, not for open-ended, generative tasks like summarization.

Evaluation Biases Another possible explanation for our results lies in evaluation biases. Unlike classification tasks where evaluation is straightforward, in summarization tasks, apart from using ROGUE, we rely on pretrained language models themselves as evaluators. Recent studies such as Li et al. (2024a) and Gu et al. (2025) highlight several concerns with the use of LLMs as judges, including various forms of bias. For example, BERTScore is implemented with general-purpose pretrained language models, which are likely affected by domain shift in our climate science setting. Similarly, recent work (Panickssery et al., 2024) suggests that LLM-based evaluators may favor outputs generated by architectures similar to their own. This could partly explain why more advanced reasoning LLMs do not show clear advantages under G-Eval, especially since the evaluator used is GPT-4o itself.

That said, it is noteworthy that DeepSeek-R1, despite likely having less architectural similarity to GPT-4o than o3-mini, achieves the best overall G-Eval scores. While this complicates the interpretation, it also suggests that other factors, such as training data or output style, may influence evaluation outcomes. Addressing all of these issues is beyond the scope of this paper, and we welcome further work to develop more robust, domain-sensitive evaluation frameworks for summarization tasks.

7 Conclusion

Our evaluation of reasoning-optimized LLMs for climate evidence synthesis reveals both promising capabilities and important limitations. These models demonstrate significant advantages in contextual confidence classification, outperforming general-purpose LLMs by 8 percentage points in accuracy and macro-F1 scores when assigning confidence levels to climate statements. This suggests potential utility as auxiliary tools for confidence assessment in scientific workflows.

However, in factual summarization tasks, reasoning LLMs show minimal and inconsistent advantages over general-purpose models. Despite their enhanced logical capabilities, they struggle with the nuanced requirements of scientific summarization when evaluated on relevance, faithfulness, confidence level assignment, which fares much worse than expert-written summaries.

These findings indicate that current reasoning LLMs can potentially contribute to specific aspects of climate evidence synthesis while highlighting the continued necessity of human expertise for summarization tasks. Future work should focus on developing specialized models for scientific synthesis and exploring human-AI collaborative frameworks that leverage the complementary strengths of both. Ultimately, a targeted approach to integrating these technologies into scientific assessment will be essential to maintain rigor while enhancing efficiency.

8 Limitations

We acknowledge that our research faces several limitations.

First, our evidence base excludes visual data such as graphs, charts, and images, which often contain critical climate information in IPCC reports. This omission potentially limits the comprehensiveness of our evaluation, as multi-modal reasoning capabilities would be necessary for complete assessment of climate evidence.

Second, we rely on prompt-based approaches without domain-specific adaptation or fine-tuning. While this allows for assessment of off-the-shelf model capabilities, it likely underestimates the potential performance of models specifically adapted to climate science terminology and reasoning patterns.

Third, our evaluation metrics for summarization tasks, despite careful design, may be susceptible to "LLM-as-judge" biases. Models evaluating other models' outputs could share fundamental limitations or biases, potentially inflating quality assessments of machine-generated summaries compared to expert evaluation.

Finally, our study represents a point-in-time assessment of rapidly evolving technologies. The performance gaps and capabilities identified may change significantly as reasoning LLMs continue to develop. Future work should address these limitations through multi-modal evidence inclusion, domain adaptation techniques, and more robust human-in-the-loop evaluation frameworks.

9 Ethical Considerations

This research evaluates LLMs on existing IPCC assessment data without involving human subjects or generating new climate recommendations. We acknowledge that AI tools for scientific synthesis raise important considerations regarding transparency, accountability, and potential automation bias. While our work demonstrates potential utility in specific tasks, we emphasize that these technologies should supplement rather than replace expert judgment in climate assessment.

10 Acknowledgment

This research was supported in part by credits from OpenAI's Researcher Access Program. We also gratefully acknowledge the NYU IT High Performance Computing (HPC) team for providing computational resources, services, and technical expertise that facilitated this work. We also thank the anonymous reviewers for their helpful feedback.

References

Thomas Bernauer and Liam F. McGrath. 2016. Simple reframing unlikely to boost public support for climate policy. *Nature Climate Change*, 6(7):680–683.

Jannis Bulian, Mike S. Schäfer, Afra Amini, Heidi Lam, Massimiliano Ciaramita, Ben Gaiarin, Michelle Chen Huebscher, Christian Buck, Niels G. Mede, Markus Leippold, and Nadine Strauss. 2024a. Assessing large language models on climate information. In *Proceedings of the 41st International Conference on Machine Learning*, volume 235 of *Proceedings of Machine Learning Research*, pages 4884–4935. PMLR.

Jannis Bulian, Mike S. Schäfer, Afra Amini, Heidi Lam, Massimiliano Ciaramita, Ben Gaiarin, Michelle Chen Hübscher, Christian Buck, Niels G. Mede, Markus Leippold, and Nadine Strauß. 2024b. Assessing large language models on climate information.

Fengxiang Cheng, Haoxuan Li, Fenrong Liu, Robert van Rooij, Kun Zhang, and Zhouchen Lin. 2025. Empowering llms with logical reasoning: A comprehensive survey.

John Cook, Naomi Oreskes, Peter T. Doran, William R. L. Anderegg, Bart Verheggen, Ed W. Maibach, J. Stuart Carlton, Stephan Lewandowsky, Andrew G. Skuce, Sarah A. Green, et al. 2016. Consensus on consensus: a synthesis of consensus estimates on human-caused global warming. *Environmental Research Letters*, 11(4):048002.

Théo Alves Da Costa, Timothée Bohe, Jean Lelong, Nina Achache, Gabriel Olympie, Nicolas Chesneau, and Natalia De la Calzada. 2024. Climateqa, ai-powered conversational assistant for climate change and biodiversity loss.

Adrian J. De-Gol, Corinne Le Quéré, Andrew J. P. Smith, et al. 2023. Broadening scientific engagement and inclusivity in IPCC reports through collaborative technology platforms. *npj Climate Action*, 2:49.

DeepSeek-AI, Daya Guo, Dejian Yang, Haowei Zhang, Junxiao Song, Ruoyu Zhang, Runxin Xu, Qihao Zhu, Shirong Ma, Peiyi Wang, Xiao Bi, Xiaokang Zhang, Xingkai Yu, Yu Wu, Z. F. Wu, Zhibin Gou, Zhihong Shao, Zhuoshu Li, Ziyi Gao, Aixin Liu, Bing Xue, Bingxuan Wang, Bochao Wu, Bei Feng, Chengda Lu, Chenggang Zhao, Chengqi Deng, Chenyu Zhang, Chong Ruan, Damai Dai, Deli Chen, Dongjie Ji, Erhang Li, Fangyun Lin, Fucong Dai, Fuli Luo, Guangbo Hao, Guanting Chen, Guowei Li, H. Zhang, Han Bao, Hanwei Xu, Haocheng Wang, Honghui Ding, Huajian Xin, Huazuo Gao, Hui Qu, Hui Li, Jianzhong Guo, Jiashi Li, Jiawei Wang, Jingchang Chen, Jingyang Yuan, Junjie Qiu, Junlong Li, J. L. Cai, Jiaqi Ni, Jian Liang, Jin Chen, Kai Dong, Kai Hu, Kaige Gao, Kang Guan, Kexin Huang, Kuai Yu, Lean Wang, Lecong Zhang, Liang Zhao, Litong Wang, Liyue Zhang, Lei Xu, Leyi Xia, Mingchuan Zhang, Minghua Zhang, Minghui Tang, Meng Li,

Miaojun Wang, Mingming Li, Ning Tian, Panpan Huang, Peng Zhang, Qiancheng Wang, Qinyu Chen, Qiushi Du, Ruiqi Ge, Ruisong Zhang, Ruizhe Pan, Runji Wang, R. J. Chen, R. L. Jin, Ruyi Chen, Shanghao Lu, Shangyan Zhou, Shanhuang Chen, Shengfeng Ye, Shiyu Wang, Shuiping Yu, Shunfeng Zhou, Shuting Pan, S. S. Li, Shuang Zhou, Shaoqing Wu, Shengfeng Ye, Tao Yun, Tian Pei, Tianyu Sun, T. Wang, Wangding Zeng, Wanjia Zhao, Wen Liu, Wenfeng Liang, Wenjun Gao, Wenqin Yu, Wentao Zhang, W. L. Xiao, Wei An, Xiaodong Liu, Xiaohan Wang, Xiaokang Chen, Xiaotao Nie, Xin Cheng, Xin Liu, Xin Xie, Xingchao Liu, Xinyu Yang, Xinyuan Li, Xuecheng Su, Xuheng Lin, X. Q. Li, Xiangyue Jin, Xiaojin Shen, Xiaosha Chen, Xiaowen Sun, Xiaoxiang Wang, Xinnan Song, Xinyi Zhou, Xianzu Wang, Xinxia Shan, Y. K. Li, Y. Q. Wang, Y. X. Wei, Yang Zhang, Yanhong Xu, Yao Li, Yao Zhao, Yaofeng Sun, Yaohui Wang, Yi Yu, Yichao Zhang, Yifan Shi, Yiliang Xiong, Ying He, Yishi Piao, Yisong Wang, Yixuan Tan, Yiyang Ma, Yiyuan Liu, Yongqiang Guo, Yuan Ou, Yuduan Wang, Yue Gong, Yuheng Zou, Yujia He, Yunfan Xiong, Yuxiang Luo, Yuxiang You, Yuxuan Liu, Yuyang Zhou, Y. X. Zhu, Yanhong Xu, Yanping Huang, Yaohui Li, Yi Zheng, Yuchen Zhu, Yunxian Ma, Ying Tang, Yukun Zha, Yuting Yan, Z. Z. Ren, Zehui Ren, Zhangli Sha, Zhe Fu, Zhean Xu, Zhenda Xie, Zhengyan Zhang, Zhewen Hao, Zhicheng Ma, Zhigang Yan, Zhiyu Wu, Zihui Gu, Zijia Zhu, Zijun Liu, Zilin Li, Ziwei Xie, Ziyang Song, Zizheng Pan, Zhen Huang, Zhipeng Xu, Zhongyu Zhang, and Zhen Zhang. 2025. Deepseek-r1: Incentivizing reasoning capability in llms via reinforcement learning. Technical report, DeepSeek-AI.

F. M. Delgado-Chaves, M. J. Jennings, A. Atalaia, J. Wolff, R. Horvath, Z. M. Mamdouh, J. Baumbach, and L. Baumbach. 2025. Transforming literature screening: The emerging role of large language models in systematic reviews. *Proceedings of the National Academy of Sciences of the United States of America*, 122(2):e2411962122.

Jiawei Gu, Xuhui Jiang, Zhichao Shi, Hexiang Tan, Xuehao Zhai, Chengjin Xu, Wei Li, Yinghan Shen, Shengjie Ma, Honghao Liu, Saizhuo Wang, Kun Zhang, Yuanzhuo Wang, Wen Gao, Lionel Ni, and Jian Guo. 2025. A survey on llm-as-a-judge.

Jie Huang and Kevin Chen-Chuan Chang. 2023. Towards reasoning in large language models: A survey. In *Findings of the Association for Computational Linguistics: ACL 2023*, pages 1049–1065, Toronto, Canada. Association for Computational Linguistics.

Lei Huang, Weijiang Yu, Weitao Ma, Weihong Zhong, Zhangyin Feng, Haotian Wang, Qianglong Chen, Weihua Peng, Xiaocheng Feng, Bing Qin, and Ting Liu. 2023. A survey on hallucination in large language models: Principles, taxonomy, challenges, and open questions. *ACM Trans. Inf. Syst.*, 43(2).

IPCC. 2010. Guidance note for lead authors of the IPCC fifth assessment report on consistent treatment of uncertainties. Technical report, Intergovernmental Panel on Climate Change. Prepared by the IPCC Cross-Working Group Meeting on Consistent Treatment of Uncertainties, Jasper Ridge, CA, USA, 6-7 July 2010.

Elphin Joe, Sai Koneru, and Christine Kirchhoff. 2024. Assessing the effectiveness of GPT-4o in climate change evidence synthesis and systematic assessments: Preliminary insights. In *Proceedings of the 1st Workshop on Natural Language Processing Meets Climate Change (ClimateNLP 2024)*, pages 251–257, Bangkok, Thailand. Association for Computational Linguistics.

Wojciech Kryscinski, Bryan McCann, Caiming Xiong, and Richard Socher. 2020. Evaluating the factual consistency of abstractive text summarization. In *Proceedings of the 2020 Conference on Empirical Methods in Natural Language Processing (EMNLP)*, pages 9332–9346, Online. Association for Computational Linguistics.

Romain Lacombe, Kerrie Wu, and Eddie Dilworth. 2023. Climatex: Do llms accurately assess human expert confidence in climate statements? In *NeurIPS 2023 Workshop on Tackling Climate Change with Machine Learning*.

Haitao Li, Qian Dong, Junjie Chen, Huixue Su, Yujia Zhou, Qingyao Ai, Ziyi Ye, and Yiqun Liu. 2024a. Llms-as-judges: A comprehensive survey on llm-based evaluation methods.

Ruiqi Li, Paige Reeves, Alasdair Tran, and Lexing Xie. 2024b. Profiling and analyzing climate change statements in IPCC reports. In *2nd AI4Research Workshop: Towards a Knowledge-grounded Scientific Research Lifecycle*.

Chin-Yew Lin. 2004. ROUGE: A package for automatic evaluation of summaries. In *Text Summarization Branches Out*, pages 74–81, Barcelona, Spain. Association for Computational Linguistics.

Yang Liu, Dan Iter, Yichong Xu, Shuohang Wang, Ruochen Xu, and Chenguang Zhu. 2023. G-eval: NLG evaluation using gpt-4 with better human alignment. In *Proceedings of the 2023 Conference on Empirical Methods in Natural Language Processing*, pages 2511–2522, Singapore. Association for Computational Linguistics.

Yinhan Liu, Myle Ott, Naman Goyal, Jingfei Du, Mandar Joshi, Danqi Chen, Omer Levy, Mike Lewis, Luke Zettlemoyer, and Veselin Stoyanov. 2019. Roberta: A robustly optimized bert pretraining approach.

Valérie Masson-Delmotte, Panmao Zhai, Anna Pirani, Sarah L. Connors, C. Péan, S. Berger, N. Caud, Y. Chen, L. Goldfarb, M. I. Gomis, M. Huang, K. Leitzell, E. Lonnoy, J. B. R. Matthews, T. K. Maycock, T. Waterfield, O. Yelekçi, R. Yu, and B. Zhou, editors. 2021. *Climate Change 2021: The Physical Science Basis. Contribution of Working Group I to the Sixth Assessment Report of the Intergovernmental Panel on Climate Change*. Cambridge University Press, Cambridge, United Kingdom and New York, NY, USA. In press.

Michael D. Mastrandrea, Katharine J. Mach, Gian-Kasper Plattner, et al. 2011. The IPCC AR5 guidance note on consistent treatment of uncertainties: a common approach across the working groups. *Climatic Change*, 108:675–691.

Amr Mohamed, Mingmeng Geng, Michalis Vazirgiannis, and Guokan Shang. 2025. Llm as a broken telephone: Iterative generation distorts information.

Gaku Morio and Christopher D Manning. 2023. An nlp benchmark dataset for assessing corporate climate policy engagement. In *Advances in Neural Information Processing Systems*, volume 36, pages 39678–39702. Curran Associates, Inc.

Veruska Muccione, Saeid Ashraf Vaghefi, Julia Bingler, et al. 2024. Integrating artificial intelligence with expert knowledge in global environmental assessments: opportunities, challenges and the way ahead. *Regional Environmental Change*, 24:121.

Sahal Shaji Mullappilly, Abdelrahman Shaker, Omkar Thawakar, Hisham Cholakkal, Rao Muhammad Anwer, Salman Khan, and Fahad Khan. 2023. Arabic mini-ClimateGPT : A climate change and sustainability tailored Arabic LLM. In *Findings of the Association for Computational Linguistics: EMNLP 2023*, pages 14126–14136, Singapore. Association for Computational Linguistics.

OpenAI. 2025. System card: O3-Mini. Technical report, OpenAI.

Arjun Panickssery, Samuel R. Bowman, and Shi Feng. 2024. Llm evaluators recognize and favor their own generations.

Yifan Peng, Justin F. Rousseau, Edward H. Shortliffe, and Chunhua Weng. 2023. AI-generated text may have a role in evidence-based medicine. *Nature Medicine*, 29(7):1593–1594.

Nick Pidgeon and Baruch Fischhoff. 2011. The role of social and decision sciences in communicating uncertain climate risks. *Nature Climate Change*, 1(1):35–41.

Hans-Otto Pörtner, Debra C. Roberts, Melinda Tignor, Elvira S. Poloczanska, Katja Mintenbeck, Andrés Alegría, Morgan Craig, Stefanie Langsdorf, Sina Löschke, Vincent Möller, Andrew Okem, and Bardhyl Rama, editors. 2022. *Climate Change 2022: Impacts, Adaptation, and Vulnerability. Contribution of Working Group II to the Sixth Assessment Report of the Intergovernmental Panel on Climate Change*. Cambridge University Press, Cambridge, UK and New York, NY, USA.

Tobias Schimanski, Jingwei Ni, Mathias Kraus, Elliott Ash, and Markus Leippold. 2024. Towards faithful and robust llm specialists for evidence-based question-answering.

P. R. Shukla, Jim Skea, Raphael Slade, Alaa Al Khourdajie, Renée van Diemen, David McCollum, Minal Pathak, Shreya Some, Purvi Vyas, Roger Fradera, Malek Belkacemi, Amrita Hasija, Giovanna Lisboa, Suvadip Luz, and Juliette Malley, editors. 2022. *Climate Change 2022: Mitigation of Climate Change. Contribution of Working Group III to the Sixth Assessment Report of the Intergovernmental Panel on Climate Change*. Cambridge University Press, Cambridge, UK and New York, NY, USA.

Richard C. J. Somerville and Susan Joy Hassol. 2011. Communicating the science of climate change. *Physics Today*, 64(10):48–53.

Dominik Stammbach, Jingwei Ni, Tobias Schimanski, Kalyan Dutia, Alok Singh, Julia Bingler, Christophe Christiaen, Neetu Kushwaha, Veruska Muccione, Saeid A. Vaghefi, and Markus Leippold, editors. 2024. *Proceedings of the 1st Workshop on Natural Language Processing Meets Climate Change (ClimateNLP 2024)*. Association for Computational Linguistics, Bangkok, Thailand.

John D. Sterman. 2011. Communicating climate change risks in a skeptical world. *Climatic Change*, 108:811–826.

Jiankai Sun, Chuanyang Zheng, Enze Xie, Zhengying Liu, Ruihang Chu, Jianing Qiu, Jiaqi Xu, Mingyu Ding, Hongyang Li, Mengzhe Geng, Yue Wu, Wenhai Wang, Junsong Chen, Zhangyue Yin, Xiaozhe Ren, Jie Fu, Junxian He, Wu Yuan, Qi Liu, Xihui Liu, Yu Li, Hao Dong, Yu Cheng, Ming Zhang, Pheng Ann Heng, Jifeng Dai, Ping Luo, Jingdong Wang, Ji-Rong Wen, Xipeng Qiu, Yike Guo, Hui Xiong, Qun Liu, and Zhenguo Li. 2024. A survey of reasoning with foundation models.

Huy Quoc To, Ming Liu, and Guangyan Huang. 2024. Towards efficient large language models for scientific text: A review.

C. W. van Eck. 2023. The next generation of climate scientists as science communicators. *Public Understanding of Science*, 32(8):969–984.

Daniel Van Veen, Cornelia Van Uden, Leah Blankemeier, Jean-Benoit Delbrouck, Ali Aali, Christian Bluethgen, Anuj Pareek, Michela Polacin, Eduardo Pontes Reis, Alena Seehofnerová, Nikita Rohatgi, Praveer Hosamani, William Collins, Nikhil Ahuja, Curtis P. Langlotz, Jason Hom, Sergios Gatidis, John Pauly, and Akshay S. Chaudhari. 2023. Clinical text summarization: Adapting large language models can outperform human experts. *Research Square*, pages rs.3.rs–3483777. Preprint.

Bin Wang, Chao Xu, Xiaomeng Zhao, Linke Ouyang, Fan Wu, Zhiyuan Zhao, Rui Xu, Kaiwen Liu, Yuan

Qu, Fukai Shang, Bo Zhang, Liqun Wei, Zhihao Sui, Wei Li, Botian Shi, Yu Qiao, Dahua Lin, and Conghui He. 2024. Mineru: An open-source solution for precise document content extraction.

Nicolas Webersinke, Mathias Kraus, Julia Anna Bingler, and Markus Leippold. 2022. Climatebert: A pretrained language model for climate-related text.

Jason Wei, Xuezhi Wang, Dale Schuurmans, Maarten Bosma, Brian Ichter, Fei Xia, Ed H. Chi, Quoc V. Le, and Denny Zhou. 2022. Chain-of-thought prompting elicits reasoning in large language models. In *Proceedings of the 36th International Conference on Neural Information Processing Systems*, NIPS '22, Red Hook, NY, USA. Curran Associates Inc.

Fengli Xu, Qianyue Hao, Zefang Zong, Jingwei Wang, Yunke Zhang, Jingyi Wang, Xiaochong Lan, Jiahui Gong, Tianjian Ouyang, Fanjin Meng, Chenyang Shao, Yuwei Yan, Qinglong Yang, Yiwen Song, Sijian Ren, Xinyuan Hu, Yu Li, Jie Feng, Chen Gao, and Yong Li. 2025. Towards large reasoning models: A survey of reinforced reasoning with large language models.

Haopeng Zhang, Philip S. Yu, and Jiawei Zhang. 2025. A systematic survey of text summarization: From statistical methods to large language models. *ACM Comput. Surv.* Just Accepted.

Tianyi Zhang, Varsha Kishore, Felix Wu, Kilian Q. Weinberger, and Yoav Artzi. 2020. Bertscore: Evaluating text generation with bert. In *International Conference on Learning Representations*.

A Appendix: Prompt Details

Following DeepSeek's recommendation for data cleaning and general analysis tasks,[8] we set temperate at 1 for DeepSeek-R1 (and also GPT-4o). For o3-mini, we set reasoning effort to be "medium".
Reference Only Prompt is adapted from Lacombe et al. (2023). Contextual prompts (zeroshot and fewshot) and summarization prompt are produced with the help of meta-prompts by OpenAI.[9] G4 Eval Prompt is adapted from Liu et al. (2023) and OpenAI.[10]

Reference Only Prompt

You are a knowledgeable climate science assistant trained to assess the confidence level associated with various statements about climate change. You will be presented with a statement about climate science, climate impacts, or climate change mitigation retrieved from the IPCC AR6 WGI, WGII, or WGIII assessment reports.
You will be given the subsection header from where the statement is from. Climate scientists have evaluated that statement as low confidence, medium confidence, or high confidence, based on evidence (type, amount, quantity, consistency) and agreement among their peers.
What is their confidence level? Respond *only* with one of the following words: 'low', 'medium', 'high'.

Reference: {WG}, {header}
Statement: {conclusion_statement}
Output:

Contextual Prompt

Analyze the provided scientific text *evidence excerpt* and predict the <MASKED> confidence level of the *conclusion statement* based on contextual evidence.
Avoid external knowledge. Rely only on the provided text. Base your evaluation on the type, quantity, consistency, and agreement of presented evidences.
Respond *only* with one of the following words: 'low', 'medium', 'high'.

Example One
Evidence Excerpt:
{evidence_excerpt}
Conclusion Statement:
{conclusion_statement}
Output:
{true_confidence}

Example Two
Evidence Excerpt:
{evidence_excerpt}
Conclusion Statement:
{conclusion_statement}
Output:
{true_confidence}

Example Three
Evidence Excerpt:
{evidence_excerpt}
Conclusion Statement:
{conclusion_statement}
Output:
{true_confidence}

Input:
Evidence Excerpt:
{evidence_excerpt}
Conclusion Statement:
{conclusion_statement}
Output:

[8] https://api-docs.deepseek.com/quick_start/parameter_settings

[9] https://platform.openai.com/docs/guides/prompt-generation

[10] https://cookbook.openai.com/examples/evaluation/how_to_eval_abstractive_summarization

Summarization Prompt

You are a scientific analyst summarizing key findings from scientific literature. Given a passage of scientific evidence, synthesize the information concisely while preserving quantitative details, uncertainty assessments, and key conclusions.

Guidelines:
1. Focus on the core scientific claims, ensuring clarity and accuracy.
2. Include key findings with numerical data and confidence levels when appropriate.
3. Be concise, your answer should not be longer than one paragraph.
4. Avoid speculation. Use only the provided information; exclude external knowledge.
5. Use precise and neutral language.

Example Input: {evidence_excerpt}

Example Output: {conclusion}

Input: {evidence_excerpt}

Output:

G4 Eval Prompt

Scientific Conclusion Evaluation You are an expert evaluator assessing the quality of LLM-generated scientific conclusions. Your task is to evaluate how well a model has synthesized scientific literature according to specific criteria. For each submission, you will be provided with:
1. The original scientific passage
2. The LLM-generated conclusion
3. The expected guidelines for the conclusion
Evaluation Criteria (Score each on a scale of 1-5):
{criteria}
Evaluation Process: {steps}
Now evaluate: Original Passage: {passage}
LLM-Generated Conclusion: {conclusion}
{guideline_section}
Your evaluation must follow this exact format: Evaluation:
-Relevance: Score: X/5
-Faithfulness: Score: X/5
-Confidence Level Appropriateness: Score: X/5

G4 Eval Prompt - Relevance

Relevance
* 5: Perfectly captures the core scientific findings and key quantitative details
* 4: Identifies most important findings but misses minor details
* 3: Captures some key findings but omits several important elements
* 2: Focuses primarily on peripheral information rather than central findings
* 1: Fails to identify the main scientific findings

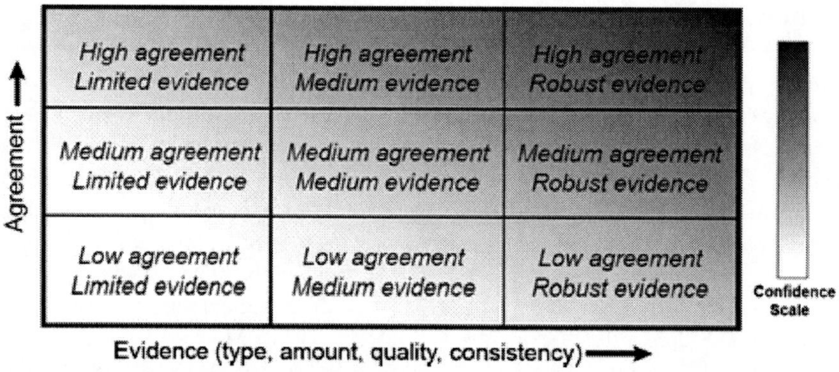

Figure 5: Confidence Evaluation Matrix from (Mastrandrea et al., 2011)

G4 Eval Prompt - Faithfulness

Faithfulness
* 5: Completely faithful to the original text with no misrepresentations or distortions
* 4: Largely faithful with only minor inaccuracies that don't affect the core meaning
* 3: Generally faithful but contains some misrepresentations of moderate importance
* 2: Contains significant misrepresentations or fabricated information
* 1: Fundamentally misrepresents the scientific content or contradicts the original text

G4 Eval Prompt - Confidence Level Appropriateness

Confidence Level Appropriateness
* 5: All confidence levels expressed in conclusion statement strictly follow from scientific text
* 4: Contain confidence level statements with minor inaccuracies or somewhat dubious nature
* 3: Preserves some uncertainty statements but omits or misrepresents others
* 2: Significantly understates or overstates confidence in findings
* 1: Completely misrepresents or omits critical uncertainty statements and confidence levels

B Appendix: Figures from IPCC AR6

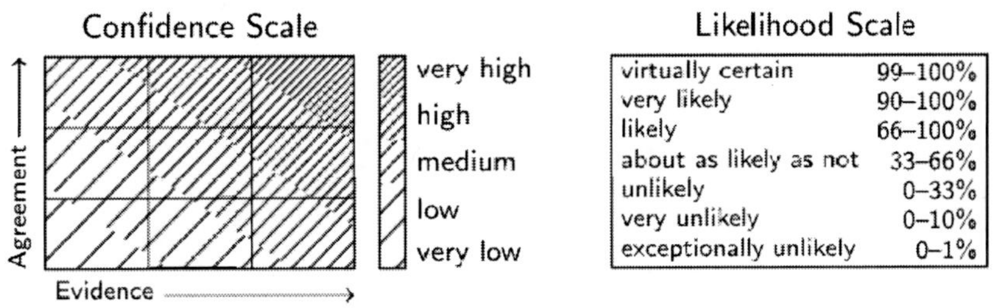

Figure 6: Confidence and likelihood scales for communicating degree of certainty in key findings of the IPCC AR5, adapted from (Mastrandrea et al., 2011)

Chapter 3 Human Influence on the Climate System

human influence on historical blocking activity. The *low confidence* statements are due to the limited number of studies available. The shift of the Southern Hemisphere jet is correlated with modulations of the SAM (Section 3.7.2). There is *medium confidence* in model performance regarding the simulation of the extratropical jets, storm track and blocking activity, with increased resolution sometimes corresponding to better performance, but important shortcomings remain, particularly for the Euro-Atlantic sector of the Northern Hemisphere. Nonetheless, synthesizing across Sections 3.3.3.1–3.3.3.3, there is *high confidence* that CMIP6 models capture the general characteristics of the tropospheric large-scale circulation.

3.3.3.4 Sudden Stratospheric Warming Activity

Sudden stratospheric warmings (SSWs) are stratospheric weather events associated with anomalously high temperatures at high latitudes persisting from days to weeks. Section 2.3.1.4.5 discusses the definition and observational aspects of SSWs. SSWs are often associated with anomalous weather conditions, for example, winter cold spells, in the lower atmosphere (e.g., Butler et al., 2015; Baldwin et al., 2021).

Seviour et al. (2016) found that stratosphere-resolving CMIP5 models, on average, reproduce the observed frequency of vortex splits (one form of SSWs) but with a wide range of model-specific biases. Models that produce a better mean state of the polar vortex also tend to produce a more realistic SSW frequency (Seviour et al., 2016). The mean sea level pressure anomalies occurring in CMIP5 model simulations when an SSW is underway, however, differ substantially from those in reanalyses (Seviour et al., 2016). Unlike stratosphere-resolving models, models with limited stratospheric resolution, which make up more than half of the CMIP5 ensemble, underestimate the frequency of SSWs (Osprey et al., 2013; J. Kim et al., 2017). Taguchi (2017) found a general underestimation in CMIP5 models of the frequency of 'major' SSWs (which are associated with a break-up of the polar vortex), an aspect of an under-representation in those models of dynamical variability in the stratosphere. Wu and Reichler (2020) found that finer vertical resolution in the stratosphere and a model top above the stratopause tend to be associated with a more realistic SSW frequency in CMIP5 and CMIP6 models.

Some studies find an increase in the frequency of SSWs under increasing greenhouse gases (e.g., Schimanke et al., 2013; Young et al., 2013; J. Kim et al., 2017). However, this behaviour is not robust across ensembles of chemistry-climate models (Mitchell et al., 2012; Ayarzagüena et al., 2018; Rao and Garfinkel, 2021). There is an absence of studies specifically focusing on simulated trends in SSWs during recent decades, and the short record and substantial decadal variability yields *low confidence* in any observed trends in the occurrence of SSW events in the Northern Hemisphere winter (Section 2.3.1.4.5). Such an absence of a trend and large variability would also be consistent with a recent reconstruction of SSWs extending back to 1850, based on sea level pressure observations (Domeisen, 2019), although this time series has limitations as it is not based on direct observations of SSWs.

In summary, an anthropogenic influence on the frequency or other aspects of SSWs has not yet been robustly detected. There is *low confidence* in the ability of models to simulate any such trends over the historical period because of large natural interannual variability and also due to substantial common biases in the simulated mean state affecting the simulated frequency of SSWs.

3.4 Human Influence on the Cryosphere

3.4.1 Sea Ice

3.4.1.1 Arctic Sea Ice

The AR5 concluded that 'anthropogenic forcings are *very likely* to have contributed to Arctic sea ice loss since 1979' (Bindoff et al., 2013), based on studies showing that models can reproduce the observed decline only when including anthropogenic forcings, and formal attribution studies. Since the beginning of the modern satellite era in 1979, Northern Hemisphere sea ice extent has exhibited significant declines in all months with the largest reduction in September (see Section 2.3.2.1.1, and Figures 3.20 and 3.21 for more details on observed changes). The recent Arctic sea ice loss during summer is unprecedented since 1850 (*high confidence*), but as in AR5 and SROCC there remains only *medium confidence* that the recent reduction is unique during at least the past 1000 years due to sparse observations (Sections 2.3.2.1.1 and 9.3.1.1). CMIP5 models also simulate Northern Hemisphere sea ice loss over the satellite era but with large differences among models (e.g., Massonnet et al., 2012; Stroeve et al., 2012). The envelope of simulated ice loss across model simulations encompasses the observed change, although observations fall near the low end of the CMIP5 and CMIP6 distributions of trends (Figure 3.20). CMIP6 models on average better capture the observed Arctic sea ice decline, albeit with large inter-model spread. Notz et al. (2020) found that CMIP6 models better reproduce the sensitivity of Arctic sea ice area to CO_2 emissions and global warming than earlier CMIP models although the models' underestimation of this sensitivity remains. Davy and Outten (2020) also found that CMIP6 models can simulate the seasonal cycle of Arctic sea ice extent and volume better than CMIP5 models. For the assessment of physical processes associated with changes in Arctic sea ice, see Section 9.3.1.1.

Since AR5, there have been several new detection and attribution studies on Arctic sea ice. While the attribution literature has mostly used sea ice extent (SIE), it is closely proportional to sea ice area (SIA; Notz, 2014), which is assessed in Chapters 2 and 9 and shown in Figures 3.20 and 3.21. Kirchmeier-Young et al. (2017) compared the observed time series of the September SIE over the period 1979–2012 with those from different large ensemble simulations which provide a robust sampling of internal climate variability (CanESM2, CESM1, and CMIP5) using an optimal fingerprinting technique. They detected anthropogenic signals which were separable from the response to natural forcing due to solar irradiance variations and volcanic aerosol, supporting previous findings (Figure 3.21; Min et al., 2008b; Kay et al., 2011; Notz and Marotzke, 2012; Notz and Stroeve, 2016). Using selected CMIP5 models and three independently derived sets of observations, Mueller et al. (2018) detected fingerprints from greenhouse gases, natural, and other anthropogenic forcings simultaneously in the September Arctic SIE over

Figure 7: Example Section from IPCC AR6

Author Index

Association for Computational Linguistics
209 N. Eighth Street
Stroudsburg, Pennsylvania 18360

ISBN 979-8-3313-2405-6

CURRAN ASSOCIATES INC.
proceedings.com